T0224853

Integrative and Interdisciplinary Aspects of Intermetallics

MATERIALS RESEARCH SOCIETY
SYMPOSIUM PROCEEDINGS VOLUME 842

Integrative and Interdisciplinary Aspects of Intermetallics

Symposium held November 29–December 1, 2004, Boston, Massachusetts, U.S.A.

EDITORS:

Michael J. Mills
Ohio State University
Columbus, Ohio, U.S.A.

Haruyuki Inui
Kyoto University
Kyoto, Japan

Helmut Clemens
Montanuniversität Leoben
Leoben, Austria

Chong-Long Fu
Oak Ridge National Laboratory
Oak Ridge, Tennessee, U.S.A.

Materials Research Society
Warrendale, Pennsylvania

CAMBRIDGE UNIVERSITY PRESS
Cambridge, New York, Melbourne, Madrid, Cape Town,
Singapore, São Paulo, Delhi, Mexico City

Cambridge University Press
32 Avenue of the Americas, New York NY 10013-2473, USA

Published in the United States of America by Cambridge University Press, New York

www.cambridge.org
Information on this title: www.cambridge.org/9781107409057

Materials Research Society
506 Keystone Drive, Warrendale, PA 15086
http://www.mrs.org

© Materials Research Society 2005

This publication is in copyright. Subject to statutory exception
and to the provisions of relevant collective licensing agreements,
no reproduction of any part may take place without the written
permission of Cambridge University Press.

This publication has been registered with Copyright Clearance Center, Inc.
For further information please contact the Copyright Clearance Center,
Salem, Massachusetts.

First published 2005
First paperback edition 2013

Single article reprints from this publication are available through
University Microfilms Inc., 300 North Zeeb Road, Ann Arbor, MI 48106

CODEN: MRSPDH

ISBN 978-1-107-40905-7 Paperback

Cambridge University Press has no responsibility for the persistence or
accuracy of URLs for external or third-party internet websites referred to in
this publication, and does not guarantee that any content on such websites is,
or will remain, accurate or appropriate.

CONTENTS

Preface xv

Materials Research Society Symposium Proceedings xvi

IRON ALUMINIDES

* Strengthening of Iron Aluminide Alloys for High-Temperature
Applications 3
    Martin Palm, André Schneider, Frank Stein, and Gerhard Sauthoff

Effect of Alloying Elements (Ga, Ge, Si) on Pseudoelasticity in
Fe₃Al Single Crystals 15
    Hiroyuki Y. Yasuda, Takashi Kase, and Yukichi Umakoshi

Pseudoelasticity of DO₃ Ordered Monocrystalline Fe₃Al 21
    S. Kabra, H. Bei, D.W. Brown, M.A.M. Bourke, and
    E.P. George

Development of High Temperature Creep Resistance in
Fe-Al Alloys 27
    D.G. Morris, M.A. Muñoz-Morris, and C. Baudin

Microstructure and Mechanical Properties of Fe-Ni-Mn-Al Alloys 35
    M.W. Wittmann, I. Baker, J.A. Hanna, and P.R. Munroe

Thermomechanical Treatment of a Fe₃Al Alloy 41
    Joachim Konrad, Stefan Zaefferer, André Schneider,
    Georg Frommeyer, and Dierk Raabe

Optimization of Precipitation for the Development of
Improved Wrought Fe₃Al-Based Alloys 47
    Satoru Kobayashi, Stefan Zaefferer, and André Schneider

Effect of Excess Vacancies on Antiphase Domain Growth
in Fe₃Al 53
    Y. Koizumi, T. Hagiwara, Y. Minamino, and N. Tsuji

*Invited Paper

Microstructure, Mechanical Properties and Wear Resistance
of Fe-Al Based Alloys With Various Alloying Elements 59
Han-Sol Kim, In-Dong Yeo, Tae-Yeub Ra, and Won-Yong Kim

NICKEL ALUMINIDES

⅛ Microstructures and Mechanical Properties of NiAl-Mo
Composites 67
H. Bei and E.P. George

Microstructure and Thermo-Mechanical Behavior of NiAl Coatings 73
G. Dehm, J. Riethmüller, P. Wellner, O. Kraft, H. Clemens,
and E. Arzt

Quantitative Analysis of γ Precipitates in Cyclically
Deformed Ni_3(Al,Ti) Single Crystals Using Magnetic
Technique 79
Yukichi Umakoshi, Hiroyuki Y. Yasuda, and
Toshifumi Yanai

Microstructures in Cold-Rolled Ni_3Al Single Crystals 85
Kyosuke Kishida, Masahiko Demura, and
Toshiyuki Hirano

Crystal Structure, Phase Stability and Plastic Deformation
Behavior of Ti-Rich Ni_3(Ti,Nb) Single Crystals With Various
Long-Period Ordered Structures 91
Koji Hagihara, Tetsunori Tanaka, Takayoshi Nakano,
and Yukichi Umakoshi

Effect of Fe Alloying on the Vacancy Formation in Ni_3Al 97
E. Partyka, R. Kozubski, W. Sprengel, and H.-E. Schaefer

TITANIUM ALUMINIDES

An Electron Microscope Study of Mechanical Twinning in
a Precipitation–Hardened TiAl Alloy 103
Fritz Appel

Increase in γ/α_2 Lamellar Boundary Density and Its Effect
on Creep Resistance of TiAl Alloy 109
Kouichi Maruyama, Jun Matsuda, and Hanliang Zhu

Effect of Long-Term Aging and Creep Exposure on the
Microstructure of TiAl-Based Alloy for Industrial
Applications...115
    Juraj Lapin, Mohamed Nazmy, and Marc Staubli

Creep Behavior and Microstructural Stability of Ti-46Al-9Nb
With Different Microstructures...121
    S. Bystrzanowski, A. Bartels, H. Clemens, R. Gerling,
    F.-P. Schimansky, and G. Dehm

Internal Friction of a High-Nb Gamma-TiAl-Based Alloy
With Different Microstructures...127
    M. Weller, H. Clemens, G. Dehm, G. Haneczok,
    S. Bystrzanowski, A. Bartels, R. Gerling, and E. Arzt

Investigation of the Structure and Properties of Hypereutectic
Ti-Based Bulk Alloys..133
    Dmitri V. Louzguine-Luzgin, Larissa V. Louzguina-Luzgina,
    and Akihisa Inoue

TEM Characterization of Fatigue Tested Lamellar
Ti-44Al-8Nb-1B...139
    H. Jiang, D. Hu, and I.P. Jones

On the Effects of Interstitial Elements on Microstructure and
Properties of Ternary and Quaternary TiAl Based Alloys......................145
    Jean-Pierre Chevalier, Mélanie Lamirand, and
    Jean-Louis Bonnentien

The Correlation of Slip Between Adjacent Lamellae in TiAl.................151
    G. Molénat, A. Couret, M. Sundararaman, J.B. Singh,
    G. Saada, and P. Veyssière

A Study of the Deformation Behavior of Lamellar γ-TiAl
by Numeric Modeling..157
    T. Schaden, F.D. Fischer, H. Clemens, F. Appel, and
    A. Bartels

Effect of B Addition on Thermal Stability of Lamellar
Structure in Ti-47Al-2Cr-2Nb Alloys..163
    Y. Yamamoto, N.D. Evans, P.J. Maziasz, and C.T. Liu

Compressive Creep Behavior in Coarse Grained Polycrystals
of Ti₃Al and Its Dependence on Binary Alloy Compositions...................169
    Tohru Takahashi, Yuki Sakaino, and Shunzi Song

8 Micro Fracture Toughness Testing of TiAl Based
   Alloys With a Fully Lamellar Structure 175
   K. Takashima, T.P. Halford, D. Rudinal, Y. Higo, and
   M. Takeyama

High-Temperature Environmental Embrittlement of
Thermomechanically Processed TiAl-Based Intermetallic
Alloys With Various Kinds of Microstructures..181
   T. Takasugi, Y. Hotta, S. Shibuya, Y. Kaneno, H. Inoue,
   and T. Tetsui

Experimental Studies and Thermodynamic Simulation of
Phase Transformations in γ-TiAl Based Alloys...187
   Harald F. Chladil, Helmut Clemens, Harald Leitner,
   Arno Bartels, Rainer Gerling, Wilfried T. Marketz, and
   Ernst Kozeschnik

Massive Transformation in High Niobium Containing
TiAl-Alloys 193
   A. Bartels, S. Bystrzanowski, H. Chladil, H. Leitner,
   H. Clemens, R. Gerling, and F.-P. Schimansky

The Creep Behaviors of Two Fine-Grained XD TiAl Alloys
Produced by Similar Heat Treatments 199
   Hanliang Zhu, Dongyi Seo, Kouichi Maruyama, and
   Peter Au

Creep of TiAl Alloys at 750°C Under Moderate Stress...205
   Alain Couret and Joel Malaplate

Abnormal Deformation Behavior in Polysynthetically-
Twinned TiAl Crystals With A and N Orientations—
An AFM Study 211
   Yali Chen and David P. Pope

Rationalization of the Plastic Flow Behavior of
Polysynthetically-Twinned (PST) TiAl Crystals Based on
Slip Mode Observation Using AFM and Schmid's Law...217
   Yali Chen and David P. Pope

Texture Development During Hot Extrusion of a Gamma
Titanium Aluminide Alloy 223
   M. Oehring, F. Appel, H.-G. Brokmeier, and U. Lorenz

Fatigue Testing of Microsized Samples of γ-TiAl Based Material 229
    Timothy P. Halford, Kazuki Takashima, and Yakichi Higo

Microstructure-Property Relationships of Two Ti₂AlNb-Based Intermetallic Alloys: Ti-15Al-33Nb(at.%) and Ti-21Al-29Nb(at.%) 235
    Christopher J. Cowen, Dingqiang Li, and Carl J. Boehlert

Phase Transformation in Orthorhombic Ti₂AlNb Alloys Under Severe Deformation 241
    B.A. Greenberg, N.V. Kazantseva, and V.P. Pilugin

Impact Properties of Hot-Worked Gamma Alloys With BCC β-Ti Phase..247
    Kentaro Shindo, Toshimitsu Tetsui, Toshiro Kobayashi,
    Shigeki Morita, Satoru Kobayashi, and Masao Takeyama

Effects of Long-Period Superstructures on Plastic Properties in Al-Rich TiAl Single Crystals..253
    Takayoshi Nakano, Koutaro Hayashi, Yukichi Umakoshi,
    Yu-Lung Chiu, and Patrick Veyssière

TEM Analysis of Long-Period Superstructures in TiAl Single Crystal With Composition Gradient.................................259
    S. Hata, K. Shiraishi, N. Kuwano, M. Itakura,
    Y. Tomokiyo, T. Nakano, and Y. Umakoshi

Influence of Micro-Alloying on Oxidation Behavior of TiAl 265
    Michiko Yoshihara and Shigeji Taniguchi

SILICIDES

The Effects of Substitutional Additions on Tensile Behavior of Nb-Silicide Based Composites 273
    Laurent Cretegny, Bernard P. Bewlay, Ann M. Ritter,
    and Melvin R. Jackson

Effect of Microstructure and Zr Addition on the Crystallographic Orientation Relationships Among Phases Related to the Eutectoid Decomposition of Nb₃Si in Near Eutectic Nb-Si Alloy 279
    Seiji Miura, Kenji Ohkubo, and Tetsuo Mohri

Microstructures of LENS™ Deposited Nb-Si Alloys 285
  Ryan R. Dehoff, Peter M. Sarosi, Peter C. Collins,
  Hamish L. Fraser, and Michael J. Mills

Microstructures and Mechanical Properties in Ni_3Si-Ni_3Ti-
Ni_3Nb-Based Multi-Intermetallic Alloys 291
  T. Takasugi, K. Ohira, and Y. Kaneno

Experimental Investigation and Thermodynamic Modeling
of Phase Equilibria in the Hf-Ti-Si System...297
  Y. Yang, B.P. Bewlay, M.R. Jackson, and Y.A. Chang

Role of Microstructure in Promoting Fracture and Fatigue
Resistance in Mo-Si-B Alloys 303
  J.J. Kruzic, J.H. Schneibel, and R.O. Ritchie

⊗ A Bond-Order Potential Incorporating Analytic
  Screening Functions for the Molybdenum Silicides 309
  Marc J. Cawkwell, Matous Mrovec, Duc Nguyen-Manh,
  David G. Pettifor, and Vaclav Vitek

High Temperature Oxidation Behavior of Al Added
Mo/Mo_5SiB_2 In Situ Composites...315
  Akira Yamauchi, Kyousuke Yoshimi, and Shuji Hanada

Nucleation of (Mo) Precipitates on Dislocations During
Annealing of a Mo-Rich Mo_5SiB_2 Phase 321
  Nobuaki Sekido, Ridwan Sakidja, and John H. Perepezko

FUNCTIONAL INTERMETALLICS

* Advanced TEM Investigations on Ni-Ti Shape Memory
  Material: Strain and Concentration Gradients Surrounding
  Ni_4Ti_3 Precipitates..329
  Dominique Schryvers, Wim Tirry, and Zhiqing Yang

* Sub-Nano and Nano-Structures of Hydrides of $LaNi_5$ and
  its Related Intermetallics..339
  Etsuo Akiba, Kouji Sakaki, and Yumiko Nakanura

*Invited Paper

Transformation Behavior of TiNiPt Thin Films Fabricated Using Melt Spinning Technique 347
Tomonari Inamura, Yohei Takahashi, Hideki Hosoda,
Kenji Wakashima, Takeshi Nagase, Takayoshi Nakano,
Yukichi Umakoshi, and Shuichi Miyazaki

Factors for Controlling Martensitic Transformation Temperature of TiNi Shape Memory Alloy by Addition of Ternary Elements 353
Hideki Hosoda, Kenji Wakashima, Shuichi Miyazaki,
and Kanryu Inoue

Shape Memory Effect Through L1$_0$-fcc Order-Disorder Transition 359
K. Tanaka

Effect of Heat Treatment Conditions on Multistage Martensitic Transformation in Aged Ni-Rich Ti-Ni Alloys 365
Minoru Nishida, Toru Hara, Yasuhiro Morizono,
Mitsuhiro Matsuda, and Kousuke Fujishima

Application of the CSL Model to Deformation Twin Boundary in B2 Type TiNi Compound 371
Minoru Nishida, Mituhiro Matsuda, Yasuhiro Morizono,
Towako Fujimoto, and Hideharu Nakashima

Coarsening Behavior and Coercivity in L1$_0$-Ordered Intermetallic Fe-Pd Ferromagnets With Equiaxed Grain Morphology 377
Jörg M.K. Wiezorek and Anirudha R. Deshpande

Comparison of Temperature Driven Ordering in Bulk Foil and Thin Film of L1$_0$ Ordered FePd 383
Chaisak Issro, Wolfgang Püschl, Wolfgang Pfeiler,
Bogdan Sepiol, Peter F. Rogl, William A. Soffa,
Manuel Acosta, and Véronique Pierron-Bohnes

Formation of Defect Structures During Annealing of Cold-Deformed L1$_0$-Ordered Equiatomic FePd Intermetallics 389
Anirudha R. Deshpande and Jörg M.K. Wiezorek

Magnetic Properties of E2$_1$-Base Co$_3$AlC and the Correlation With the Ordering of Carbon Atoms and Vacancies 395
Yoshisato Kimura, Fu-Gao Wei, Hideyuki Ohtsuka,
and Yoshinao Mishima

Structural Properties and Magnetic Behavior in the Pseudobinary Alloys CoFe-CoAl 401
Nobutoshi Tadachi, Hiroki Ishibashi, and
Mineo Kogachi

Directional Thermoelectric Performance of Ru_2Si_3 407
Benjamin A. Simkin, Yoshinori Hayashi, and
Haruyuki Inui

Mechanical Aspects of Structural Optimization in a Bi-Te Thermoelectric Module for Power Generation 413
Yujiro Nakatani, Reki Takaku, Takehisa Hino,
Takahiko Shindo, and Yoshiyasu Itoh

Effects of Ga- and In-Doping on the Thermoelectric Properties in Ba-Ge Clathrate Compounds 419
Norihiko L. Okamoto and Haruyuki Inui

Crystal Structure and Thermoelectric Properties of Al-Containing Re Silicides 425
Eiji Terada, Min-Wook Oh, Dang-Moon Wee, and
Haruyuki Inui

Characterization and Catalytic Properties of Ni_3Al for Hydrogen Production From Methanol 431
Ya Xu, Satoshi Kameoka, Kyosuke Kishida,
Masahiko Demura, An-pang Tsai, and Toshiyuki Hirano

OTHER INTERMETALLICS

* **Complex Intermetallic Compounds: Defects, Disordering, Details** 439
W. Sprengel, F. Baier, K. Sato, X.Y. Zhang, and
H.-E. Schaefer

Effect of Heat Treatments on Microstructure of Rapidly Solidified TiCo Ribbons 449
Kyosuke Yoshimi, Akira Yamauchi, Ryusuke Nakamura,
Sadahiro Tsurekawa, and Shuji Hanada

*Invited Paper

In Situ Observation of Surface Relief Formation and Disappearance During Order-Disorder Transition of Equi-Atomic CuAu Alloy Using Laser Scanning Confocal Microscopy 455
Seiji Miura, Hiroyuki Okuno, Kenji Ohkubo, and Tetsuo Mohri

Phase Equilibria and Lattice Parameters of Fe_2Nb Laves Phase in Fe-Ni-Nb Ternary System at Elevated Temperatures 461
Masao Takeyama, Nobuyuki Gomi, Sumio Morita, and Takashi Matsuo

Sputtered Coatings Based on the Al_2Au Phase 467
Christian Mitterer, Helmut Lenhart, Paul H. Mayrhofer, and Martin Kathrein

Solidification Processing and Fracture Behavior of RuAl-Based Alloys 473
Todd Reynolds and David Johnson

Formation and Morphology of Kurnakov Type $D0_{22}$ Compound in Disordered fcc γ-(Ni, Fe) Matrix Alloys 479
Akane Suzuki and Masao Takeyama

Parameters of Dislocation Structure and Work Hardening of Ni_3Ge 485
N.A. Koneva, Yu.V. Solov'eva, V.A. Starenchenko, and E.V. Kozlov

Identification of the Chirality of Intermetallic Compounds by Electron Diffraction 491
S. Fujio, H. Sakamoto, K. Tanaka, and H. Inui

Mechanical Behavior of a Pt-Cr Jewelry Alloy Hardened by Nano-Sized Ordered Particles 497
Kamili M. Jackson, Miyelani P. Nzula, Silethelwe Nxumalo, and Candace I. Lang

MODELING OF INTERMETALLICS

First-Principles Study of Structural and Defect Properties in FeCo Intermetallics 505
M. Krcmar, C.L. Fu, and J.R. Morris

Dislocation Structure, Phase Stability and Yield Stress Behavior of Ultra-High Temperature L1$_2$ Intermetallics: Combined First Principles-Peierls-Nabarro Approach.....511
Oleg Y. Kontsevoi, Yuri N. Gornostyrev, and Arthur J. Freeman

Atomic, Electronic, and Magnetic Structure of Iron-Based Sigma-Phases.....517
Pavel A. Korzhavyi, Bo Sundman, Malin Selleby, and Börje Johansson

B2 Phases and Their Defect Structures: Part I. Ab Initio Enthalpy of Formation and Enthalpy of Mixing in the Al-Ni-Pt-Ru System.....523
Sara Prins, Raymundo Arroyave, Chao Jiang, and Zi-Kui Liu

B2 Phases and Their Defect Structures: Part II. Ab Initio Vibrational and Electronic Free Energy in the Al-Ni-Pt-Ru System.....529
Raymundo Arroyave, Sara Prins, and Zi-Kui Liu

Thermal Analysis of Relaxation Processes of Supersaturated Vacancies in B2-Type Aluminides.....535
Ryusuke Nakamura, Kyosuke Yoshimi, Akira Yamauchi, and Shuji Hanada

Ab Initio Calculation of Point Defect Energies and Atom Migration Profiles in Varying Surroundings in L1$_2$-Ordered Intermetallic Compounds.....541
Doris Vogtenhuber, Jana Houserova, Walter Wolf, Raimund Podloucky, Wolfgang Pfeiler, and Wolfgang Püschl

Author Index.....547

Subject Index.....551

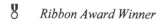 *Ribbon Award Winner*

🏆 *Trophy Award Winner*

xiv

PREFACE

These proceedings contain papers presented at Symposium S, "Integrative and Interdisciplinary Aspects of Intermetallics," held November 29-December 1 at the 2004 MRS Fall Meeting in Boston, Massachusetts. The overarching theme of this symposium is the critical importance of microstructure and defects on the physical and mechanical properties of ordered intermetallics. These compounds have great promise for a variety of diverse and important applications such as high temperature structural materials, "smart" shape memory alloys, hydrogen storage media, thermoelectric power sources, and magnetic applications. The integrative aspects of this topic were highlighted to encourage a common platform for the presentation of intermetallics research that may have very different technological backgrounds and objectives, yet in all cases require a sound understanding of the elementary processes that govern structural and functional properties—from the atomistic through microstructural scales. The symposium highlighted the tremendous progress in the field with respect to processing and properties, as well as improved fundamental insights through novel experimentation and modeling activities. Symposium S brought together researchers from twenty countries, with particularly strong representation from Asian and European laboratories.

The papers are grouped by topical area, including iron aluminides, nickel aluminides, titanium aluminides, silicides, functional intermetallics (for shape memory, thermoelectric, magnetic and hydrogen storage applications) and more generic modeling of intermetallic compounds.

The organizers would like to thank Boehler Schmiedetechnik, JEOL, Kyoto University, Oak Ridge National Laboratory, and Shimazdu Co., Ltd. for their kind support of the symposium.

Michael J. Mills
Haruyuki Inui
Helmut Clemens
Chong-Long Fu

March 2005

MATERIALS RESEARCH SOCIETY SYMPOSIUM PROCEEDINGS

Volume 807— Scientific Basis for Nuclear Waste Management XXVII, V.M. Oversby, L.O. Werme, 2004, ISBN: 1-55899-752-0

Volume 808— Amorphous and Nanocrystalline Silicon Science and Technology—2004, R. Biswas, G. Ganguly, E. Schiff, R. Carius, M. Kondo, 2004, ISBN: 1-55899-758-X

Volume 809— High-Mobility Group-IV Materials and Devices, M. Caymax, E. Kasper, S. Zaima, K. Rim, P.F.P. Fichtner, 2004, ISBN: 1-55899-759-8

Volume 810— Silicon Front-End Junction Formation—Physics and Technology, P. Pichler, A. Claverie, R. Lindsay, M. Orlowski, W. Windl, 2004, ISBN: 1-55899-760-1

Volume 811— Integration of Advanced Micro- and Nanoelectronic Devices—Critical Issues and Solutions, J. Morais, D. Kumar, M. Houssa, R.K. Singh, D. Landheer, R. Ramesh, R. Wallace, S. Guha, H. Koinuma, 2004, ISBN: 1-55899-761-X

Volume 812— Materials, Technology and Reliability for Advanced Interconnects and Low-k Dielectrics—2004, R. Carter, C. Hau-Riege, G. Kloster, T-M. Lu, S. Schulz, 2004, ISBN: 1-55899-762-8

Volume 813— Hydrogen in Semiconductors, N.H. Nickel, M.D. McCluskey, S. Zhang, 2004, ISBN: 1-55899-763-6

Volume 814— Flexible Electronics 2004—Materials and Device Technology, B.R. Chalamala, B.E. Gnade, N. Fruehauf, J. Jang, 2004, ISBN: 1-55899-764-4

Volume 815— Silicon Carbide 2004—Materials, Processing and Devices, M. Dudley, P. Gouma, P.G. Neudeck, T. Kimoto, S.E. Saddow, 2004, ISBN: 1-55899-765-2

Volume 816— Advances in Chemical-Mechanical Polishing, D. Boning, J.W. Bartha, G. Shinn, I. Vos, A. Philipossian, 2004, ISBN: 1-55899-766-0

Volume 817— New Materials for Microphotonics, J.H. Shin, M. Brongersma, F. Priolo, C. Buchal, 2004, ISBN: 1-55899-767-9

Volume 818— Nanoparticles and Nanowire Building Blocks—Synthesis, Processing, Characterization and Theory, O. Glembocki, C. Hunt, C. Murray, G. Galli, 2004, ISBN: 1-55899-768-7

Volume 819— Interfacial Engineering for Optimized Properties III, C.A. Schuh, M. Kumar, V. Randle, C.B. Carter, 2004, ISBN: 1-55899-769-5

Volume 820— Nanoengineered Assemblies and Advanced Micro/Nanosystems, J.T. Borenstein, P. Grodzinski, L.P. Lee, J. Liu, Z. Wang, D. McIlroy, L. Merhari, J.B. Pendry, D.P. Taylor, 2004, ISBN: 1-55899-770-9

Volume 821— Nanoscale Materials and Modeling—Relations Among Processing, Microstructure and Mechanical Properties, P.M. Anderson, T. Foecke, A. Misra, R.E. Rudd, 2004, ISBN: 1-55899-771-7

Volume 822— Nanostructured Materials in Alternative Energy Devices, E.R. Leite, J-M. Tarascon, Y-M. Chiang, E.M. Kelder, 2004, ISBN: 1-55899-772-5

Volume 823— Biological and Bioinspired Materials and Devices, J. Aizenberg, C. Orme, W.J. Landis, R. Wang, 2004, ISBN: 1-55899-773-3

Volume 824— Scientific Basis for Nuclear Waste Management XXVIII, J.M. Hanchar, S. Stroes-Gascoyne, L. Browning, 2004, ISBN: 1-55899-774-1

Volume 825E—Semiconductor Spintronics, B. Beschoten, S. Datta, J. Kikkawa, J. Nitta, T. Schäpers, 2004, ISBN: 1-55899-753-9

Volume 826E—Proteins as Materials, V.P. Conticello, A. Chilkoti, E. Atkins, D.G. Lynn, 2004, ISBN: 1-55899-754-7

Volume 827E—Educating Tomorrow's Materials Scientists and Engineers, K.C. Chen, M.L. Falk, T.R. Finlayson, W.E. Jones Jr,, L.J. Martinez-Miranda, 2004, ISBN: 1-55899-755-5

Volume 828— Semiconductor Materials for Sensing, S. Seal, M-I. Baraton, N. Murayama, C. Parrish, 2005, ISBN: 1-55899-776-8

Volume 829— Progress in Compound Semiconductor Materials IV—Electronic and Optoelectronic Applications, G.J. Brown, M.O. Manasreh, C. Gmachl, R.M. Biefeld, K. Unterrainer, 2005, ISBN: 1-55899-777-6

Volume 830— Materials and Processes for Nonvolatile Memories, A. Claverie, D. Tsoukalas, T-J. King, J. Slaughter, 2005, ISBN: 1-55899-778-4

Volume 831— GaN, AlN, InN and Their Alloys, C. Wetzel, B. Gil, M. Kuzuhara, M. Manfra, 2005, ISBN: 1-55899-779-2

Volume 832— Group-IV Semiconductor Nanostructures, L. Tsybeskov, D.J. Lockwood, C. Delerue, M. Ichikawa, 2005, ISBN: 1-55899-780-6

MATERIALS RESEARCH SOCIETY SYMPOSIUM PROCEEDINGS

Volume 833— Materials, Integration and Packaging Issues for High-Frequency Devices II, Y.S. Cho, D. Shiffler, C.A. Randall, H.A.C. Tilmans, T. Tsurumi, 2005, ISBN: 1-55899-781-4

Volume 834— Magneto-Optical Materials for Photonics and Recording, K. Ando, W. Challener, R. Gambino, M. Levy, 2005, ISBN: 1-55899-782-2

Volume 835— Solid-State Ionics—2004, P. Knauth, C. Masquelier, E. Traversa, E.D. Wachsman, 2005, ISBN: 1-55899-783-0

Volume 836— Materials for Photovoltaics, R. Gaudiana, D. Friedman, M. Durstock, A. Rockett, 2005, ISBN: 1-55899-784-9

Volume 837— Materials for Hydrogen Storage—2004, T. Vogt, R. Stumpf, M. Heben, I. Robertson, 2005, ISBN: 1-55899-785-7

Volume 838E—Scanning-Probe and Other Novel Microscopies of Local Phenomena in Nanostructured Materials, S.V. Kalinin, B. Goldberg, L.M. Eng, B.D. Huey, 2005, ISBN: 1-55899-786-5

Volume 839— Electron Microscopy of Molecular and Atom-Scale Mechanical Behavior, Chemistry and Structure, D. Martin, D.A. Muller, E. Stach, P. Midgley, 2005, ISBN: 1-55899-787-3

Volume 840— Neutron and X-Ray Scattering as Probes of Multiscale Phenomena, S.R. Bhatia, P.G. Khalifah, D. Pochan, P. Radaelli, 2005, ISBN: 1-55899-788-1

Volume 841— Fundamentals of Nanoindentation and Nanotribology III, D.F. Bahr, Y-T. Cheng, N. Huber, A.B. Mann, K.J. Wahl, 2005, ISBN: 1-55899-789-X

Volume 842— Integrative and Interdisciplinary Aspects of Intermetallics, M.J. Mills, H. Clemens, C-L. Fu, H. Inui, 2005, ISBN: 1-55899-790-3

Volume 843— Surface Engineering 2004—Fundamentals and Applications, J.E. Krzanowski, S.N. Basu, J. Patscheider, Y. Gogotsi, 2005, ISBN: 1-55899-791-1

Volume 844— Mechanical Properties of Bioinspired and Biological Materials, C. Viney, K. Katti, F-J. Ulm, C. Hellmich, 2005, ISBN: 1-55899-792-X

Volume 845— Nanoscale Materials Science in Biology and Medicine, C.T. Laurencin, E. Botchwey, 2005, ISBN: 1-55899-793-8

Volume 846— Organic and Nanocomposite Optical Materials, A. Cartwright, T.M. Cooper, S. Karna, H. Nakanishi, 2005, ISBN: 1-55899-794-6

Volume 847— Organic/Inorganic Hybrid Materials—2004, C. Sanchez, U. Schubert, R.M. Laine, Y. Chujo, 2005, ISBN: 1-55899-795-4

Volume 848— Solid-State Chemistry of Inorganic Materials V, J. Li, M. Jansen, N. Brese, M. Kanatzidis, 2005, ISBN: 1-55899-796-2

Volume 849— Kinetics-Driven Nanopatterning on Surfaces, E. Wang, E. Chason, H. Huang, G.H. Gilmer, 2005, ISBN: 1-55899-797-0

Volume 850— Ultrafast Lasers for Materials Science, M.J. Kelley, E.W. Kreutz, M. Li, A. Pique, 2005, ISBN: 1-55899-798-9

Volume 851— Materials for Space Applications, M. Chipara, D.L. Edwards, S. Phillips, R. Benson, 2005, ISBN: 1-55899-799-7

Volume 852— Materials Issues in Art and Archaeology VII, P. Vandiver, J. Mass, A. Murray, 2005, ISBN: 1-55899-800-4

Volume 853E—Fabrication and New Applications of Nanomagnetic Structures, J-P. Wang, P.J. Ryan, K. Nielsch, Z. Cheng, 2005, ISBN: 1-55899-805-5

Volume 854E—Stability of Thin Films and Nanostructures, R.P. Vinci, R. Schwaiger, A. Karim, V. Shenoy, 2005, ISBN: 1-55899-806-3

Volume 855E—Mechanically Active Materials, K.J. Van Vliet, R.D. James, P.T. Mather, W.C. Crone, 2005, ISBN: 1-55899-807-1

Volume 856E—Multicomponent Polymer Systems—Phase Behavior, Dynamics and Applications, K.I. Winey, M. Dadmun, C. Leibig, R. Oliver, 2005, ISBN: 1-55899-808-X

Volume 858E—Functional Carbon Nanotubes, D.L. Carroll, B. Weisman, S. Roth, A. Rubio, 2005, ISBN: 1-55899-810-1

Volume 859E—Modeling of Morphological Evolution at Surfaces and Interfaces, J. Evans, C. Orme, M. Asta, Z. Zhang, 2005, ISBN: 1-55899-811-X

Volume 860E—Materials Issues in Solid Freeforming, L. Jayasinghe, L. Settineri, A.R. Bhatti, B-Y. Tay, 2005, ISBN: 1-55899-812-8

Volume 861E—Communicating Materials Science—Education for the 21st Century, S. Baker, F. Goodchild, W. Crone, S. Rosevear, 2005, ISBN: 1-55899-813-6

Prior Materials Research Society Symposium Proceedings available by contacting Materials Research Society

Iron Aluminides

Mater. Res. Soc. Symp. Proc. Vol. 842 © 2005 Materials Research Society   S1.7

Strengthening of iron aluminide alloys for high-temperature applications

Martin Palm, André Schneider, Frank Stein and Gerhard Sauthoff
Max-Planck-Institut für Eisenforschung GmbH, Max-Planck-Str. 1, 40237 Düsseldorf, Germany

ABSTRACT

An overview is given on materials developments of ferritic and Fe_3Al-based iron aluminium alloys with strengthening precipitate phases for high-temperature applications currently underway at the Max-Planck-Institut für Eisenforschung GmbH (MPIE). The development of high-temperature alloys for structural applications is to be focussed on optimisation of strength, creep and corrosion resistance at high temperatures and sufficient ductility at lower temperatures. This is discussed with respect to recent studies on Fe-Al-based alloys with strengthening precipitates, such as κ-phase Fe_3AlC_x, MC-carbides, Laves phase, and the B2-ordered intermetallic phase NiAl. The following alloy systems have been investigated: Fe-Al-X (X=C, Ti, Ta, Mo, Zr), Fe-Al-Ti-Nb, Fe-Al-Ni-Cr, and Fe-Al-M-C (M=Ti, V, Nb, Ta).

The investigations have been focussed on the microstructure, constitution, mechanical properties, and high-temperature corrosion behaviour of Fe-Al-based alloys with Al contents ranging from 10 to 30 at. %.

INTRODUCTION

Iron aluminide-based alloys have a considerable potential for structural applications at elevated temperatures as they show a good corrosion resistance at high temperatures, have a low density compared with other iron-based alloys and they may be manufactured and processed with existing equipment [1-5]. However, in order to qualify as structural materials, strength and creep resistance at high-temperatures have to be improved and appropriate strategies have been discussed [1, 4, 6-10]. These strategies include strengthening by solid-solution hardening, coherent or incoherent precipitates, dispersoids, and/or ordering.

To what extent these measures are possible in certain Fe-Al-X(-Y) systems and how they affect other important properties such as ductility at low temperatures and corrosion resistance is not well understood. Therefore a variety of ternary and higher-order Fe-Al systems have been investigated at the MPIE with respect to basic mechanical properties and oxidation resistance.

SOLID-SOLUTION HARDENING

Few data are available on the effect of solid-solution hardening in iron aluminium alloys at high temperatures [11]. This is at least partly due to the fact that even at high temperatures the solid solubility of many elements in the Fe-Al phases is restricted to a few atomic percent or less. In figure 1 the 0.2%-proof stress of some Fe-Al-X alloys at 800 °C is shown in dependence of the concentration of X. All alloys under consideration are single-phase with ordered B2 structure at this temperature (see Table I). A marked influence of the yield stress anomaly can be ruled out as this temperature is well above of that for the anomaly for the respective alloys.

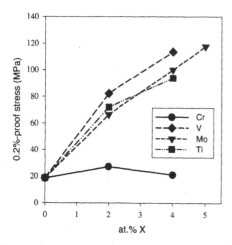

Figure 1. 0.2%-proof stress (in compression; $10^{-4}s^{-1}$ deformation rate) at 800 °C for as-cast Fe-26Al (all compositions given are in atomic percent) with additions of 2 and 4 at.% X (X=Cr, V, Mo,Ti) [12] and Fe-28Al-5Mo [13].

According to figure 1 alloying with V, Mo and Ti leads to a marked increase of the proof stress while adding Cr shows no effect at this temperature. The data do not indicate that the increase is related to the atomic size misfit as has been reported for solid-solution hardening at room temperature [14]. Also for B2-ordered alloys with 35 at.% Al a significant effect of solid-solution hardening on the proof stress has been reported for various additions of third elements [15]. Regarding the influence of solid-solution hardening on creep only few data do exist [11]. The data indicate that the creep resistance at elevated temperatures may increase by the addition of substitutional solutes and that especially Mo has a marked effect [16].

It is concluded that solid-solution hardening contributes to strengthening of Fe-Al-based alloys at high temperatures, but is restricted by the solubility of the alloying element in the Fe-Al phases which varies widely depending on the particular solute.

YIELD STRESS ANOMALY

Both intermetallic phases, Fe₃Al and FeAl, show an anomalous dependence of the yield stress on temperature. After an initial decrease between 100 and 300 °C the yield stress increases and shows a relative maximum at intermediate temperatures. The shape of the yield stress curve as a function of temperature is sensitively influenced by the alloy composition, the strain rate and the thermal pre-treatment (see e.g. [2, 17-20]). Since the temperature of the yield stress maximum as well as the yield stress at the maximum itself decrease with decreasing strain rate, the yield stress anomaly has a more pronounced effect on the yield stress than on creep resistance.

For binary Fe₃Al the maximum of the yield stress anomaly has been found at about 500 to 570 °C depending on the thermal pre-treatment of the alloy [21]. Because this temperature is near the temperature T_c for the D0₃/B2 transition, the yield stress anomaly has been associated

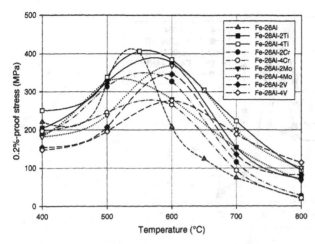

Figure 2. 0.2%-proof stress (in compression; $10^{-4}s^{-1}$ deformation rate) in dependence on temperature for as-cast Fe-26Al with additions of 2 and 4 at.% X (X=Cr, V, Mo, Ti).

with this transition [21-23], but more recent studies indicate that there is no direct correlation [24, 25].

The effect of alloying additions on the yield stress anomaly is illustrated in figures 2 and 3. In figure 2 the compressive 0.2%-proof stress is shown for a binary alloy and several ternary alloys containing 26 at.% Al (experimental details are given in [25]). Depending on temperature, all these alloys are single-phase DO_3- or B2-ordered. The transition temperature T_c has been determined by DTA (Table I). Apparently all alloying elements under consideration affect the temperature of the maximum of the yield stress anomaly in different ways. Addition of Cr only

Table I. Transition temperatures T_c for the DO_3 to B2 and B2 to A2 transition as determined by DTA (*: estimated according to [26]; x: B2 at the melting temperature).

Alloy composition (at.%)	Temperature of the yield stress maximum (°C)	T_c for DO_3/B2 (°C)	T_c for B2/A2 (°C)
Fe-26Al	540	546	829
Fe-26Al-2Ti	580	669	970
Fe-26Al-4Ti	560	773	1055
Fe-26Al-2Cr	550	551	822
Fe-26Al-4Cr	560	553	827
Fe-26Al-2Mo	510	610	873
Fe-26Al-4Mo	600	666	894
Fe-26Al-2V	600	635	975
Fe-26Al-4V	600	698	1045
Fe-26.9Al-15.4Ti	700	1150*	x
Fe-25.8Al-21.9Ti	700	1200*	x

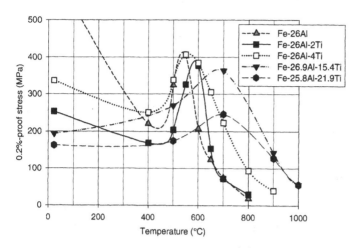

Figure 3. 0.2%-proof stress (in compression; $10^{-4}s^{-1}$ deformation rate) in dependence on temperature for Fe-26Al with various additions of Ti. Alloys with 0, 2 and 4 at.% are as-cast while those with 15.4 and 21.9 at.% Ti have been annealed at 1000 °C for 96 h for equilibration to avoid second phases [27].

slightly increases the temperature, the more the more Cr is added. Ti has a somewhat stronger effect, but the maximum of the yield stress anomaly is more increased by adding 2 at.% than by adding 4 at.% Ti. By alloying with 2 at.% Mo the maximum is even shifted to lower temperatures while an increase of about 50 K is observed when 4 at.% Mo are added. About the same increase is produced by V, but independent from the added amount. In a preceding study it was found that for Fe-26Al-4Ti-2X (X=Cr, Mo, V) alloys all X elements increase the temperature of the maximum of the yield stress anomaly to the same extent [25].

The comparison of the temperatures for the individual maxima of the yield stress anomaly with T_c for the $DO_3/B2$ transition (Table I) reveals that apparently no direct relation between them exists as had been suggested before [22]. For some alloys, e.g. Fe-26Al and Fe-26Al-2Cr, both temperatures are about the same, while for others T_c may be much higher than the temperature of the maximum of the yield stress anomaly, e.g. T_c for the $DO_3/B2$ transition for Fe-26Al-4Ti is about 200 K above that of the yield stress maximum. The same behaviour was found for the Fe-26Al-4Ti-X alloys [25]. It is noted that for all alloys discussed here the maximum of the yield stress anomaly is always observed in the DO_3 stability range or at about the $DO_3/B2$ transition but not within the stability range of B2.

In many Fe-Al-X systems the solid solubility of X in the binary Fe-Al phases is limited. A notable exception is the Fe-Al-Ti system where up to 25 at.% Ti can be dissolved in Fe_3Al [28]. Because Ti belongs to those elements which substitute for Fe on ½ ½ ½, the 4(b) Wyckoff site of the lattice [29], the DO_3-type ordering is maintained. Figure 3 shows the compressive proof stress of several single-phase Fe-26Al alloys with varying additions of Ti. With increasing Ti content the yield stress curves are shifted towards higher temperatures. Therefore, alloying with Ti leads to a marked increase of the yield stress at high temperatures. E.g. for a deformation rate of

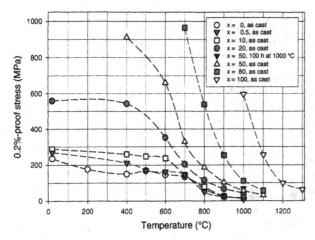

Figure 4. 0.2%-proof stress (in compression; 10^{-4}s^{-1} deformation rate) in dependence on temperature for Fe-10Al-Zr with varying volume fractions (x) of Laves phase (λ) [30].

10^{-4} s^{-1} a yield stress of 100 MPa is observed at about 670 °C for Fe-26Al and Fe-26Al-2Ti, at about 790 °C for Fe-26Al-4Ti and at about 940 °C for Fe-26.9Al-15.4Ti and Fe-25.8Al-21.9Ti. Apparently this increase does not depend on the Ti content in a linear way. It seems that the effect also comes from a decrease of the maximum yield stress which is accompanied by a broadening of the curve as it flattens.

It is noted that the shape of the yield stress curve is also influenced by annealing. For Fe-25.8Al-21.9Ti annealing at 400 °C prior to testing at this temperature led to a decrease in yield stress [27]. But it seems that such a low-temperature anneal does not affect the yield stress at high temperatures [25].

It is concluded that dissolving various ternary elements in Fe-Al alloys can have a varying influence on the yield stress anomaly, which is not yet understood. Again this effect is limited by the restricted solid solubility of many of these third elements in the Fe-Al phases. Only for systems where a considerable solid solubility exists and where T_c for the DO$_3$/B2 transition is increased thereby, a considerable contribution for increasing the strength at high temperatures can be expected.

INCOHERENT PRECIPITATES

Because the solid solubility for the X component is limited in many Fe-Al-X systems, various precipitates may be generated by alloying, e.g. other intermetallic phases or carbides. For a study of the sole effect of such precipitates, the Fe-Al-Zr system is well suited because nearly no solid solubility for Zr exists in the Fe-Al phases [31]. Alloying Fe-Al with up to 30 at.% Al with Zr results in the precipitation of a Laves phase (λ). Figure 4 shows the effect of varying volume fractions of λ on the yield stress of two-phase α (disordered A2) + λ alloys with a constant Al content of 10 at.% [30]. Already a volume fraction of 0.5% increases the yield stress, though a

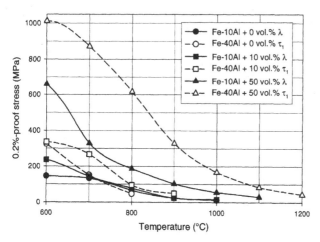

Figure 5. 0.2%-proof stress (in compression; $10^{-4}s^{-1}$ deformation rate) at high temperatures of as-cast Fe-10Al-xZr and Fe-40Al-xZr alloys with varying volume fractions of Laves phase (λ) or $(Fe,Al)_{12}Zr$ (τ_1), respectively.

marked increase is only observed for alloys with more than 10 vol.% of Laves phase. For these alloys not only an increase of the yield stress is seen at high temperatures but also at low and ambient temperatures. The increase in strength is paralleled by increased brittleness which already leads to fracture of the specimens when the load is applied (as indicated by missing data in figure 4).

The same behaviour is observed for alloys containing 20 and 40 at.% Al. The alloys with 40 at.% Al differ from those described above as the matrix is B2 ordered and the strengthening precipitates are $(Fe,Al)_{12}Zr$ (τ_1). According to figure 5 the alloys with precipitates of τ_1 have a higher yield stress at high temperatures than those with λ. As the alloys without any precipitates show about the same yield stress for Fe-10Al and Fe-40Al, the higher yield stress of the τ_1 containing alloys may be attributed to these precipitates and/or differences in the microstructure (figure 6). Detailed studies of the high-temperature strength have also been carried out at the MPIE for other Fe-Al alloys with strengthening incoherent precipitates of intermetallic phases, i.e. Fe-Al-Ta [32, 33], Fe-Al-Ti [27], Fe-Al-Ti-X (X = Cr, Nb, Mo) [25] and Fe-Al-Mo [13, 34].

Besides intermetallic phases, carbides may also act as strengthening phases. In the Fe-Al-C system a cubic perovskite-type κ-phase Fe_3AlC_x exists within an extended temperature and composition range [35]. The strengthening effect of Fe_3AlC_x at high temperatures has been studied in detail at MPIE [36-39]. For alloys with Al contents between 23 and 29 at.% no strengthening effect at high temperatures was observed when up to 3 at.% C, corresponding to a volume fraction of up to about 10 vol.% of Fe_3AlC_x, were added [39, 40]. For alloys with 21 to 30 at.% Al a marked increase in high-temperature hot hardness, yield stress and creep resistance is observed when the carbon content is raised to 6 to 11 at.% C, corresponding to volume fractions between 45 and 100 vol.% of Fe_3AlC_x [36, 37]. As has been described above for precipitates of intermetallic phases, the ductility at low and ambient temperatures decreases also with increasing volume fractions of Fe_3AlC_x. As the compressive yield stress for an alloy with 45

 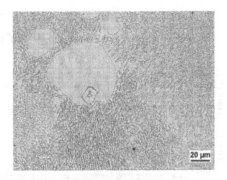

Figure 6. Light optical micrographs of (a) Fe-10Al-12.5Zr with 50 vol.% Laves phase (λ) and (b) Fe-40Al-4Zr with 50 vol.% (Fe,Al)$_{12}$Zr (τ_1). Fe-10Al-12.5Zr comprises a microstructure of primary dendrites of λ (light phase) within the λ + A2 phase eutectic, Fe-40Al-4Zr shows primary B2 (light phase) within the τ_1 + B2 phase eutectic.

vol.% Fe$_3$AlC$_x$ is about 100 MPa at 800 °C [41], it appears that this phase has a less strengthening effect than λ and τ_1 in Fe-Al-Zr alloys (figure 5), though other factors such as differences in microstructures may also account for this difference.

The combined effect of precipitating a Laves phase and additional carbides has also been studied. For Fe-Al-C-X (X=Ti, Nb, Ta) no marked coarsening of the precipitates was observed after annealing for 1 month at 1000 °C indicating the relative stability of these microstructures at high temperatures [42]. Though the volume fraction of the precipitates is only 2 to 4 vol.%, a slight increase of the yield strength was found at 800 °C compared to the solid solution strengthened alloys shown in figure 1, e.g. Fe-26Al-2Nb-1C and Fe-26Al-2Ta-1C showed a compressive yield strength of 100 and 110 MPa [43].

It is concluded that Fe-Al-based alloys can be considerably strengthened at high temperatures by precipitation of second phases. As the increase in strength is usually paralleled by a decrease in ductility at low temperatures, the use of this effect may be restricted.

COHERENT PRECIPITATES

In some Fe-Al-X systems, in which an appropriate miscibility gap exists and the lattice mismatch between the coexisting phases is sufficiently small, coherent precipitates may form. This is the case in the Fe-Al-Ni system, where a miscibility gap between disordered α-(Fe,Al) (A2) and ordered (Ni,Fe)Al (B2) exists. Other examples are the Fe-Al-Ti system where coherent α-(Fe,Al) (A2) + Fe$_2$TiAl (L2$_1$) alloys can be obtained [11] or the Fe-Al-V system, where coherent α-(Fe,Al) (A2) + Fe$_2$VAl (L2$_1$) as well as coherent (Fe,V)Al (B2) + Fe$_2$VAl (L2$_1$) alloys can be generated [44].

Only for the Fe-Al-Ni system some mechanical properties of such coherent microstructures have been studied before [45, 46]. These studies revealed that the fine-scaled microstructures with the hard (Ni,Fe)Al being either the matrix or the precipitate phase have a strong strengthening effect. These studies have now been extended to quaternary Fe-Al-Ni-Cr alloys

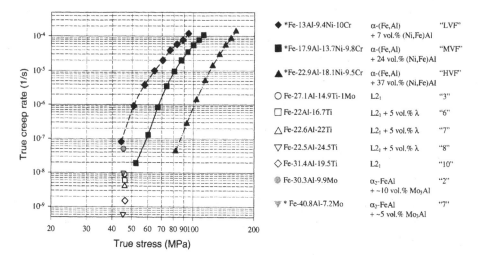

Figure 7. Stress dependence of secondary creep rate (in compression) at 750 °C for various alloys. Data for Fe-Al-Ni-Cr alloys were evaluated by step-wise increasing the load while all other data have been determined at constant load. Compositions are analysed or nominal (*) ones. Designations given in "" are those in the respective references (Fe-Al-Ni-Cr [47], Fe-Al-Ti(-Mo) [27], Fe-Al-Mo [13]). All alloys are as-cast except for Fe-Al-Ni-Cr which were solution-treated and annealed with a two-stage anneal as described in [47].

[47]. Cr has been added in view of improving the oxidation resistance and the current emphasis is on evaluating the creep behaviour while simultaneously maintaining a sufficient ductility at low and ambient temperatures.

Figure 7 shows the creep behaviour of three different Fe-Al-Ni-Cr alloys with low (LVF), medium (MVF) and high volume fractions (HVF) of (Ni,Fe)Al, which corresponds to a volume fraction of 7, 24, and 37 vol.%, respectively, at 900 °C. With increasing volume fractions of precipitate there is a marked increase of the creep resistance. The alloys were solution treated at 1200 °C for 24 h in order to ensure a homogeneous distribution of the alloying elements and to avoid segregation effects. The alloys were subsequently aged in air by a two-step anneal at 900 °C for 48 h followed by 300 h at 750 °C. By this heat treatment the volume fraction of fine-scaled secondary precipitates is considerably reduced. Thereby the brittle-to-ductile transition temperatures (BDTT) are maintained at 400 to 500 °C [47].

Figure 7 also shows secondary creep rates for some as-cast Fe-Al-Ti(-Mo) and Fe-Al-Mo alloys. Details of the preparation of these alloys as well as other mechanical properties are given in [27] and [13], respectively. The Fe-Al-Ti alloys are $DO_3(L2_1)$-ordered at 750 °C. They show a higher creep resistance than B2-ordered Fe-Al-Mo alloys with comparable or slightly higher volume fractions of precipitates. The secondary creep resistance of the Fe-Al-Ti and Fe-Al-Mo alloys corresponds to that of the LVF and MVF Fe-Al-Ni-Cr alloys, but the BDTT of the former alloys is between 700 and 900 °C for the Fe-Al-Ti alloys [27] and 600 to 700 °C for the Fe-Al-Mo alloys [13]. Whether the BDTT can be improved for the Fe-Al-Mo and Fe-Al-Ti alloys, has

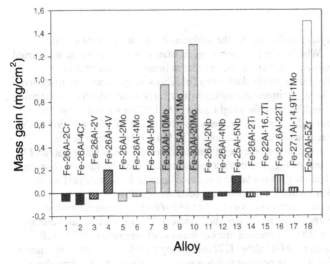

Figure 8. Weight gains during compression testing in air at 800 °C for various Fe-Al-based alloys with respect to Fe-26Al (Weight gain of Fe-26Al: 0.1 mg/cm^2 at 800 °C).

to be tested. Nevertheless the creep resistances already achieved are quite promising for Fe-Al-based alloys.

ORDERING

The high creep resistance of the Fe-Al-Ti alloys is attributed to the effect of the $D0_3(L2_1)$-ordered matrix. A number of elements like Ti [29], Mo, Ta and W [48] substitute preferentially for Fe on the 4(b) Wyckoff sites in $D0_3$-ordered Fe_3Al. Complete substitution of Fe on this site leads to the Heusler-type Fe_2XAl $L2_1$-structure. As discussed above, the solid solubility for many X elements is low in Fe_3Al and therefore substitution can only occur to a certain extent. An exception is the Fe-Al-Ti system where up to 25 at.% Fe can be substituted, i.e. the stoichiometric compound Fe_2TiAl can be achieved [28].

The effect of $L2_1$-ordering on yield stress and creep resistance at high temperatures has already been discussed in connection with figures 3 and 7. For Fe-Al-Ti alloys it has been found that the creep resistance and the yield stress at high temperatures increase while the strength at room temperature decreases with progressive atomic order [27], i.e. ordering has a positive effect on these properties. However, the BDTT for all investigated alloys have been found to be rather high, i.e. between 675 and 900 °C, though data for the ratio of plastic to elastic work indicate an increased ductility with increased order.

OXIDATION

For binary Fe-Al alloys it has been found that the oxidation behaviour improves with increasing Al content [1, 49]. Whether the high corrosion resistance of these alloys is maintained after alloying with other elements, which are added in order to improve the mechanical properties at high temperatures, has to be checked. For a first check of the short-term oxidation behaviour, the weight gains during compression testing in air may be used, which refer to testing times of about 45 min. In figure 8 the specific weight gains at 800 °C for various alloys are shown in comparison with that of Fe-26Al. As has been reported before [49, 50] addition of Cr even improves the oxidation resistance. Small additions of V, Mo and Nb at least do not deteriorate the oxidation resistance, though the oxidation resistance decreases when 4 at.% of V or 5 at.% of Mo or Nb are added. At high Mo contents a fast oxidation of the alloys is observed. Alloying with Ti does not affect the high oxidation resistance up to high contents [27] while marked additions of Zr are detrimental for the oxidation resistance [30].

For some of the alloys the oxidation behaviour has been verified by thermogravimetric measurements. They confirm the trends which have been indicated by the weight gains during compression testing in air. For 900 °C parabolic rate constants k_p of the order of 10^{-13} to 10^{-14} $g^2cm^{-4}s^{-1}$ were found for various Fe-Al-Ti alloys [27] confirming their high oxidation resistance at least up to additions of 17 at.% Ti. In contrast to this it was found for Fe-Al-Zr alloys that a k_p of 10^{-13} to 10^{-14} $g^2cm^{-4}s^{-1}$ at 900 °C is only observed for Zr contents of up to about 1 at.% [30], while higher Zr contents lead to a marked reduction of the oxidation resistance. First thermogravimetric measurements for Fe-Al-Mo alloys with Mo contents of 5 at.% or more indicate that these alloys show no parabolic but an unfavourable hyperbolic growth of the oxide scale.

Regarding the oxidation behaviour it can be concluded that alloying may have a strong influence on the oxidation resistance. While in many of the investigated alloys additions up to a few atomic percent did not show a marked effect on the oxidation behaviour larger contents can have a marked effect and even change the oxidation behaviour in an unfavourable way. In any case the oxidation behaviour should be evaluated when alloying additions are considered.

CONCLUSIONS

Fe-Al-base alloys are highly promising for high-temperature applications because of their outstanding oxidation and corrosion resistance, comparatively high thermal conductance and low thermal expansion, strengthening through atomic ordering (with a transition from A2 structure, i.e. bcc disorder, to ordered DO_3 structure and B2 structure with increasing Al content), solid-solution hardening and manifold possibilities of precipitate hardening through ternary alloying. A proper selection of the ternary alloying elements with optimisation of alloy content and processing for microstructure optimisation allows for balancing high-temperature strength, low-temperature ductility and high-temperature corrosion resistance in view of specific possible applications.

REFERENCES

1. D. Hardwick and G. Wallwork, *Rev. High-Temp. Mater.* **4**, 47 (1978).
2. K. Vedula, "FeAl and Fe$_3$Al", *Intermetallic Compounds Vol. 2, Practice*, ed. J.H. Westbrook and R.L. Fleischer (John Wiley & Sons, Chichester, 1995) pp. 199-209.
3. C.G. McKamey, "Iron Aluminides", *Physical Metallurgy and Processing of Intermetallic Compounds*, ed. N.S. Stoloff and V.K. Sikka (Chapman & Hall, New York, 1996) pp. 351-391.
4. N.S. Stoloff, *Mater. Sci. Eng.* **A258**, 1 (1998).
5. K. Natesan and P.F. Tortorelli in *Int. Symp. on Nickel and Iron Aluminides: Processing, Properties, and Applications*, edited by S.C. Deevi, P.J. Maziasz, V.K. Sikka and R.W. Cahn (ASM International, 1997) pp. 235-242.
6. D.G. Morris and M.A. Morris, *Mater. Sci. Eng.* **A239**, 23 (1997).
7. D.G. Morris, *Intermetallics* **6**, 753 (1998).
8. G. Sauthoff, *Intermetallics* **8**, 1101 (2000).
9. D.G. Morris and M.A. Munoz-Morris, *Revista de Metalurgia* **37**, 230 (2001).
10. A. Bahadur, *Mater. Sci. Technol.* **19**, 1627 (2003).
11. M. Palm, *Intermetallics*, accepted for publication.
12. F. Stein, A. Schneider and G. Frommeyer, presented at *"Discussion Meeting on the Development of Innovative Aluminium Alloys"* MPI für Eisenforschung, Düsseldorf, March 9th 2004.
13. M. Eumann, M. Palm and G. Sauthoff *Intermetallics*, **12**, 625 (2004).
14. J.H. Schneibel, E.P. George, E.D. Specht and J.A. Horton, *Mater. Res. Soc. Symp. Proc.* **364**, 73 (1995).
15. M.G. Mendiratta, S.K. Ehlers, D.M.Dimiduk, W.R. Kerr, S. Mazdiyasni and H.A. Lipsitt, *Mater. Res. Soc. Symp. Proc.* **81**, 393 (1987).
16. D.H. Sastry and R.S. Sundar in *Int. Symp. on Nickel and Iron Aluminides: Processing, Properties, and Applications*, edited by S.C. Deevi, P.J. Maziasz, V.K. Sikka and R.W. Cahn (ASM International, 1997) pp. 123-144.
17. I. Baker and E.P. George, *Mater. Res. Soc. Symp. Proc.* **552**, KK4.1.1. (1999).
18. D.J. Schmatz and R.H. Bush, *Acta metall.* **16**, 207 (1968).
19. D.G. Morris, P. Zhao and M.A. Munoz-Morris, *Mater. Sci. Eng.* **A297**, 256 (2001).
20. K. Yoshimi, S. Hanada and M.H. Yoo, *Acta metall. mater.* **43**, 4141 (1995).
21. N.S. Stoloff and R.G. Davies, *Acta metall.* **12**, 473 (1964).
22. R.S. Diehm and D.E. Mikkola, *Mater. Res. Soc. Symp. Proc.* **81**, 329 (1987).
23. U. Prakash, R.A. Buckley, H. Jones and C.M. Sellars, *ISIJ Int.* **31**, 1113 (1991).
24. W. Schröer, C. Hartig and H. Mecking, *Z. Metallkde.* **84**, 294 (1993).
25. F. Stein, A. Schneider and G. Frommeyer, *Intermetallics* **11**, 71 (2003).
26. I. Ohnuma, C.G. Schön, R.Kainuma, G. Inden and K. Ishida, *Acta mater.* **46**, 2083 (1998).
27. M. Palm and G. Sauthoff, *Intermetallics* **12**, 1345 (2004).
28. M. Palm, G. Inden and N. Thomas, *J. Phase Equilibria* **16**, 209 (1995).
29. G. Athanassiadis, G. Le Caer, J. Foct and L. Rimlinger, *Phys. Stat. Sol.* **40a**, 425 (1977).
30. F. Stein, M. Palm and G. Sauthoff, *Intermetallics*, accepted for publication.
31. F. Stein, G. Sauthoff and M. Palm, *Z. Metallkde.* **96**, 469 (2004).

32. D. D. Risanti and G. Sauthoff in *PRCIM-5 Advanced Materials and Processing* edited by Z.Y. Zhong, H. Saka, T.H. Kim, E.A.Holm, Y.F.Han and X.S. Xie (Trans Tech Publications, Zürich, 2004) pp. 865-868.
33. D. D. Risanti and G. Sauthoff, *Intermetallics*, accepted for publication.
34. M. Eumann, M. Palm and G. Sauthoff, *Steel Res. Int.* **75**, 62 (2004).
35. M. Palm and G. Inden, *Intermetallics* **3**, 443 (1995).
36. I. Jung and G. Sauthoff, *Z. Metallkde.* **80**, 490 (1989).
37. W. Sanders and G. Sauthoff, *Intermetallics.* **5**, 361 (1997).
38. W. Sanders and G. Sauthoff, *Intermetallics* **5**, 377 (1997).
39. A. Schneider, L. Falat, G. Sauthoff and G. Frommeyer, *Intermetallics*, accepted for publication.
40. A. Schneider and G. Sauthoff, *Steel Res. Int.* **75**, 55 (2004).
41. W. Wunnike-Sanders, "*Verformungsverhalten der Perowskitphasen im System Fe-Ni-Al-C*", thesis (RWTH Aachen, 1993) pp. 1-123.
42. A. Schneider, L. Falat, G. Sauthoff and G. Frommeyer, *Intermetallics* **11**, 443 (2003).
43. L. Falat, A. Schneider, G. Sauthoff and G. Frommeyer, *Intermetallics*, accepted for publication.
44. T. Maebashi, T. Kozakai and M. Doi, *Z. Metallkde.* **95**, 1005 (2004).
45. I. Jung, "*Untersuchung des Verformungsverhaltens ferritischer zweiphasiger Fe-Ni-Al-Legierungen mit großen Anteilen der intermetallischen (Fe,Ni)Al-Phase bei hohen Temperaturen*", thesis (RWTH Aachen, 1987) pp. 1-72.
46. I. Jung and G. Sauthoff, *Z. Metallkde.* **80**, 484 (1989).
47. C. Stallybrass, A. Schneider and G. Sauthoff, *Intermetallics*, accepted for publication.
48. L. Anthony and B. Fultz, *Acta metall. mater.* **43**, 3885 (1995).
49. R. Prescott and M.J. Graham, *Oxidat. Met.* **38**, 73 (1992).
50. Z.G. Zhang and Y. Niu in *PRCIM-5 Advanced Materials and Processing* edited by Z.Y. Zhong, H. Saka, T.H. Kim, E.A.Holm, Y.F.Han and X.S. Xie (Trans Tech Publications, Zürich, 2004) pp. 685-688.

Mater. Res. Soc. Symp. Proc. Vol. 842 © 2005 Materials Research Society   S1.10

Effect of Alloying Elements (Ga, Ge, Si) on Pseudoelasticity in Fe₃Al Single Crystals

Hiroyuki Y. Yasuda[1,2], Takashi Kase[2] and Yukichi Umakoshi[2]
[1]Research Center for Ultra-High Voltage Electron Microscopy, Osaka University, 7-1, Mihogaoka, Ibaraki, Osaka 567-0047, Japan
[2]Department of Materials Science and Engineering, Graduate School of Engineering, Osaka University, 2-1, Yamada-oka, Suita, Osaka 565-0871, Japan

ABSTRACT

Pseudoelasticity in Fe₃Al single crystals doped with a small amount of Ga, Ge and Si was investigated focusing on the antiphase boundary (APB) energy and the ordered domain structure. Single crystals of Fe-23at%Al and Fe-21at%Al-2at%X (X=Ga, Ge, Si) were grown by a floating zone method. In Fe-23at%Al single crystals, superpartial dislocations with Burgers vector (b) of 1/4<111> moved dragging APB during loading, while APB pulled back the superpartials during unloading. This resulted in giant pseudoelasticity regardless of martensitic transformation and the recoverable strain was about 5%. Ga addition was found to be effective in increasing the recovery strain compared with Fe-23at%Al crystals. In contrast, both Ge and Si additions decreased the amount of shape recovery. Stress at which the shape recovery started, was increased by Ga, Ge and Si additions. This means the APB energy increased by the additions, since the surface tension of APB pulling back the superpartials increases with increasing the energy. On the other hand, the frictional stress of the superpartials with b=1/4<111> increased significantly by Ge or Si doping due to solid solution hardening, though the stress of Ga-doped crystals was almost the same as that of the binary crystals. Higher frictional stress of Ge- and Si-doped crystals made the reversible motion of the superpartials difficult, resulting in the small recovery ratio. Ordered domains with displacement vector (R) of 1/4<111> in Fe-23at%Al and Fe-Al-Ga alloys were observed to be small, less than 100nm. In contrast, Ge and Si additions increased the domain size to more than 500nm. Since the domain boundaries with R=1/4<111> played an important role in the individual motion of the superpartials with b=1/4<111>, the fine domain structure was found to be favorable for giant pseudoelasticity in Fe₃Al single crystals. Ga addition increased the APB energy following the superpartials and kept the domain size small, resulting in the increase in recovery strain.

INTRODUCTION

Pseudoelasticity is a phenomenon by which strain is recovered during unloading. A giant pseudoelasticity generally results from a martensitic transformation; stress-induced martensitic transformation during loading and the reverse transformation during unloading [1]. However, DO₃-ordered Fe₃Al single crystals where martensitic transformation never occurred were found to demonstrate pseudoelasticity [2-4]. In Fe₃Al with the DO₃ structure, a superlattice dislocation with Burgers vector (b) of <111> generally dissociates into four superpartial dislocations with b=1/4<111> bound by antiphase boundary (APB) [5]. In contrast, the superpartial dislocations moved individually dragging nearest-neighbour APB (NNAPB) whose displacement vector (R) is 1/4<111> [2-4]. The NNAPB pulled back the superpartials during unloading resulting in the giant pseudoelasticity. The recoverable strain of Fe₃Al single crystals was approximately 5%, comparable with that of Ti-Ni alloys [1].

The amount of the strain recovery showed a maximum at an Al concentration of 23at% [3]. In contrast, little strain recovery was observed in Fe-28at%Al single crystals. In Fe-23at%Al crystals, superpartials with b=1/4<111> moved individually dragging NNAPB, while four superpartials bound by NNAPB and next-nearest neighbour APB (NNNAPB) moved in a group in Fe-28at%Al. There was also a big difference in ordered domain structure between Fe-23at%Al and Fe-28at%Al crystals; the domain size decreased with a decrease in Al content. When the superpartials with b=1/4<111> pass through B2-type domain boundaries with R=1/4<111>, high-energy-APB was created at the intersection between the superpartial and the domain boundaries. So, the superpartials may start to move individually not to increase the intersection area [3]. Moreover, The APB at the intersection pulled back the superpartials more strongly during unloading since the energy was higher than that of normal APB. Thus, the ordered domain structure played an important role in the pseudoelasticity in Fe_3Al single crystals; fine domain structure was found to be favourable for the giant pseudoelasticity [3].

A backward stress originated from APB increases with increasing APB energy. In contrast, an increase in ordered domain size results in a decrease in shape recovery. Therefore, control of both APB energy and ordered domain size may enhance the pseudoelasticity in Fe_3Al single crystals; addition of third elements is effective in controlling both of them. Fe_3Ga, Fe_3Ge and Fe_3Si are also known to have the DO_3 structure. According to Fe-Ga, Fe-Ge and Fe-Si binary phase diagrams [6,7], the ordering temperatures of these compounds are higher than that of Fe_3Al. This means Ga, Ge and Si additions may increase the APB energy of Fe_3Al. In addition, we have found giant pseudoelasticity appeared in Fe_3Ga single crystals and the APB energy was higher than that of Fe_3Al crystals [8]. In this paper, we report the effect of third element, Ga, Ge and Si on the pseudoelasticity in Fe_3Al single crystals focusing on the APB energy and the ordered domain structure.

EXPERIMENTAL PROCEDURE

Since Ga, Ge and Si are believed to preferentially substitute for Al atoms [9], Al+X (X=Ga, Ge, Si) concentration should be kept constant in order to examine the effect of the doping. Fe-23at%Al and Fe-21at%Al-2at%X (X=Ga, Ge, Si) single crystals were grown by the floating zone method at a growth rate of 5mm/h. After homogenization at 1373K for 48h, the single crystal rods were slowly cooled to room temperature at a cooling rate of 80K/h. The compression specimens with [$\bar{1}$49] loading axis were cut from the single crystals by spark machining. The Schmid factor for ($\bar{1}$01)[111] primary slip system is 0.5. Compression tests were performed in air at room temperature at a constant cross-head speed of 0.05mm/min corresponding to an initial strain rate of 1.67×10^{-4}/s. After the compression, the specimens were subsequently unloaded at the same cross-head speed. The shape recovery of the specimens during unloading was checked from the stress-strain curve and the dimension of the specimens. The recovery ratio (r) was defined as follows

$$r = \frac{\varepsilon_p - \varepsilon_r}{\varepsilon_p} \times 100 \tag{1}$$

where ε_p and ε_r are the applied plastic strain and the residual strain, respectively. Ordered domain structure and deformation substructure of the crystals was observed by a transmission electron microscopy (TEM).

RESULTS and DISCUSSION

Pseudoelasticity in Fe₃Al single crystals doped with Ga, Ge and Si

Figure 1 shows stress-strain curves of Fe-23at%Al and Fe-21at%Al-2at%X single crystals. These crystals were compressed to ε_p=5.0% and subsequently unloaded. Fe₃Al single crystals doped with 2at%Ge and Si exhibit higher yield stress and smaller strain recovery than Fe-23at%Al crystals. In contrast, the plastic strain applied to Fe-23at%Al and Fe-21at%Al-2at%Ga single crystals is perfectly recovered during unloading. It should be noted that recovery-start stress of Fe-21at%Al-2at%Ga single crystals is higher than that of Fe-23at%Al crystals. Moreover, Fe₃Al single crystals doped with 2at%Ge and Si also exhibit higher recovery-start stress than Fe-23at%Al crystals, though the recovery ratio of the doped crystals is less than 50%. Recovery ratios of these crystals are plotted against applied plastic strain in Fig.2. Ga-doped crystals demonstrate higher recovery ratio than the others. In particular, the recovery ratio of Fe-Al-Ga crystal at more than ε_p=7.5% is higher than that of Fe-Al binary crystals. Thus, Ga addition is effective in the enhancement of pseudoelasticity in Fe₃Al single crystals. From a simple estimation, if mobile dislocation density is 10^{12}m^{-2}, an average distance of dislocation motion is calculated to be 400μm at ε_p=5%. Thus, a recoverable strain of 5% in Fe-Al and Fe-Al-Ga single crystals seems to be too large to be achieved by dislocation motion. However, Kubin et al. [2] directly observed the superpartials pulled back by NNAPB. Therefore, the mechanism of pseudoelasticity in Fe₃Al single crystals needs further discussion.

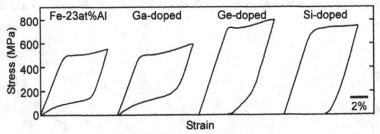

Figure 1. Stress-strain curves of Fe-Al-X (X=Ga, Ge, Si) single crystals compressed to ε_p=5.0%.

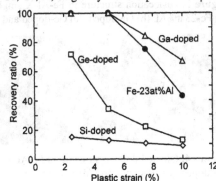

Figure 2. Recovery ratio of Fe-Al-X (X=Ga, Ge, Si) single crystals as a function of plastic strain.

Dislocation and ordered domain structures in Fe-Al-X single crystals

Figure 3 shows dislocation structure in Fe-Al and Fe-Al-X single crystals compressed to ε_p =5.0% and subsequently unloaded. Interestingly, dislocation densities in Fe-Al and Fe-Al-Ga single crystals are quite low, though these crystals were deformed to ε_p=5.0%. Furthermore, dislocation loops elongated along [111] direction are frequently observed in Fe-Al and Fe-Al-Ga crystals, as shown in Fig.3 (a) and (b). In the magnified weak beam image, the loop-like dislocations were confirmed to be uncoupled superpartials with b=1/4[111]. From a contrast analysis with reflection vector (g) of 020, NNAPB with R=1/4[111] exists inside the dislocation loops. During the TEM observation, the loops sometimes shrank and subsequently disappeared. So, the superpartials were pulled back by NNAPB during unloading, resulting in the giant pseudoelasticity in Fe-Al and Fe-Al-Ga single crystals. In contrast, highly dense dislocations are observed in Ge- and Si-doped crystals as shown in Fig.3 (c) and (d), though the superpartials with b=1/4[111] are also uncoupled, similar to Fe-Al and Fe-Al-Ga crystals. It is also noted that dislocation walls aligned nearly parallel to [1$\bar{2}$1] and [010] directions are formed in Ge- or Si-doped crystals. Since [1$\bar{2}$1] direction is the edge orientation of superpartials with b=1/4[111], pileup of the superpartials at the edge dipoles leads to the formation of [1$\bar{2}$1] walls. On the other hand, [010] walls may be created by the interaction between primary ($\bar{1}$01)[111] and secondary (101)[$\bar{1}$11] slips. These dislocation walls may suppress the backward motion of the superpartials during unloading, resulting in low recovery ratio of Ge- and Si-doped crystals.

Ordered domain structure in Fe-Al and Fe-Al-X crystals is shown in Fig.4. There are two types of ordered domain boundaries in the D0$_3$ phase; B2-type with R=1/4<111> and D0$_3$-type

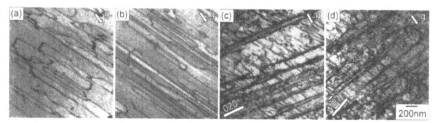

Figure 3. Dislocation structure in Fe-Al-X single crystals compressed to ε_p=5.0%; g=202. (a) Fe-23at%Al, (b) Ga-doped, (c) Ge-doped and (d) Si-doped.

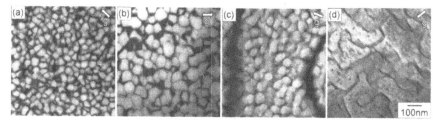

Figure 4. Ordered domain structure in Fe-Al-X single crystals; g=020. (a) Fe-23at%Al, (b) Ga-doped, (c) Ge-doped and (d) Si-doped.

with R=1/2<111>. The contrast of the domain boundaries is known to depend on the ordered reflection [4]; B2-type boundary is visible with both g=111 and g=020, while DO_3-type one is out of contrast with g=020. There exist numerous small domains less than 100nm in the binary and Ga-doped crystals as shown in Fig. 4 (a) and (b). Both B2 and DO_3-type domain boundaries are densely introduced in the crystals since some domain boundaries are in contrast with g=020. In contrast, a few B2-type domain boundaries can be seen in Ge-doped crystals (Fig.4 (c)). Furthermore, there is no B2-type domain boundary in the observed area in Si-doped crystals as shown in Fig.4 (d). Since interaction between B2-type domain boundaries and superpartials with b=1/4[111] plays an important role in the pseudoelasticity in Fe_3Al single crystals [3], fine domain structure with B2-type boundaries is favourable for the pseudoelasticity [3]. Coarse domain structure and low density of B2-type domain boundaries in Ge- and Si-doped crystals resulted in the low recovery ratio. It is also noted that spherical bcc phase is observed to precipitate in Si-doped crystals as shown in Fig.4 (d). The precipitation of bcc phase leads to the localized slip in Si-doped crystals and accelerate the formation of the dislocation walls, resulting in little strain recovery in the crystals. On the other hand, in Ga-doped crystals, the size of ordered domains surrounded by B2-type boundaries kept to be small, which results in giant pseudoelasticity similar to the binary crystals.

Stress acting on superpartial dislocation in Fe-Al-X single crystals

When the pseudoelasticity caused by surface tension of APB appears in Fe_3Al single crystals, the critical resolved shear stress (τ_y) during loading and the recovery-start shear stress (τ_r) during unloading are given by

$$\tau_y = \tau_0 + \tau_b \qquad (2)$$
$$\tau_r = \tau_b - \tau_0 \qquad (3)$$

where τ_0 is the frictional stress of superpartials with b=1/4[111] and τ_b is the backward stress acting on the superpartials. Since τ_b is mainly originated from NNAPB following the superpartial, τ_b increases with increasing the energy of NNAPB. By substituting each parameter, τ_0 and τ_b were calculated and plotted in Fig.5. τ_0 remarkably increases by Ge or Si doping, while τ_0 of Ga-doped crystals is similar to that of Fe-Al binary crystals. An increase in τ_0 by Ge or Si doping may result from solid solution hardening, since the atomic sizes of Ge and Si are quite smaller than that of Al. The increase of τ_0 is disadvantageous for the pseudoelasticity since the superpartial with b=1/4[111] must smoothly move forth and back during loading and unloading to exhibit shape recovery. On the other hand, τ_b increases by adding any element, especially by Ge and Si additions. This means that the energy of NNAPB was

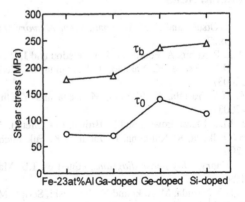

Figure 5. τ_0 and τ_b in Fe-Al-X single crystals deformed to ε_p=5.0%.

increased by addition of Ga, Ge and Si. However, Ge or Si doping was unfavourable for the pseudoelasticity in Fe_3Al single crystals. Both Ge and Si additions increase not only τ_b but also τ_0 and the ordered domain size. This leads to the small strain recovery in Ge- and Si-doped crystals. In contrast, Ga addition increased the energy of NNAPB and kept the domain size small, resulting in the enhancement of the pseudoelasticity.

CONCLUSIONS

The effect of Ga, Ge and Si addition on pseudoelasticity in Fe_3Al single crystals was investigated. The following conclusions were reached.
(1) Ga-doped Fe_3Al single crystals showed higher recovery ratio than Fe-Al binary crystals, while addition of Ge and Si was not effective in the enhancement of the pseudoelasticity.
(2) Ga addition increased the energy of NNAPB dragged by superpartials with b=1/4[111] and kept the domain size small, resulting in the increase of recovery strain.
(3) Ge and Si doping increased the frictional stress of the superpartials, the energy of NNAPB and size of ordered domains surrounded by B2-type boundaries. As a result, the strain recovery of Ge- and Si-doped crystals was small compared with that of Fe-Al binary crystals.

ACKNOWLEDGMENTS

This work was supported by a Grant-in Aid for Scientific Research from the Ministry of Education, Culture, Sports, Science and Technology of Japan. Part of this work was carried out at the Strategic Research Base "Handai Frontier Research Center" supported by the Japanese Government's Special Coordination Fund for Promoting Science and Technology. This work was also supported by "Priority Assistance of the Formation of Worldwide Renowned Centers of Research - The 21st Century COE Program (Project: Center of Excellence for Advanced Structural and Functional Materials Design)" from the Ministry of Education, Sports, Culture, Science and Technology of Japan.

REFERENCES

1. K. Otsuka and C. M. Wayman, *Shape Memory Materials*, (Cambridge University Press, 1998), pp.27.
2. L. P. Kubin, A. Fourdeux, J. Y. Guedou and J. Rieu, Phil. Mag., **A46**, 357 (1982).
3. H.Y. Yasuda, K. Nakano, T. Nakajima, M. Ueda and Y. Umakoshi, Acta Mater., **51**, 5101 (2003).
4. H.Y. Yasuda, K. Nakano, M. Ueda and Y. Umakoshi, Mater. Sci. Forum, **426-432**, 1801 (2003).
5. M. J. Marcinkowski and N. Brown, Acta Metall., **9**, 764 (1961).
6. O. Ikeda, R. Kainuma, I. Ohnuma, K. Fukamichi and K. Ishida, J. Alloys Comp., **347**, 198 (2002).
7. *Binary Alloy phase diagrams*, edited by T.B. Massalski (Materials Park, ASM International, 1990)
8. H.Y. Yasuda, M. Aoki and Y. Umakoshi, Scripta Mater., in press.
9. S. Zuqing, Y. Wangyue, S. Lizhen, H. Yuanding, Z. Baisheng and Y. Jilian, Mater. Sci. Eng., **A258**, 69 (1998).

Mater. Res. Soc. Symp. Proc. Vol. 842 © 2005 Materials Research Society   S1.9

PSEUDOELASTICITY OF $D0_3$ ORDERED MONOCRYSTALLINE Fe_3Al

S. Kabra,[1] H. Bei,[1,3] D. W. Brown,[2] M.A.M. Bourke,[2] and E. P. George, [1,3]

[1]The University of Tennessee, Department of Materials Science and Engineering, Knoxville, TN 37996
[2]Los Alamos National Laboratory, Los Alamos, NM 87545
[3]Oak Ridge National Laboratory, Metals and Ceramics Division, Oak Ridge, TN 37831

ABSTRACT

Pseudoelasticity in monocrystalline Fe_3Al (23 at.% Al) was investigated by room-temperature mechanical testing along the <418> tensile and compressive axes. In tension, up to ~10% strain is recoverable whereas only ~5% strain is recoverable in compression. Straight, parallel, surface step lines were seen to appear/disappear as the specimens were pseudoelastically loaded/unloaded. In contrast, in the plastic region ($\varepsilon > 10\%$), wavy slip lines appeared on the specimen surfaces which did not disappear upon unloading. In-situ neutron diffraction was performed during compressive straining and the intensities of several diffraction peaks increase/decrease reversibly during loading/unloading. These changes are consistent with a deformation twin which produces large crystal rotations. They could also be indicative of a phase transformation. Unfortunately, we were able to sample only a limited range of 2θ in the present investigation and, within this range, none of the new peaks that appeared during the pseudoelastic deformation were disallowed peaks for the $D0_3$ crystal structure. Therefore we are unable at this time to distinguish between the two possible mechanisms, twinning and phase transformation.

INTRODUCTION

Iron aluminides based on Fe_3Al have attractive properties for structural applications at intermediate temperatures, including good oxidation resistance and low density, e.g. [1]. However, a property that has gone relatively unnoticed is its room-temperature pseudoelasticity. Pseudoelasticity in single crystal Fe_3Al was reported by Guedou et. al. [2] as early as 1976; however, there have been relatively few papers since then regarding this phenomenon [2-9]. Briefly, pseudoelasticity is observed only for compositions in the range 21-29 at.% Al [2] and for $D0_3$ rather than B2 ordering [3]. Up to ~5% compressive strain is recoverable, but there is a strong orientation dependence with full recovery possible only for deformation along directions near the <419> that maximize the Schmid factor for the (101)[111] system, and significantly less recovery as one deviates from this orientation [8]. When the alloy is deformed at 77 K, the imposed strain does not recover upon unloading but does so only after the specimen warms up to room temperature [2,3], indicating that Fe_3Al exhibits both shape memory behavior and pseudoelasticity.

At first, twinning was suggested as a possible mechanism for pseudoelasticity [3], but no supporting experimental evidence was offered. Subsequently, an APB-dragging mechanism was postulated [4], which seemed to be supported by in-situ transmission electron microscopy observations showing reversible motion of the leading partial dislocation during loading/unloading [4,7-9]. However, it is difficult to see how large amounts of strain can be produced by the APB mechanism if only the leading partial moves while the trailing partial remains fixed. On the other hand, if the entire dislocation moves (and multiplies), enough strain can certainly be generated, but then there is no restoring force to drive the strain recovery during unloading.

In the most widely studied system that exhibits both pseudoelasticity and shape memory, NiTi, a stress-induced martensitic transformation is known to be responsible for pseudoelasticity [e.g., 10]. While there have been no reports of such a transformation in Fe$_3$Al, it may be because, to our knowledge, there have been no attempts to perform in-situ structural characterization during deformation.

In this study we first investigate pseudoelasticity in tension (since prior experiments were done in compression) and compare our results with the earlier compression data. We also performed in-situ neutron diffraction during compressive loading/unloading to determine whether any structural transformations occur during the pseudoelastic deformation.

EXPERIMENTAL DETAILS

Alloys of composition Fe-23Al (all compositions in at.%) were arc melted, drop-cast, and directionally solidified in an optical floating zone furnace to produce a <100> single crystal. This single crystal was oriented and cut normal to <418> and used to seed additional single crystals having the <418> growth direction. The <418> crystals were homogenized at 1100°C for 48 h and furnace cooled at 80°C/h to maximize DO_3 order [9].

Dogbone shaped specimens having a 1 × 2 × 24 (mm) gage section were tensile tested along <418> direction at room temperature at a constant crosshead speed of 0.001 mm/s, which corresponded to an engineering strain rate in our specimens of 5×10^{-5} s^{-1}.

In-situ neutron diffraction experiments were performed on the SMARTS diffractometer at the Lujan Center for Neutron Scattering at the Los Alamos National Laboratory [11]. Compressive loading-unloading cycles were performed on cylindrical specimens 6 mm in diameter and 14.4 mm long with their compression axes along <418>. At various points on the stress-strain curve, the cross-head motion was stopped (for ~ 3 minutes) to collect diffraction data.

RESULTS AND DISCUSSION

Figure 1 shows tensile stress-strain curves of Fe$_3$Al exhibiting almost complete recovery of applied strains up to ~10% and plastic (unrecoverable) deformation beyond that. There is no indication of any hardening in tension, unlike in compression (Fig. 2) where there is a small amount of hardening evident in the pseudoelastic regime (Fig. 2). Another difference between tension and compression is that significantly more strain is recovered in tension than in compression. The elastic to pseudoelastic transition occurs at a stress of ~500 MPa in both tension and compression and there is a large loading/unloading hysteresis.

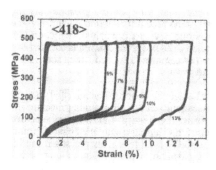

Figure 1 Tensile stress-strain curves of monocrystalline Fe$_3$Al.

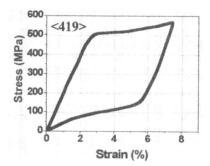

Figure 2 Compressive stress-strain curve of monocrystalline Fe$_3$Al [9].

Most of the applied strain (~97%) was recovered immediately upon unloading [Fig. 3 (a)]. The remaining strain recovered in a time-dependent manner, as shown in the inset of Fig. 3 (a) for 10% applied strain, and in Fig. 3 (b) for other strains. Straight, parallel, step lines appeared on the sample surfaces in the pseudoelastic region [Fig. 4 (a)], and their density increased with increasing strain. Upon unloading from the pseudoelastic region these lines disappeared completely [Fig. 4 (b)].

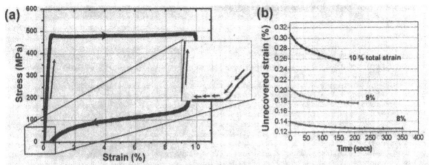

Figure 3 (a) Tensile stress-strain curve showing that ~97% of the applied strain of 10% is recovered instantaneously upon unloading with the remainder recovered in a time-dependent manner; (b) time dependence of strain recovery after unloading.

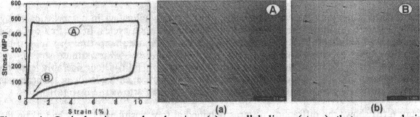

Figure 4. Optical micrographs showing (a) parallel lines (steps) that appear during pseudoelastic deformation, and (b) the complete disappearance of these surface steps upon unloading.

When specimens were loaded in tension to strains more than 10%, a large part of the strain was unrecoverable upon unloading (Fig. 1). This result can be related to the two types of surface features that appear during loading. Figure 5 (a) shows parallel step like features (Type 1 lines) that appear on the surface of the specimen when it is deformed in the pseudoelastic region [similar to Fig. 4 (a)]. After ~10% strain, wavy slip lines (Type 2 lines) appear on the surface [Fig. 5 (b)]. Upon unloading from ~13% strain, most, but not all, of the Type 1 lines disappear, whereas all the Type 2 lines remain [Fig. 5 (c)]. Clearly, Type 1 lines are associated with the recoverable part of the strain while Type 2 lines are slip lines associated with plastic deformation. Only ~4% strain is recovered when the sample is unloaded from 13% strain, whereas essentially all of the strain is recovered when the sample is unloaded from 10% strain (within the pseudoelastic regime). This indicates that a major part of the pseudoelastic (recoverable) strain is trapped once irreversible plastic deformation commences.

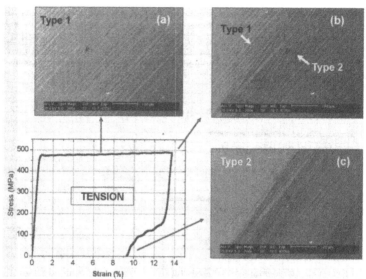

Figure 5 SEM micrographs showing (a) surface steps (Type 1 lines) in the pseudoelastic region, (b) wavy slip lines (Type 2 lines) beyond ~ 10% applied strain, and (c) Type 2 lines remaining after unloading (many, but not all, of the Type 1 lines are gone).

Neutron diffraction patterns were obtained during loading/unloading in compression along <418>. Figure 6 (a) shows the stress-strain curves for two such cycles. In the first cycle the specimen was deformed in the pseudoelastic regime (~3% maximum strain), while in the second cycle the specimen was taken into the plastic regime (~6% maximum strain). As mentioned before, the recoverable strain in compression [Fig. 6 (a)] is considerably less than that in tension (Fig. 1). Neutron diffraction patterns were obtained while the cross-head was stopped at different positions [some of which are indicated by arrows in Fig. 6 (a)].

Figure 6 (b) shows the diffraction patterns obtained at four positions (1, 2, 3 and 4) on the stress-strain curve. Position 1 lies well within the elastic region of the stress-strain curve and the diffraction pattern at this position is essentially the same as that in the unloaded condition (Position 3). Large changes in peak intensity with respect to the initial pattern are evident in the pseudoelastic region (Position 2). Some peaks increase in intensity, while others decrease. Two of the notable peaks are (422) and (844): they have zero intensity in position 1 but have significant intensity at position 2. In fact, the (422) peak goes from zero intensity before straining (1), to being the most intense peak in the pseudoelastic region (2), and back to zero intensity in the unloaded condition (3). Such changes in peak intensity are indicative of large crystal rotations, which may be caused by stress-induced twinning or a change in crystal symmetry due to a stress-induced phase transformation. Unfortunately, the detector configuration in our current experiments allowed us to sample only a limited range of 2θ. Within this range, none of the new peaks that appeared at position 2 are disallowed peaks for the $D0_3$ crystal structure. Therefore we are unable at this time to distinguish between the two possible mechanisms. Additional experiments are planned using more complete detector coverage that will allow us to measure a larger portion of the standard stereograph of the material. Figure 6 (b) also shows a spectrum obtained at position 4, (i.e. after the specimen was unloaded from the plastic region). Unlike what was observed at position 3, the intensities of the (422) and (844) peaks do not go to zero at position 4, indicating that a significant portion of the previously recoverable (pseudoelastic) strain gets trapped and becomes unrecoverable once plastic deformation commences, consistent with the behavior of the Type 1 and 2 surface features discussed earlier (Fig. 5).

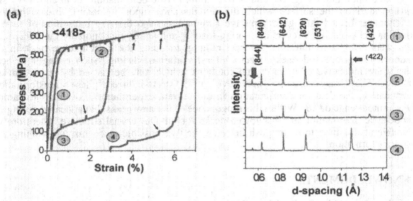

Figure 6 (a) Compressive stress-strain curves of Fe_3Al single crystal, (b) neutron diffraction patterns obtained at different locations (1, 2, 3, and 4) on the stress-strain curves.

Figure 7 (a) shows the variation of the (422) peak intensity as a function of the applied strain. In the elastic region ($\varepsilon < 0.5\%$), the (422) peak is not present. It makes its first appearance in the pseudoelastic region, and thereafter increases in intensity almost linearly with increasing strain. Upon unloading, its intensity decreases with decreasing strain, following to the loading curve but with a small hysteresis. The normalized (422) peak intensity can also be plotted as a function of applied stress, and compared with the stress-strain behavior as shown in Fig. 7 (b). There is an almost one-to-one correlation between the two curves.

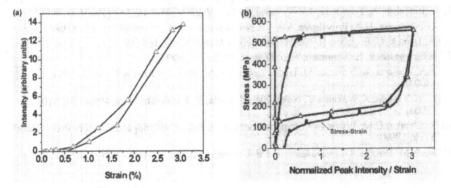

Figure 7 (a) Peak intensity of (422) reflections as a function of applied strain; (b) Variation of the normalized (422) peak intensity with stress compared to the pseudoelastic stress-strain behavior.

CONCLUSIONS

Single crystal Fe_3Al can recover up to 10% strain in tension along <418>, almost twice as much as previously shown to be recoverable in compression. In conjunction with this recoverable strain, straight, parallel step lines appeared on the specimen surface, which disappeared upon unloading. Beyond ~10% strain, plastic deformation set in and wavy slip lines appeared on the surface which did not disappear upon unloading. In-situ neutron diffraction performed during compressive straining showed that the intensities of several diffraction peaks changed reversibly during loading/unloading, most notably the (422) peak which went from zero intensity before straining, to being the most intense peak in the pseudoelastic region, and back to zero intensity after unloading. Such changes in peak intensity are indicative of large crystal rotations, which may be caused by stress-induced twinning. They may also be indicative of a stress-induced phase transformation. Unfortunately, the detector configuration in our current experiments allowed us to sample only a limited range of 2θ. Within this range, none of the new peaks that appeared during the pseudoelastic deformation are disallowed peaks for the DO_3 crystal structure. Therefore we are unable at this time to distinguish between the two possible mechanisms, twinning and phase transformation.

ACKNOWLEDGMENTS

Research sponsored by the Division of Materials Sciences and Engineering, Office of Basic Energy Sciences, U. S. Department of Energy, under Contract DE-AC05-00OR22725 with UT-Battelle, LLC. This work also benefited from the use of the Los Alamos Neutron Science Center (LANSCE) at the Los Alamos National Laboratory funded by the US Department of Energy under Contract W-7405-ENG-36.

REFERENCES

1. C.G. McKamey, J.H. DeVan, P.F. Tortorelli and V.K. Sikka. *J. Mater. Res.* 6, 1779 (1991).
2. J.Y. Guedou, M. Paliard, J. Rieu, *Scripta Met.* 10, 631 (1976).
3. J.Y. Guedou, J. Rieu, *Scripta Met.* 12, 927 (1978).
4. L.P. Kubin, A. Fourdeux, J.Y. Guedou, J. Rieu, *Phil. Mag. A* 46, 357 (1982).
5. G.I. Nosova, N.A. Polyakova, Ye. Ye. Novikova, *Phys. Met. Metall.* 61, 151 (1986).
6. A. Brinck, C. Engelke, H. Neuhäuser, *Scipta Mater.* 37(5), 569 (1997).
7. E. Langmaack, E. Nembach, *Phil. Mag. A* 79, 2359 (1999).
8. H.Y. Yasuda, K. Nakano, M. Ueda, Y. Umakoshi, *Mat. Sci. Forum* 426-432, 1801 (2003).
9. H.Y. Yasuda, K. Nakano, T. Nakajima, M. Ueda, Y. Umakoshi, *Acta Mater.* 51, 5101 (2003).
10. K. Otsuka, C.M. Wayman, "Shape memory materilas", Cambridge University Press, New York, 1998.
11. M. A.M. Bourke, D. C. Dunand, and E. Ustundag, *Applied Physics A* 74, s1707 (2002).

Mater. Res. Soc. Symp. Proc. Vol. 842 © 2005 Materials Research Society   S1.8

Development of high temperature creep resistance in Fe-Al alloys

D.G. Morris[1], M.A. Muñoz-Morris[1] and C. Baudin[2]
[1]Department of Physical Metallurgy, CENIM, CSIC, Avenida Gregorio del Amo 8, E-28040 Madrid, Spain.
[2]Instituto de Ceramica y Vidrio, CSIC, Campus de Cantoblanco, Camino de Valdelatas, E-28049, Madrid, Spain

ABSTRACT

Most of the studies aimed at the development of creep-resisting Fe-Al intermetallics have been oriented at application temperatures of the order of 500-650°C, where these materials may compete with conventional stainless steels. The Fe-Al intermetallics are, however, particularly excellent in their oxidation and corrosion resistances at temperatures of the order of 1000°C, where Chromium-Nickel steels are no longer able to withstand the aggressive environments. This presentation is part of a study aimed at the development of good creep resistance at such high temperatures.

Studies of a variety of cast Fe_3Al-base alloys, strengthened by solution or precipitate/dispersoid-forming alloying additions, are reported. The alloys show good strength from room temperature to about 500°C, but thereafter strength falls rapidly as thermally-activated deformation processes become operative. Solution additions are capable of producing good low temperature strength, but do not contribute significantly to creep strength at very high temperatures (above 700°C). Precipitation hardening has been examined in Nb-containing alloys, where Fe_2Nb Laves precipitates form at intermediate temperatures. These materials show good strength up to about 700°C, but at higher temperatures the fine precipitates coarsen excessively. Strengthening in the intermediate temperature range varies depending on whether the solute is precipitated prior to high temperature testing or concurrent with this.

INTRODUCTION

Most of the work carried out to date on the high temperature or creep strength of iron aluminides has examined strengthening at temperatures of the order of 500-650°C [1-8]. While the corrosion and oxidation resistances of iron aluminides remain excellent up to much higher temperatures, considerable improvement of flow and creep strength at such temperatures is needed [9]. The present study examines the high temperature flow behaviour of several Fe_3Al-based intermetallics, and analyses their behaviour in terms of the strengthening mechanisms operating.

EXPERIMENTAL DETAILS

The alloys reported here are a solution-hardened alloy designated 28CrSi [composition (atomic percentage throughout) Fe-28%Al-5%Cr-1%Si] and a Fe-25%Al-2%Nb alloy examined in two heat-treated states – solutionised by a 1h anneal at 1300°C and water quench [designated 25-2-I], and precipitation hardened by a subsequent anneal of 1h at 900°C [designated 25-2-II].

Compression testing was carried out on cylindrical specimens of these alloys at temperatures from room temperature to 900°C at a standard strain rate of 2×10^{-4}/s, with several additional tests carried out at 750°C at different strain rates (from 10^{-3}/s to 10^{-6}/s). Initial microstructures, as well as deformation structures after high temperature testing, were examined by scanning and transmission electron microscopy.

RESULTS
Starting microstructures

The 28CrSi alloy had no second phase particles since the various alloying additions remained in solution (apart from a small amount of Nb and B in the form of coarse particles), and a grain size of about 250μm. The solutionising treatment given to 25-2-I had dissolved approximately half the Nb, with the remainder present in the form of coarse C14 Laves particles, generally arranged on grain boundaries [10], as illustrated in Fig. 1a. Further ageing at 900°C led to the precipitation of most of the Nb as precipitates about 0.25μm in size, again of C14 Laves phase, which were uniformly distributed throughout the grains, see Fig. 1b. The grain size was again about 250μm.

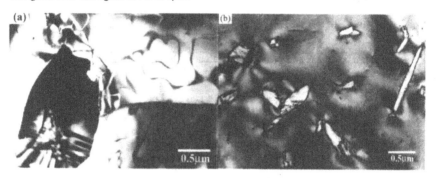

Figure 1: Initial microstructures of (a) alloy 25-2-I; and (b) alloy 25-2-II: (a) shows a coarse Laves particle with faulting on its basal plane at a grain boundary between Fe_3Al grains showing domain boundary contrast. (b) shows Laves precipitates formed on ageing.

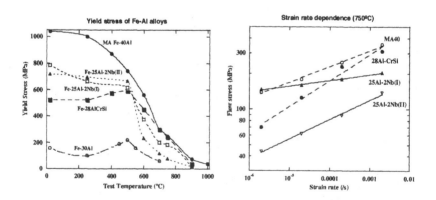

Figure 2: Variation of flow stress with test temperature. Strain rate 2×10^{-4}/s.

Figure 3: Strain rate dependence of flow stress at 750°C.

Mechanical testing

The most important results of compression testing are shown in Figs. 2 and 3. It should be noted that for tests carried out at temperatures below about 500-600°C there was strong work hardening after initial yielding, and hence the flow stress recorded varies with the strain level considered, while for temperatures of about 700°C and above there was no work hardening, and the flow stress remained constant after the onset of plasticity. Both figures show, for comparison, results on testing a mechanically-alloyed Fe-40Al alloy (see reference [11] for more details on structure and mechanical properties), and Fig. 2 shows also some data on the high temperature strength of a binary Fe-30Al alloy for comparison [12]. Clearly the binary Fe-30Al alloy is extremely weak at all test temperatures, while the mechanically-alloyed Fe-40Al alloy is very strong at low temperatures but loses much of its strength by 900-1000°C. This loss of strength may be related to the very fine grain size (below 1μm) and the thermally activated movement of dislocations over the fine oxide particles [11]. In addition, while this mechanically-alloyed material is very strong at high strain rates, even at 750°C (see Fig. 3), the variation of flow stress with strain rate is moderately high (stress exponent n in Table 1 is moderate), and the flow stress required for slow strain rate deformation at 750°C, i.e. under creep conditions, becomes low.

Considering that creep at these temperatures is determined by climb-controlled dislocation annihilation, we should expect a stress exponent of creep rate of about 4-5 [13]. Considering also that dispersed particles introduce an internal stress resistance (σ_0) which subtracts from the applied stress (σ), creep will then be determined by the effective stress, $\sigma_{eff} = \sigma - \sigma_0$, and the creep equation relating strain rate and stress can be written as:

$$\varepsilon = B(\sigma - \sigma_0)^4.$$

The reason for the high apparent exponents of Table 1 is the presence of these internal stress resistances. Accordingly we have analysed our creep data in terms of the equation above using a stress exponent of 4. (Similar results, and equally good fit to experimental data are obtained using instead a stress exponent of 5, but no further discussion of this point will be given here.)

Figs. 2 and 3, in conjunction with the analyses reported in Table 1, provide a very interesting comparison of the roles of solution and particle additions on strengthening. The solution hardened 28CrSi alloy shows moderate flow strength at temperatures up to about 500°C, Fig. 2, but at higher temperatures is among the strongest of all the alloys tested here. However, as seen in Fig. 3, and reflected in the low values of the stress exponent (n) and the internal resistance stress (σ_0) in Table 1, the flow stress varies strongly with strain rate, such that the flow stress for very slow strain rates is the lowest for all the alloys tested.

The pre-aged 25-2-II shows reasonable strength at low temperature, but above 600°C softens to very low flow stress values, Fig. 2. As seen in Fig. 3, and also deduced in Table 1

Table 1. Analysis of strain rate sensitivity of flow stress at 750°C: strain rate range 10^{-3}/s to 10^{-6}/s. Experimental flow stress is shown for a strain rate of 2×10^{-4}/s, and an extrapolated value, based on equations $\varepsilon = A\sigma^n$ and $\varepsilon = B(\sigma - \sigma_0)^4$, for a strain rate of 2×10^{-7}/s.

Alloy	Value n, from $\varepsilon = A\sigma^n$	Value σ_0, from $\varepsilon = B(\sigma - \sigma_0)^4$	Flow stress, at 2×10^{-4}/s	Flow stress, at 2×10^{-7}/s
25-2-I	25	135	180	142
25-2-II	5.7	30	88	35
28CrSi	3.5	-	240	32
MA40	6.9	106	250	110

from the low σ_0 and n values, there is a strong variation of flow stress with strain rate such that the flow stress becomes very low at low strain rates (35MPa at a strain rate of 2×10^{-7}/s).

Finally, the initially solutionised 25-2-I shows good strength to about 500°C, and thereafter reasonable strength retention up to temperatures of about 800°C. Fig 3 shows a good strength at 750°C to low strain rates, reflected in the high values of stress exponent (n) and internal threshold stress (σ_0) in Table 1. Such variation of flow stress with strain rate means that this material has the best strength at strain rates of 2×10^{-7}/s and below.

Deformation microstructures after high temperature straining

Fig. 4 shows microstructures after deformation at various temperatures. Deformation was carried out to a strain of about 2% at a rate of 2×10^{-4}/s, and the sample then rapidly cooled.

Fig. 4a shows the dislocation structure found in alloy 28CrSi deformed at 750°C, showing how the dislocations are randomly arranged, with no clear sign of their collection as cell or subgrain structures. All dislocations are single, which is not surprising since the material is either completely disordered or weakly B2-ordered at the test temperature.

Fig. 4b shows the equivalent dislocation structure found in alloy 25-2-II after deformation at 750°C, where the dislocations are seen to have collected together as loose subgrain boundaries which are pinned at the many relatively fine Laves precipitate particles. These particles, previously shown in Fig. 1b, have not changed in size or number during the relatively short time of testing at 750°C. The rearrangement of dislocations into subgrain boundaries is a reflection of the greater dislocation mobility at 750°C in this alloy than in the 28CrSi alloy, also reflected in the lower flow stress at this temperature, as seen in Fig. 2.

Figs. 4c-g illustrate dislocation structures in alloy 25-2-I deformed at various temperatures. After deformation at room temperature, Fig. 4c, pairs of partial dislocations forming superdislocations (imperfect for the DO_3 ordered material), are seen, which tend to pile up against the relatively coarse Laves particles. The Laves particles themselves do not appear to deform, consistent with their much greater strength. Deformation at a somewhat higher temperature, 500°C, (see Figs. 4d-e, which show a bright field and weak beam image, respectively) leads to the decoupling of dislocations into single dislocations, each of which appears to move independently of each other. They are in fact still loosely coupled by a B2 type (nearest-neighbour violation) antiphase boundary, as confirmed when imaging using a suitable superlattice vector. The coupling between the two partials is sufficiently weak, however, that the pairs of dislocations show a large and highly variable separation. Deformation at 750°C leaves curled single dislocations inside the material, with incipient precipitation on the dislocations, see Figs. 4f-g. The precipitation at this stage is fine, of size about 50-80nm, and appears as fringed objects on the dislocations when imaged close to the Bragg condition for the diffracting beam. Studies elsewhere [14,15] suggest that such fringed precipitates are a metastable $L2_1$ phase, forming as a precursor to the appearance of the C14 Laves phase. Deformation at higher temperatures, 800-900°C, leads to dislocation structures which are very similar to those found in alloy 25-2-II, namely that the single dislocations have accummulated into loose (at 800°C) or well-defined (at 900°C) subgrain structures, with the subgrain interiors relatively free of dislocations.

DISCUSSION

The following discussion will examine quickly the origins of the high temperature behaviour of the various alloys tested here. Strength at lower temperatures, room temperature to 500°C or so, is strongly affected by a multitude of factors – such as Peierls forces (hence the superdislocations found after room temperature deformation of alloy 25-2-I in Fig. 4c are relatively straight), vacancies if present, cross-slip locking, etc. – as well as solution and particle hardening, and will not be discussed further here.

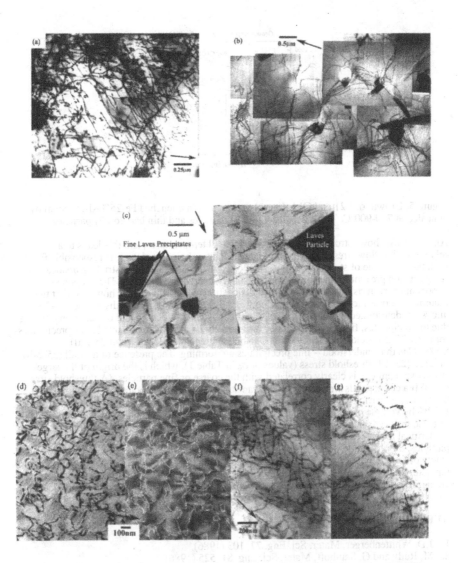

Figure 4: (a) Randomly arranged dislocations in alloy 28CrSi after deformation at 750°C (zone axis near [110], g vector arrowed of type 220); (b) dislocations forming subgrain arrangement tied to Laves particles in alloy 25-2-II deformed at 750°C (zone axis near [110] and g vector arrowed of type 220); (c) superdislocations held up at Laves particles in alloy 25-2-I deformed at room temperature (zone axis near [001] and g vector arrowed of type 220); (d-e) (Bright field-Dark field) curled, single dislocations in alloy 25-2-I deformed at 500°C (zone axis near [111] and g vector of type 220); (f-g) fine precipitation on dislocations in alloy 25-2-I deformed at 750°C (zone axis near [110] and g vector of type 220).

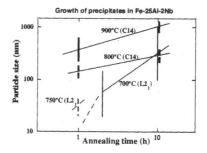

Growth of precipitates in Fe-25Al-2Nb

Figure 5: Growth of L21 and C14 (Laves) precipitates in quenched Fe-25%Al-2%Nb alloy annealed at 700-900°C. Thick bars refer to C14 particles and thin bars to $L2_1$ particles.

Alloy 28CrSi shows strong solution hardening at all temperatures, and this leads to a relatively high flow stress to high temperatures. This solution hardening is responsible for both the low value of stress exponent (n) reported in Table 1, and for restricting dislocation recovery and preventing the formation of a subgrain structure (Fig. 4a). The low stress exponent characterising solute-controlled dislocation movement means, however, that the material fast becomes weak at low strain rates. High temperature strengths of both 25-2 alloys are determined by the absence of significant solute hardening and by the strengthening due to precipitates, fine $L2_1$ precipitates for alloy 25-2-I at 750°C or coarse Laves precipitates for alloy 25-2-II (and 25-2-I at 900°C). The better strength of 25-2-I (than 25-2-II) at 600-800°C°C is thus understood – fine precipitates are forming. The presence of a small (25-2-II) or large (25-2-I) threshold stress (values of σ_0 in Table 1), which is the origin of the large stress exponent (n), is clearly correlated with the coarse or fine particles. (A threshold stress of 30MPa corresponds well with the Orowan stress produced by abour 2% by volume of 0.3μm Laves particles; a threshold stress of 135MPa corresponds well with the Orowan stress for 2% by volume of 60nm $L2_1$ particles.) It is the good strength and good stress sensitivity of these materials that translates into good high temperature strength at low strain rates.

Critical to the development of good creep strength is a uniform dispersion of fine, stable particles. Solidification processes play a major role in determining the initial distribution of solute or second phase particles. Precipitation and coarsening processes determine to which temperature given phases can be stable, with Fig. 5 showing how precipitates in alloy 25-2 are hardly sufficiently stable above 700°C. Further studies examining structures produced by solidification and the stability of alternative phases are underway.

REFERENCES

1. J.D. Whittenberger, Mater. Sci. Eng. **77**, 103 (1986).
2. M. Rudy and G. Sauthoff, Mater. Sci. Eng. **81**, 525 (1986).
3. C.G. McKamey, J.H. DeVan, P.F. Tortorelli, and V.K. Sikka, J. Mater. Res. **6**, 1779 (1991).
4. C.G. McKamey, P.J. Maziasz, and J.W. Jones, J. Mater. Res. **7**, 2089 (1992).
5. D.G. Morris, M. Nazmy, and C. Noseda, Scripta Metall. Mater. **31**, 173 (1994).
6. J.A. Jimenez and G. Frommeyer, Mater. Sci. Eng. **220A**, 93 (1996).

7. W.J. Zhang, R.S. Sundar and S.C. Deevi, Intermetallics **12**, 893 (2004).
8. P. Kratochvil, J. Pesicka, J. Hakl, T. Vlasak and P. Hanus, J. Alloys and Compounds **378**, 258 (2004).
9. D.G. Morris, M.A. Muñoz-Morris and J. Chao, Intermetallics **12**, 821 (2004).
10: D.G. Morris, M.A. Muñoz-Morris and C. Baudin, Acta Mater. **52**, 2827 (2004).
11. M.A. Muñoz-Morris, C. Garcia Oca and D.G. Morris, Acta Mater. **51**, 5187 (2003).
12. C. Hartig, MH Yoo, M. Koeppe and H. Mecking, Mater. Sci. Eng. **258A**, 59 (1998).
13. J. Weertman, J. Appl. Phys. **28**, 362 (1957); Trans. AIME **218**, 207 (1960).
14. D.M. Dimiduk, M.G. Mendiratta, D. Banerjee and H.A. Lipsitt, Acta Metall. **36**, 2947 (1988).
15. D.G. Morris, L.M. Requejo and M.A. Muñoz-Morris, submitted to Intermetallics.

Mater. Res. Soc. Symp. Proc. Vol. 842 © 2005 Materials Research Society　　　　　　　S5.17

Microstructure and Mechanical Properties of Fe-Ni-Mn-Al Alloys

M.W. Wittmann I. Baker, J.A. Hanna, Thayer School of Engineering, Dartmouth College, Hanover, NH 03755-8000, USA, P. R. Munroe, Electron Microscope Unit, University of New South Wales, Sydney, NSW 2052, Australia

Abstract

In an attempt to produce a two-phase alloy consisting of a L2$_1$–structured (Fe,Ni)$_2$MnAl-based phase in either a B2 or b.c.c. matrix, seven Fe-Ni-Mn-Al alloys were cast. Transmission electron microscopy (TEM) of the as-cast alloys revealed a range of microstructures including single phase L2$_1$, a f.c.c./B2 eutectic, and alternating, coherent 10-60 nm wide ordered and disordered b.c.c. rods aligned along <100>. A description of the phases, including chemical compositions and hardnesses is presented.

Introduction

Incorporating L2$_1$ precipitates into a B2 matrix has been shown to be an effective method for producing an alloy with excellent high temperature strength. For example, an alloy consisting of L2$_1$-structured Ni$_2$AlTi precipitates in a B2 NiTi matrix has far superior creep properties relative to the matrix phase alone [1], and Ni$_2$AlTi particles in a NiAl matrix results in better creep properties than either of the constituent phases [2-4]. The drawback with these strongly-ordered nickel-based alloys is that they are quite brittle. By comparison, there has been only limited work on the less strongly ordered Fe$_2$AlX compounds. A recent investigation of the mechanical properties of Fe$_2$AlMn single crystals revealed that they could exhibit some room temperature tensile ductility (ε_f~6%) and an increasing yield strength with increasing temperature to 800 K [5]. Since Ni$_2$MnAl also adopts the L2$_1$ structure, we explored the possibility of forming a multi-phase alloy containing an L2$_1$ phase by substituting Ni for Fe in Fe-rich Fe$_2$MnAl. To this end, seven alloys were cast with concentrations of Fe, Ni, Mn, and Al ranging from 15-35 at. %. One alloy of interest, Fe-15Ni-25Mn-25Al, was also given a 115 hour anneal at 823 K in an attempt to increase the L2$_1$ ordering. Transmission electron microscopy (TEM) including energy dispersive x-ray spectrometry (EDS) was used to characterize the resulting microstructure in the as cast, and annealed condition, and room temperature hardness measurements were performed to survey the mechanical properties.

Experimental

Seven alloys, see Table 1 for compositions, were arc melted in a water-cooled copper mold under argon from constituent elements that were of 99.9% purity or better. Ingots were flipped and melted 4 times to ensure mixing. Slices were cut from the as-cast samples from which 3 mm dia. disks were cut using an EDM. The disks were mechanically polished using 200 grit SiC paper to a thickness of 200 μm and subsequently electropolished in 30% nitric acid in methanol at -20°C using a Tenupol 5, at 12V and 100 mA. TEM analysis was performed using either a Tecnai FEG F20 or Philips CM200 operated at 200kV, both equipped with EDS. Hardness measurements were performed at room temperature using a Leitz MINIload tester with a Vickers-type indenter using a 200g load and a 12 s drop time. Reported values are the average of at least 5 measurements. Heat treatments were performed in air at 823 K.

Table 1: Summary of alloy data

Nominal Composition	Phases Present	Composition of Phases (at%) Fe:Ni:Mn:Al	Hardness (VPN)
1a - $Fe_{30}Ni_{20}Mn_{15}Al_{35}$	B2 and f.c.c.	B2=9:34:20:37 f.c.c.=46:12:41:1	434±10
1b - $Fe_{30}Ni_{20}Mn_{20}Al_{30}$	$L2_1$, B2, and b.c.c.	#	456±10
1c - $Fe_{30}Ni_{20}Mn_{25}Al_{25}$	B2 and b.c.c.	B2 = 13:33:14:40 b.c.c.=55:8:20:17	501±17
1d - $Fe_{30}Ni_{20}Mn_{30}Al_{20}$	B2 and b.c.c.	B2=12:33:19:35 b.c.c.= *	445±15
1e - $Fe_{30}Ni_{20}Mn_{35}Al_{15}$	$L2_1$	Single Phase	310±5
1f - $Fe_{25}Ni_{25}Mn_{25}Al_{25}$	B2 and b.c.c. b.c.c. (precipitates)	B2=9:36:15:40 b.c.c. (spinodal)=* b.c.c. precip. = 46:8:32:14	437±15
1g - $Fe_{35}Ni_{15}Mn_{25}Al_{25}$	$L2_1$, B2, and b.c.c.	#	534±12

\# - Individual phases too small for chemical analysis.
* - Accurate chemical information not available due to preferential etching of the b.c.c. phase.

Results

Figures 1a-g are TEM micrographs of the seven as-cast alloys. The crystal structures of the phases were determined by electron diffraction techniques. A characteristic [$1\bar{1}0$] diffraction pattern, showing reflections consistent with $L2_1$ ordering is shown in Figure 2. The microstructure and crystal structures of each of the alloys are outlined below:

Alloy 1a: Eutectic microstructure of alternating B2 and f.c.c., ~60 nm wide and 2μm long lamellae.

Alloy 1b: Diffraction patterns indicate $L2_1$ ordering with the microstructure showing mottled contrast which is more clearly seen in Figure 3. This suggests the alloy is in the early stages of spinodal decomposition in which little phase separation has occurred. Comparison with alloys 1a and 1c, whose compositions bracket alloy 1b, and which do not show a spinodal microstructure, and a well-established spinodal respectively, suggests that the composition of alloy 1c falls near the edge of the coherent spinodal phase field. As a result, the temperature at which decomposition occurs is much lower, resulting in slower diffusion and less phase separation. Alternatively, it is also possible that diffusion is slower due to $L2_1$ ordering resulting in less phase separation.

Alloy 1c: Interconnected array of alternating, coherent B2 and b.c.c. phases having a wavelength of 60-80 nm, which probably formed by spinodal decomposition.

Alloy 1d: Interconnected array of alternating, coherent B2 and b.c.c. phases having a wavelength of 60-80 nm, which again probably formed by spinodal decomposition.

Alloy 1e: Single phase $L2_1$ structure.

Alloy 1f: Multi-phase microstructure consisting of large, ~5 μm b.c.c. precipitates (arrowed in Figure 1f) in a matrix of fine alternating b.c.c. and B2 phases of wavelength ~120 nm. The similarities in size and morphology of the alternating phases with those seen in alloys 1c and 1d suggest that they form by spinodal decomposition. It is unclear why the large, Fe-rich b.c.c. precipitates form in this alloy and not in alloy 1c which has a higher concentration. One possibility is that the casting was not completely homogeneous.

Alloy 1g: Diffraction patterns indicate $L2_1$ ordering with the microstructure showing tweed contrast with a wavelength of 10-20 nm. This suggests that the alloy is two-phase with one phase having the $L2_1$ structure. A dark field (DF) image taken along <110> using a {200} B2 superlattice (Figure 4a) showed that the tweed contrast arises from alternating B2 (bright) and b.c.c. (dark) rods, which are aligned along <100>. A similarly oriented DF image taken using a {111} $L2_1$ superlattice reflection (Figure 4b) showed that the regions with $L2_1$ ordering (bright) are less well defined. It is concluded that there are regions of $L2_1$ ordering within the B2 ordered rods. This microstructure probably also formed by spinodal decomposition.

A summary of phase's present, chemical compositions and as cast hardnesses is given in Table 1. No cracking was observed when making hardness measurements suggesting all of the alloys have some ductility. In making the TEM samples, the b.c.c. phase was preferentially etched, making chemical analysis of this phase using EDS problematic and, thus, some compositions of this phase are not reported. From those available, it can be seen that in alloys with a spinodal microstructure the elements partition into a Ni and Al rich B2 phase and a Fe and Mn rich b.c.c. phase. While it cannot be determined conclusively from morphology alone if a microstructure results from spinodal decomposition, the regular interval and interconnectedness of the phases in alloys 1b, 1c, 1d, 1f, and 1g suggests that is how they formed.

Spinodal alloys have been of interest for their high strength which is believed to result from periodic strain fields arising from composition fluctuations [6]. Similar results can be seen in this investigation in which the alloys having a spinodal microstructure have a hardness ranging from 437-534 VPN whereas the single phase alloy (alloy 1e) has a hardness of 310 VPN. The same strengthening mechanism is believed to apply in these alloys.

Two alloys (1b and 1g) were found to have a microstructures consisting of an $L2_1$ phase in a B2, or b.c.c. matrix. Alloy 1g, the hardest alloy investigated, most clearly showed this structure with a tweed structure of ~10 nm wide, alternating ordered and disordered rods. In an attempt to increase the $L2_1$ ordering this alloy was annealed for 115 hours at 823 K. This temperature is slightly below the $L2_1$-B2 transition temperature of 898 K for Fe_2MnAl [5], but somewhat above the aging temperature of 673 K used for Ni_2MnGa [7]. A dark field image showing the microstructure of the annealed alloy is shown in Figure 5a with a [110] selected area diffraction pattern in Figure 5b which exhibits $L2_1$ ordering. The wavelength of the initial microstructure has increased to ~50 nm and numerous elongated incoherent precipitates have

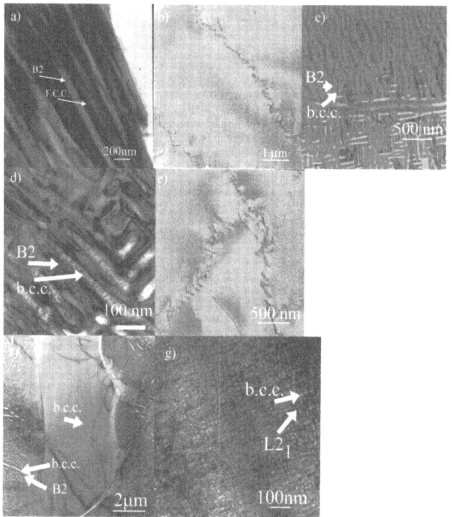

Figure 1: Bright field TEM images of alloys 1a-1g.

formed. The chemical composition of these precipitates was determined to be 44.8 ± 0.3% Fe, 43±1% Mn, and 12.7±0.9%Al which is similar to the composition of the f.c.c. phase in alloy 1a, and falls on the inside edge of the γ phase field of the Fe-Mn-Al ternary [8]. The L2$_1$ reflections in Figure 5 have become very weak suggesting that continued segregation of the elements between the three phases has resulted in a decrease in L2$_1$ ordering. More detailed investigations into the formation of the aged microstructure, including mechanical tests to characterize age-hardening, high-temperature strength, and deformation mechanisms are currently underway.

Figure 2: Characteristic [110] zone axis diffraction pattern for alloys 1b, 1e and 1g showing L2₁ ordering.

Figure 3: Dark field TEM image of alloy 1b using a {002} superlattice reflection showing mottled contrast.

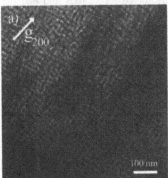

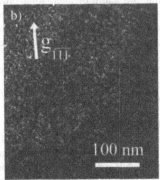

Figure 4: Dark field TEM images of as cast alloy 1g taken using a) B2 and b) L2₁ superlattice reflections.

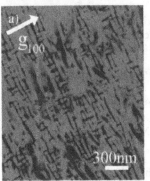

Figure 5: a) Dark field image taken along <100> using a B2 superlattice reflection of alloy 1g annealed for 115 h at 823 K showing coarsening of the original microstructure and formation of elongated precipitates. b) [110] diffraction pattern showing only weak L2₁ reflections.

CONCLUSIONS

Seven Fe-Ni-Mn-Al alloys were cast and their microstructures and hardnesses determined. It was found that:

1. A microstructure comprised of alternating, coherent ordered and disordered b.c.c. rods with wavelengths ranging from 10-80 nm, aligned along <100> exists for a range of alloy compositions near the composition $(FeNi)_2MnAl$. Based on the regular interval and interconnectedness of the phases, the microstructure is attributed to spinodal decomposition.
2. Chemical analysis revealed that in alloys with a spinodal microstructure the elements tended to segregate into a Fe and Mn-rich b.c.c. phase and Ni and Al-rich B2 phase.
3. Alloys which have a multi-phase spinodal microstructure show good hardness in the as cast condition with hardness values ranging from 445-539 VPN whereas a single phase alloy with similar composition had a hardness of 310 VPN.
4. Two alloys having compositions of Fe-20Ni-20Mn-30Al and Fe-15Ni-25MN-25Al were found to have microstructures incorporating a $L2_1$ phase into a b.c.c. matrix. The second of these had a microstructure of alternating ordered and disordered ~8nm rods and had the highest hardness of the alloys investigated. From dark field images using a $L2_1$ superlattice reflection, the rods showed only partial $L2_1$ ordering. Annealing at 823 K for 115 hours resulted in coarsening of the as cast structure, formation of elongated precipitates, and a decrease in $L2_1$ ordering.
5. A eutectic structure of b.c.c. and f.c.c. lamellae resulted for the alloy composition Fe-20Ni-15Mn-35Al. The hardness of this alloy was similar to those having a spinodal microstructure.

ACKNOWLEDGEMENTS

This research was supported by NIST grant 60NANB2D0120 and NSF grant DMR0314209. The views and conclusions contained herein are those of the authors and should not be interpreted as necessarily representing official policies, either expressed or implied, of the National Science Foundation, the National Institute of Standards and Technologies, or the U.S. Government.

REFERENCES

1. Koizumi, Y., Ro, Y., Nakazawa, S., and Harada, H., Materials Science and Engineering A-Structural Materials Properties Microstructure and Processing, **223**, 1997, 36-41.
2. Polvani, R.S., Tzeng, W.S., and Strutt, P.R., Metallurgical Transactions A-Physical Metallurgy and Materials Science, **7**, 1976, 33-40.
3. Strutt, P.R., Polvani, R.S., and Ingram, J.C., Metallurgical Transactions A-Physical Metallurgy and Materials Science, **7**, 1976, 23-31.
4. Strutt, P.R. and Kear, P.R., Proceedings of the Material Research Society, **39**, 1985, 279-292.
5. Wittmann, M., Munroe, P.R., and Baker, I., Philosophical Magazine-Structure and Properties of Condensed Matter, **84**, 2004, 3169-3194.
6. Cahn, J.W., Acta Metallurgica, **11**, 1963, 1275.
7. Fujita, A., Fukamichi, K., Gejima, F., Kainuma, R., and Ishida, K., Applied Physics Letters, **77**, 2000, 3054-3056.
8. Schmatz, D.J., Transactions of the Metallurgical Society of AIME, **215**, 1959, 112-114.

Mater. Res. Soc. Symp. Proc. Vol. 842 © 2005 Materials Research Society   S5.18

Thermomechanical Treatment of a Fe₃Al alloy

Joachim Konrad[1,2], Stefan Zaefferer[2], André Schneider[1], Georg Frommeyer[1], Dierk Raabe[2]
[1]Materials Technology, [2]Microstructure Physics and Metal Forming
Max-Planck-Institut fuer Eisenforschung GmbH
Max-Planck-Str.1
40237 Duesseldorf

ABSTRACT

A binary Fe₃Al alloy is investigated with respect to hot and warm rolling behavior and microstructural as well as microtextural modifications. Rolling has been performed in the A2 and B2 order regimes. The differences in microstructure are investigated. The performed texture analysis reveals the differences in hot and warm rolling textures depending on the hot rolling temperature. On the basis of microtexture investigations by means of electron backscatter diffraction (EBSD) differences concerning orientation gradients and sub-grain structures are found. A model of combined order-related and non-order related effects is proposed explaining the observed material behavior. The results are used for process modification.

INTRODUCTION

Fe₃Al-based alloys are regarded as promising for high temperature applications in corrosive atmospheres. Generally Fe₃Al shows higher strength compared to disordered Fe-Al alloys and good corrosion resistance in oxidizing and sulphidising atmospheres. At room temperature the material has DO₃ crystal structure, at temperatures of 546 °C (for the binary alloy Fe 26at.%Al, [1]) it transforms into a B2 structure and at 829 °C into a disordered A2 structure. The long-range order at low temperatures leads to a lack of room temperature ductility which is attributed to the low mobility of superdislocations present in the DO₃-ordered state and the difficulty of cross-slip due to the generation of anti-phase boundaries [2]. Thermomechanical treatment of Fe₃Al-based alloys is regarded as important method to overcome their intrinsic brittleness. Sun et al [3] introduced a concept of thermomechanical treatment in the temperature range of the B2-ordered phase. The process, consisting of hot and warm rolling steps and heat treatments below the recrystallization temperature, leads to an improvement of room temperature ductility [4]. This increased ductility is considered to be based on residual single fold dislocations created during hot deformation [4] and on strain induced disordering [5]. McKamey[6], in contrast, attributed the increased ductility of recovered (but not recrystallized) material to the reduction of environmental embrittlement - an extrinsic effect attributed to reactions of the environmental water vapor with the aluminum atoms - due to the elongated grain structure. Finally, the crystallographic texture of the recovered material may also have an influence on the ductility [4]. In the project on which we report here it is investigated whether a thermomechanical treatment in the A2 region (i.e. above 829 °C [1]) where the material is completely disordered [7] could further improve the warm rolling behavior as well as the mechanical properties at room temperature. At these temperatures thermally activated dynamic processes such as recovery and recrystallization are more important than in the B2 region also due to the effect that dislocations no longer appear as superdislocations and therefore show a much improved mobility. As a result of annihilation of dislocations the amount of remaining single-fold dislocations at room temperature may be reduced. This, however, also leads to a lower driving force for recrystallization and may therefore assist the retention of a recovered microstructure.

EXPERIMENT

A Fe-26 at.% Al alloy was selected to study the forming behavior of the binary base alloy without the influence of ternary alloying elements. The alloy was produced in an induction furnace under argon atmosphere. The melt was cast in copper moulds and cooled down with rates of about 10 K/s. The cast block hat the dimensions 85 x 30 x 200 mm³. Grain size distribution was measured in the as-cast state by means of light optical microscopy (LOM)

From data analysis of the deformation simulation in Konrad et al [8] a change in the hot deformation behavior was found between 800 and 900 °C which is assumed to be due to the order transformation occurring between these temperatures. It is supposed that this change should also influence the evolution of microstructure and microtexture during hot rolling at these temperature. Therefore rolling experiments at 800 °C and 900 °C were carried out. The total deformation of 1.67 (80 %) was achieved in six passes with constant relative height reduction of 0.28. The subsequent warm rolling was performed using samples of each hot rolling temperature to investigate differences in deformation behavior caused by the different pretreatment. The rolling at 700°C (B2) was performed in 20 passes to a total deformation of 1.81 resulting in a sheet of 1mm thickness. After the last pass of hot and warm rolling the samples were water quenched.

Microstructural characterization of these samples was carried out on transverse sections using automatic crystal orientation mapping (ACOM) in a JSM 6500F (JEOL) SEM equipped with an OIM electron backscatter diffraction (EBSD) system. The results are presented in inverse pole figure of the normal direction (ND) and kernel average misorientation maps. The latter are constructed by calculating the kernel average misorientation of every measurement point up to it's third nearest neighbors. The texture components are visualized in $\varphi_2 = 45°$ sections of the Euler space for the ODF calculated by the discrete binning method.

RESULTS

The as-cast microstructure is characterized by a coarse and inhomogeneous grain structure. Most of the material consists of columnar grains, where the column axis indicates the solidification direction. These grains show a sharp <100> fiber texture with the <100> direction parallel to the solidification direction [9]. The average grain size is 2.48 mm (circular diameter).

The six pass hot rolling produces a crack free hot band at both temperatures 800 °C and 900 °C. Figure 1 shows SEM images of the samples' cross sections observed with the backscatter electron (BSE) detector which provides crystal orientation contrast. Figure 1a shows the sample rolled at 800 °C in the B2-ordered state. Large orientation gradients in the grain interiors can be seen. Grain boundaries show a seam of strong gradients. The sample rolled at 900 °C (disordered A2, Fig. 1b) also shows orientation gradients. In contrast to the sample rolled at 800 °C, they are in general not continuous but appear in form of discrete subgrains. These subgrains indicate stronger recovery or the onset of dynamic recrystallization (DRX).

The $\varphi_2 = 45°$ sections of the Euler space for the ODF of samples rolled at 800°C and 900°C are shown in Figure 2. The dominating texture components are visible. The γ-fiber component (<111> ‖ normal direction) and the α-fiber component (<110> ‖ rolling direction) are components that are typically found in the microstructure of deformed b.c.c. Fe alloys. The third component, the cube component {100}<001>, in contrast, is rather unusual for these kind of alloys. It is assumed that it is a remainder of the strong <001> fiber texture of the starting material. While the sample rolled at 800°C reveals a relatively strong α-fiber component and retained cube and rotated cube components, the γ-fiber component is rather weak. In contrast the sample rolled at 900°C shows a stronger γ-fiber component and weaker α-fiber, retained cube and rotated cube components. Figure 3 shows the inverse pole figures of the normal direction in orientation maps calculated from EBSD data. It is clearly visible that rolling at 800°C generates a pancake structure of deformed grains with more pronounced edges at triple points and straighter grain boundaries aligned parallel to normal direction than rolling at 900°C where bulged grain boundaries and larger grains can be seen. Kernel average misorientation maps of the same area are presented in Figure 4. These maps do not show long distance orientation gradients but bring out short range orientation changes and cell or subgrain boundaries. The sample rolled at 800 °C shows strong orientation gradients close to grain boundaries. At 900 °C, in contrast, the orientation gradients are less pronounced. In the grain interiors cell like structures start to evolve and some large low misorientation subgrains become visible.

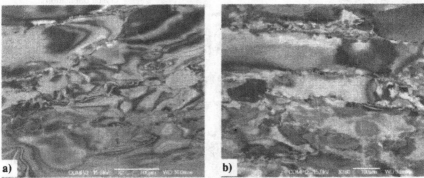

Figure 1: SEM images applying the BSE detector, i.e. orientation contrast. a) hot rolled at 800°C and b) at 900°C Fe26at.%Al to a total deformation of 1.67.

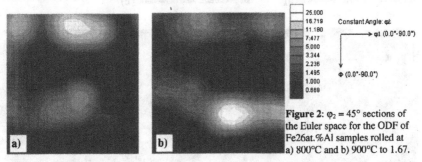

25.000
16.719
11.180
7.477
5.000
3.344
2.236
1.495
1.000
0.669

Constant Angle: φ2

→ φ1 (0.0°-90.0°)

Φ (0.0°-90.0°)

Figure 2: φ₂ = 45° sections of the Euler space for the ODF of Fe26at.%Al samples rolled at a) 800°C and b) 900°C to 1.67.

Warm rolling at 700°C produces a crack free sample only from starting material hot rolled at 900°C (A2). The B2 (800°C) hot rolled sample breaks during the final pass by a crack in rolling direction. Figure 5 shows the cross sections of both samples. The BSE images do not reveal significant differences in microstructure. Both samples show characteristic layer like grain morphology. Strong orientation gradients in grains as well as close to grain boundaries.

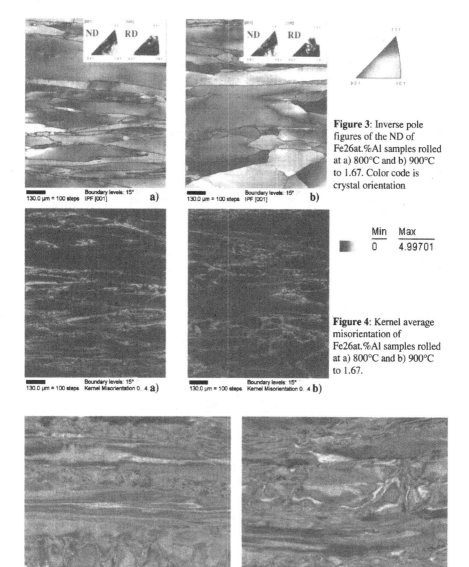

Figure 3: Inverse pole figures of the ND of Fe26at.%Al samples rolled at a) 800°C and b) 900°C to 1.67. Color code is crystal orientation

Min	Max
0	4.99701

Figure 4: Kernel average misorientation of Fe26at.%Al samples rolled at a) 800°C and b) 900°C to 1.67.

Figure 5: SEM images applying the BSE detector, i.e. orientation contrast. a) hot rolled at 800°C and b) at 900°C Fe26at.%Al to a total deformation of 1.67 and both warm rolled at 700°C to additional 0.83.

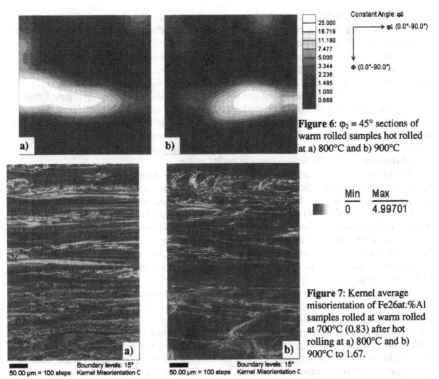

Figure 6: $\varphi_2 = 45°$ sections of warm rolled samples hot rolled at a) 800°C and b) 900°C

Min	Max
0	4.99701

Figure 7: Kernel average misorientation of Fe26at.%Al samples rolled at warm rolled at 700°C (0.83) after hot rolling at a) 800°C and b) 900°C to 1.67.

Boundary levels: 15°
50.00 μm = 100 steps Kernel Misorientation C

Boundary levels: 15°
50.00 μm = 100 steps Kernel Misorientation C

Some grains show characteristic orientation gradients which divide some of the flat grains into necklace like structures. These substructures seem to be somewhat larger in the 900°C hot rolled sample than in the 800°C sample. The $\varphi_2 = 45°$ sections in Figure 6 show that in both warm rolled samples the three texture components found in the hot rolled state are again present. While the texture of the sample hot rolled at 900°C is strongly dominated by the γ-fiber component (maximum at (111) <0-11>) the texture of the sample hot rolled at 800°C shows a significantly stronger α-fiber component and a different intensity maximum of the γ-fiber component (at (111) <6-71>). Figure 7 shows the kernel average misorientation of both samples revealing significantly stronger orientation gradients especially close to grain boundaries in the sample hot rolled at 800°C. They also show a smaller grain size in the sample hot rolled at 800°C which corresponds in size to the subgrain structure of the sample hot rolled at 900°C.

CONCLUSIONS

The SEM images of the hot rolled samples (Fig. 1) show typical necklace structures at grain boundaries as it is usually observed in early stages of DRX. For both temperatures the necklace structure is about equally developed. This observation shows that in the chosen strain rate - temperature regime softening occurs by dynamic processes as has been assumed for the hot compression test results [8]. The sample deformed at the higher temperature, additionally shows strong recovery visible in the form of subgrain and dislocation cell formation while the lower temperature sample shows strong continuous deformation gradients particularly in front of grain boundaries. Two effects may contribute to these differences. First the higher temperature allows for quicker recovery, thus orientation gradients (which consist of geometrically necessary dislocations) may separate into subgrains which then finally become nuclei for dynamic or post-dynamic recrystallization [10]. Second, the slip of single-fold dislocations, in contrast to superdislocations, may easier allow for cross-slip and activation of multiple slip regimes, thus reducing the occurrence of orientation gradients at obstacles, such as grain boundaries. On the other hand, superdislocations which are more strongly confined to their slip planes may lead to dislocation pile-ups in front of obstacles which is visible in form of sharp orientation gradients. The presented results may be interpreted as another confirmation of the importance of superdislocation motion for deformations in the B2 region.

The influence of non order related temperature effects on the microtexture evolution during hot rolling is well known. Recovery is enhanced and thus stored energy in terms of dislocation density is reduced while cellular dislocation structures and subgrains evolve. The results of the microtexture analysis (Fig. 4) show the applicability of the model to the studied alloy. Expected order effects aggravate this behavior by hindering cross-slip and recovery.

Texture is influenced by the more pronounced recovery in the A2 rolled sample because in ferritic b.c.c. alloys the dynamic and static recrystallization leads to a preference of the γ-fiber due to its high nucleation rate [11] while the rolling deformation leads to a preference of the α-fiber component.

During subsequent warm rolling the effects of the pretreatment temperature are inherited. The higher amount of piled up dislocation in front of grain boundaries, visible as strong orientation gradients, in the B2 hot rolled sample leads to a stronger additional pile up during warm rolling. It can be stated that the warm rolling in the B2 order regime leads to sharp orientation gradients close to grain boundaries due to the described lack of cross slip activity. In combination with a hot rolling temperature in the B2 regime generating similar microtexture features and a reduced amount of formability, this leads to catastrophic failure during warm rolling. The inherited differences in texture also support this model of the influence of hot rolling temperature on the warm rolling behavior.

For the development of the thermomechanical process hot rolling in the A2 order regime at low temperatures seems to be recommendable.

REFERENCES

[1] F. Stein, A. Schneider and G. Frommeyer, *Intermetallics* **11**, 71 (2003).
[2] H. J. Leamy, F. X. Kayser and M. J. Marcinkowski, *Phil. Mag.* **20**, 763 (1969).
[3] Z. Q. Sun, Y. D. Huang, W. Y. Yang and G. L. Chen, in *High temperature ordered intermetallic alloys V*, edited by I. Baker, R. Darolia, J. D. Whittenberger and M. H. Yoo, (Mater. Res. Symp. Proc. **288**, Pittsburgh, PA, 1993) p. 885.
[4] Y. D. Huang, W. Y. Yang, G. L. Chen and Z. Q. Sun, *Intermetallics* **9**, 331 (2001).
[5] M. M. Dadras and D. G. Morris, *Scripta Metall. Mater.* **28**, 1245 (1993).
[6] C. G. McKamey and D. H. Pierce, *Scripta Metall. Mater.* **28**, 1173 (1993).
[7] S. M. Kim and D. G. Morris, *Acta mater.* **46**, 2587 (1998).
[8] J. Konrad, S. Zaefferer, A. Schneider, D. Raabe and G. Frommeyer, *Intermetallics* (acc. 2005).
[9] J. Konrad, A. Günther, S. Zaefferer and D. Raabe D. *GeNF Experimental Report* **1**, 183 (2003).
[10] R. A. Petkovic, M. J. Luton, J. J. Jonas, *Acta metal.* **27**, 1633 (1979).
[11] B. Hutchinson and P. Bate, *Proc. of IF Steels 2003*, ISIJ, Tokyo, 337 (2003).

Mater. Res. Soc. Symp. Proc. Vol. 842 © 2005 Materials Research Society   S5.19

Optimisation of Precipitation for the Development of Improved Wrought Fe₃Al-based Alloys

Satoru Kobayashi [a, b], Stefan Zaefferer [a], André Schneider [b]
[a] Dept. of Microstructure Physics and Metal Forming, [b] Dept. of Materials Technology,
Max-Planck-Institute for Eisenforschung, Max-Planck-Str. 1, D-40237 Düsseldorf, GERMANY.

ABSTRACT

Effect of TiC precipitates on the kinetics of static recrystallisation has been studied by using a Fe-26Al-5Cr (at%) single-phase (α:A2/B2/D0$_3$) alloy and two-phase (α+TiC) alloys with different amounts of TiC precipitates. Based on the results, a desirable thermo-mechanical processing is proposed for the development of wrought Fe₃Al-based alloys with strengthening MC carbides.

In the alloys with a high amount of TiC, needle-like TiC precipitates with 1-10 μm in length formed during air-cooling after homogenisation. Hot deformations with such large precipitates cause inhomogeneous deformation around the particles, leading to particle stimulated nucleation (PSN) and hence accelerate recrystallisation.

The occurrence of PSN is harmful for the embrittlement problem, i.e. ductility drastically decreases when recrystallisation occurs, but useful for grain refinement. The following process is proposed to accomplish grain refinement, strengthening by precipitates and avoidance of the embrittlement: hot deformation with a large amount of precipitates to make grain refinement possible by using PSN, followed by hot deformation with a small amount of precipitates near α single-phase region and a subsequent heat treatment to obtain fine precipitates. The fine particles would also act to pin the boundaries of growing grains, thus leading to extended recovery rather than recrystallisation. This process is difficult to carry out in the (Fe-26Al-5Cr)-TiC system because the temperature necessary to enable precipitation is very high and the kinetics is quick. The precipitation temperature is significantly decreased by replacing TiC by VC or MoC.

INTRODUCTION

Fe₃Al-based alloys with bcc structures (α: disordered A2 and B2- or D0$_3$ ordered) have been considered as a structural material for high-temperature applications due to their resistance to high-temperature oxidation and sulphidation [1, 2]. Serious unsolved problems are the poor high-temperature strength, creep resistance and the limited low-temperature ductility. This paper proposes a thermo-mechanical processing to develop wrought Fe₃Al-based alloys with improved high-temperature strength as well as low-temperature ductility.

An effective method to improve high-temperature strength is to introduce fine precipitates. Recent investigations demonstrated that precipitates such as carbides, borides and Laves phase are effective for strengthening even at very high temperatures around 800°C [3-6]. The limited ductility at low temperatures can be improved by careful thermo-mechanical treatments to achieve deformed and well-recovered states. The occurrence of recrystallisation, however, drastically reduces ductility [7]. Consequently, in order to accomplish both strengthening by precipitates and improving low-temperature ductility, it is of importance to understand precipitation behaviour and its effect on recrystallisation since precipitates in deformation process greatly affects the recovery and recrystallisation.

In this study, precipitation of TiC and its effect on the kinetics of static recrystallisation have been studied by using a Fe-26Al-5Cr (at%) single-phase (α:A2/B2/D0$_3$) alloy and two-phase (α+TiC) alloys with different amounts of TiC precipitates. Based on these results, a desirable thermo-mechanical process is proposed for the development of wrought Fe$_3$Al-based alloys with strengthening MC carbides.

EXPERIMENTAL PROCEDURE

Figure 1 shows a vertical section along the tie-line between α phase with Fe-26Al-5Cr (at%) and TiC phase, which was determined in another paper [8]. A Fe-26Al-5Cr single-phase alloy and two different two-phase alloys containing Ti and C, which are on the same tie-line, were used in this study. Hereafter the Fe-26Al-5Cr alloy is referred to as base alloy and the Ti and C containing alloys are designated according to their Ti and C concentrations.

The alloys were prepared from 3N purity iron, 4N aluminium, 3N chromium, 4N titanium and 3N carbon as 2 kg ingots by induction melting performed in argon atmosphere. The surface of the ingots was machined and these were cut to a size of 56 x 27 x 80 mm. The specimens were homo-genised in the α single-phase temperature region followed by air-cooling. The homogenisation temperatures are shown for each alloy by circles in Figure 1. Detailed procedures are described in another paper [9].

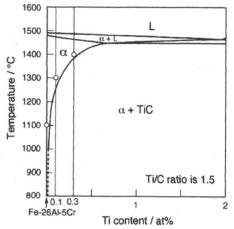

Figure 1. Vertical section along the tie-line between α (Fe-26Al-5Cr) and TiC [8].

The homogenised specimens were hot rolled to a reduction of 65% in four passes without intermediate heat treatments. Subsequently the samples were heat-treated at 900 °C up to 1h. The heat-treated samples were cut into halves and the cross section was used for macro and microstructure observations. Microstructure observations were performed by means of optical microscopy (OM) and scanning electron microscopy (SEM) using a backscattered electron detector.

Additionally, Fe-26Al-5Cr based alloys with additions of V, C or Mo, C were prepared with 3N vanadium and molybdenum chips in the same way as for the other samples. The alloys were heat-treated in the temperature range between 800°C and 1400°C in order to determine the solubility range of the α single-phase regions.

RESULTS & DISCUSSIONS

Kinetics of TiC precipitation

Kinetics of TiC precipitation was investigated by isothermal aging experiments in the α+TiC two-phase region after homogenisation in the α single-phase region. Figure 2 shows determined TTT diagram for the TiC precipitation from the solid. The 0.1Ti-0.07C alloy has C curve with a nose at 1000°C and about 30 seconds. In the 0.3Ti-0.2C alloy, TiC precipitates form even during water quenching, indicating C curve exists at a very short time range. The homogenised samples consist of needle-like TiC precipitates with the length of less than 1 μm in the 0.1Ti-0.07C alloy but 1-10 μm in the 0.3Ti-0.2C alloy.

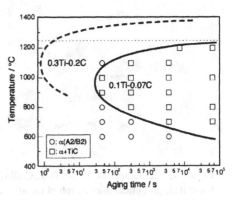

Figure 2. TTT diagram for the precipitation of TiC in the 0.1Ti-0.07C alloy (solid line) and 0.3Ti-0.2C alloy (broken line).

Effect of precipitates on recrystallisation

The homogenised samples were hot-rolled at 800°C and subsequently aged at 900°C. Area fractions recrystallised were determined by using OM and the change in the area fraction with annealing at 900°C is shown in Figure 3. It is noted that the kinetics of recrystallisation is slightly retarded in the 0.1Ti-0.07C but drastically promoted in the 0.3Ti-0.2C alloy. This acceleration of recrystallisation in the 0.3Ti-0.2C alloy is attributed to an increase in the deformation heterogeneity which may act as nucleation site (particle stimulated nucleation, PSN). Figure 4(a) shows deformation zones close to precipitates in the 0.3Ti-0.2C alloy after hot rolling. The needle-like precipitates are TiC particles. A

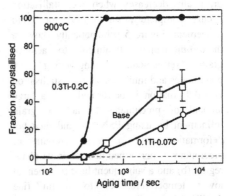

Figure 3. Change in fraction recrystallised with annealing at 900°C for the hot rolled samples.

sharp orientation contrast is clearly seen around the large TiC, demonstrating that intense deformation occurred around precipitates. It is therefore reasonable to assume that the intense deformation led to nucleation of grains around the particles (Figure 4(b)), resulting in a promotion of recrystallisation.

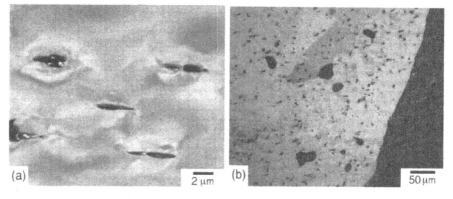

Figure 4. Micrographs of the 0.3Ti-0.2C alloy; (a) backscattered electron image of the alloy hot-rolled, (b) optical micrograph of the alloy annealed at 900°C for 5 min.

Desirable thermo-mechanical processing

The occurrence of PSN is harmful for the embrittlement problem, i.e. ductility drastically decreases when recrystallisation occurs completely, but useful for grain refinement. Figure 5 schematically shows a thermo-mechanical treatment to achieve grain refinement, strengthening by precipitates and inhibition of recrystallisation. Hot deformation is performed with a large amount of precipitates to achieve grain refinement by using PSN (a), and then hot deformation with a small amount of precipitates is performed near α single-phase region (b) and a subsequent heat treatment at lower temperature (c) to obtain fine precipitates. The fine particles would also act to pin the boundaries of growing grains, thus leading to extended recovery rather than recrystallisation. In the (Fe-26Al-5Cr)-TiC

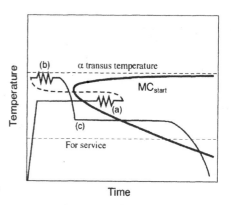

Figure 5. Schematic illustration showing a desirable thermo-mechanical processing to develop wrought Fe₃Al-based alloys with improved high-temperature strength as well as low-temperature ductility

system this sequence is difficult to realise because the temperature necessary to enable precipitation is very high and the kinetics is quick. Therefore the question arises, which alloy system exhibits an α single-phase region extending towards lower temperature, enabling the desirable thermo-mechanical treatment.

α single-phase regions in vertical sections through the phase diagrams between Fe-26Al-5Cr and M-40C (M= V, Mo) was experimentally determined. Figure 6 shows the determined α single-phase regions of these systems superimposed with the vertical section of the (Fe-26Al-5Cr)-TiC system. In both, the VC and MoC containing systems the solubility limit of the α single-phase region is more than twice as large as that in -TiC system. It is remarkable that in the (Fe-26Al-5Cr)-MoC system the phase boundary between α and α+MC is shifted by about 400°C towards lower temperatures compared to the system with TiC at 0.5M and 0.33C, for example. For achieving the desired thermo-mechanical processing characteristics, as discussed in the previous chapter, -MoC and -VC systems should be effective due to their extended single-phase regions.

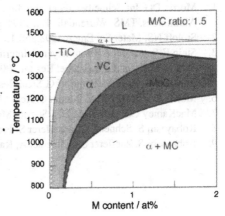

Figure 6. Experimentally determined α single-phase regions in the vertical sections between Fe-26Al-5Cr and MC with the M/C ratios of 1.5; (a) M= V, (b) M= Mo.

SUMMARY

A Fe-26Al-5Cr single-phase (α) alloy and two different two-phase (α+TiC) alloys with different volume fractions of TiC have been hot-rolled, and the kinetics of static recrystallisation has been examined. Based on the results, a desirable thermo-mechanical process is proposed for the development of wrought Fe₃Al-based alloys with strengthening MC carbides. The following results were obtained;

1. In the alloys with a high amount of TiC, α single-phase region exists at around 1400°C and the kinetics of TiC precipitation is very quick.
2. Needle-like TiC precipitates with 1-10 μm in length formed after homogenisation cause particle stimulated nucleation (PSN) and hence accelerate recrystallisation.
3. In order to achieve improved high-temperature strength as well as low-temperature ductility, the following thermo-mechanical treatment is proposed: hot deformation with a large amount of precipitates to make grain refinement possible by PSN, followed by hot deformation with a small amount of precipitates and a subsequent heat treatment to obtain fine precipitates.
4. This process could be possible to carry out in the (Fe-26Al-5Cr)-MoC and -VC systems due to the relatively low precipitation temperature.

ACKNOWLEDGMENT

The authors would like to thank Mr. K. Markmann and U. Tezins of the Max-Planck-Institute for Iron Research for their help in the experiments.

REFFERENCES

1. Morris DG. In: Schneibel JH et al, editor. Processing, Properties and Applications of Iron Aluminides, TMS, Warrendale, PA, 1994. p. 3.
2. Stoloff NS. Mater Sci Eng A 1998;258:1.
3. Stein F, Schneider A, Frommeyer G. Intermetallics 2003;11:71.
4. Schneider A, Sauthoff G. Steel Res Int 2004;75:55.
5. Eumann M, Palm M, Sauthoff G. Steel Res Int 2004;75:62.
6. Morris DG, Morris MA, Baudin C. Acta Mater 2004;52:2827.
7. MacKamey CG and Pierce DH. Scripta Metall 1993;28:1173.
8. Kobayashi S, Schneider A, Zaefferer S, Frommeyer G, Raabe D. to be submitted.
9. Kobayashi S, Zaefferer S, Schneider A, Raabe D, Frommeyer G. accepted to Intermetallics.

Mater. Res. Soc. Symp. Proc. Vol. 842 © 2005 Materials Research Society   S5.20

Effect of Excess Vacancies on Antiphase Domain Growth in Fe$_3$Al

Y. Koizumi, T. Hagiwara, Y. Minamino, and N. Tsuji
Department of Adaptive Machine Systems, Osaka University
2-1 Yamada-oka, Suita, Osaka 565-0871, Japan

ABSTRACT

The growth of the D0$_3$-type antiphase domain (APD) in Fe$_3$Al was investigated focusing on the effect of excess vacancies that were introduced during the quenching process from the disordered state. The variation in the APD size exhibited considerable deviation from the conventional "parabolic growth law" in the early stage of APD growth. This variation was numerically calculated on the assumption that the migration of the APD boundaries was enhanced by non-equilibrium excess vacancies and the vacancy concentration decreased during the isothermal annealing for the APD growth. The calculated variations in the APD size could be successfully fitted to the experimental results in cases with quenching temperatures (T_q) of 873 K or 1073 K, but not when T_q was 1273 K. The APD growth in the latter case was much slower than the expected growth derived from the calculation. This discrepancy was attributed to the rapid decrease in the vacancy concentration due to vacancy clustering since a significant amount of dotted contrasts were observed in TEM image of only the specimen quenched from 1273K.

INTRODUCTION

The control of APD is getting more important because the dramatic change in properties or novel phenomena related with APDs in intermetallic compounds have been found [1, 2]. Generally, the variation in the APD size during isothermal annealing is considered to follow the parabolic growth law [3], which is expressed by the following equation:

$$l^2 - l_0^2 = kt \qquad (1)$$

where l_0 is the initial APD size, k is the coefficient of the growth rate, and t is the annealing time. However, according to our study on APD growth in Ti$_3$Al, the APD growth deviates from the parabolic growth in the initial stage. This deviation was explained by taking the k as a function of time in the following differential equation, from which the law is derived when k is constant [4]:

$$dl/dt = k/(2l) \qquad (2)$$

The k was described as a function of time by taking the variations in the mobility and the energy of APD boundaries (APDBs) accompanying the changes in the vacancy concentration and the LRO inside domains into account. Consequently, it was revealed that the excess vacancies existing in the initial stage are primarily responsible for the deviation. This means that the APD growth depends on the concentration of vacancies introduced during the quenching; and it is implied that the APD growth is affected by the condition of heat treatment for the disordering. This may be verified by investigating the APD growth in cases with various quenching temperatures for the disordering. However, such types of experiment are difficult to perform by using Ti$_3$Al because of the narrow range of temperature for the disordered phase of Ti$_3$Al [5].

On the other hand, Fe$_3$Al has wide temperature range of disordered state. While it has D0$_3$-type ordered structure below approximately 800K, B2-type partly disordered structure between about 800K and 1200K, and disordered bcc structure at higher temperatures [5].

Because of the wide ranges of temperature for the disordering, it is possible to investigate the APD growth in cases with various quenching temperatures. However, even the time-series data of the APD growth in Fe$_3$Al based on the observation by TEM, not to mention the effect of the quenching temperature on APD growth, are not available, although quite limited number of data on APD growth by TEM or unreliable estimation by XRD are available [6].

In analyzing the effect of excess vacancies on APD growth, it is necessary to estimate the variation in the vacancy concentration and the long range order (LRO) inside APD, but some difficulties exist in the estimation. With increasing number of the unknown parameters which determines the rate of APD growth, such as APB energy and diffusion coefficient, APD growth data at increasing number of temperature are required [7]. However, it is anticipated that the variation in LRO can be estimated by measuring the resistivity, which is considered to vary depending on the LRO [8].

The aim of this study is to evaluate the influence of the excess vacancy on the APD growth by investigating the growth of the D0$_3$-type APD in Fe$_3$Al by TEM and the variation in the resistivity in the cases with different quenching temperatures.

EXPERIMENTAL PROCEDURES

An ingot of Fe$_3$Al compound with nominal compositions of Fe-28at%Al was prepared by melting high-purity metals of 99.9mass%Fe and 99.99mass%Al in an argon arc furnace. It was annealed at 1523K for 3.6x10^3 s for homogenization. Chips of approximately 3x3x1 mm^3 were cut from the ingot. The chips were hold at 1273K (A2 phase) for 600s or at a temperature of 873K and 1073K (B2 phase) for 3.6x10^{3}s, and brine-quenched. Prior to the annealing at 873K, the chips were annealed at 1073K for 3.6x10^{3}s in order to annihilate the B2-type APDBs. They were isothermally annealed at 673 K, 723 K, and 773 K for 5x10^2-5x10^5 s. Thin foils for the observation by TEM were prepared from the chips by mechanical and electrolytic polish using the twin-jet method with a solution of 33vol% nitric acid-67vol% methanol at 248 K.

The APDs were observed in the dark field by TEM with 111 or 222 superlattice reflection. The average linear intersection lengths of APDBs with more than two hundred intersections were measured for each specimen. There are two types of APDBs in the D0$_3$ structure with different displacement vectors of R_1=1/4<111> and R_2=1/2<100>. The displacement vectors of APDBs were determined by comparing the images by the two reflections. APDBs with the R_1 are visible with any of the two reflections while those with the R_2 are visible only with 111 reflection.

Specimens for resistivity measurement with the size of size 1.1x1.1x40 mm^3 were cut out from the rest of the homogenized ingot. They were executed the disordering heat treatments and then polished with emery papers. Ni lead wires with a diameter of 0.3 mm were spotwelded to the specimens for the measurement by the potentiometric method. The measurements were carried out at 77K, with the specimens in liquid nitrogen. The variations in the resistivity during the ordering were measured by means of repeating the annealing and the measurement alternatively using identical specimens. A salt bath was used in order to bring the specimens quickly to the desired temperature of isothermal ordering. The method for calculating the APD size variation is identical with the previous study [4], and the equations used for calculating the resistivity are given in the literature [7]. The equations and parameters used for the calculation are not shown here, due to the limitation of space. They will be shown in another paper in the near future.

RESULTS & DISCUSSION

Figure 1 shows, by way of example, dark field images with 111 reflection of specimens quenched from 873K and annealed for various period at 723 K. Although both the B2-type and D0₃-type APDBs can exhibit their images with the 111 reflection, no APDB was observed in the as-quenched specimen (Fig. 1(a)). This implies that the B2-type APDs were annihilated during the annealing at 1073K prior to the quenching from 873K and no D0₃-type APD was formed during the quenching. After annealing for 5×10^3 s, the APDs were formed, and the APDs size

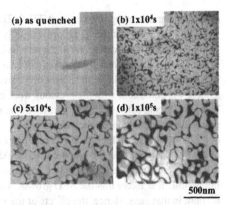

Fig.1 TEM images APDs in Fe-28at%Al quenched from 873K and annealed at 673K for various period.

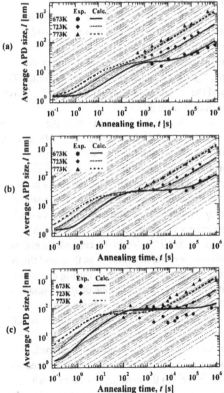

Fig.2 Variations in average APD size during annealing at various temperatures after quenched from (a) 873K, (b) 1073K, and (c) 1273K.

increased with increasing annealing time.

Figure 2 shows the variations in the APD sizes during annealing at various temperatures after the quenching. Figure 2(a) shows the variation in the case where the quenching temperature (T_q) is 873 K. With the increase in the annealing temperature (T_a), the APD sizes of the same annealing time increases, which implies a higher growth rate of the APD at higher temperatures, as is the normal cases. Provided that the APD growth follows the parabolic growth law, the APD size varies along a line parallel to the thin dotted lines in the figure. In the case of T_a=773K, data points run parallel to the thin lines, while in the case of T_a=673K and T_a=723K, the slopes in the early periods are smaller than those for the parabolic growth law. The deviation is more remarkable at the lower temperature of 673K. Figure 2(b) shows the variation in the APD size in the case of T_q=1073K. The deviation from the parabolic growth law is remarkable in comparison with the case of T_q=873 K. Particularly at 673 K, it appears that the APD size is almost constant during the time period of 5×10^3s - 5×10^4s. Figure 2(c) shows the variation in the APD sizes in the case of T_q=1273 K. The deviation is more remarkable and is recognized even at 773K. The APD size variation for the time period of 10^2s–10^3s suggests that the APDs grow rapidly before the plateau.

Figure 3 shows the variations in the resistivity. Figure 3(a) shows the variation in the case where T_q=873 K. The resistivity greatly decreased within a time of 10^2s at the all T_a s. At lower T_a s, the variations in the initial stages are slower, but the final amount of change is larger. This tendency is consistent with the previous work conducted at lower annealing temperatures [8]. Figures 3(b) and (c) show the variations in the resistivity in the cases of T_q=1073K and 1273K, respectively. In both cases, the variations are qualitatively similar to that in the case of T_q=873K, but somewhat faster. Thick lines in Fig.2 and Fig.3 indicate the calculated variations. In the cases of T_q=873K and 1073K, the variations in the APD size and the resistivity could be calculated so that they agree with the experimental data simultaneously, but in the case of T_q=1273K, it was impossible. It is likely that the APD growth is affected by something which is not taken into account in that case. Hence, the effects of the variations in vacancy concentration (C_V) and LRO

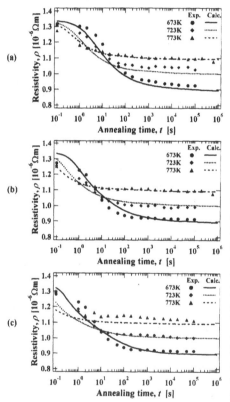

(η) on APD growth are analyzed for the case of T_q=1073K, by way of example.

Figure 4(a) and (b) show the variations in C_v and η, respectively. The C_V varies toward the equilibrium value for each annealing temperature (T_a) from the equilibrium value for the T_q; η approaches the equilibrium for the T_a and the variations are higher at higher T_a. This is quite reasonable.

Figure 5(a), (b), (c), and (d) show the variations in the boundary mobility (M), driving force (γl), growth rate (r), and l calculated using the variations in C_V and η at 623 K as an example of the analysis. The variations in the cases of no excess vacancy, and of no variation in η are shown for comparison. In the case of no excess vacancies, M is a constant with a value of 7×10^{-22} m^2/s; in the case where excess vacancies exist, the value is as thousand times larger than this initial stage. During the initial stage, due to a smaller value of η, γ is small. With the increment of η within a few seconds of annealing, γ and γl also increase. However, in the case of no excess vacancies, r is in the order of 10^{-12} m/s (10^{-3} nm/s), which explains why the variation in l cannot be recognized in the figure. On the contrary, in the case where excess vacancies exist, the variation in l can be recognized in the figure, since the r is as high as 10^{-9} m/s. When η approaches the equilibrium

Fig.3 Variations in resistivity during annealing at various temperatures after quenched from (a) 873K, (b) 1073K, and (c) 1273K.

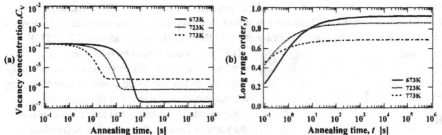

Fig.4 Calculated variations in (a) vacancy concentration and (b) long range order parameter during annealing at various temperatures after quenched from 1073K.

and appears to be constant at $t \cong 10s$, although M decreases slowly with the decrement of C_V, it is still as large as over 10^{-20} m²/s, and accordingly r is also large, and thus the APD growth is still remarkable. Subsequently, as the C_V decreases approaching the equilibrium, M decreases, and thus the r also decreases. When C_V reaches the equilibrium at approximately $t \cong 10^3$ s, the curve of the APD size variation appears to be almost flat. At this point, r is smaller than that in the case of no excess vacancies. This can be explained as follows. M and γ remain nearly identical in both cases of with and without excess vacancy. In the case with excess vacancies, l has already become large, and therefore, γ/l has become smaller than that in the case of no excess vacancy. Following this at $t \cong 10^3$ s, M and γ appear to be constant at their equilibrium values. The variation in the APD size follows the parabolic growth law in which the size at this point is taken as the initial size.

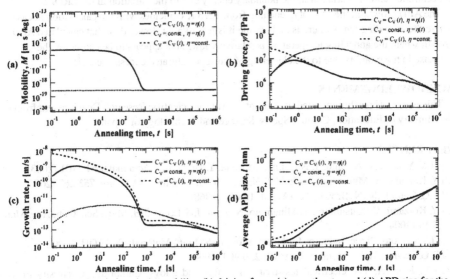

Fig.5 Calculated variations in (a) mobility, (b) driving force, (c) growth rate, and (d) APD size for the case of annealing at 673K after quenching from 1073K. Variations for the cases of no excess vacancy (C_V=const.) and of no variation in LRO (η=const.) are indicated together for comparison.

Figure 6 shows the bright-field image of the specimen quenched from 1273K and annealed at 673K for 10^4s where the APD growth was slower than expected from the calculation. Such types of contrasts were not observed in the case of T_a=723K or 773K. The condition of T_q=1273K and

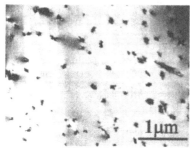

T_a=673K is accompanied by the highest concentration of excess vacancy in the present study. This implies that the dotted contrasts were formed by rearrangement or clustering of the excess vacancies. If the vacancies are consumed for the formation of the vacancy clusters, the vacancy concentration effective for APDB migration decreases with a resulting decrease in the APD growth rate. This appears to explain the reason for the delay. More detailed experiments and observations are required to elucidate the character of the dotted contrasts and their influence on the APD growth.

Fig.6 Bright field image of Fe_3Al quenched from 1273K and annealed at 673K for 10^4s.

CONCLUSION

The higher the quenching temperature is, the faster the APD growth in the initial stage is, because of an increase in the mobility of APDBs due to the excess vacancies. Particularly, in the case where the annealing temperature for ordering is low, the effect is remarkable.

The APD growth after quenching at 873 K or 1073 K can be described by calculation in consideration of the variation in the boundary mobility due to the variation in the vacancy concentration and the variation in the boundary energy due to the variation in the LRO.

The APD growth in the case of quenching from 1273 K coincides with the calculation in cases where the annealing temperature is high, while it is remarkably slower than that calculated APD growth where the annealing temperature is relatively low. This appears to be caused by the decrease in the mobility due to the clustering of the supersaturated excess vacancies.

ACKNOWLEDGEMENTS

This work was partly supported by a Grant-in-Aid for Scientific Research Development from the Ministry of Education, Culture, Sports, Science and Technology of Japan.

REFERENCES

[1] H.Y. Yasuda, K. Nakano, T. Nakajima, M.Ueda, Y. Umakoshi, Acta mater **51** 5101 (2003)

[2] Y. Koizumi, Y. Minamino, N. Tsuji, T. Nakano, Y. Umakoshi, *MRS Sympo* **753** 267 (2003)

[3] S.G Cupschalk, N. Brown, Acta Metal. **16** 657 (1968)

[4] Y. Koizumi, H. Katsumura, Y. Minamino, N. Tsuji, J.G Lee and H. Mori, Sci. Tech. Adv. Mat. **5** 19 (2004).

[5] H. Okamoto, Phase diagrams for binary alloys (ASM int., Materials Park, OH, 2000).

[6] R.G Davies, Trans. Metal. Soc. AIME **230** 903 (1964).

[7] P.L. Rossiter, The electrical resistivity of metals and alloys, Cambridge University Press (1987)

[8] R. Feder and R.W. Cahn, Philo. Mag. A **5** 343 (1960).

Mater. Res. Soc. Symp. Proc. Vol. 842 © 2005 Materials Research Society

Microstructure, Mechanical Properties and Wear Resistance of Fe-Al Based Alloys with Various Alloying Elements

Han-Sol Kim, In-Dong Yeo, Tae-Yeub Ra and Won-Yong Kim
Advanced Materials R&D Center, Korea Institute of Industrial Technology,
#994-32, Dongchun-dong, Yeonsu-ku, Incheon 406-254, South Korea

ABSTRACT

We report on microstructure, mechanical properties and wear resistance of Fe-Al based alloys with various alloying elements. The microstructures were examined using optical and scanning electron microscopy (SEM) equipped with energy dispersive X-ray spectroscope (EDS). Two types of alloys were prepared using vacuum arc melting; one is Fe-28Al based alloys (DO_3 structured) with and without alloying elements such as Mo and Zr. The other one is Fe-35Al based alloys (B2 structured) produced with same manner. For both types of alloys, equiaxed microstructures were observed by the addition of Mo, while dendritic structures were observed by the Zr addition. These microstructural features were more evinced with increasing the content of alloying elements. Concerning the mechanical properties and wear resistance, Fe-35Al based alloys with or without Mo addition superior to Fe-28Al based alloys especially in the high temperature region.

INTRODUCTION

Iron aluminides are of interest for high temperature structural materials in the replacement of highly alloyed heat resistant steels or stainless steels, because they have many attractive characteristics, such as excellent resistance to oxidation and corrosion in hazardous environments, relatively low density, and low material cost [1]. However, their insufficient ductility at room temperature due to insufficient number of slip systems and therefore cleavage fracture tendency has hindered extended applications in the various fields of industries [2, 3]. Thus most of studies for Fe-Al based alloys have been carried out in enhancement of room temperature ductility as well as high temperature strength and creep resistance. Relatively poor studies have been performed on wear resistance at room temperature and at ambient temperatures in Fe-Al based intermetallic alloys. We have prepared two types of alloys with different alloying elements (Mo and Zr) and different structures (DO_3 and B2) in order to understand the effect of microstructure on mechanical properties and wear resistance.

EXPERIMENTAL DETAILS

Various alloy ingots were prepared by arc melting on a water-cooled copper hearth under an Ar gas atmosphere with a non-consumable tungsten electrode. The ingots were re-melted several times to ensure chemical homogeneity. The nominal compositions for all of the samples studied here are listed in Table 1. Samples for metallographic observations were mounted and mechanically polished with SiC paper and Al_2O_3 particles with water. Scanning electron microscopy (SEM) equipped with energy dispersive X-ray spectroscope (EDS) was used in determining the chemical composition of each constituent phase.

Mechanical properties were evaluated by micro-vickers hardness and tensile tests. The

Table 1. Nominal compositions of Fe-Al based alloys used in this study.

Alloy	Chemical composition (at.%)				
	Al	Cr	B	Mo	Zr
A0	28	2	0.2	-	-
AM1	28	2	0.2	0.5	-
AM3	28	2	0.2	2.0	-
AZ1	28	2	0.2	-	0.5
AZ3	28	2	0.2	-	2.0
B0	35	2	0.2	-	-
BM1	35	2	0.2	0.5	-
BM3	35	2	0.2	2.0	-
BZ1	35	2	0.2	-	0.5
BZ3	35	2	0.2	-	2.0

micro-vickers hardness was measured under a load of 4.9N at room temperature. For tensile tests, the specimens were sectioned from as-cast buttons by electro-discharge machining. These specimens were tested in air at an initial strain rate of $1.0 \times 10^{-3} \text{s}^{-1}$ at 773K.

Tribological properties were evaluated by using a ball-on-disk type tribo-meter with alumina balls as the counter material. In order to retain uniform test conditions, a new ball was used for each test. The tests were conducted at a sliding speed of 0.1m/s to a total distance of 100m under a load of 10N. The tests were carried out at room temperature and 673K. The average wear track profile was integrated to get cross-section area of wear track, and the wear volume was calculated by multiplying the area by the average track length. The friction force was measured using a load cell and the friction coefficient was determined from the friction force and the load.

RESULTS AND DISCUSSION

Microstructure

Optical micrographs of as-cast Fe-Al based alloys studied are shown in figure 1. In DO_3 Fe_3Al based alloys with Mo addition (b and c in the figure 1), the microstructures consisting of equiaxed feature of grains and small dispersoids within grains and at grain boundaries are observed over the whole microstructure. The apparent volume fraction of dispersoids increases with increasing Mo content as compared in figure 1(b) and (c).

In B2 FeAl-based alloys with Mo addition (g and h in the figure 1), the microstructures consisting of equiaxed feature of grains and needle-like dispersoids within grains and at grain boundaries are seen. In the sample with 2at.% Mo these needle-like fine dispersoids are disappeared, while dispersoids are formed along the grain boundaries, as shown in figure 1(h).

In the samples with Zr addition, the microstructures observed are different to the other alloys. Zr addition to Fe-Al based alloys causes the formation of a dendrite microstructure with a residual eutectic phase, as shown in figure 1(d), (e), (i) and (j). This can be explained by limited solubility range of Zr in DO_3 and B2 structured Fe-Al intermetallic phases [4]. The residual eutectic phase has been known as $Zr(Fe, Al)_2$ Laves phase [5] that strengthen Fe-Al based alloy [6]. The volume fraction of residual eutectic phase drastically increases with increasing Zr content in both DO_3 and B2 type Fe-Al based alloys. This propensity for microstructural formation is more evinced in B2-structured alloys than DO_3-structured ones. Figure 2 shows the

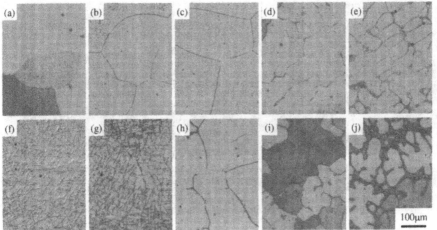

Figure 1. Optical micrographs of D0$_3$ structured alloys; (a) A0, (b) AM1, (c) AM3, (d) AZ1 and (e) AZ3, and B2 structured alloys; (f) B0, (g) BM1, (h) BM3, (i) BZ1 and (j) BZ3.

volume fraction of dispersoids for the present alloys studied. The volume fraction of dispersoids was measured using computer-aided image analyzing system. In both D0$_3$ and B2 structured alloys, the volume fraction of dispersoids formed at grain boundaries increases with increasing content of alloying elements.

In order to identify the dispersoids, SEM/EDS analysis was done for some alloys. SEM micrographs with EDS results are shown in figure 3. From the SEM/EDS results, it can be confirmed that as follows: the dispersoids of alloy AM3 contain higher Mo and Cr content and smaller content of Al compared to the chemical composition of the matrix phase (figure 3(a)), the eutectic phase in the alloy AZ3 contain high Al and Zr content (figure 3(b)), and the dispersoids in alloy BM1 contain slightly higher Al content (about ~5 at.%) than the matrix phase but the other differences in chemical composition aren't detected (figure 3(c)).

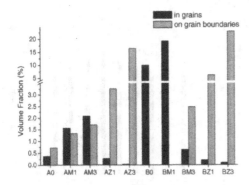

Figure 2. Apparent volume fraction of dispersoids in the present alloys.

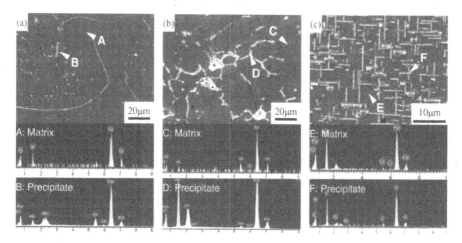

Figure 3. SEM micrographs and corresponding EDS results of the three alloys; (a) AM3, (b) AZ3 and (c) BM1.

Mechanical properties

Micro-hardness and tensile properties of the present alloys are shown in figure 4. The hardness value increases with increasing content of the alloying elements for both types of DO_3 and B2 structured alloys. This may be attributed to the volume fraction of dispersoids as well as their structure type, shape and size. Within the experimental range investigated, it is found that Zr is more effective alloying element to increase hardness than Mo for both structured alloys. This may be associated with microstructural change from equiaxed structure with fine dispersoids to dendritic structure with a residual eutectic phase. Concerning the tensile results tested at 773K, no discernable change in ultimate tensile strength is obtained, while fracture strain decreases with transition element alloying such as Zr and Mo in DO_3 structured alloys. This

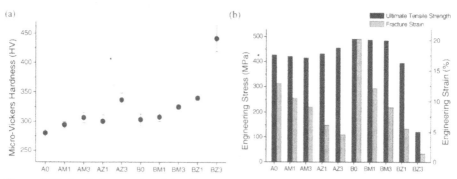

Figure 4. Mechanical properties of the alloys: (a) microhardness at room temperature (b) ultimate tensile strength and fracture strain obtained by tensile test at 773K.

result may indicate that fracture strain is sensitive to the size, shape and distribution of dispersoids in D0$_3$ structured Fe-Al based alloys. In B2 structured alloys, we have observed that both ultimate tensile strength and fracture strain decreases with an addition of transition elements such as Mo and Zr. On the basis of stress-strain behavior, it is recognized that the decrease in ultimate tensile strength is closely related to the brittle fracture of the B2 structured alloys tested.

Wear resistance

Figure 5 shows the results of wear rate and friction coefficient for B2 and D0$_3$ structured alloys. At room temperature, B2 structured alloys show lower friction coefficient than that for the D0$_3$ structured alloys. This is well correlated with the hardness results. Basically, low friction coefficient can be obtained in highly alloyed samples to show the effect of transition element alloying for both B2 and D0$_3$ structured alloys. Similar tendency is seen for both types of samples tested at 673K. With respect to the wear rate B2 structured alloy without containing transition elements appeared to display lower wear rate than D0$_3$ structured alloy without containing transition elements at room temperature. Interestingly, BZ3 alloy shows the higher wear rate than the other alloys even though its friction coefficient is the minimum among the alloys examined. This may be due to a weak phase boundaries between B2 or D0$_3$ ordered matrix phases and Laves dispersoids formed as a residual eutectic phase. Thus it can be suggested that finely distributed dispersoids in the microstructure is responsible for lower wear rate at room

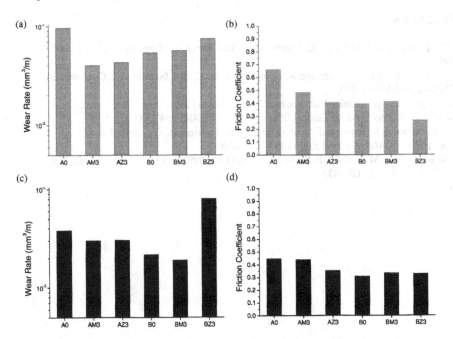

Figure 5. Tribological properties of experimental alloys tested at room temperature((a) and (b)) and at 673K((c) and (d)): (a, c) wear rate (b, d) friction coefficient

temperature. At 673K, besides alloy BZ3, wear rate decreases with an addition of alloying elements for Fe-Al based alloys tested. This result is in contrast to the result obtained at room temperature. One plausible explanation is that thermally activated process may compensate the brittleness due to a weak interface boundary between matrix and dispersoids resulting in an improvement in wear resistance. Nevertheless the poor wear resistance of the alloy BZ3 is believed to lack of ductility at 673K. Conclusively, microstructure, alloying elements as well as crystal structure would be significant to improve the wear resistance at room temperature and at ambient temperature in the present Fe-Al based alloys.

CONCLUSION

The microstructure, mechanical properties and wear behavior were investigated in Fe-Al based alloys with different alloying elements and different order structures (D0$_3$ and B2). The microstructure of the alloys containing Mo and Zr exhibits dispersoids in the D0$_3$ or B2 matrix. The volume fraction, shape and distribution of the dispersoids strongly depend on the Mo and Zr contents. At the given condition, besides BZ3 alloy, B2 structured alloys exhibit better wear resistance than D0$_3$ structured alloys. In B2 structured alloys, finely dispersed microstructure has a better effect to enhance the wear resistance than the dendritic microstructure with a residual eutectic phase. Within the experimental range investigated, the B2 structured alloys with or without Mo addition show good both tensile properties and wear resistance at high temperature.

REFERENCES

1. C.G. McKamey, J.H. DeVan, P.E. Tortorelli and V.K. Sikka, *J. Mater. Res.* **6**, 1779-1805 (1991).
2. O. Tassa, C. Testani, J. Lecoze and A. Lefort, *Proc. Int. Symp. Intermetallic Compounds* **JIMIS-6**, 573-577 (1991).
3. M.A. Crimp and K. Vedula, *Mat. Sci. and Eng.* **78**, 193 (1986).
4. F. Stein, G. Sauthoff and M. Palm, *J. Phase Equilib.* **23**, 480-494 (2002).
5. A. Wasilkowska, M. Bartsch, F. Stein, M. Palm, K. Sztwiertnia, G. Sauthoff and U. Messerschmidt, *Mat. Sci. and Eng.* **A380**, 9-19 (2004).
6. A. Wasilkowska, M. Bartsch, F. Stein, M. Palm, G. Sauthoff and U. Messerschmidt, *Mat. Sci. and Eng.* **A381**, 1-15 (2004).

Nickel Aluminides

Mater. Res. Soc. Symp. Proc. Vol. 842 © 2005 Materials Research Society

S1.2

MICROSTRUCTURES AND MECHANICAL PROPERTIES OF NiAl-Mo COMPOSITES

H. Bei[1,2] and E. P. George[1,2]
[1]The University of Tennessee, Department of Materials Science and Engineering, Knoxville, TN 37996
[2]Oak Ridge National Laboratory, Metals and Ceramics Division, Oak Ridge, TN 37831

Abstract

In-situ composites consisting of ~14 vol.% continuous Mo fibers embedded in a NiAl matrix were produced by directional solidification in a xenon-arc-lamp, floating-zone furnace. The fiber spacing and size were controlled in the range 1-2 μm and 400-800 nm, respectively, by varying the growth rate between 80 and 20 mm/h. Electron back-scatter diffraction patterns from the constituent phases revealed that the growth directions and interface boundaries exhibited the following orientation relationships: $\langle 100 \rangle_{NiAl} // \langle 100 \rangle_{Mo}$ and $\{011\}_{NiAl} // \{011\}_{Mo}$. The temperature dependence of the tensile strength and ductility were investigated and the NiAl-Mo composite was found to be both stronger and have a lower ductile-brittle transition temperature than the unreinforced NiAl matrix.

Introduction

The use of NiAl as a structural material suffers from two major drawbacks: poor ductility/fracture toughness at room temperature and low strength/creep resistance above 600°C. Attempts have been made to toughen NiAl by combining it with a ductile metal [1-11]. For example, Johnson et al. [4] obtained a room-temperature fracture toughness of ~20 MPa√m in a composite alloy of composition NiAl – 34 at.% (Cr, Mo), and Misra et al. [11] obtained a fracture toughness of ~14 MPa√m in a NiAl – 9 at.% Mo eutectic alloy. Both these values are significantly higher than the room-temperature fracture toughness of monolithic NiAl single crystals (~6 MPa √m [12]).

In this study, well-aligned microstructures of NiAl-Mo, devoid of any cellular or dendritic regions, were produced by directional solidification under carefully controlled conditions in a high-temperature optical floating zone furnace having a relatively steep temperature gradient (~30 K/mm). This furnace has been used by us previously to produce well-aligned eutectic microstructures of Cr-Cr₃Si and V-V₃Si alloys over a wide range of growth conditions [13-16].

Experimental Procedures

Alloys having the nominal composition Ni – 45.5Al – 9Mo (at.%) were arc melted, drop cast, and directionally solidified in a high-temperature optical floating zone furnace in flowing argon gas at growth rates of 20-80 mm/h and a fixed rotation rate of 60 rpm. Additional details of the processing conditions are given elsewhere [17]. Total weight losses after melting and casting were less than 0.05%, so nominal (starting) compositions are used throughout this paper.

Representative samples were cut from the directionally solidified (DS) rods along the transverse and longitudinal directions, polished using standard metallographic techniques, etched with a solution of 80% hydrochloric and 20% nitric acids, and examined by optical, electron, and orientation imaging microscopy.

Dogbone-shaped specimens were machined from the DS rods and tensile tested parallel to the fiber direction in vacuum at 25-1000°C and an engineering strain rate of 4.2 × 10^{-3} s^{-1}. Fracture surfaces were examined in a scanning electron microscope (SEM).

Results and Discussion

Figure 1 shows the well-aligned fibrous microstructure of the NiAl-Mo eutectic obtained by directional solidification at 40 mm/h and 60 rpm where the matrix is the NiAl phase and the continuous fibers are Mo (solid-solution). X-ray microprobe analysis revealed that the NiAl matrix contained essentially no Mo (< 0.1 at.%) and had the off-stoichiometric composition Ni-45.2Al, whereas the Mo fibers contained all three elements and had the composition Mo-10.1Al-3.9Ni (all compositions given in at.%). The volume fraction of the Mo fibers was determined to be 0.141.

The arrangement of the Mo rods in the transverse section is approximately hexagonal [Fig.1 (c)], which is the pattern that naturally results at the points of intersection of equal-radius circles drawn around the Mo fibers with each such point equidistant from its neighbors [Fig. 1 (d)]. During steady-state growth, the radius of the circles is determined by the diffusion distance, which in turn is determined by the time available for diffusion (or the directional solidification rate); i.e., the faster the growth rate, the smaller the circles and hence the closer the fiber spacing.

The shape of the individual Mo rods in the transverse section is square rather than circular [Fig. 1(c)], indicating a highly anisotropic NiAl-Mo interfacial energy. Consistent with this, electron backscatter diffraction patterns recorded from the constituent phases at different locations within the as-grown composite revealed the following orientation relationships. The NiAl-Mo interface boundaries were parallel to the {011} planes in both NiAl and Mo, whereas, the growth direction was <100> in both phases.

The {011} planes observed in this study are in agreement with those reported previously [11]. These are the closest-packed planes in the BCC and B2 crystal structures of Mo and NiAl [18], and are probably selected during eutectic growth as inter-phase boundaries to minimize the boundary energy. However, the <100> growth directions observed here are not the same as those reported before [11] where they were found to deviate by ~15-20° from <100>. A possible reason for this difference is that a well-aligned microstructure was obtained in this study whereas a cellular microstructure was observed earlier [11]. When the microstructure is well-aligned, all the Mo fibers are normal to the transverse section (on which the microstructural observations are made). In contrast, the Mo fibers in the cellular structures intersect the transverse section at different angles making it difficult to unambiguously determine the growth direction.

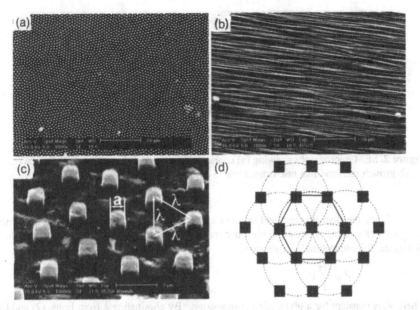

Figure 1. SEM micrographs showing well-aligned rod-like microstructure of the NiAl-Mo eutectic alloy: (a) transverse section, (b) longitudinal section, and (c) higher magnification view of transverse section; (d) schematic of growth mechanism resulting in hexagonal arrangement of Mo fibers.

The fiber spacing, defined as the average distance on a transverse section between the centers of adjacent Mo fibers [λ in Fig. 1(c)], and the fiber size, defined as the average edge length of the square cross-sections [a in Fig. 1 (c)], were measured over a range of growth rates, R, from 20 to 80 mm/h at a fixed rotation rate of 60 rpm. Figure 2 shows examples of microstructures obtained at growth rates of 20 and 80 mm/h.

Jackson and Hunt [19] obtained the following relationship between growth rate R and spacing λ by considering the balance between the diffusion required for phase separation and the energy required for interphase boundary formation:

$$\lambda^2 R = \frac{a^R}{Q^R} = C_l,\qquad(1)$$

where Q^R and a^R (and, therefore, C_l) are constants related to the magnitudes of the liquidus slopes at the eutectic temperature, the composition difference between the two phases, their volume fractions, the solid-liquid interface energies of two phases, and the liquid–solid interface shape. Here, C_l is assumed to have a fixed value over the range of experimental conditions investigated.

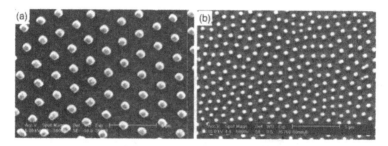

Figure 2. SEM micrographs showing (a) coarse (20 mm/h growth rate), and (b) fine (80 mm/h growth rate) rod-like microstructures in our DS NiAl-Mo alloy.

Since the fiber arrangement on the transverse section is hexagonal, it can be shown that the fiber size (a) is related to the fiber spacing (λ) and the volume fraction of the fibers (V_f) by the following simple expression:

$$a^2 = \frac{\sqrt{3}}{2} V_f \lambda^2,\tag{2}$$

where V_f is constant for a given alloy composition. By eliminating λ from Eqns. (2) and (3) we obtain the following relationship between fiber size and growth rate:

$$a^2 R = C_2.\tag{3}$$

where C_2 is another constant. Therefore, both fiber size and spacing vary inversely as the square root of growth rate [Eqns. (1) and (3)].

Figure 3 is a plot of the experimentally determined values of λ and a versus the reciprocal of the square root of growth rate, $\sqrt{R}$.

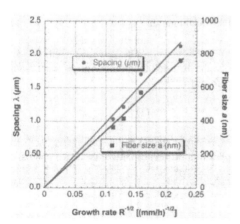

Figure 3. Effects of growth rate on Mo fiber spacing and size in the DS NiAl-Mo alloy investigated in this study.

Consistent with Eqns. (1) and (3) the following linear fits were obtained:

$$\lambda R^{1/2} = 9.76 \ \mu m \cdot mm^{1/2} / h^{1/2}$$
$$a R^{1/2} = 3.42 \ \mu m \cdot mm^{1/2} / h^{1/2} \qquad (4)$$

From Eqn. (4), the ratio of fiber spacing to size, λ/a, is found to be 2.85. When this value is substituted into Eqn. (2), the volume fraction of Mo fibers is calculated to be 14.2%, which is in excellent agreement with the measured value of 14.1%.

Tensile tests were performed parallel to the Mo fibers at room temperature and various elevated temperatures to investigate the mechanical properties of the NiAl-Mo composite directionally solidified at 80 mm/h and 60 rpm [microstructure shown in Fig 2 (b)]. Figure 4 is a plot of the 0.2% off-set yield strength and ductility of the composite as a function of temperature. Also included in this figure are the yield strength and ductility of <100>-oriented NiAl having the same composition as the matrix in the composite. The composite has a higher yield strength and lower DBTT than the NiAl matrix.

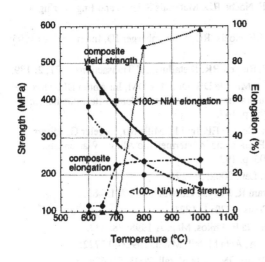

Figure 4. Temperature dependence of the yield strength and ductility of NiAl -Mo composite compared to those of <100> NiAl having the same composition as the matrix in the composite.

Conclusions

Directional solidification in an optical floating zone furnace was used to produce well-aligned rod-like microstructures of NiAl-Mo eutectic alloys. The structures consist of 14 vol.% Mo (solid-solution) fibers having a square cross section embedded in an intermetallic NiAl matrix. The composition of the fibers is Mo-10.1Al-3.9Ni and that of the matrix Ni-45.2Al (< 0.1 Mo). The growth direction is <100> in both the NiAl matrix and the Mo fibers, and the NiAl/Mo interface boundaries are parallel to the {011} planes in both phases. Fiber spacing and size decrease inversely as the square root of the growth rate, and range from ~1 to 2 μm and ~400 to 800 nm, respectively, for growth rates of 80 to 20 mm/h. When compared to <100> NiAl single crystals having the same composition as the matrix, the NiAl-Mo composite has higher yield strength and lower DBTT.

Acknowledgment

This research was sponsored by the Division of Materials Sciences and Engineering, Office of Basic Energy Sciences, U. S. Department of Energy, under contract DE-AC05-00OR22725 with UT-Battelle, LLC.

References

[1]   Misra A, Wu ZL, Kush MT, Gibala R, Materials Science and Engineering A 1997; 239-240: 75.

[2]   Subramanian PR, Mendiratta MG, Miracle DB, Metallurgical and Materials Transactions A 1994; 25: 2769.

[3]   Joslin SM, Chen XF, Oliver BF, Noebe RD, Materials Science and Engineering A 1995; 196: 9.

[4]   Johnson DR, Chen XF, Oliver BF, Noebe RD, Whittenberger JD, Intermetallics 1995; 3; 99.

[5]   Cline HE, Walter JL, Lifshin E, Russell RR, Metallurgical Transactions 1971; 2: 189.

[6]   Subramanian R, Mendiratta MG, Miracle DB, Dimiduk DM, In: Anton DL, Martin PL, Miracle DB, McMeeking R, editors. Intermetallic Matrix Composites. Pittsburgh: Materials Research Society 1990. p. 147.

[7]   Frommeyer G, Rahlbauer R, In: George EP, Inui H, Mills MJ, Eggeler G, editors. Defect Properties and Related Phenomena in Intermetallic Alloys. Warrendale: Materials Research Society 2003. p. 193.

[8]   Milenkovic S, Caram R, Mater. Lett. 2002; 55: 126.

[9]   Milenkovic S, Coelho AA, Caram R, J Cryst. Growth 2000; 211: 485.

[10]   Walter JL, Cline HE, Metall. Trans. 1970; 1: 1221

[11]   Misra A, Wu ZL, Kush MT, Gibala R, Philos. Mag. A 1998; 78: 533.

[12]   Chang KM, Darolia R, Lipsitt HA, Acta Metall. Mater. 1992; 40 2722.

[13]   Bei H, George EP, Kenik EA, Pharr GM, Acta Metall. 2003; 51: 6241.

[14]   Bei H, George EP, Pharr GM. Intermetallics 2003; 11: 283

[15]   Bei H, George EP, Kenik EA, Pharr GM, Z. Metallkunde 2004: in press.

[16]   George EP, Bei H, Serin K, Pharr GM, Mater. Sci. Forum 2003; 426-432: 4579.

[17]   Bei H, George EP, Acta Mater. 2005; 53: 69.

[18]   Villars P, Calvert LD, "Pearson's Handbook of Crystallographic Data for Intermetallic Phases", American Society for Metals, Metals Park, 1985.

[19]   Jackson KA, Hunt JD. Trans. Metall. Soc. AIME 1966; 236: 1129.

Mater. Res. Soc. Symp. Proc. Vol. 842 © 2005 Materials Research Society S1.3

Microstructure and Thermo-Mechanical Behavior of NiAl Coatings

G. Dehm[1], J. Riethmüller[1], P. Wellner[1], O. Kraft[2], H. Clemens[3], E. Arzt[1]

[1]Max-Planck-Institut für Metallforschung, 70569 Stuttgart, Germany
[2]Institut für Materialforschung II, Forschungszentrum Karlsruhe and Institut für Zuverlässigkeit von Bauteilen und Systemen, Universität Karlsruhe (TH), 76344 Karlsruhe, Germany
[3]Department of Physical Metallurgy and Materials Testing, Montanuniversität Leoben, 8700 Leoben, Austria

ABSTRACT

In this study the thermo-mechanical behavior of a commercial Pt containing NiAl coating deposited on a Ni-base superalloy is compared with a Ni-rich NiAl coating sputter-deposited on a Si substrate. Both types of coatings possess high tensile room temperature stresses after thermal straining. The Pt-NiAl coating shows negligible plasticity as a result of solid solution and dispersion strengthening. In contrast, for the NiAl coatings on Si noticeable plasticity can be obtained if the film thickness exceeds the sub-micrometer range.

INTRODUCTION

NiAl is applied as a protective high temperature coating for turbine blades used in engines [1]. In order to improve the oxidation resistance commercial NiAl coatings contain several at% Pt. Pt substitutes the Ni atoms of the ordered β-NiAl phase and suppresses the formation of ternary oxides which are less stable than a dense Al_2O_3 scale [2-4]. While the mechanical behavior of bulk NiAl has been extensively studied over the last decades little information is known about the mechanical behaviour of NiAl-based coatings.

In this paper we report on thermal straining results of a commercial Pt containing NiAl coating deposited on a Ni-base superalloy. Since the microstructure of commercial coatings strongly depends on the processing route and chemical composition of the Ni-base superalloy additional straining experiments were performed on a model system. The model system consists of NiAl coatings sputter deposited onto a single crystal Si substrate with an amorphous diffusion barrier between the coating and the substrate to prevent interdiffusion. The deformation behavior of the coatings is characterized on the basis of stress-temperature cycles and correlated with microstructural and dimensional effects.

NiAl DEPOSITION AND STRESS MEASUREMENT

Industrial NiAl coatings containing approximately 7 at% Pt were grown on (100) oriented Ni-base superalloy single crystals in a two step process. Initially, a several μm thick Pt layer was electro-deposited on the substrate and annealed at elevated temperatures in order to stimulate Ni-Pt interdiffusion. Subsequently, Al was deposited by chemical vapour deposition. The Al diffuses into the Ni-Pt layer and forms a Pt alloyed NiAl coating. The commercial Ni-base superalloy substrate contained additions of Cr, Mo, W, Co, Ta, and Re.

For microstructural and mechanical comparison with the industrial Pt-NiAl coatings pure NiAl films were grown as a model system by magnetron-sputtering at nominally room temperature. (100) oriented single-crystal Si substrates coated with a 50nm thick amorphous SiN$_x$ diffusion barrier served as substrates. After deposition, the samples were annealed at 600 °C for 1 h without breaking vacuum in the ultrahigh vacuum chamber of the sputtering system to obtain a stable microstructure.

The microstructure of all coatings was analysed using X-ray diffraction, optical microscopy, focussed-ion beam microscopy (FIB), scanning electron microscopy (SEM) and transmission electron microscopy (TEM). Chemical compositions of the coatings and individual phases were qualitatively determined using energy-dispersive X-ray spectroscopy (EDS) in the SEM and TEM. Further details of the microstructural studies are reported in reference [5].

The biaxial stresses of the pure NiAl coatings were measured in a nitrogen rich atmosphere using the substrate-curvature technique [5,6]. Residual stresses in the film lead to a bending of the film/substrate composite which is measured with an optical system and analyzed as described in [5,6]. Measuring the curvature C and using the biaxial elastic modulus of the substrate M_s and the thicknesses of the film h_f and the substrate h_s, the film stress σ_f can be calculated according to [6]:

$$\sigma_f = \frac{M_s\, h_s^2}{6\, h_f}\, C\,. \tag{1}$$

In order to obtain absolute stress values for NiAl coatings on Si substrates we measured the curvature of the bare substrate prior to NiAl deposition and subtracted this value from the composite curvature. This procedure results in an accuracy of ± 5 % for absolute film stresses. In contrast, for Pt-NiAl coatings on Ni-base superalloy substrates, the presence of a diffusion layer prevents an absolute stress determination by this technique, since the evolution of curvature during thermal cycling contains contributions of the coating and the diffusion layer. However, the absolute stress of the Pt-NiAl coating was determined at room temperature using the $\sin^2\Psi$ method. Based on this value the stress-temperature values during thermal cycling can be estimated assuming a constant curvature of the diffusion layer/substrate couple. A detailed description of the substrate-curvature technique and of the $\sin^2\Psi$ method are given in references [5,6] and [6].

RESULTS AND INTERPRETATION

Pt-NiAl coating on a Ni-base superalloy

Optical microscopy revealed an average Pt-NiAl grain size of 75 ± 15 µm with the grains extending over the complete coating thickness of ~45 µm. In this region a chemical composition of ~40 at% Ni, 40-45 at% Al and ~7 at% Pt was measured by EDS linescans in the SEM (figure 1a). A corresponding cross-sectional image recorded using secondary electrons in the FIB is presented in figure 1b. XRD measurements indicate that the coating consists solely of randomly oriented β-NiAl grains exhibiting a CsCl-structure (figure 2). Detailed TEM investigations of the coating are in agreement with SEM and XRD observations. The NiAl grains contain several at% Pt in solid solution (hereafter named (PtNi)Al) and form the matrix of the coating. Additionally, TEM reveals the presence of sub-micrometer sized particles in the coating. The particles are occluded in the (PtNi)Al grains (see figure 3) and consist mainly of refractory metals which must

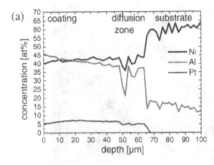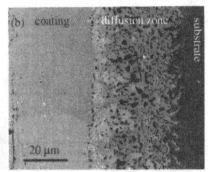

Figure 1. (a) EDS linescan across the coating and diffusion zone into the Ni-base superalloy. The coating contains approximately 40 at% Ni, 40-45 at% Al and 7 at% Pt. (b) FIB image of a cross-section reveals columnar NiAl grains in the coating and a ~40 μm wide diffusion zone.

have diffused from the Ni-base superalloy substrate into the coating. Adjacent to the coating a ~40 μm thick diffusion zone has formed, which contains a large number of micrometer-sized particles (figure 1b). Chemical analyses performed in the SEM indicate that these particles are carbides and oxides.

Two samples with similar coating thickness and chemical composition were cycled between room temperature and 750 °C at heating and cooling rates of 4-6 K/min and the changes in curvature were recorded every 10K. Since the curvature of the substrate was not known the initial stress of the coating was determined by the $\sin^2\Psi$ method using the {211} (PtNi)Al reflections at $2\theta = 81.3$ °. A Poisson ratio of $\nu = 0.307$ was assumed for the (PtNi)Al phase and a Youngs modulus E=126GPa was deduced from unloading segments of nanoindentation experiments following the procedure of [7]. Using these numbers a room temperature stress of 3.3 ± 0.2 GPa was determined by the $\sin^2\Psi$ method. Thermo-mechanical cycles of a Pt-NiAl coating are shown in figure 4. In addition to the measured changes in curvature, stress values based on the room temperature stress obtained by the $\sin^2\Psi$ method are indicated in figure 4.

The stress values were calculated from the curvature changes under the simplifying assumption that the curvature of the substrate/diffusion zone couple remains unchanged during thermal cycling. Upon first heating the tensile stresses of the coating relax as indicated by the negative change in curvature. At temperatures between 550 °C and 700 °C a positive change in curvature is observed in the first heating cycle which might result from grain growth, diffusion and/or surface oxidation although thermal cycling was performed in a N_2 atmosphere. During cooling tensile stresses develop in the coating. Subsequent thermal cycles reveal a narrow curvature (stress) - temperature hysteresis indicating that very little plasticity occurred in the coating.

NiAl film on Si substrate

Details of the microstructure and thermal stress evolution of the magnetron-sputtered NiAl films are described in [5] and are briefly described in this paragraph. After magnetron sputtering and annealing at 600 °C, the Ni-rich NiAl films with thicknesses ranging between 0.2

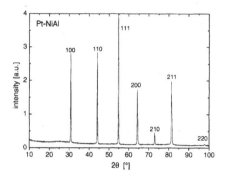

Figure 2. θ-2θ scan of the Pt-NiAl coating. The diffraction peaks arise from β-NiAl.

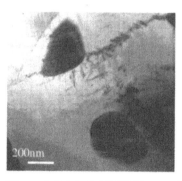

Figure 3. Bright-field TEM image of particles in the (PtNi)Al matrix, which consist of refractory metals.

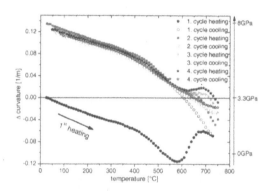

Figure 4. Curvature-temperature curves for the first four cycles between room temperature and 750 °C. The curvature changes are measured by a laser optical system. X-ray diffraction studies revealed that the coating is initially under a tensile stress of 3.3 ± 0.2 GPa. Based on this value stresses between -1 GPa and ~8 GPa were estimated to have developed in the coating during thermal cycling assuming negligible stresses in the diffusion layer.

and 3 μm revealed a polycrystalline microstructure with equiaxed NiAl grains. The median grain size was usually smaller than the film thickness and increased with increasing film thickness (figure 5). Chemical analysis revealed a concentration of 45-50 at% Al. Similarly to the Pt-NiAl coating all magnetron-sputtered NiAl films solely consisted of β–NiAl as observed by X-ray diffraction. TEM investigations of 3.0 and 0.2 μm thick films (figure 5) revealed no additional phases in the NiAl films. Dislocations were found in the 3 μm thick film, mainly in the larger grains, while the 0.2 μm thick film was devoid of dislocations.

The stress evolution of the NiAl films was studied during thermal cycling to 700 °C by the substrate-curvature technique. In figure 6 the stress-temperature curves of the first and the second cycle of a 0.4 μm thick film (Al content of 48.5 at%) are shown. Upon first heating from room temperature, the initially tensile stress of 1600 MPa decreased linearly with increasing temperature at a rate of 2.7 MPa/K up to about 400 °C and a biaxial tensile stress of 600 MPa. On further heating, an accelerated stress relaxation with an increased rate of up to 5.5 MPa/K occurred resulting in a deviation from the linear stress-temperature behavior. This increase is

Figure 5. TEM micrographs of (a) 0.2 μm and (b) 3 μm thick NiAl films on Si substrate (48.5at% Al). The grain size increases with increasing film thickness.

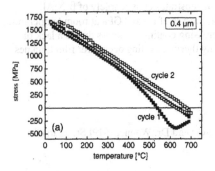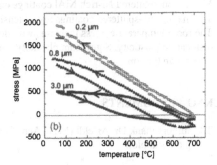

Figure 6. (a) First and second stress temperature cycles of a 0.4 μm thick NiAl film reveal a change in slope upon the first heating cycle. In subsequent cycles the 0.4 μm thick film deforms mainly elastically. (b) Comparison of the stress-temperature curves for different film thicknesses. Thicker films possess larger stress hysteresis and lower room temperature flow stresses indicating stronger dislocation plasticity.

associated with constrained diffusional creep [8], a stress relaxation mechanism where atoms diffuse from the coating surface via the grain boundaries into the interior in order to relax tensile stresses [5,8]. Above 500 °C the film stress became compressive and eventually reached compressive stress values up to 400 MPa, which then relaxed to about 250 MPa. During cooling the film stress became tensile again and increased linearly with decreasing temperature at a rate of 2.8 MPa/K. At room temperature the initial tensile stress of 1600 MPa before thermal cycling was regained. The second and subsequent temperature cycles of the 0.4 μm thick film coincide with the first cycle during heating to 400 °C (see figure 6). However, no change in stress slope was observed upon heating to 600 °C. It is believed that residual oxygen during thermal cycling led to the formation of a passivation layer on the surface shutting down constrained diffusional creep. Subsequent stress-temperature cycles were almost identical to the second cycle.

The effect of film thickness on the stress evolution is shown in figure 7 for 0.2, 0.8 and 3.0 μm thick NiAl films with Al contents of 48.5 at%. It is clearly seen that the room temperature stresses increase with decreasing film thickness. Additionally, the width of the stress-temperature hysteresis decreases with decreasing film thickness, indicating less dislocation plasticity for thinner films. This finding is in agreement with TEM observations [5].

SUMMARY

- A commercial Pt-NiAl high temperature coating consists of β-NiAl with Pt in solid solution. Additionally, submicrometer-sized particles of refractory metals have formed in the coating. Interdiffusion during coating deposition caused the formation of a diffusion zone.
- The commercial coating is under a high biaxial tensile stress at room temperature. Upon heating the stresses relax. Thermal straining experiments reveal predominantly elastic deformation for the Pt-NiAl coatings on the Ni-base superalloy.
- Magnetron-sputtered Ni-rich NiAl coatings on Si substrates consist solely of β-NiAl.
- The magnetron-sputtered coatings reveal tensile stresses of up to 2 GPa at room temperature. The room temperature stress increases with decreasing coating thickness due to a lack of dislocation plasticity. Significant plasticity during thermal cycling occurs for film thickness larger than 0.8 μm.

ACKNOWLEDGMENTS

The authors thank Dr. Affeldt (MTU Aero Engines) and Dr. Wagner (MPI Stuttgart) for providing the samples.

REFERENCES

1. E. E. Affeldt, Adv. *Eng. Mat.* **2**, 811 (2000).
2. M. J. Stiger, N. M. Yanar, M. G. Topping, and F.S. Meier, *Z. Metallkde* **90**, 1069 (1999).
3. A. L. Purvis, B. M. Warnes, *Surface & Coatings Technology* **146-147**, 1 (2001).
4. M. W. Chen, R. T. Ott, T. C. Hufnagel, P. K. Wright, and K. J. Hemker, *Surface & Coatings Technology* **163-164**, 25 (2003).
5. P. Wellner, G. Dehm, O. Kraft, and E. Arzt, *Z. Metallkde* **95**, 769 (2004).
6. W. D. Nix, *Met. Trans. A* **20**, 2217 (1989).
7. G. M. Pharr, W. C. Oliver, and F. R. Brotzen, *J. Mater. Res.* **7**, 613 (1992).
8. H. Gao, L. Zhang, W. D. Nix, C. V. Thompson, and E. Arzt, *Acta Mat.* **47**, 2865 (1999).

Quantitative Analysis of γ Precipitates in Cyclically Deformed Ni3(Al,Ti) Single Crystals Using Magnetic Technique

Yukichi Umakoshi[1], Hiroyuki Y. Yasuda[1,2] and Toshifumi Yanai[1]

[1]Department of Materials Science and Engineering, Graduate School of Engineering, Osaka University, 2-1, Yamada-oka, Suita, Osaka 565-0871 Japan
[2]Research Center for Ultra-High Voltage Electron Microscopy, Osaka University, 7-1, Mihogaoka, Ibaraki, Osaka 565-0047 Japan

ABSTRACT

Ni-18at.%Al-4at.%Ti single crystals containing small γ precipitates in γ' matrix were prepared by aging at 1073K after solution treatment from γ' single phase region. The size, shape and volume fraction of the γ precipitates were examined during aging and deformation by means of changes in spontaneous magnetization and coercive force. Cyclic deformation behavior of these crystals was discussed through change in the morphology of γ precipitates using magnetic technique.

INTRODUCTION

Magnetic properties are known to be influenced by plastic deformation. For example, Pt_3Fe is antiferromagnetic in the fully ordered $L1_2$ structure while it shows strong ferromagnetism after rolling because the atomic environment and magnetic couplings change by cold working [1]. Magnetic anisotropy induced by deformation in some ferromagnetic alloys is mainly caused by atomic arrangements around an anti-phase boundary (APB) and the interaction between the internal stress field around dislocations and spontaneous magnetization. The type, density and distribution of lattice defects such as dislocations and planar faults can be examined using deformation induced magnetic anisotropy. Dislocations and APBs in ordered and disordered Ni_3Fe single crystals fatigued were examined by the magnetic anisotropy of permeability and high-field susceptibility resulting in making clear the mechanism of cyclic softening and cyclic hardening [2].

Recently, we tried to predict a nondestructive lifetime of fatigued Fe-40at%Al alloy by magnetic measurement [3]. Atomic rearrangement near APBs in ordered FeAl alloys results in a paramagnetic-to-ferromagnetic transition. In fatigued Fe-40at%Al spontaneous magnetization increased with increasing quantity of APB ribbons and tubes. Since spontaneous magnetization rapidly increased due to the increase in APB tubes at the final stage of fatigue and fracture occurred, the remaining lifetime can be evaluated from the onset of an abrupt increase in magnetization.

Ni-based superalloys are composed of γ' precipitates with the $L1_2$ structure and disordered f.c.c. γ matrix. The size and distribution of the precipitates strongly influence mechanical properties of the alloys. Change in morphology of the γ' precipitates such as subdivision or dissolution by shear is important to examine cyclic deformation behavior. Since γ' phase shows weak ferromagnetism, significant change in magnetic properties is not expected by deformation. In contrast, γ phase of Ni-rich solid solution shows strong ferromagnetism and its deformation behavior can be examined using magnetic technique.

In this paper, we report magnetic observation of γ precipitates in γ'-Ni3(Al,Ti) single crystals during annealing and cyclic deformation focusing on changes in size, shape and volume fraction of the precipitates. Much attention is placed on not only subdivision but dissolution of the γ precipitates in γ' matrix during cyclic deformation.

EXPERIMENTAL PROCEDURE

Single crystals of γ'-$Ni_3(Al,Ti)$ were grown from ingots of Ni-18at.%Al-4at.%Ti alloy by a floating zone method. The crystals were homogenized at 1423K for 168h. Fatigue specimens (7mmx2x2mm^2 at gauge) whose loading axis is [149] were cut from the single crystals by spark machining. The Schmid factor for the primary (111)[$\bar{1}$01] slip is 0.5. Specimens for Vickers hardness and compression tests were also prepared.

Specimens were quenched into ice brine from 1423K after annealing for 2h to obtain γ' single phase. The quenched specimens were aged at 1073K for appropriate periods, and small γ phase precipitated in γ' matrix. Compression and fatigue tests were carried out at room temperature at strain rate of $1.67\times10^{-4}s^{-1}$ and $3.0\times10^{-4}s^{-1}$, respectively. Fatigue tests were carried out in tension/compression mode at total strain amplitude of 0.2% to 10^3 or 10^4 cycles.

The magnetization process of deformed and undeformed specimens was measured by a vibrating sample magnetometer in the temperature range between 298K and 4.2K. An external magnetic field was applied parallel to the loading axes of compressed and fatigued specimens.

The morphology of γ precipitates was observed by TEM focusing on changes in volume fraction, size and shape of the precipitates. The deformation substructure and the change in γ precipitates during deformation were also examined by TEM.

RESULTS AND DISCUSSION

Fig.1 shows change in Vickers hardness of Ni-18at.%Al-4at.%Ti alloy during aging at 1073K. The hardness increases with increasing aging time and results in the decrease after reaching a maximum at 3h. The change in hardness is closely related to the precipitation of γ phase in γ' matrix. According to TEM observation, both size and volume fraction of γ precipitates increase with increasing aging time up to 3h but after showing a peak the precipitates are coarsened maintaining a constant volume fraction. Spherical precipitates were observed in the peak-aged specimen as shown in Fig.2 (a) but the shape of the precipitates changed from spherical to plate-like in the over-aged specimen at 100h as shown in Fig.2. Although the microstructure is not shown here, numerous spherical γ precipitates with average size of 12nm maintain coherency with γ' matrix in the peak-aged specimen due to the coffee beam contrast around the precipitates. In contrast, coarse plate-like precipitates in the over-aged specimen showed fringe contrast with different reflection of g=220 and their habit plane was {100} to minimize the elastic energy at the interface

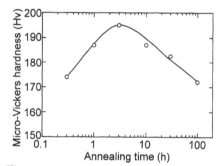

Figure 1. Change in Vickers hardness of Ni-18at.%Al-4at.%Ti alloy annealed at 1073K with annealing time.

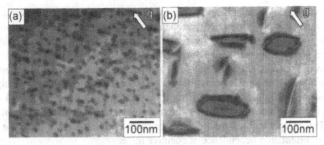

Figure 2. TEM micrographs of Ni-18at.%Al-4at.%Ti alloys annealed at 1073K for 3h (a) or 100h (b); g=110.

between γ precipitates and γ' matrix.

Fig.3 shows changes of spontaneous magnetization and coercive force in Ni-18at.%Al-4at.%Ti single crystals during aging at 1073K. The spontaneous magnetization monotonically increases with increasing annealing time, while coercive force rapidly increases after gradual increase at initial annealing time up to 3h and then reached a constant around 100h. Since γ' matrix shows weak ferromagnetism, the increase in spontaneous magnetization is due to the precipitation of γ phase showing ferromagnetism. Although the increase in spontaneous magnetization at an initial stage of aging is caused by change in volume fraction of ferromagnetic γ precipitates, the further increase at the aging time longer than 3h cannot be explained only by the volume fraction. From EDS analysis, the relative concentration of Ni to (Al,Ti) in quenched γ' phase was 3.77.

In the peak-aged (3h) specimen, the relative concentrations in γ precipitates and γ matrix were 4.70 and 3.63, respectively, while they changed to 5.06 and 3.37 in the over-aged (100h) specimen. At the longer aging, Ni concentration in ferromagnetic γ precipitates increased. Therefore, change in spontaneous magnetization during aging is closely related to changes in volume fraction and Ni concentration of γ precipitates.

The coercive force showed a gradual increase at the initial stage of aging followed by a rapid increase at over-aged period. Total magnetic energy (E_T) of a precipitate is known to be composed of magnetostatic energy, magnetocrystalline anisotropy energy, shape magnetic anisotropy energy. From a condition of $\partial^2 E_T / \partial\varphi^2 = 0$ where φ is the angle between easy axis and an external field direction, coercive force is derived. The coercive force is strongly

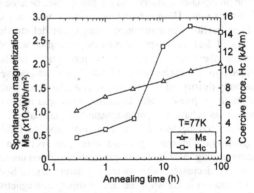

Figure 3. Change in spontaneous magnetization and coercive force of Ni-18at.%Al-4at.%Ti alloys annealed at 1073K with annealing time. Magnetic measurement was done at 77K.

influenced by shape magnetic anisotropy energy. In the case of spherical precipitates, magnetic spins aligned randomly along easy axis of each precipitates in a demagnetization process after magnetization resulting in small coercive force. In contrast, when an external field is applied to plate-like precipitates, magnetic spins are aligned parallel to the field. In a demagnetization process, however, the spins align preferentially along the longitudinal axis of the precipitates and large coercive force is obtained. According to TEM observation of the γ precipitates during aging, shape of the precipitates changed from spherical to plate-like after the peak-aging time of 3h. The rapid increase in coercive force is due to the shape magnetic anisotropy of the plate-like precipitates. The morphology of the precipitates such as the aspect ratio can be examined by measurement of orientation dependence of coercive force.

The cyclic hardening behavior of as-quenched and aged Ni-18at.%Al-4at.%Ti single crystals was examined in the previous paper [4]. The maximum stress rapidly increased with increasing number of cycles and was saturated at about 500 cycles. Although the maximum stress of the as-quenched or over-aged specimen was lower than that of the peak-aged specimen at the initial stage of cyclic deformation, the saturation stress of the over-aged specimen was almost the same as that of the peak-aged specimen. The over-aged and peak-aged specimens demonstrated the strongest and weakest cyclic hardening among three specimens, respectively. As a result, the saturation stress of the as-quenched specimen was the lowest.

Effect of plastic deformation on spontaneous magnetization and coercive force in peak-aged and over-aged specimens is shown in Table 1. Since effect of cyclic deformation on magnetic properties of weak ferromagnetic γ' phase was negligible [5], changes in spontaneous magnetization and coercive force come from γ precipitates in γ' matrix. Although even 10% compression strain induced no significant spontaneous magnetization for both aged specimens, the spontaneous magnetization rapidly decreased by cyclic deformation and small spontaneous magnetization at 10^4 cycles suggests the disappearance of ferromagnetism. In compression little change in spontaneous magnetization is in good agreement with no significant changes in the morphology and the volume fraction of γ precipitates. During cyclic deformation subdivision of the precipitates occurred and the precipitates became small by very frequent to-and-fro motion of dislocations. However, the precipitates were not dissolved in γ' matrix and the initial volume fraction was maintained during cyclic deformation.

Rapid decrease in spontaneous magnetization was observed in both fatigued specimens. The average diameters of spherical γ precipitates in the peak-aged and over-aged specimens fatigued to 10^4 cycles were 15 and 22nm, respectively. These nano-scale precipitates were randomly distributed in γ' matrix. In a nano-scale particle single magnetic domain only exists. Although γ phase is ferromagnetic, each spin is randomly oriented in γ' matrix and the γ precipitates behave as a paramagnetic-like material [6].

After cooling to 4.2K under a small applied field of 100Oe, temperature dependence of magnetization in peak-aged and over-aged specimens fatigued to 10^4 cycles was measured during heating to 298K The magnetization showed unusual increase at low temperatures and then gradually decreased with increasing temperature. Such unusual temperature dependence of magnetization is known to be due to superparamagnetism which occurs in materials containing nano-particles [6]. The volume of each particle can be calculated from the blocking temperature obtained from the magnetization-temperature curve. The detail will be presented elsewhere [7].

In Table 1 coercive force of the over-aged specimen is larger than that of the peak-aged specimen because of high shape magnetic anisotropy energy of plate-like precipitates in the over-aged specimen. The coercive force of the over-aged specimen decreases by deformation

Table I. Effect of fatigue deformation on spontaneous magnetization and coercive force at 77K in Ni-18at.%Al-4at.%Ti alloys annealed at 1073K.

	peak-aged (3h)			over-aged (100h)		
	undeformed	10^3 cycles	10^4 cycles	undeformed	10^3 cycles	10^4 cycles
Spontaneous magnetization, Ms ($\times 10^{-2}$Wb/m^2)	3.18	2.10	1.63	3.83	2.85	1.70
Coercive force, Hc (kA/m)	2.79	4.71	1.22	15.59	10.68	5.86

since the precipitates are cut by shear and the shape changes from plate-like to spherical.

During cyclic deformation large plate-like precipitates are cut and number of small particles increases. Therefore, the saturation stress of the over-aged specimen finally reaches the same value of the peak-aged specimen at 10^4 cycles. The coercive force in the peak-aged specimen decreases after showing a slight increase during deformation. The increase is due to change in shape from spherical to ellipsoidal by shear. At 10^4 cycles precipitates become spherical nano-particles and the coercive force decreases.

When a nonmagnetic material contains ferromagnetic precipitates, the deformation behavior in the material is examined and the fatigue life can be estimated by monitoring change in magnetic properties of ferromagnetic precipitates.

CONCLUSIONS

Deformation behavior of Ni-18at%Al-4at%Ti single crystals was examined using magnetic technique focusing on changes in spontaneous magnetization and coercive force in γ precipitates in γ' matrix, and the following conclusions were reached.

1. Vickers hardness of the crystal increased followed by decrease after showing a maximum peak during aging at 1073K. The size and volume fraction of γ precipitates increased with annealing time up to a hardness peak, while coarsening of the precipitates occurred but the volume fraction maintained constant in the over-aging range.

2. Spontaneous magnetization increased with annealing time. The change in magnetization is due to an increase in volume fraction of γ precipitates. The increase after peak-aging time is caused by Ni-enrichment in the precipitates during aging.

3. Shape change of γ precipitates can be examined by coercive force. Rapid increase of coercive force in the over-aged range is caused by the shape change of the precipitates from spherical to plate-like.

4. Cyclic deformation behavior can be monitored by changes in spontaneous magnetization and coercive force. The γ precipitates are cut by shear during cyclic deformation. The precipitates become spherical nano-particles by strong cyclic deformation and superparamagnetism appears.

By monitoring changes in spontaneous magnetization and coercive force, deformation behavior can be examined.

ACKNOWLEDGEMENTS

This work was supported by a Grant-in-Aid from the Japanese Ministry of Education, Culture, Sports, Science and Technology, and also by "Priority Assistance of the Formation of Worldwide Renowned Centers of Research - The 21st Century COE Program (Project: Center of Excellence for Advanced Structural and Functional Materials Design)" from the same Ministry.

REFERENCES

1. V.R. Hahn and E. Kneller, Z. Metallkunde, **49**, 426(1958).
2. H.Y. Yasuda, A. Sasaki and Y. Umkoshi, J. Appl. Phys., **88**, 5909 (2003).
3. H.Y. Yasada, R. Jimba and Y. Umakoshi, Scripta Mater., **48**, 589 (2003).
4. H.Y. Yasuda, R. Jimba and Y. Umakoshi, Acta Mater., **50**, 161 (2002).
5. H.Y. Yasuda, A. Sasaki, Y. Umakoshi and S. Takahahi, Scripta Mater., **44**, 581 (2001).
6. K. Yakushiji, S. Mitani, K. Takanashi, J.-G. Ha and H. Fujimori, J. Magn. Magn. Mater., **212**, 75 (2000).
7. H.Y. Yasuda, T. Yanai and Y. Umakoshi, submitted.

Mater. Res. Soc. Symp. Proc. Vol. 842 © 2005 Materials Research Society   S5.22

Microstructures in Cold-Rolled Ni₃Al Single Crystals

Kyosuke Kishida, Masahiko Demura, and Toshiyuki Hirano
Materials Engineering Laboratory, National Institute for Materials Science,
1-2-1 Sengen, Tsukuba, Ibaraki 305-0047, JAPAN

ABSTRACT

Microstructure evolution during cold rolling of binary stoichiometric Ni_3Al single crystals was examined by the optical (OM) and transmission electron microscopy (TEM). In the case of the <001> initial RD, the banded structure is formed. Inside each matrix band, the localized shear deformations occur alternately on two {111} planes. In addition, huge amounts of widely extended superlattice intrinsic stacking faults (SISFs) are observed from relatively early stage of cold rolling. The occurrence of the localized shear deformation is considered to be controlled by the SISFs since they must be strong obstacles for the dislocation motion on the other glide plane. The extensive formation of the SISFs is therefore considered to be one of the most important microstructural features which control the cold rolling behavior of Ni_3Al.

INTRODUCTION

Considerable amount of studies on the deformation microstructures in Ni_3Al-base intermetallic alloys have been carried out so far, however, main interests were the dislocation structures near the yield point mainly because the anomalous yield behavior is one of the most attractive properties of Ni_3Al [1,2]. In contrast, the deformation microstructures after large plastic deformation such as cold rolling have rarely been studied by transmission electron microscopy (TEM) [3,4]. Recently we have carried out systematical studies on cold rolling of binary stoichiometric Ni_3Al single crystals with various initial crystal orientations and revealed that the cold rolling behaviors are strongly dependent on the initial crystal orientations, especially on the initial rolling direction (RD) [5]. Very thin cold-rolled foils (approximately 20μm in thickness) with good quality in shape can be obtained only when the initial RD is close to <001>, whereas macroscopic curving or cracking hinders high cold reduction for samples with the other initial RD [5,6]. Such an anisotropic rolling behavior must be closely related to the internal deformation microstructure developed during cold rolling. It is therefore necessary to investigate the characteristics of the microstructures in the cold rolled thin foils for understanding the details of the texture formation mechanisms. In the present study, we report the results of the TEM observations of cold-rolled Ni_3Al single crystals.

EXPERIMENTAL PROCEDURES

Single crystalline rods of binary stoichiometric Ni_3Al were grown by optical floating zone methods [7]. The sample (#47-1) used in this study possessed approximate initial orientations of (530)[001] [5]. All cold rolling processes were performed at room temperature without intermediate annealing. Details of the cold rolling were described in our previous papers [7].

Microstructures of the samples were characterized by optical microscopy (OM) and TEM. For OM observations, three orthogonal sections were cut from each cold rolled sample. Surfaces were mechanically polished and then chemically etched in a Marble reagent (5g of copper sulfate pentahydrate, 20cm^3 of hydrochloric acid and 20cm^3 of water) at room temperature. TEM experiments were carried out in a Philips CM200 operated at 200kV. Both the rolling plane (RD-TD plane) and the longitudinal (RD-ND) sections of the cold-rolled foils were prepared for the TEM observations. They were mechanically thinned and polished by a standard double-jet electropolishing method in a solution of perchloric acid, n-butyl alcohol and methanol (ratio 1:5:10) at 233K. In the case of the longitudinal sections of very thin foils below 100μm in thickness, the ion-milling machine was used for the final thinning process.

RESULTS

Banded structure

Figure 1 shows optical microstructures of three orthogonal sections in a 54% cold-rolled Ni$_3$Al single crystal. As previously reported [4], so-called banded structure is clearly observed in all sections and each component matrix band develops as a thin plate inclined to the rolling plane. Orientation difference of fine traces seen in each matrix band indicates that two types of differently oriented matrix bands are developed during cold rolling. The formation of the banded structure results in the clear texture split as shown in the {220} pole figure.

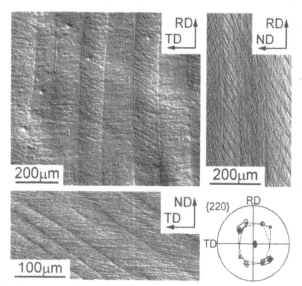

Figure 1. Optical microstructures of three orthogonal sections of a 54% cold-rolled Ni$_3$Al single crystal with the banded structure and corresponding {220} pole figure.

Localized shear deformation

Figure 2a shows deformation microstructures observed in a rolling plane section of a 54% cold-rolled sample containing the transition region between two adjacent matrix bands. Diffraction analysis confirms that the ND of each matrix band is approximately parallel to [110] with some deviation less than 10°, which is consistent with the result of the pole figure measurement. In each matrix band, striped contrast is observed to be extended unidirectionally along [$\bar{1}$10]. Microstructural features relating to the striped contrast are therefore expected to be viewed edge on by the projection along the [$\bar{1}$10] of the matrix band. Such an observation condition can be achieved by using the longitudinal (RD-ND) section. Figure 2b shows a bright field image observed along [$\bar{1}$10] of a matrix band. It is clearly seen from the figure, the microstructure contains many parallelograms in various sizes, both edges of which are actually viewed edge on being virtually parallel to two {111} planes, namely (111) and ($\bar{1}$1$\bar{1}$). The inset of the figure 2(b) is selected area electron diffraction (SAD) pattern taken from the central region in the figure 2(b) with approximately 600nm in diameter containing many parallelograms. The SAD pattern does not contain any additional spots indicating that the crystal orientations of all parallelograms are macroscopically the same. Therefore, the observed structure cannot be interpreted as the fcc-type {111}<11$\bar{2}$> deformation twins. Orientation difference between the adjacent parallelograms is measured to be less than 2° by the microdiffraction experiments. Details of the microdiffraction analysis will be published in elsewhere [8].

As marked by some arrows in figure 3a, one set of edges of parallelograms on (111) are largely shifted at intersections with another set of edges on ($\bar{1}$1$\bar{1}$), which indicates the occurrence of intensive localized shear on ($\bar{1}$1$\bar{1}$) planes. Both edges of the parallelograms are thus

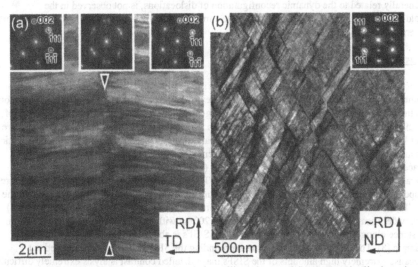

Figure 2. Deformation microstructures in (a) the rolling plane and (b) the longitudinal sections of a 54% cold-rolled Ni₃Al single crystal.

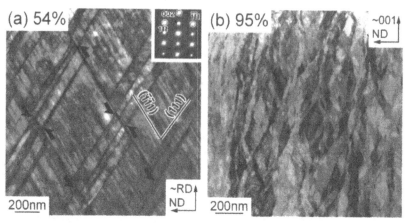

Figure 3. Bright field images of Ni₃Al single crystals cold-rolled to (a) 54% and (b) 95% thickness reductions.

considered to be developed by the localized shear deformations on two {111} planes. Further cold reduction makes the size of the parallelograms much smaller as seen in the 95% cold-rolled foils (figure 3b). It can be thus concluded that the cold deformation of the Ni₃Al single crystals mainly proceeds through the alternate occurrence of the localized shear deformations on two {111} planes. It should be noted that any apparent equiaxed cell or sub-grain structure, which is generally related to the dynamic reconfiguration of dislocations, is not observed in the cold-rolled Ni₃Al single crystals in contrast to some conventional metals with high stacking fault energy such as Al and Ni.

Stacking faults

Although no additional diffraction spots are seen, strong streaks are observed in the SAD patterns in figure 2b and 3a. In most cases, they extend along one of two {111} reflecting vectors in the $[\bar{1}10]$ diffraction pattern. Such streaks are generally considered to reflect the existence of planar defects being parallel to the incident beam. However, the edges of the parallelograms are not likely to be the dominant reasons for the streaks since they mostly extend along one direction. In an example shown in figure 3a, streaks are seen parallel to $(\bar{1}\bar{1}1)$ reflecting vector and fine traces of the planar defects on the corresponding $(\bar{1}\bar{1}1)$ plane are observed inside the parallelograms. Weak-beam and high-resolution TEM analyses reveal that the planer defects are superlattice intrinsic stacking faults (SISFs) (figure 4). As clearly seen in figure 4a, width of the SISFs are not uniform. Further TEM analyses confirm that the widths of the SISFs along [112] are widely scattered ranging from 4nm to above 200nm with an average of 35nm and an average distance along the direction perpendicular to the SISFs is about 10nm in the 54% cold rolled sample [8]. The types of the dislocations surrounding the SISF have not been clarified yet because extremely high amounts of the SISFs make detailed contrast analysis extremely difficult. It is obvious that such extensive formation of the SISFs is responsible for the absence of sub-grain structure inside the parallelograms. In addition, the development of the parallelograms

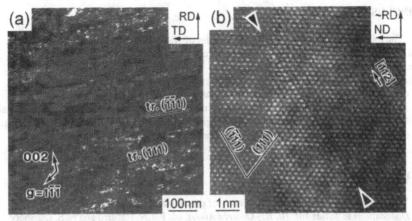

Figure 4. (a) Weak beam and (b) high-resolution TEM images of the superlattice intrinsic stacking faults (SISFs) in a 54% cold-rolled Ni₃Al single crystal. Arrowheads in (b) indicate the location of the SISF.

is considered to be controlled by the SISFs since they may act as strong obstacles for the dislocation motion on the other glide plane [9]. Thus, it can be concluded that the extensive formation of the SISFs is one of the most important microstructural features in the cold rolled Ni₃Al.

DISCUSSION

As described above, the characteristics of the microstructures of the cold-rolled Ni₃Al with the banded structure can be summarized as the extensive development and accumulation of the planar defects such as the SISFs and the planes of the localized shear deformations on two sets of {111} planes. Previously, we discussed the deformation mechanisms for the cold rolled Ni₃Al with the banded structure by assuming that the deformations are carried out with simultaneous activation of two {111}<110> slip systems in each matrix band [5].The shear planes observed in this study are fully consistent with those deduced on the basis of the macroscopic information such as OM observations, macroscopic shape change and the texture evolution during the cold rolling [5]. It is therefore likely to be reasonable to conclude that all observed shear deformation play the same role as the {111}<101> slip systems.

In addition, we have found that the simultaneous slip activation on more than three slip planes is difficult during heavy cold deformation of Ni₃Al single crystals [5]. It is apparent that such limitations on the slip activation originate from the planar nature of the stored deformations, i.e. slip activation become more difficult with increasing the number of the planes with planar defects. Since all planar nature of deformation is related to the SISF formation, it would be possible to reduce the anisotropy of the Ni₃Al single crystals by increasing the SISF energy. For this purpose, alloying of proper ternary elements and/or warm-rolling should be considered.

CONCLUSIONS

The microstructures in the cold-rolled samples of Ni_3Al with the banded structure were studied by the OM and TEM as a function of the cold reduction. The main results are summarized as follows.

1. The cold deformation of the Ni_3Al single crystals mainly proceeds through the alternate occurrence of the localized shear deformations on two {111} planes.
2. The SISFs are extensively developed and accumulated in the cold rolled Ni_3Al single crystals. Widths of the SISFs are widely scattered ranging from 4nm to above 200nm with an average of 35nm and an average distance is about 10nm in the 54% cold rolled sample.

REFERENCES

1. Y.Q. Sun and P.M. Hazzledine, "Geometry of dissociation glide in L1$_2$ γ'- phase", *Dislocations in Solids Vol. 10: L1$_2$ Ordered Alloys*, ed. F.R.N. Nabarro and M.S. Duesbery (Elsevier, Amsterdam, 1996), chapter 49, pp. 27-68.
2. P. Veyssiere and G. Saada, "Microscopy and plasticity of the L1$_2$ γ' phase", *Dislocations in Solids Vol. 10: L1$_2$ Ordered Alloys*, ed. F.R.N. Nabarro and M.S. Duesbery (Elsevier, Amsterdam, 1996), chapter 53, pp. 253-441.
3. J. Ball and G. Gottstein, *Intermetallics,* **1**, 171 (1993).
4. S.G. Chowdhury, R.K. Ray, A.K. Jena, *Mater. Sci. Engng. A*, **246**, 289 (1998).
5. K. Kishida, M. Demura, Y. Suga and T. Hirano, *Philos. Mag.*, **83**, 3029 (2003).
6. M. Demura, K. Kishida, Y. Suga, M. Takanashi, and T. Hirano, *Scripta Mater.*, **47**, 267 (2002)
7. M. Demura, Y. Suga, O. Umezawa, K. Kishida, E. P. George, and Hirano, T., *Intermetallics*, **9**, 157 (2001).
8. K. Kishida, M. Demura and T. Hirano, unpublished work.
9. F.R.N. Nabarro, Z.S. Basinski and D.B. Holt, *Advances in Physics*, **13**, 193 (1964).

Mater. Res. Soc. Symp. Proc. Vol. 842 © 2005 Materials Research Society   S5.26

Crystal Structure, Phase Stability and Plastic Deformation Behavior of Ti-rich $Ni_3(Ti,Nb)$ Single Crystals with Various Long-Period Ordered Structures.

Koji Hagihara, Tetsunori Tanaka, Takayoshi Nakano and Yukichi Umakoshi
Department of Materials Science and Engineering & Handai Frontier Research Center,
Graduate School of Engineering, Osaka University,
2-1, Yamada-oka, Suita, Osaka 565-0871, Japan

ABSTRACT

In Ni-Ti-Nb ternary system, there are some geometrically close-packed (GCP) phases with long-period stacking sequences of a close-packed plane (CPP). Among them, our focus is on the $Ni_3(Ti_{0.90}Nb_{0.10})$ crystals with Pb_3Ba-type rhombohedral structure with nine-fold stacking sequence. Compression tests were conducted using the single crystals and the temperature and orientation dependences of plastic deformation behavior were investigated in comparison with those of DO_{24}-Ni_3Ti crystals with the four-fold stacking sequence. The K-W locking of screw dislocation was found to occur not only in the compounds such as Ni_3Al and Ni_3Ti with a relatively small unit cell, but also even in complex compounds with longer-period stacking structures by slip on the common CPP in the GCP structures.

INTRODUCTION

The Ni_3X-type intermetallic compounds (X= Al, Ti, Sn, Nb, V, etc.) are known to crystallize in various geometrically close-packed (GCP) crystal structures: $L1_2$ for Ni_3Al, DO_{24} for Ni_3Ti, DO_{19} for Ni_3Sn, DO_a for Ni_3Nb and DO_{22} for Ni_3V, respectively. The crystal structures of those GCP phases are composed of the common close-packed plane (CPP), which corresponds to the {111}in fcc and (0001) in hcp unit cell. Their crystal structures are therefore characterized by the two features focusing on the CPP; the stacking sequence of CPP and the ordered arrangements of X atoms on the CPP. Among the Ni_3X-type compounds, much attention has been paid to the $L1_2$-Ni_3Al as an important strengthening component in Ni-based superalloys. They show superior mechanical properties including the anomalous strengthening behavior due to the Kear-Wilsdorf (K-W) locking of screw dislocations [1]. Recently, we found that the yield stress anomaly (YSA) appears not only in $L1_2$-Ni_3Al [1] but also in other GCP compounds with a relatively small unit cell, such as DO_a-Ni_3Nb [2], DO_{24}-Ni_3Ti [3] and DO_{19}-Ni_3Sn [4]. Therefore, they are also expected to act as a strengthening phase in Ni-based superalloys and multi-phase composite alloys [5-7].

Recently, we investigated the phase stability of ternary Ni_3X-type compounds in a Ni-Ti-Nb system. We fabricated single crystals of $Ni_3(Ti, Nb)$ with various different Nb content by the floating zone (FZ) method, and investigated their microstructures. Figure 1 shows the selected area electron diffraction (SAED) patterns in the various $Ni_3(Ti_{1-x}Nb_x)$ single crystals (x=0, 0.03, 0.05 and 0.10) taken from the $[2\bar{1}\bar{1}0]$ direction. All of the SAED patterns are obviously different each other and the periodicity of spots along the c-axis lengthens with the Nb addition. This indicates that their crystal structures changed from the four-fold DO_{24} structure of Ni_3Ti to the different structures with longer-stacking sequence of CPP depending on the quantity of Nb content [8].

To date there has been few reports of the plastic behavior of such compounds with the complex stacking sequence. Since the atomic arrangements on the non-close packed planes are too complicated, we assumed in the previous paper that the K-W locking is unlikely to occur in those complex compounds [9]. In this study we focused on the $Ni_3(Ti_{0.90}Nb_{0.10})$ crystals with Pb_3Ba-type rhombohedral structure. Compression tests were conducted using the single crystals and the temperature dependences of plastic deformation behavior, operative slip system and dislocation structure were compared with those of DO_{24}-structured Ni_3Ti crystals with the four-fold stacking sequence of the CPP, with particular focus on the YSA.

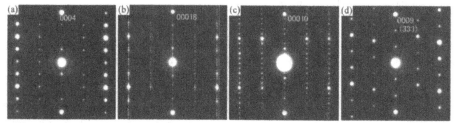

Figure 1. Selected area electron diffraction patterns observed along the $[2\bar{1}\bar{1}0]$ direction in the various $Ni_3(Ti_{1-x}Nb_x)$ single crystals (a) x=0, (b) x=0.03, (c) x=0.05 and (d) x=0.10, respectively. The stacking sequence of CPP varies from (a) four-fold (DO_{24}) to (b) eighteen-fold, (c) ten-fold and (d) nine-fold by the addition of different amount of Nb.

EXPERIMENTAL PROCEDURE

A master ingot with a nominal composition of $Ni_3(Ti_{0.9}Nb_{0.1})$ was prepared by arc-melting. Single crystals were grown by the FZ method in an NEC SC-35HD furnace at a growth rate of $5mmh^{-1}$ under a high-purity argon gas flow. The crystal structure of $Ni_3(Ti_{0.9}Nb_{0.1})$ is schematically drawn in Fig.2(a). The crystal is composed of nine-fold stacking of the common CPP, and the Ti(Nb) atoms are arranged in a triangular fashion on the CPP. We denote this structure as a "nine-layered structure" hereafter. In the rhombohedral lattice system, the crystal orientation can also be indexed by the hexagonal notation shown in Fig. 2(b), and this aids our understanding of the crystal geometry.

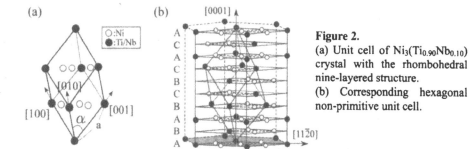

Figure 2.
(a) Unit cell of $Ni_3(Ti_{0.90}Nb_{0.10})$ crystal with the rhombohedral nine-layered structure.
(b) Corresponding hexagonal non-primitive unit cell.

The crystal obtained was a single-phase single-oriented crystal, but it contained two variants of matrix and twin-related phases. They existed with plate-like morphology, whose boundaries lay on the (111) close-packed plane. The volume fraction of the twin-variant was about the same as that of the matrix-variant. Rectangular specimens with dimensions of approximately $2 \times 2 \text{mm}^2 \times 5\text{mm}$ were prepared by electro-discharge machining from the as-grown crystal. Two orientations denoted by A and B were chosen as the loading axes, which are depicted on a stereographic projection in Fig. 3. In the hexagonal notation, the orientations A and B are comparable with those previously tested in Ni$_3$Ti single crystals [3]. The Schmid factors for the $(1\bar{1}00)[11\bar{2}0]$ prism slip and $(0001)[2\bar{1}\bar{1}0]$ basal slip are approximately 0.5 at A and B orientation, respectively. Taking the crystal geometry into consideration, it is believed that the existence of twin-variant may not give a significant additional influence on the plastic behavior. Compression tests were performed at a nominal strain rate of 1.7×10^{-4} s^{-1} in a temperature range from room temperature (RT) to 1200°C in a vacuum. Thin foils were cut from the deformed specimens, and dislocation structures were observed in a TEM (JEOL-3010) operated at 300kV.

RESULTS

Figure 3 shows the temperature dependence of the yield stress (0.2% offset stress) in Ni$_3$(Ti$_{0.90}$Nb$_{0.10}$) single crystals deformed at A and B orientations. The yield stresses show strong temperature dependence at both loading orientations. At the A orientation, the yield stress exhibits relatively high value at RT, and this gradually decreases with increasing temperature. However, the yield stress shows a slight increase with increasing temperature from 300°C to 500°C. Subsequently, the yield stress rapidly decreases above the peak temperature of 500°C. At the B orientation, the yield stress exhibits a considerably low value at RT, but it rapidly increases with rising temperature, reaching a maximum stress peak at around 700°C. The yield stress at the peak temperature is six times higher than that at RT. It should be

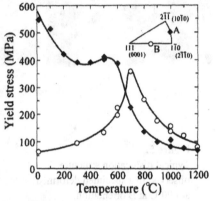

Figure 3. Temperature dependence of the yield stress in Ni$_3$(Ti$_{0.90}$Nb$_{0.10}$) single crystals deformed at A and B orientations.

noted that the profile of yield stress-temperature curves at A and B orientations are both very similar to those observed in Ni$_3$Ti single crystals with the DO$_{24}$ structure [3], despite the difference in the crystal structure. In this paper, we have discussed only the deformation behavior at B orientation in which the operation of slip on CPP is expected. Details of the deformation behavior at A orientation will be described elsewhere.

Figure 4 (a) and (b) show the slip traces on the side surfaces of the B-oriented specimens deformed at 500°C and 800°C, respectively. Slip traces parallel to (111) were clearly observed at 500°C where strong YSA appeared. The (111) slip corresponds to the (0001) basal slip in the hexagonal notation, the slip on the CPP. The slip traces were fairly fine and homogeneously distributed in the specimens. Above the peak temperature of YSA, however, the (111) slip trace disappeared and only {100} slip traces were observed accompanied by rapid decrease in yield stress, as shown in Fig.4(b). For the {100} slip, the Burgers vector of dislocations was confirmed to be <$01\bar{1}$> by TEM observation using the conventional $g \cdot b$ contrast analysis. The {100}<$01\bar{1}$> slip corresponds to the {$10\bar{1}1$}<$\bar{1}2\bar{1}0$> pyramidal slip in the hexagonal notation. As reported in the previous paper [3], the {$10\bar{1}0$} prism slip is operative in addition to the (0001) basal slip in Ni$_3$Ti at high temperatures in this orientation, while the {$10\bar{1}1$} pyramidal slip was identified in Ni$_3$(Ti$_{0.90}$Nb$_{0.10}$) crystals owing to the change in crystal structure. This change of operative slip systems from the {$10\bar{1}0$} prism slip in Ni$_3$Ti to the {$10\bar{1}1$} pyramidal slip in Ni$_3$(Ti$_{0.90}$Nb$_{0.10}$) was also detected at A orientation.

Figure 4.
Slip traces observed in specimens deformed at B orientation at different temperatures of (a) 500°C and (b) 800°C. The subscript M and T indicate that the slips in the matrix and twin-variant phases, respectively.

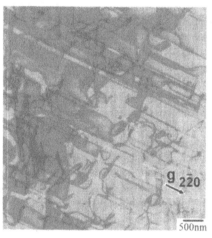

Figure 5.
Typical dislocation structure observed in a specimen deformed at 600°C at B orientation. Straight screw dislocations with the Burgers vector parallel to [$1\bar{1}0$] were dominantly seen. Beam // [111]*, g=$2\bar{2}0$.

Dislocation structure in the B-oriented specimen was examined at temperatures where strong YSA appeared. Figure 5 shows the bright field image of dislocations in a specimen deformed at 600°C. In this temperature region, most dislocations align straight along the [$1\bar{1}0$] direction. These straight dislocations are connected to each other by large bowed-out edge segments (superkinks). At the superkinks, the dislocations are often dissociated in a wide distance into two superpartials on the (111) slip plane. In contrast, no clear dissociation is visible at the straight segments. The straight dislocations lose their contrast when imaged with g=22$\bar{4}$. This indicates that the Burgers vector of straight dislocations are parallel to [1$\bar{1}$0]; they have a screw character, and the [1$\bar{1}$0] screw dislocations were not dissociated on the (111) slip plane. However, it was confirmed from the tilting observation that the dissociation of superdislocation occurs on the other non-slip plane. From the weak-beam observation of dislocations on the (001) plane, the [1$\bar{1}$0] dislocations at screw segments was confirmed to be dissociated with a separation distance of 6nm. It is therefore concluded that the dissociation mode of [1$\bar{1}$0] dislocation at screw segments are as follows;

$$[1\bar{1}0] \rightarrow 1/2[1\bar{1}0] + APB + 1/2[1\bar{1}0]$$

This APB-type dissociation occurs not on the slip plane but on the (001) plane via microscopic cross-slip process in the strong YSA temperature region. This observed dislocation configuration is analogous to the K-W locking configuration of screw dislocation observed in some GCP compounds [1-4].

DISCUSSION

{100}<01$\bar{1}$> and {111}<1$\bar{1}$0> slips were identified to be operative in $Ni_3(Ti_{0.90}Nb_{0.10})$ crystals. They correspond to the {10$\bar{1}$1}<$\bar{1}$2$\bar{1}$0> pyramidal slip and (0001)<2$\bar{1}$$\bar{1}$0> basal slip in the hexagonal notation, respectively. The substitution of 10%Ti by Nb in Ni_3Ti induced change in the crystal structure from the four-layered DO_{24} structure to the nine-layered structure. The change in crystal structure hindered the operation of {10$\bar{1}$0} prism slip ({2$\bar{1}$$\bar{1}$} slip in the rhombohedral cell), but the {10$\bar{1}$1} pyramidal slip ({100} slip) was activated instead in $Ni_3(Ti_{0.90}Nb_{0.10})$.

It is well known that the Peierls stress for the motion of dislocation increases with decreasing d/b value, where d is the plane-distance and b is the length of Burgers vector [10]. Namely, the narrower the distance of slip plane, the harder is the motion of dislocation. Concerning with the (001) slip plane in the nine-layered structure, although the packing density of atom on the plane is very low (31.3%), the plane-distance is relatively wide and is comparable to those of other slip systems. In addition, the constituent atoms are periodically arranged with the same distance on the {001} plane in the nine-layered structure, while that is not true on the {11$\bar{2}$} prism plane. This is one of the reasons why the {001} pyramidal slip was activated instead of the {11$\bar{2}$} prism slip in the nine-layered crystal.

Strong YSA appeared by slip on {111}<1$\bar{1}$0>. The characteristic dislocation structures of the straight screw dislocations and the dissociation of superdislocation on non-glide plane of {001} were observed in the temperature region where YSA occurred. These are very similar to those observed in some GCP compounds in which the YSA occurs by the K-W locking of screw dislocations [1-4]. The main driving force for the K-W locking is known to be the anisotropy of APB energy depending on the plane. We estimated the energies of APBs formed by the motion of 1/2<1$\bar{1}$0> superpartial dislocation in the nine-layered structure as a function of ordering energy. APB formed on the (111) slip plane must contain the disturbance of the first nearest neighbor interaction, and hence the energy of APB is believed to be high. In contrast, when APB is formed on a particular (001) plane, on which only Ni atoms are arranged, the atomic interaction at the first nearest neighbor is not disturbed by the formation of APB. Therefore, the energy of APB formed on (001) is supposed to be much lower than those on the other planes. This indicates that the difference in APB energies between the primary (111) slip plane and the {001} plane can provide the driving force for locking of screw dislocations. Indeed, the dissociation of superdislocation into two $1/2[1\bar{1}0]$ superpartials was confirmed not on (111) slip plane but on (001) by the weak-beam observation.

In the long-period ordered GCP structures with the complex stacking sequence, the arrangement of atoms on the prism and pyramidal planes is complicated and the atomic packing density on those planes is very low. In the previous study, therefore, it was supposed that the K-W locking accompanied with the microscopic cross-slipping of screw dislocations is unlikely to occur in those compounds [9]. However, the present result clearly demonstrates that the K-W locking can occur even in compounds with long-period ordered structures and where the strong YSA behavior appears.

CONCLUSIONS

The plastic deformation behavior of $Ni_3(Ti_{0.90}Nb_{0.10})$ crystals with Pb_3Ba-type rhombohedral structure with nine-fold stacking of the close-packed plane was examined. $\{100\}<01\bar{1}>$ slip and $\{111\}<1\bar{1}0>$ slip were identified to be operative depending on the crystal orientation. Strong yield stress anomaly appeared by slip on the close-packed plane of $\{111\}<1\bar{1}0>$. The result indicates that K-W locking of screw dislocation occurs not only in the compounds such as Ni_3Al and Ni_3Ti with relatively small unit cells based on fcc, hcp or dhcp, but also even in complex compounds with long-period stacking structures by slip on the common close-packed plane in the GCP structures.

ACKNOWLEDGEMENTS

This work was supported by a Grant-in-Aid from the Japanese Ministry of Education, Culture, Sports, Science and Technology, and also by "Priority Assistance of the Formation of Worldwide Renowned Centers of Research - The 21st Century COE Program (Project: Center of Excellence for Advanced Structural and Functional Materials Design)" from the same Ministry. A part of this work was carried out at the Strategic Research Base "Handai Frontier Research Center" supported by the Japanese Government's Special Coordination Fund for Promoting Science and Technology.

REFERENCES

1. M.Yamaguchi and Y. Umakoshi, Prog. Mater. Sci. **34**, 1 (1990).
2. K. Hagihara, T. Nakano and Y. Umakoshi, Acta Mater. **48**, 1469 (2000).
3. K. Hagihara, T. Nakano and Y. Umakoshi, Acta Mater. **51**, 2623 (2003).
4. K. Hagihara, T. Nakano and Y. Umakoshi, Scripta Mater. **48**, 577 (2003).
5. K. Tomihisa, Y. Kaneno and T. Takasugi, Intermetallics **12**, 317 (2004).
6. Y. Nunomura, Y. Kaneno, H. Tsuda and T. Takasugi, Intermetallics **12**, 389 (2004).
7. K. Hagihara, T. Nakano and Y. Umakoshi, Sci. Tech. Adv. Mater. **3**, 193 (2002).
8. J. H. N. Van Vucht, J Less-common metals, **11**, 308 (1966).
9. K. Hagihara, T. Nakano and Y. Umakoshi, Mat. Res. Soc. Symp. Proc. **357**, 753 (2003).
10. J. P. Hirth and J. Lothe, "Theory of Dislocation (second edition)", Krieger publishing company, Malabar Florida, p.237.

Mater. Res. Soc. Symp. Proc. Vol. 842 © 2005 Materials Research Society

Effect of Fe Alloying on the Vacancy Formation in Ni$_3$Al

E. Partyka[1,2], R. Kozubski[1], W. Sprengel[2] and H.-E. Schaefer[2]
[1]Jagiellonian University, M. Smoluchowski Institute of Physics, 30-059 Kraków, Poland.
[2]Stuttgart University, Institute of Theoretical and Applied Physics, D-70550 Stuttgart, Germany.

ABSTRACT

The influence of Fe alloying on the formation of thermal vacancies in Ni$_{75}$Al$_{25-x}$Fe$_x$ has been specifically studied by high-temperature, positron lifetime spectroscopy (PLS). The results are consistent with a previously reported decrease of the activation energy for the vacancy formation upon alloying of Ni$_{75}$Al$_{25}$ with Fe derived from residual isochronal resistometry (REST): an indirect method for vacancy studies.

INTRODUCTION

The ductility of the intermetallic compound Ni$_{75}$Al$_{25}$ which exhibits an ordered L1$_2$ superstructure almost up to the melting point ($T_m = 1660$ K) [1] can be improved drastically by alloying with Fe. The resulting ordered L1$_2$ phase Ni$_{75}$Al$_{25-x}$Fe$_x$ is stable within a wide Fe concentration range ($0 \leq x \leq 25$). However, the temperature of the order-disorder transformation decreases as the Fe content increases [2].

A resistometric study of the ordering kinetics in Ni$_{75}$Al$_{25-x}$Fe$_x$ showed that the destabilization of the L1$_2$ phase is accompanied by a decrease of the activation energy E_A for ordering [3]. Taking into account the fact that ordering at a fixed temperature occurs due to the atomic mechanism of the migration of vacancies which are present in equilibrium concentration, the overall activation energy E_A for ordering comprises the vacancy formation energy E_F and the vacancy migration energy E_M expressed as $E_A = E_F + E_M$. For a more profound understanding of the stability and the ordering kinetics of the L1$_2$ superstructure it is therefore important to determine both values separately.

A first comprehensive study of the ordering kinetics in Ni$_{75}$Al$_{25-x}$Fe$_x$ by means of quasi-residual resistometry (REST) has been reported recently [4]. By applying the Schulze-Lücke formalism [5] for the analysis of the resulting isochrones of the resistivity measurements an estimation of the vacancy formation energy E_F and the migration energy E_M is possible. It has been found that the destabilization of the L1$_2$ superstructure by Fe is accompanied by a decrease of the vacancy formation energy E_F whereas the vacancy migration energy shows no dependence on the Fe-content (see Fig.1).

These results correlate very well with a simple theoretical model [6] where the equilibrium configuration of a ternary L1$_2$ ordered A$_3$B$_{1-x}$C$_x$ system is described within a Bragg-Williams approximation predicting a decrease of the vacancy formation energy E_F with an increase of a C admixture which destabilizes L1$_2$ superstructure.

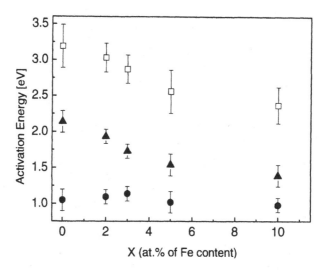

Fig.1. Fe- concentration dependence of the vacancy formation energy E_F (▲), the vacancy migration energy E_M (●) and the resulting activation energy E_A (□) for the ordering kinetics as derived from the analysis of quasi-residual resistometry [4] .

The aim of the present study is to investigate the vacancy formation energy E_F in Fe doped Ni_3Al - as deduced from resistometry and modelling – by means of positron lifetime measurements which allow a specific detection of vacancy-type defects

EXPERIMENTAL

A cylinder with an axial bore hole and a cover were prepared from a $Ni_{75}Al_{15}Fe_{10}$ single crystal. A $^{58}CoCl_2$ solution was deposited inside and annealed in a hydrogen atmosphere yielding a metallic ^{58}Co positron source with an activity of $1 \times 10^6 Bq$ which subsequently was diffused into the specimen during a 24 h anneal at 1273 K. The cylinder with the cover was hermetically sealed in Nb container by electron beam welding and heated in an electron beam furnace for measuring the temperature dependence of the positron lifetime. For the positron lifetime measurement a fast-slow γ-γ coincidence spectrometer equipped with BaF_2 scintillators with a time resolution of 264 ps was used. The positron lifetime components were numerically determined by means of standard techniques from positron lifetime spectra with a total number of approximately 1.5×10^6 coincidence counts [7].

RESULTS AND DISCUSSION

Numerical analysis of the positron lifetime spectra resulted in two time constants τ_0, τ_v and their intensities I_0 and I_v characteristic for positrons annihilated in the free delocalised state and in the vacancy-trapped state, respectively.

Figure 2 shows the reversible sigmoidal temperature dependence of the mean positron lifetime $\bar{\tau} = I_0\tau_0 + I_v\tau_v$ indicating the formation of thermal vacancies. For lower temperatures up to 1000K a weak, linear temperature dependence of the free positron lifetime is visible which can be expressed as $\tau_f(T) = \tau_{f,0}(1+\alpha T)$. Above 1100 K a strong increase of the mean positron lifetime $\bar{\tau}$ occurs due to positron trapping and annihilation at thermally formed vacancies. Saturation trapping of positrons sets in at about 1550 K.

The mean positron lifetime according to the two-state trapping model [8] is given by

$$\bar{\tau} = \tau_f \frac{1+\sigma C_v\tau_v}{1+\sigma C_v\tau_f} \tag{1}$$

with the vacancy concentration

$$C_v = \exp\left(\frac{S_F}{k_B}\right)\exp\left(\frac{-E_F}{k_BT}\right). \tag{2}$$

By a fit of Eq. (1) to the experimental data and taking into account the effect of thermal expansion on τ_f we obtain a value for the vacancy formation energy $E_F = (1.5 \pm 0.2)$ eV and $\sigma\exp(S_F/k_B) = (4.4 \pm 9.3) \times 10^{15}$ s^{-1} in Ni$_{75}$Al$_{15}$Fe$_{10}$. This value of E_F is lower than $E_F = (1.81 \pm 0.08)$ eV determined for Ni$_{75}$Al$_{25}$ by Badura-Gergen and Schaefer [9] by employing the same method of positron lifetime spectroscopy.

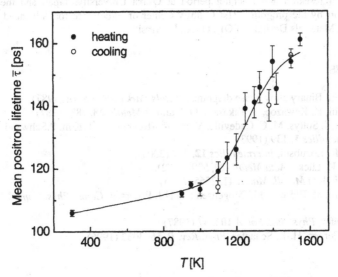

Fig. 2. Reversible S-shaped curve of the temperature dependence of the mean positron lifetime $\bar{\tau}$ in Ni$_{75}$Al$_{15}$Fe$_{10}$.

It should be pointed that thermal vacancy formation in $Ni_{75}Al_{15}Fe_{10}$ is detected in the disordered phase above the order-disorder transition temperature $T_c = 880K$ whereas in the former measurement on Ni_3Al [9] vacancy formation is detected in the ordered phase. From this we conclude that the vacancy formation enthalpy E_F as specifically detected by positron lifetime studies is slightly lowered upon Fe doping, in qualitative agreement with the results of the resistivity studies [4].

The measurement of the Fe concentration dependence of E_F in $Ni_{75}Al_{25-x}Fe_x$ combined with modelling (e.g. within the Bragg-Williams approximation) may be the basis for the determination of the two-body bond energies for nearest-neighbor Fe-Fe, Fe-Ni and Fe-Al pairs – in analogy to the earlier work [9].

CONCLUSION

Investigations of $Ni_{75}Al_{25-x}Fe_x$ by means of high-temperature positron lifetime spectroscopy showed that the vacancy formation energy (E_F) is Fe dependent and decreases with an increase of the Fe concentration. The result confirms previous studies performed by means of residual isochronal resistometry .

ACKNOWLEDGEMENTS

One of the authors (E.P.) appreciates financial support by the EU-Marie Curie Training Site Program (HPMT-GH-01-00224-02). H.-E. S. acknowledges the hospitality of Prof. Y. Shirai during his 2004 research and teaching period at Osaka University, Japan and the generous financial support by the program "21st Century Center of Excellence for Advanced Structural and Functional Materials Design (CEO)", Osaka University.

REFERENCES

1. .TB Massalski. Binary alloy phase diagrams. *Metals Park (OH) ASME* (1987).
2. N. Masahashi, H. Kawazoe, T. Takasugi, O. Izumi, *Z Metall.* **78**, 788 (1987).
3. R. Kozubski, J. Sołtys, M. C. Cadeville, V. Pierron-Bohnes, T. H. Kim, P.Schwander, et al. *Intermetallics* **1**, 139 (1993).
4. E. Partyka, R. Kozubski, *Intermetallics* **12**, 213 (2004).
5. A. Schulze, K. Lücke, *Acta Metall.* **20**, 529 (1972).
6. R. Kozubski, *Acta Metall. Mater.* **41**, 2565 (1993).
7. P. Kirkegaard, M. Eldrup, O.E. Mogensen and N.J. Pedersen, *Comp. Phys. Comm.* **23**, 307 (1081).
8. H.- E. Schaefer, *Phys. Stat. Sol. A* **102**, 47 (1987).
9. K. Badura-Gergen, H.-E. Schaefer, *Phys. Rev. B*, **56**, 3032 (1996).

Titanium Aluminides

An Electron Microscope Study of Mechanical Twinning in a Precipitation–Hardened TiAl Alloy

Fritz Appel
Institute for Materials Research, GKSS Research Centre Geesthacht,
Geesthacht, D-21502, Germany

ABSTRACT

The high temperature capability of TiAl alloys can be significantly improved by the implementation of precipitation reactions. From the engineering viewpoint the challenge is to establish the hardening mechanism without compromising desirable low temperature properties such as ductility and toughness. With this perspective in mind, twin structures were investigated in a TiAl alloy containing Ti_3AlC carbides, in order to identify mitigation strategies for overcoming precipitation induced embrittlement. The study is focussed on the nucleation and immobilization of twins at precipitates. The results are discussed in the context of the plastic anisotropy and fracture behaviour of TiAl alloys.

INTRODUCTION

In an attempt to improve the high temperature strength of TiAl alloys, several studies on precipitation hardening have been performed [1-3]. The strengthening effect critically depends on the size and dispersion of the particles. In this respect, carbides appear to be beneficial as the optimum dispersion can be achieved by homogenization and ageing procedures. The precipitation reaction gives rise to a marked age hardening; suitable particle dispersions can reduce creep rates over several orders of magnitude, when compared with the homogenized reference state. However, the attributes that are desirable for high temperature service might be counter to low temperature ductility and damage tolerance. TiAl alloys suffer from brittle fracture that persist up to relatively high temperatures of 700-800 °C. The brittleness is mainly associated with the existence of low index cleavage planes [4] and generally low dislocation mobility [5]. Any further increase of the glide resistance by precipitation hardening can therefore be extremely harmful for the fabrication and handling of components. As described in a detailed study of Yoo and Fu [4], deformation of γ(TiAl) can also be provided by mechanical twinning along $1/6<11\overline{2}]\{111\}$. Thus, the mechanism provides auxiliary slip systems and plays an important role in alloy design strategies for mitigating the problems associated with the poor damage tolerance of the material. While the literature abounds with the results on twin structures in TiAl, there has been a striking lack of systematic work reported on the interaction of mechanical twins with precipitates. To some extend this due to the atomic scale of the processes and the difficulties associated with the complexity of defect structures that are developed in precipitation-hardened material. This imbalance of information is addressed in the present paper by a transmission electron microscope (TEM) study of twin structures in a precipitation hardened TiAl alloy, utilising conventional and high resolution imaging techniques.

EXPERIMENTAL DETAILS

In a Ti-48.5Al-0.36C (at.%) alloy, Ti$_3$AlC perovskite precipitates were formed by homogenisation at 1523 K and subsequent ageing at 1023 K [3]. Compression samples of this alloy were deformed at room temperature to strain ε=3%. For the TEM observations, slices were sectioned perpendicular to the sample axis so that it was a simple matter to estimate the relative orientation of a given large grain with respect to the deformation axis. The slices were ground to 150 µm in thickness and thinned to electron transparency by twin-jet electropolishing. The thin foils were examined in a Philips CM300 microscope operated at 300 kV.

INTERACTIONS OF MECHANICAL TWINS WITH PRECIPITATES

The structural characteristics of the Ti$_3$Al precipitates formed by the above described thermal treatment are, in broad terms, similar to those described previously by Tian et al. [2]. Accordingly, the precipitates exhibit with the TiAl matrix the orientation relationship

$$[001]_p \parallel [001]_m, <100>_p \parallel <100>_m, \tag{1}$$

p and m designate the precipitate and the $\gamma \square \square \square$ matrix, respectively. There is a significant lattice mismatch between the precipitates and the γ matrix, which may be expressed as ε_m=2(a_m-a_p)/(a_m+a_p). a_m and a_p are the lattice parameters of the matrix and the precipitate. The values estimated in the plane of the interface along the a- and c-direction are ε_a=-0.057 and ε_c=-0.021, respectively. This significant anisotropy in the lattice misfit is probably the reason why the precipitates have a rod-like morphology. In the present case the precipitates had an average length (along [001]) l_p=22 nm and width (along <100>) d_p=3.3 nm. The separation distance of the precipitates along the dislocations was typically 50-100 nm. As described in a previous study [3], the glide resistance provided by the precipitates mainly relies on the long-range coherency stresses built up between the precipitates and the matrix due to the difference in the lattice constants. The precipitates are strong glide obstacles for all types and characters of perfect dislocations, which are capable, however, of being sheared by the dislocations. Orowan loops and other microstructural evidence of hardening by impenetrable obstacles were not observed. In this precipitation hardened alloy mechanical twinning is a prominent deformation mechanism as indicated by streaking along the <111> traces in the diffraction pattern. The interaction of the twinning partial dislocations with the precipitates leads to a distinct twin morphology. The twinning dislocations are strongly bowed-out and apparently penetrate the obstacle array along paths of easy movement and become immobilized at groups of unfavourably arranged particles. The twins often appear fragmented, i.e. islands of untwined regions occur. Such processes apparently locally impede twin growth and lead to a significant variation of the twin thickness.

Figure 1 is a high-resolution electron micrograph of an immobilized twin. Before being finally terminated, the thickness of the twin is reduced by three (111) planes, as indicated by the stacking sequences shown in Fig. 1b. The Burgers circuit (utilizing a stress-free coherent twin boundary to define the reference lattice) around the defect configuration flanked by the two stacking sequences is closed; indicating the net Burgers vector is zero. A formation mechanism of this type of pure ledge was first proposed for bcc crystals by Sleeswyk [6] and later on modified for other structures [7,8]. Following this concept, strain accommodation at the triple step may occur by the recombination of the twinning partial dislocations into an emissary lattice dislocation with the Burgers vector $b=3x1/6<11\,2]$. This dislocation can slip away, leaving behind a stress free ledge with almost no shear discontinuity across the interface. Indeed, close to the twin tip an isolated dislocation could be recognized (Fig. 1a). However, the resolved Burgers vector of this dislocation is $b_{res}=1/4<112]$, which is not consistent with the above mentioned emissary dislocation with Burgers vector $b=1/2<11\,2]$. It might be expected that the emissary dislocation is subjected to decomposition and dissociation reactions, which in the close neighbourhood of the twin tip can be supported by high local stresses. It should be noted that the described association of twinning partial dislocations was not a localized example, but was recognized in almost all tapered twin/matrix interfaces. The recombination reaction $3x1/6<11\,2]\rightarrow 1/2<11\,2]$ is energetically unfavourable according to Frank's rule; however, it might be supported by the high constraint stresses that are present at groups of immobilized twinning dislocations and by the fact that the emissary $1/2<11\,2]$ dislocation is a stable dislocation in the $L1_0$ structure.

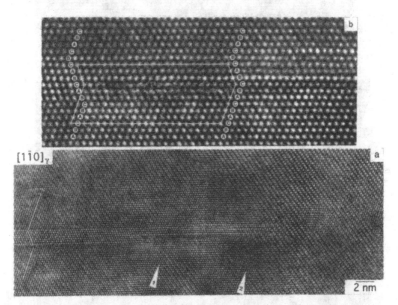

Figure 1. Immobilization of a deformation twin in a Ti-48.5Al-0.37C alloy observed after room temperature compression to strain $\varepsilon=3\%$. (a) Atomic structure of the terminated twin; note the three plane step (arrow 1) in one of the twin/matrix interfaces and the isolated dislocation at the twin tip, which can be recognized by looking along arrow 2 on the bottom of the micrograph. (b) Details of the defect structure associated with the triple step marked with arrow 1 in (a).

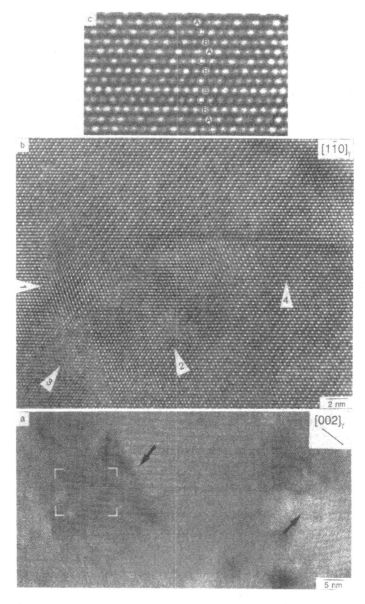

Figure 2. Heterogeneous nucleation of embryonic twins at perovskite precipitates observed in a Ti-48.5Al-0.37C alloy after room temperature compression to strain ε=3%. For details see accompanying text.

TWIN NUCLEATION at Ti₃AlC PRECIPITATES

The precipitation-hardened material exhibits an unusual propensity to profuse twinning, which warrants particular consideration. A frequent structural feature of these alloys is faulted structures, with fault planes parallel to the twin/matrix interface. The faults originate at the precipitates, as demonstrated in Fig. 2a. A complete solution to this observation is not at hand, but the formation of these planar defects is probably related to the above described misfit situation present at the precipitates. According to the theory of Frank and van der Merwe [9] the strains to fit two crystals with mismatched lattice parameters are provided partly by misfit dislocations and partly by homogeneous strains in the crystals. The precipitates are associated with dislocations that are manifested by additional $\{111\}$ planes (Fig. 2b, arrows 1 and 2). The possibility that these dislocations are matrix dislocations cannot be ruled out because the resolved Burgers vector $b_{res}=1/3<111>$ is consistent with $1/2<110]$ dislocations. However, there are also dislocations present at the precipitates that are manifested by an additional (002) plane (arrow 3). The resolved Burgers vector of these dislocations $b_{res}=1/2[001]$ is not consistent with the glide geometry of the L1₀ structure, which gives supporting evidence that these defects are misfit dislocations. This finding is contrary to the general view present in the literature [1,2], according to which perovskite precipitates are fully coherent. The discrepancy may result from a larger particle size occurring in the material investigated. The TiAl matrix surrounding the precipitates is subject to normal stresses acting in three mutually perpendicular directions, which corresponds to a state of triaxial stress. The maximal shear stresses occur on elements oriented at angles of 45° to the a- and c-axes. With growing precipitate size, the strain and surface energy of the particle can be reduced if dislocations are emitted so that the full coherency of the precipitates is lost. According to this mechanism the dislocations can be located at a significant stand-off distance from the original interface, as has been observed. Under the high constraint stresses, the interfacial dislocations may recombine so that perfect and twinning partial dislocations are released. Among the various mechanisms that may occur, the reaction

$$[00\bar{1}]\rightarrow 1/3[\bar{1}\,\bar{1}\,\bar{1}] + 2\times 1/6<11\bar{2}], \qquad (2)$$

involving two $1/2[001]$ dislocations may account for the structure shown in Fig. 2b. One of the $1/2[001]$ dislocations and the $1/3[111]$ dislocation can be recognized by a shallow view along arrows 3 and 1, respectively. The propagation of the $1/6<11\bar{2}]$ partial dislocations gives rise to the formation of the planar fault and leads to strain accommodation of the triaxial stress state. The planar defects may be described as superimposed intrinsic stacking faults, as indicated by the stacking sequence across the planar fault shown in Fig. 2c. It is tempting to speculate that such defects can easily be rearranged into embryonic twins, which may coalesce to form thicker twins. Thus, in an early stage of this growing process the twins appear fragmented. This mechanism provides a natural explanation why the twins in the precipitation-hardened material are quite irregular and often exhibit untwinned regions.

Another twin nucleation mechanism is probably associated with dissociation reactions of the dislocations [10]. The superdislocations become widely dissociated at the precipitates, which might be a consequence of the high constraint stresses described above. Twin nuclei were often found together with these dislocations, which gives rise to the speculation that the twins originate from overlapping faults trailed by the superdislocations. Investigation of these processes is the subject of ongoing work.

DISCUSSION

In TiAl alloys the von Mises criterion for a general plastic shape change is satisfied in principle because the $L1_0$ structure provides more than five independent slip systems. However, in the γ phase of two-phase alloys, glide of $<10\bar{1}]$ and $1/2<11\bar{2}]$ superdislocations is difficult because these dislocations are liable to adopt non-planar dissociations [11,12] and are subject to a high Peierls stress [13]. This leads to a significant plastic anisotropy because no glide component in the c-direction of the tetragonal unit cell is provided. In grains which are unfavourably oriented for glide of $1/2<1\bar{1}0]$ dislocations, high constraint stresses can be developed, which can easily exceed the fracture stress [5]. The required shear processes with a c-component can be accomplished by true mechanical twinning $1/6<11\bar{2}]\{111\}$. In a detailed analysis Goo [14] has actually shown that in lamellar two-phase alloys the von Mises criterion is satisfied by glide of ordinary dislocations combined with twinning, provided that all orientation variants of the γ phase are available. In this respect, the observed twin nucleation at perovskite precipitates is certainly very beneficial for the ductility of the material. Due to the high density of the precipitates a fine dispersion of mechanical twins can be produced, which is highly capable of releasing stress concentrations at constrained grains and shielding crack tips. In a separate study it was shown that twin nucleation in TiAl can also be supported by Nb additions of 5-10 at. % because the stacking fault energy is reduced [10]. Based on this concept, novel high-strength alloys with the general composition Ti-45Al-(5-10)Nb and subject to precipitation hardening were developed, which at room temperature are capable of carrying stresses in excess of 1 GPa with an appreciable plastic tensile elongation of 2 % [15].

REFERENCES

1. S. Chen, P.A. Beaven and R. Wagner, *Scripta Metall.* 26, 1205 (1992).
2. W.H. Tian, T. Sano and M. Nemoto, *Phil. Mag. A* 68, 965 (1993).
3. U. Christoph, F. Appel and R. Wagner, *Mater. Sci. Engng. A* 239-240, 39 (1997).
4. M.H. Yoo and C.L. Fu, Metall. *Mater. Trans. A* 29, 49 (1998).
5. F. Appel and R. Wagner, *Mater. Sci. Engng. R* 22, 187 (1998).
6. A. W. Sleeswyk, *Acta Metall.* 10, 803 (1962).
6. S. Mahajan and G.Y. Chin, *Acta Metall.* 22, 1113 (1974).
7. J.P. Hirth and R.W. Balluffi, *Acta Metall.* 21, 929 (1973).
8. S. Mahajan and G.Y. Chin, *Acta Metall.* 22, 1113 (1974).
9. F.C. Frank and J.H. VAN DER Merve, *Proc. R. Soc. A* 198, 216 (1949).
10. F. Appel, *Phil. Mag.*, in print.
11. G. Hug, A. Loiseau, and A. Lasalmonie, *Phil. Mag. A* 54, 47 (1986).
12. S.A. Court, V.K. Vasudevan and H.L. Fraser, *Phil. Mag. A* 61, 141(1990).
13. B.A. Greenberg, *Scripta Metall.* 23, 631 (1989).
14. E. Goo, *Scripta Mater.* 38, 1711 (1998).
15. F. Appel, J.D.H. Paul, M. Oehring, U. Fröbel, and U. Lorenz, *Metall. Mater. Trans. A* 34, 2149 (2003).

Mater. Res. Soc. Symp. Proc. Vol. 842 © 2005 Materials Research Society   S7.10

Increase in γ/α₂ Lamellar Boundary Density and its Effect on Creep Resistance of TiAl Alloy

Kouichi Maruyama, Jun Matsuda and Hanliang Zhu
Graduate School of Environmental Studies, Tohoku University
6-6-02 Aobayama, Aoba-ku, Sendai 980-8579, Japan

ABSTRACT

Several lamellar microstructures of a Ti-48 % Al alloy were made by changing heating rate in the $\alpha+\gamma$ dual phase field, and their creep properties were investigated at 1150 K. Average spacing and average length of α_2 lamellae decrease with increasing heating rate. The decrease of α_2 lamellar spacing is most effective at lower heating rate, and minimum creep rate decreases with increasing the heating rate, since a high density of γ/α_2 boundaries stabilizes lamellar microstructure during creep. On the other hand creep rate increases at high heating rate, since α_2 lamellar length becomes shorter with increasing heating rate. A reduction of creep rate by one order of magnitude is achieved at the optimum heating rate providing the best combination of narrow spacing and sufficiently long length of α_2 lamellae.

INTRODUCTION

Creep resistance of TiAl alloys depends strongly on their microstructures [1], and fully lamellar structure consisting of α_2 and γ lamellae provides the best creep resistance among typical microstructures of TiAl alloys [2]. Lamellar orientation with respect to stress axis also has a significant effect on creep resistance of lamellar TiAl alloy [3-5]. Creep rate of hard oriented PST crystals is reduced by one order of magnitude from polycrystalline aggregates consisting of randomly oriented lamellar grains. The hard oriented PST crystals have the most desirable microstructure giving the best creep resistance among various possible microstructures of TiAl alloys.

Creep resistance of the hard oriented PST crystals depends strongly on stability of their lamellar microstructures during creep deformation [5,6]. The lamellar microstructures contain γ/α_2 boundary and three types of γ/γ boundaries. Among the four types of lamellar boundaries, γ/α_2 boundary has the highest stability during creep [5-8]. A high density of γ/α_2 boundaries, in other words, a narrow spacing of α_2 lamellae is required to attain a stable lamellar microstructure and the consequent high creep resistance. It has been proved experimentally that a high density of γ/α_2 boundaries can reduce creep rate of a hard oriented PST crystal by one order of magnitude from its as-grown state [5,6]. Fabrication of a material with narrow spacing of α_2 lamellae, namely a high density of γ/α_2 boundaries is the objective of the present study.

Fine α_2 lamellar spacing is usually obtained by fast cooling after solution treatment in the α single phase field. However, PST crystals often change to polycrystals after the heat treatment probably due to nucleation of α grains whose $(0001)_\alpha$ is not parallel to the pre-existing lamellar boundary. Therefore, this

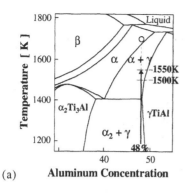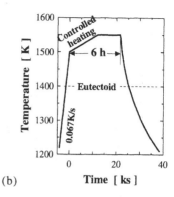

(a) **Aluminum Concentration**   (b)   **Time [ks]**

Figure 1. Secondary heat treatment procedure employed in the preset study.

conventional heat treatment is not applicable to reduction of α_2 lamellar spacing in PST crystals. Reverse phase transformation from γ to α phase occurs during heating in the $\alpha+\gamma$ dual phase field. If new α lamellae are introduced during this transformation, we can expect a decrease of α_2 lamellar spacing [9]. However, less attention has been paid to this reverse phase transformation during heating in lamellar microstructure control of TiAl alloys. In this study, effects of heating rate on the evolution of lamellar microstructure and the consequent creep property are studied with a Ti-48 % Al alloy.

EXPERIMENTAL PROCEDURE

The material used in the present study is a polycrystalline Ti-48 mol% Al alloy. The alloy was induction-skull melted, and then hot extruded at 1473 K to 80 % reduction of area. The extruded material was solution treated at 1700 K in the α single phase field and then furnace cooled at an initial cooling rate of about 0.5 Ks^{-1}. For modifying the as-grown lamellar microstructure, the material was subjected to another heat treatment. Its details are explained in the next section. Grain size was about 1 mm. Compression creep specimens with size of 2×2×3 mm were cut with a diamond saw and mechanically polished before the heat treatment. Compression creep tests were conducted at 1150 K and a constant stress of 316 MPa in an argon gas atmosphere.

RESULTS AND DISCUSSION

Heat treatment concept

A fully lamellar microstructure produced by the solution treatment is named as as-grown hereafter. The as-grown microstructure was subjected to additional heat

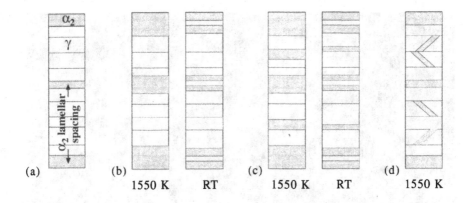

Figure 2. Schematic illustration of microstructural evolution during heating at different heating rates to 1550 K and the subsequent cooling to room temperature: (a) As-grown, (b) slow heating, (c) fast heating, and (d) very fast heating.

treatment in the α+γ dual phase field (Fig.1). According to the phase diagram of Ti-Al system (Fig.1(a)), volume fraction of α phase changes quickly during heating from 1500 to 1550 K. Therefore, heating rate was controlled in this temperature range. Samples were first heated to 1500 K at a rate of 0.067 Ks^{-1}, and then to 1550 K at a controlled heating rate ranging from 1.7×10^{-3} Ks^{-1} to 0.067 Ks^{-1}. They were held at 1550K until the total duration from 1500 K reaches 6 h, and thereafter furnace cooled to room temperature. At the heating rate of 1.7×10^{-3} Ks^{-1}, the cooling was started after 10 h. The heat treatment was carried out in a vacuum of 5×10^{-4} Pa.

The phase transformation from γ to α proceeds either by growth of pre-existing α lamellae (at a low driving force, Fig.2(b)) or by precipitation of α lamellae (at a high driving force, Fig.2(c) (d)) [10,11]. In the case of α phase precipitation (0001)$_\alpha$ should be parallel to one of the four {111} habit planes of γ phase. A large driving force of the phase transformation, in other words, a high heating rate is necessary to nucleate α lamellae during the heating. Further transformation from α to γ phase takes place during the cooling from 1550 K. When cooling rate is slow enough the newly formed α lamellae may disappear during the cooling. However, the α to γ transformation is partly accomplished by precipitation of γ plates in α lamellae at the relatively high cooling rate of about 0.3 Ks^{-1} in the present heat treatment. If this is the case we can expect reduction of α$_2$ lamellar spacing after the heat treatment.

Lamellar microstructure

The lamellar microstructures made by the heat treatment are shown in Fig.3 together with the as-grown microstructure. In the heat treated microstructures (Fig.3(b) and (c)) α$_2$ lamellar spacing decreases and density of γ/α$_2$ lamellar boundaries increases obviously as compared to those in the as-grown microstructure

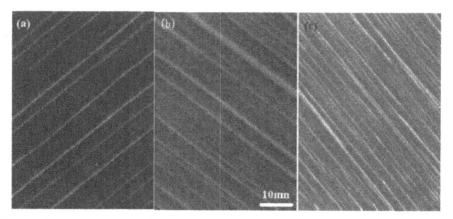

Figure 3. SEM micrographs of Ti-48 % Al alloy after the secondary heat treatment with various heating rates in the α+γ phase field: (a) As-grown, (b) 1.7×10^{-3} Ks^{-1}, and (c) 0.067 Ks^{-1}. Bright and dark bands are α_2 and γ phases, respectively.

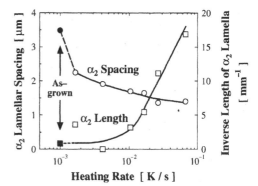

Figure 4. Effect of heating rate on average spacing and inverse length of α_2 lamellae.

(Fig.3(a)). The (0001) habit plane of newly precipitated α_2 lamellae can be parallel to any of the four {111} habit planes of γ phase. However, all the α_2 lamellae are parallel to the pre-existing ones up to the heating rate of 0.067 Ks^{-1}, suggesting preferential nucleation of the α lamellae parallel to the pre-existing ones. The following two explanations may be applicable to the experimental finding. Since there was α_2 phase along γ/γ lamellar boundaries before forming the γ/γ boundaries, Al concentration is lower along the γ/γ boundaries. Therefore, the driving force for the nucleation of α lamellae along γ/γ boundaries is essentially large. There is lattice misfit between α and γ phases and tensile plane stress is acting in γ lamellae. The tensile stress assists the nucleation of α lamellae parallel to the pre-existing ones. As seen in Fig.4, α_2 lamellar spacing after the heat treatment decreases with increasing the heating rate because of the high chance of nucleation of α lamellae.

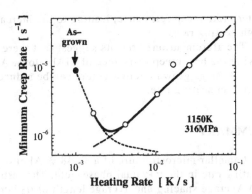

Figure 5. Effect of heating rate on minimum creep rate at 1150 K and 316 MPa.

Another point to be noted is the decrease in average length of α_2 lamellae with the increase of heating rate (Fig.4). As evident in Fig.3(c), some α_2 lamellae terminate within lamellar colonies at the high heating rate. Nuclei of α lamellae form and lengthen during heating to 1550 K. The lengthening stops at the location in which the super saturated Ti has consumed by precipitation of another α lamella. Therefore, the termination of two adjoining α_2 lamellae occurs when they meat during their lengthening. Since nucleation rate of α lamellae is high at fast heating rate, average length of α_2 lamellae decreases with increasing heating rate. Another noticeable change among the microstructures at different heating rates is thickness of α_2 lamellae. Some coarse α_2 lamellae over 1.5 μm are found at the lower heating rates (Fig.3(b)), suggesting preferential thickening of α lamellae during the heating. On the other hand, α_2 lamellae become thinner at the high heating rates (Fig.3(c)) suggesting preferential nucleation of α lamellae at the high heating rates.

Influence of heating rate on creep resistance

The effect of heating rate on minimum creep rate of the Ti-48 % Al alloy is described in Fig.5. The minimum creep rate first decreases, and then increases with increasing the heating rate. The variation of the minimum creep rate can be explained in terms of the changes in α_2 lamellar morphology with the heating rates. The lamellar microstructure contains γ/α_2 and γ/γ boundaries. During creep deformation γ/α_2 boundary usually has higher stability than γ/γ boundaries, since migration of γ/α_2 boundary requires long range diffusion. It has been confirmed experimentally [5,6] that a high density of γ/α_2 boundaries improves stability of lamellar microstructure and increases creep resistance of lamellar TiAl alloys. Therefore, the decrease of α_2 lamellar spacing (Fig.4) can explain the decrease of minimum creep rate with increasing the heating rate. On the other hand α_2 lamellae become shorter at the fast heating rate (Fig.4). Annihilation of dislocation is a rate controlling process of creep deformation, and γ/α_2 lamellar boundaries act as obstacle to the annihilation process. Since the annihilation proceeds fast at the end points of α_2 lamellae, discontinuous α_2 lamellae is less effective in strengthening than continuous ones. A high density of discontinuous α_2 lamellae more significantly deteriorate creep resistance, leading to the increase in minimum creep rate at the high heating rates.

The aforementioned results suggest that there is an appropriate heating rate giving the best creep resistance of TiAl alloys. As seen in Fig.5, the minimum creep rate of the as-grown microstructure can be reduced by one order of magnitude at the optimum heating rate.

SUMMARY

Lamellar microstructure of a Ti-48 % Al alloy was controlled by changing heating rate in the $\alpha+\gamma$ dual phase field. The results are summarized as follows:
(1) Average spacing and average length of α_2 lamellae decrease with increasing heating rate.
(2) At lower heating rate, minimum creep rate decreases with increasing the heating rate due to the decrease of α_2 lamellar spacing. This can be explained by high stability of lamellar microstructure with a high density of γ/α_2 boundaries.
(3) Creep rate increases at high heating rate, since α_2 lamellae become discontinuous with increasing the heating rate. Discontinuous α_2 lamellae are less effective in preventing recovery during creep.
(4) A reduction of creep rate by one order of magnitude was achieved at the optimum heating rate giving the best combination of narrow spacing and sufficiently long length of α_2 lamellae.

ACKNOWLEDGEMENT

Financial supports from the Ministry of Education, Science, Sports and Culture, Japan (Grant no. 15360361) and from the 21st century COE program of Materials Research Center, Tohoku University are gratefully acknowledged.

REFERENCES

1. K. Maruyama, R. Yamamoto, H. Nakakuki, and N. Fujitsuna. *Mater. Sci. Eng.* **A239-240**, 419 (1997).
2. J. Beddoes, W. Wallace, and L. Zhao, *Inter. Mater. Rev.* **40**, 197 (1995).
3. G. Wegmann, and K. Maruyama, *Phil. Mag.* **A80**, 2283 (2000).
4. H.Y. Kim, G. Wegmann, and K. Maruyama, *Mater. Sci. Eng.* **A329-331**, 790 (2002).
5. K. Maruyama, H.Y. Kim, and H. Zhu, *Mater. Sci. Eng.* **A387-389**, 910 (2004).
6. H.Y. Kim, J. Matsuda, and K. Maruyama, *Metall. Mater. Inter.* **9**, 255 (2003).
7. H.Y. Kim, and K. Maruyama, *Acta Mater.* **51**, 2191 (2003).
8. H.Y. Kim, and K. Maruyama, *Metall. Mater. Trans.* **34A**, 2191 (2003).
9. R.V. Ramanujan, *Acta Metall. Mater.* **42**, 2313 (1994).
10. X.D. Zhang, T.A. Dean, and M. H. Loretto, *Acta Metall. Mater.* **42**, 2035 (1994).
11. R.V. Ramanujan, *Inter. Mater. Rev.* **45**, 217 (2000).

Mater. Res. Soc. Symp. Proc. Vol. 842 © 2005 Materials Research Society   S7.11

Effect of Long-Term Aging and Creep Exposure on the Microstructure of TiAl-Based Alloy for Industrial Applications

Juraj Lapin[1], Mohamed Nazmy[2], and Marc Staubli[2]
[1]Institute of Materials and Machine Mechanics, Slovak Academy of Sciences, Racianska 75, SK-831 02 Bratislava, Slovak Republic
[2]ALSTOM Ltd., Department of Materials Technology, CH-5401 Baden, Switzerland

ABSTRACT

The effect of long-term aging and creep exposure on the microstructure of a cast TiAl-based alloy with nominal chemical composition Ti-46Al-2W-0.5Si (at.%) was studied. The aging experiments were performed at temperatures between 973 and 1073 K for various times ranging from 10 to 14000 h in air. Constant load tensile creep tests were performed at applied stresses ranging from 150 to 400 MPa and at temperatures between 973 and 1123 K up to 25677 h. During aging and creep testing the $\alpha_2(Ti_3Al)$-phase in the lamellar and feathery regions transforms to the $\gamma(TiAl)$-phase and fine needle-like B2 precipitates. Microstructural instabilities lead to a softening of the alloy. The effect of this softening on long-term creep resistance is negligible at temperatures of 973 and 1023 K.

INTRODUCTION

Low density, high melting temperature, good elevated-temperature strength and modulus retention, high resistance to oxidation and excellent creep properties of TiAl-based alloys make them potential candidate structural materials for various applications in the gas turbine and automotive industry. In recent years a particular interest was devoted to an alloy with nominal chemical composition Ti-46Al-2W-0.5Si (at.%), which was designated as ABB-2 [1-6]. As shown in our previous studies [2,4], the microstructure of large cast turbine blades from the ABB-2 alloy is not homogenous and changes from fully or nearly lamellar in the vicinity of the blade surface to duplex one in the central part. Such microstructural variations affect significantly local mechanical properties of the cast components [2,4]. Although several studies were also published on the microstructure of the ABB-2 alloy [4-6], information about microstructural stability of large cast components during long-term aging and creep testing are still lacking in the literature. Therefore, evaluation of long-term microstructural stability of samples prepared from different regions of cast components is of great practical interest.

The aim of this paper is to investigate the effect of long-term aging and creep exposure on the microstructure of a cast TiAl-based alloy for industrial applications. Duration of aging and creep experiments was exceptionally long comparing to existing literature data that has been published for emerging class of TiAl-based alloys.

EXPERIMENTAL DETAILS

Aging and creep experiments were conducted on the ABB-2 alloy with the chemical composition given in Table I. The as-received material was subjected to a hot isostatic pressing

Table I. Chemical composition of the ABB2 alloy (at.%).

Material	Ti	Al	W	Si	Cu	C	Fe	N	O	H
Component 1	Bal	46.61	1.88	0.49	0.006	0.034	0.038	0.057	0.177	0.060
Component 2	Bal	46.23	2.00	0.48	0.001	0.017	0.015	0.038	0.087	0.081
Component 3	Bal	46.88	1.96	0.53	0.004	0.024	0.041	0.032	0.193	0.040

at 1533 K under a pressure of 172 MPa for 4 h, which was followed by solution annealing at 1623 K for 1 h and gas fan cooling. The heat treatment was accomplished by stabilization annealing at 1273 K for 6 h and furnace cooling to room temperature.

Samples for aging experiments with dimensions of 8x8x12 mm were cut from the central part of the Component 2 by electro-spark machining. Selected positions in the component assured reproducible microstructure in all samples before aging. Aging experiments were performed at temperatures between 973 and 1073 K for various times ranging from 10 to 14000 h in air. After each aging step the samples were cooled to room temperature at a cooling rate of 2 Ks^{-1}. The Vickers microhardness measurements were performed at a load of 0.42 N on polished and slightly etched surfaces.

Cylindrical creep specimens with gauge diameters of 4, 7 and 10 mm were cut from the cast components by electro spark machining. The longitudinal axis of the creep specimens was parallel to the longitudinal axis of the components. Constant load tensile creep tests were performed at applied stresses ranging from 150 to 400 MPa and at temperatures of 973, 1023, 1033, 1073 and 1123 K up to 25677 h in air. The test temperature was monitored with two thermocouples touching the specimen gauge section and held constant within ± 1 K for each individual test. The specimen displacement was measured using a high-temperature extensometer attached to the ledges of the creep specimen. The acquisition of time-elongation data was accomplished by a computer.

The microstructure evaluation was performed by optical microscopy, scanning electron microscopy (SEM), transmission electron microscopy (TEM), and energy-dispersive X-ray spectroscopy. The size of coexisting phases was determined by computerized image analysis.

RESULTS AND DISCUSSION

Microstructure before aging and creep testing

Fig. 1 shows the typical microstructure of the samples before aging and creep testing. The microstructure consists of lamellar, feathery and γ-rich regions. The lamellar regions are composed of continuous α_2 (ordered Ti_3Al phase with DO_{19} crystal structure) and γ (ordered TiAl phase with $L1_0$ crystal structure) lamellae. Numerous nanometer-scale B2 (ordered Ti-based solid solution) and Ti_5Si_3 precipitates were observed at the apparently smooth α_2/γ interfaces [5]. The discontinuous α_2 lamellae contain fine needle-like particles with average width of 25 nm and length of 200 nm, which were identified to belong to B2-phase. Fine B2 precipitates, formed at the α_2/γ interfaces before creep testing, improve the creep resistance of the alloy through hindering interface dislocation mobility, reducing generation of γ matrix dislocations from lamellae and preventing dislocations from passing through lamellar interfaces. In the case of the

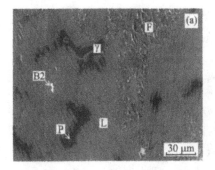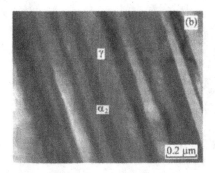

Figure 1. (a) Backscattered SEM micrograph showing the microstructure of the components before aging and creep testing. L - lamellar region; F - feathery region; P - Ti$_5$Si$_3$ precipitates. (b) TEM micrograph showing continuous α_2 and γ lamellae in the lamellar region.

ABB-2 alloy, the strengthening effect of the B2 precipitates is enhanced by co-precipitation of fine Ti$_5$Si$_3$ particles along the lamellar interfaces with a similar effect on the dislocation mobility. Some lamellar regions contain large blocky type B2 particles with an average size of about 20 μm. These particles contain elongated particles, which belong to the γ-phase and very fine Ti$_5$Si$_3$ precipitates [4,7]. The feathery region is composed of the γ-phase, irregular α_2 lamellae, B2 particles and Ti$_5$Si$_3$ precipitates. The γ-rich region is composed of the γ matrix and coarse spherical Ti$_5$Si$_3$ precipitates.

Effect of long-term aging on microstructure and microhardness

The microstructure of the aged samples observed by SEM was similar to that shown in Fig. 1. However, TEM observations showed that the as-received microstructure is unstable. During aging the α_2-phase in the lamellar and feathery regions transforms to the γ-phase and needle-like precipitates, as seen in Figs. 2a and 2b. The needle-like precipitates were identified to belong to B2-phase [5]. On the contrary to the results recently published for the samples aged at temperatures between 1173 and 1273 K for various times ranging from 24 to 800 h [8], no notable globularization of the α_2-phase and B2 precipitates within the lamellar or feathery regions was observed. The driving forces for microstructural changes in the ABB-2 alloy arise from changes of the phase equilibria. The $\gamma+\alpha_2+$B2 microstructure formed during thermal treatment is inherently unstable and transforms to a stable $\gamma+$B2 during long-term aging [6,8]. However, this process is very slow at the temperatures of the engineering interest due to low diffusivity of tungsten. As shown by Larson et al. [9], additions of tungsten stabilize α_2 lamellae against dissolution during aging. Fig. 3a shows the variation of the Vickers microhardness of the lamellar, feathery and γ-rich regions with the aging time. The microhardness of the lamellar and feathery regions decreases with increasing aging time. It is clear from Fig. 3b that the softening is faster in the lamellar regions where the total microhardness decrease is higher. All microhardness values of the γ-rich regions fall within the error of the experimental measurements without any definite evolution with the aging time. The kinetics of the softening process is analyzed assuming equation in the form

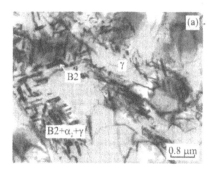 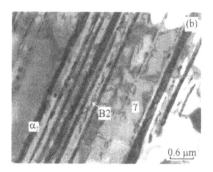

Figure 2. TEM micrograph showing the transformation the α_2 lamellae to needle-like B2 precipitates and γ-phase: (a) Feathery region of the Component 2 after aging at 973 K for 7300 h; (b) Lamellar region of the Component 2 after aging at 1023 K for 14000 h.

$$\Delta HV_m = k_0 t^m \exp\left(-\frac{Q}{RT}\right) \qquad (1)$$

where ΔHV_m is the microhardness decrease, k_0 is a material constant, t is the aging time, m is the time exponent and Q is the activation energy for softening. The microhardness decrease ΔHV_m is calculated as a difference between the microhardness of the specific region before aging and the microhardness after aging for a given time. The regression analysis of the experimental data revealed that the average time exponent is $m = 0.5$ for the lamellar regions and $m = 0.33$ for the feathery ones. Similar time exponents close to 0.5 for lamellar regions and close to 0.33 for feathery ones are also measured after aging at temperatures of 1023 and 1073 K. The activation energy calculated from an Arrhenius plot is $Q_L = 224 \pm 18$ kJ/mol and $Q_F = 193 \pm 17$ kJ/mol.

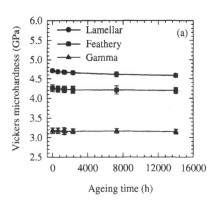 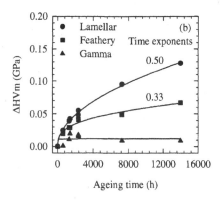

Figure 3. (a) Variation of the Vickers microhardness with the aging time at 973 K; (b) Dependence of the microhardness decrease ΔHV_m on the aging time at 973 K.

These values are lower than those determined by Muñoz-Morris et al. [8] of 349 and 415 kJ/mol for lamellar and globular regions, respectively. Main mechanisms leading to softening of the alloy are the diffusion controlled transformations of the α_2-phase, which are connected with changes of volume fraction, size and distribution of α_2 segments, B2, and Ti_5Si_3 precipitates within the lamellae. The precipitation strengthening is a complex problem including several effective mechanisms, which can operate simultaneously with various contributions to a final strength of the material [10]. The softening process in the ABB-2 alloy is very slow at 973 K. Assuming operating time of a blade of 30000 h, the softening produced by microstructural instabilities represents only 0.2 and 0.09 GPa for lamellar and feathery structure, respectively.

Effect of creep exposure on microstructure

Microstructural observations revealed that the applied stress accelerates coarsening of needle-like B2 particles. While the aging at 973 K for 14000 h result in an increase of initial mean width of B2 precipitate from 25 to 64 nm and mean length from 200 to 510 nm in unloaded button heads of the creep specimen, the mean width of the precipitates was measured to be 85 nm and mean length was 710 nm in the gauge region. Fig. 4a shows the typical deformation microstructure of the creep specimen in the lamellar region where the α_2 lamellae decomposed to the γ-phase and needle-like B2 precipitates [2,11,12]. The dislocations in the γ-matrix tend to be elongated in the screw orientation and appear to be frequently pinned along their lengths. The dislocation segments form local cusps associated with tall jogs [13,14]. Depending on the crystallographic orientation of needle-like B2 precipitates in the γ-matrix, some precipitates are easily passed by dislocations generated in the matrix. However, there are many B2 precipitates, which are effective obstacles to dislocation motion. Fig. 4b shows the typical deformation microstructure in the lamellar region with remaining α_2 lamellae. Lamellar γ/α_2 interfaces effectively constrain deformation to individual γ-lamellae with little evidence for direct transmission of dislocations under applied creep conditions. Strong interactions of ordinary dislocations with fine Ti_5Si_3 precipitates are observed [15]. The deformation microstructure in the γ-rich region is also characterized by pinned ordinary dislocations and by formation of prismatic dislocation loops [13]. Coarse Ti_5Si_3 particles are observed to be effective obstacles

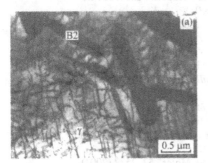 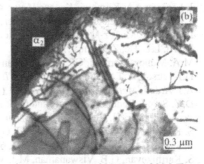

Figure 4. TEM micrographs showing deformation microstructures at 1023 K and 200 MPa: (a) Interactions of B2 precipitates with dislocations in the Component 3 after creep for 1435 h to 1.5 % strain; (b) Lamellar region of the Component 1 after creep for 25677 h to 6.7 % strain.

to dislocation motion and intensive dislocation pile-up at γ-matrix/Ti_5Si_3 interfaces was observed. Comparison of the microstructure of two creep specimens tested at 1023 K and 200 MPa prepared from the Component 1 and the Component 3 exhibiting significantly different minimum creep rates of 2.65×10^{-10} and 2.88×10^{-9} s^{-1}, respectively, showed better stability of lamellar structure in the Component 1. In the creep more resistant specimen, the ordinary dislocations are frequently pinned by fine Ti_5Si_3 precipitates formed in all regions during creep testing. In the creep less resistant specimen, the majority of silicon is bound in the coarse Ti_5Si_3 precipitates formed in the γ-rich regions and tungsten acting as solid solution strengthener [16] is partially bound in numerous coarse B2 particles formed in the lamellar regions.

CONCLUSIONS

The as-received microstructure of Ti-46Al-2W-0.5 Si (at.%) alloy is unstable during long-term aging and creep at temperatures ranging from 973 to 1123 K. The α_2-phase transforms to the γ-phase and fine needle-like B2 precipitates. The applied stresses accelerate coarsening of the needle-like B2-particles. During long-term creep testing fine Ti_5Si_3 precipitates are formed in the alloy. The microstructural instabilities lead to a softening of lamellar and feathery regions. This softening has a negligible effect on creep resistance at temperatures of 973 and 1023 K.

ACKNOWLEGMENTS

The authors acknowledge the financial support of the Slovak Grant Agency for Science under the contract VEGA 2/4166/24.

REFERENCES

1. M. Nazmy and V. Lupinc in *Materials for Advanced Power Engineering 2002*, edited by J. Lecomte-Beckers, M. Carton, F. Schubert and P.J. Ennis, (Forschungszentrum Jülich GmbH, Vol. **21**, Part I; 2002) pp. 43-56.
2. J. Lapin and M. Nazmy, *Mater. Sci. Eng. A* **380**, 298 (2004).
3. M. Nazmy and M. Staubli, U.S.Pat.#5,207,982 and European Pat.#45505 BI.
4. J. Lapin and A. Klimová, *Kovove Mater.* **41**, 1 (2003).
5. J. Lapin and T. Pelachová, *Kovove Mater.* **42**, 143 (2004).
6. I. Gil, M.A. Muñoz-Morris, and D.G. Morris, *Intermetallics* **9**, 973 (2001).
7. R. Yu, L.L. He, Z.X. Jin, J.T. Guo, H.Q. Ye, and V. Lupinc, *Scripta Mater.*, **44**, 911 (2001).
8. M.A. Muñoz-Morris, I. Gil Fernández, and D.G. Morris, *Scripta Mater.* **46**, 217 (2002).
9. D.J. Larson, C.T. Liu, and M.K. Miller, *Intermetallics*, **5**, 497 (1997).
10. A.J. Ardell, *Metall. Trans. A*, **16A**, 2131 (1985).
11. D.Y. Seo, J. Beddoes, L. Zhao, and G.H. Botton, *Mater. Sci. Eng. A*, **329-331**, 810 (2002).
12. W. Schillinger, H. Clemens, G. Dehm, and A. Bartels, *Intermetallics*, **10**, 459 (2002).
13. S. Karthikeyan, G.B. Viswanathan, P.I. Gouma, K. Vijay, V.K Vasudevan, Y.W. Kim, and M.J. Mills, *Mater. Sci. Eng. A*, **329-331**, 621 (2002).
14. S, Karthikeyan, G.B. Viswanathan, M.J. Mills, *Acta Mater.*, **52**, 2577 (2004).
15. F. Appel, *Intermetallics*, **9**, 907 (2001).
16. L. Zhao, J. Beddoes, D. Morphy, and W. Wallace: *Mater. Sci. Eng. A*, **192-193**, 957 (1995).

Creep Behavior and Microstructural Stability of Ti-46Al-9Nb with Different Microstructures

S. Bystrzanowski[a], A. Bartels[a], H. Clemens[b], R. Gerling[c], F.-P. Schimansky[c], G. Dehm[d]
[a]Materials Science and Technology, TU Hamburg-Harburg, D-21073 Hamburg, Germany
[b]Department of Physical Metallurgy and Materials Testing, Montanuniversität Leoben, A-8700 Leoben, Austria
[c]Institute for Materials Research, GKSS-Research Centre, D-21502 Geesthacht, Germany
[d]Max-Planck-Institut für Metallforschung, D-70569 Stuttgart, Germany

ABSTRACT

In this paper the creep behavior and the microstructural stability of Ti-46Al-9Nb (in at.%) sheet material were investigated in the temperature range of 700°C to 815°C. The study involves three different types of microstructure, namely fully lamellar with narrow lamellar spacing, duplex and massively transformed. Short-term creep experiments conducted at 700°C and 225 MPa confirmed that the lamellar microstructure with narrow lamellar spacing exhibits a much higher creep resistance when compared to the massively transformed and duplex ones. During long-term creep tests up to 1500 hours stress exponents (in the range of 4.4 to 5.8) and apparent activation energies (of about 4 eV) have been estimated by means of load and temperature changes, respectively. Both, stress exponents and activation energies suggest that under the applied conditions diffusion-assisted climb of dislocations is the dominant creep mechanism. The thermal stability of the different microstructures under various creep conditions has been analyzed by electron microscopy and X-ray diffraction. Our investigations revealed considerable stress and temperature induced microstructural changes which are reflected in the dissolution of the α_2 phase accompanied by precipitation of a Ti/Nb - rich phase situated at grain boundaries. This phase was identified as a ω-related phase with B8$_2$-type structure. It was shown, that in particular the duplex microstructure is prone to such microstructural instabilities.

INTRODUCTION

Since γ-TiAl based alloys are considered as structural materials for engineering applications in the temperature range of 650 to 850°C large efforts have been made to improve their oxidation and creep properties. In the frame of a new alloying strategy - based on addition of a high amount of Nb - a new class of γ-TiAl alloys has been established. These so-called TNB alloys exhibit significantly improved tensile strength, creep properties and oxidation resistance when compared to previous generations of TiAl-alloys [1-4]. At a fixed chemical composition the mechanical properties can be further improved by adjustment of appropriate microstructures, which is usually achieved by high-temperature heat treatments followed by fast cooling. However, such heat treatments lead often to constituting phases being far from thermodynamic equilibrium, which consequently makes the microstructure prone to thermal instability. The objective of the present paper was to investigate the influence of three different microstructures (duplex: DPX, massively transformed: MT, and fine spaced fully lamellar: FL) on the creep behavior of Ti-46Al-9Nb sheet material as well as to examine their thermal stability under applied creep conditions, i.e. temperatures between 700 - 815°C and stresses between 150 - 350 MPa.

EXPERIMENTAL

Sheet material with a nominal chemical composition of Ti-46Al-9Nb (at.%) was processed via a powder metallurgical (PM) route [5, 6]. The sheet shows a fine grained near-gamma (NG) microstructure as described in [7,8]. From the sheets with NG microstructure creep specimens were cut out by spark erosion and subsequently heat treated. The high-temperature heat treatments were either conducted above (in case of FL and MT microstructure) or below (in case of DPX microstructure) the α-transus temperature leading to different types of microstructures as shown in Figs. 1a-c. Heat treatment parameters and the resulting microstructural features are summarized in Table 1.

For the tensile creep experiments specimens with a gauge area of 27x5x1.2 mm³ were used. All tests were carried out in air using a tensile-creep apparatus (Mayes, TC30) with constant load equipped with a three zones furnace. During the long-term creep experiments every gradual change of stress/temperature was always accompanied by loading specimens with the stress calculated adequately to their actual cross-section. Partially, this allowed to eliminate the influence of the decreasing cross-section of the specimens. Thus, it can be assumed that the development of the creep rates observed during experiments is mainly controlled by the microstructure (or microstructural changes) and not by external factors such as reduced cross-section.

Figure 1. Investigated microstructures: (a) duplex ; (b) massively transformed; (c) fully lamellar. Left part: light-optical image; right part: SEM image in back-scattered electron mode (BSE).

Table 1. Heat treatment parameters and resulting microstructural features. Vacuum ~ 1 x 10^{-5} mbar.

Microstructure	Heat treatment	General features
Duplex	1300°C / 30 min / furnace cooling in vacuum (~100 K/min)	Fine γ grains and small lamellar $\alpha_2 + \gamma$ colonies (~ 10-15 μm)
Massively transformed	1330°C / 4 min / air cooling (>> 100 K/min)	Mostly massively transformed with some fine lamellar colonies and small γ-grains along former α grain boundaries
Fully lamellar	1360°C / 10 min / furnace cooling in vacuum (~ 100 K/min)	Large lamellar colonies (~300 μm) with fine lamellar spacing (~110 nm)

RESULTS AND DISCUSSION

Figs. 2 a-c summarize the creep behavior of Ti-46Al-9Nb sheet material with DPX, MT and FL microstructures. At 700°C and 225 MPa the FL microstructure exhibits significantly higher creep resistance than the MT and the DPX microstructure. This is reflected in the creep rates as indicated in Fig. 2a, which were calculated from the last section of the creep curves (in every case the last 50 hours have been taken into account). It should be noted that these creep rates represent rather apparent creep rates than steady state or minimum creep rates. The FL microstructure creeps 4 times slower than the MT and nearly 10 times slower than the DPX microstructure. In addition, the primary creep rate in case of the DPX microstructure is enormously high when compared to MT and FL. The pronounced primary range in case of duplex microstructure has been already noticed by Skrotzki et al. [9], however, explanations for this behavior were not reported.

Stress exponents and apparent activation energies were determined by analyzing the creep data according to:

$$\dot{\varepsilon} = S\sigma^n \exp(-Q_C / RT), \qquad (1)$$

where $\dot{\varepsilon}$ represents the steady-state or minimum creep rate, S is a structural factor, n denotes the stress exponent, and Q_C is the apparent activation energy for creep. The stress exponent was calculated from the data shown in Fig. 2b. At 700°C the n-values vary from 4.4 (for DPX at stresses of 200 - 275 MPa) to 4.5 (for MT at 275 - 350 MPa) up to 5.8 (for FL at 275 - 350 MPa). At 800°C two types of microstructures were examined, i.e. MT and FL. For these microstructures stress exponents of 5.0 and 5.6 were determined, respectively. The apparent activation energies for creep have been obtained from the Arrhenius plot shown in Fig. 2c. The Q-values for MT and FL are close to each other (3.95 and 3.91 eV, respectively), while the activation energy in case of the DPX microstructure approaches 4.25 eV. However, the value of 4.25 eV seems to be overestimated, since the creep rate taken at 745°C was determined at a specimen elongation of ~ 5%. Thus, the creep rate could have been affected by significant microstructural changes. It should be noted that the activation energies which were obtained by creep tests are in good agreement with values determined from internal friction experiments conducted on the same material [10].

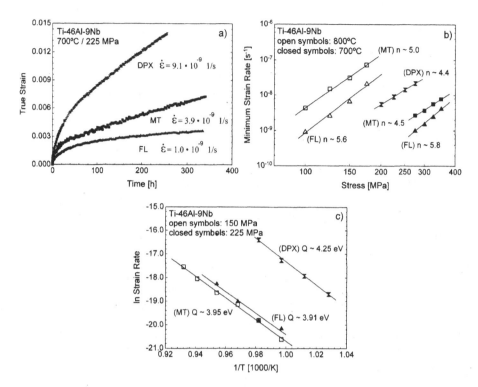

Figure 2. Creep behavior of Ti-46Al-9Nb sheet material with different microstructures: (a) creep curves; (b) minimum strain rate vs. stress; (c) ln strain rate vs. 1/T. Experimental data and the determined values for n and Q are indicated in the insets.

From both, i.e. stress exponents (between 4.4 - 5.8) and activation energy (about 4 eV) it is concluded that diffusion-assisted climb of dislocations is the major creep controlling mechanism in Ti-46Al-9Nb material with DPX, MT and FL microstructures under the applied temperature and stress conditions.

During creep all investigated microstructures showed a tendency to stress and temperature induced microstructural changes, which are reflected in the dissolution of the α_2-phase and formation of a new phase located preferentially at grain and/or colony boundaries. Figs. 3a-c show BSE images of DPX, MT and FL microstructures after long-term creep tests including stress/temperature changes. In every case small precipitations showing a white contrast are present when compared with initial microstructures in Figs. 1a-c. These precipitations have been identified as a ω-related phase with B8₂-type crystal structure. Fig. 3d shows XRD spectra as obtained from crept Ti-46Al-9Nb with DPX microstructure. Additional ω-type related peaks clearly appear in the crept material. At the same time the amount of α_2-phase is significantly reduced when compared to initial material. The presence of ω-related phases in TiAl alloys (also showing relatively high Nb contents but lower Al concentrations than our alloy) have been reported by several authors [11-13]. Bendersky et al. [12] have proposed the following transformation path for the formation of ω-phase in a Ti-37.5Al-12.5Nb alloy: B2 $\rightarrow$ ω" $\rightarrow$ ω(B8$_2$), where ω" is an intermediate trigonal phase. In our case the α_2-phase undergoes a decomposition which at first most probably leads to the formation of β/B2 phase and eventually to creation of the Ti and Nb rich ω-phase of B8$_2$ structure according to the above mentioned sequence. Simultaneously, new γ-TiAl grains enriched in Al appear in the vicinity of the ω-precipitates. They can be recognized as small dark spots, e.g. in Fig. 3a. Dissolution of the α_2-phase might occur as a result of discontinuous coarsening, which preferentially takes place at grain and colony boundaries. Therefore, the duplex microstructure exhibits the strongest tendency to such changes. This is a direct consequence of its fine-grained morphology and significant amount of small lamellar grains showing a high amount of non-equilibrium α_2-phase. The influence of the ω-phase on mechanical properties and ductility of Ti-46Al-9Nb material, however, requires further investigations.

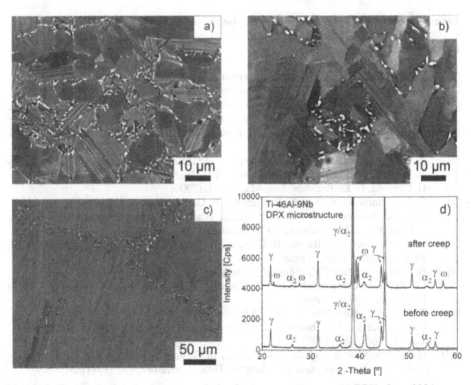

Figure 3. Changes in the microstructures during long term creep tests: (a) DPX after ~800 hrs at 225 MPa and 700-775°C; (b) MT after ~1500 hrs at 150 MPa and 730-815°C; (c) FL after ~1500 hrs at 800°C and 100-200 MPa. All pictures were taken in BSE mode; (d) X-ray diffraction spectra of DPX material before and after creep deformation (~800 hrs at 225 MPa and 700-775°C)

CONCLUSIONS

i) The creep properties of Ti-46Al-9Nb material depend strongly on the prevailing microstructure; fully lamellar microstructure exhibits superior creep resistance when compared to massively transformed and duplex ones.

ii) Stress exponents in the range of 4.4 - 5.8 and an activation energy of about 4eV imply diffusion-assisted climb of dislocations as the dominant creep mechanisms for the test conditions used in this study (700°C, 200 - 350 MPa and 800°C, 100 - 175 MPa).

iii) As a result of accelerated cooling rates the examined microstructures are not in thermodynamic equilibrium. Therefore, they are prone to temperature and stress induced microstructural instabilities which is reflected in the dissolution of excess α_2 phase and in the formation of ω-phase with $B8_2$ structure.

ACKNOWLEDGEMENTS

The authors thank Dr. H. Kestler, Plansee AG, Austria, for supplying the sheet material.

REFERENCES

[1] F. Appel, M. Oehring, R. Wagner, Intermetallics, 8, 1283, (2000)

[2] W.J. Zhang, S.C. Deevi, G.L. Chen, Intermetallics, 10, 403, (2002)

[3] M. Yoshira, Y.-W. Kim, in Gamma Titanium Aluminides; edited by Y.-W. Kim, H. Clemens, A.H. Rosenberg (TMS, Warrendale, PA, 2003) p.559

[4] R. Gerling, A. Bartels, H. Clemens, H. Kestler, F.-P. Schimansky, Intermetallics 12, 275, (2004)

[5] R. Gerling, F.-P. Schimansky, H. Clemens, in Gamma Titanium Aluminides; edited by Y.-W. Kim, H. Clemens, A.H. Rosenberg (TMS, Warrendale, PA, 2003) p.249

[6] H. Clemens, H. Kestler, Adv. Eng. Mat. 2, 551, (2000)

[7] S. Bystrzanowski, A. Bartels, H. Clemens, R. Gerling, F.-P. Schimansky, H. Kestler, G. Dehm, G. Haneczok, M. Weller, in Gamma Titanium Aluminides; edited by Y.-W. Kim, H. Clemens, A.H. Rosenberg (TMS, Warrendale, PA, 2003) p.431

[8] S. Bystrzanowski, A. Bartels, H. Clemens, R. Gerling, F.-P. Schimansky, G. Dehm, H. Kestler, Intermetallics, in print

[9] B. Skrotzki, T. Rudolf, G. Eggeler, Z. Metallkd., 90, 8, 393, (1999)

[10] M. Weller, H. Clemens, G. Dehm, G. Haneczok, S. Bystrzanowski, A. Bartels, R. Gerling E. Arzt, these proceedings

[11] T.H. Yu, C.H. Koo, Mat. Sci. Eng., A239-240, 694, (1997)

[12] L.A. Bendersky, W.J. Boettinger, B.P. Burton, F.S. Biancaniello, Acta metall. mater., Vol. 38, No.6, 931, (1990)

[13] G.L. Chen, W.J. Zhang, Z.C. Liu, S.J. Li, in Gamma Titanium Aluminides; edited by Y.-W. Kim, D.M. Dimiduk, M.H. Loretto (TMS, Warrendale, PA, 1999) p.371

Mater. Res. Soc. Symp. Proc. Vol. 842 © 2005 Materials Research Society   S7.13

Internal Friction of a High-Nb Gamma-TiAl-Based Alloy with Different Microstructures

M. Weller[1], H. Clemens[2], G. Dehm[1], G. Haneczok[3] , S. Bystrzanowski[4], A. Bartels[4], R. Gerling[5], and E. Arzt[1]

[1]Max-Planck-Institut für Metallforschung, Heisenbergstr. 3, D-70569 Stuttgart, Germany
[2]Dept. of Physical Metallurgy and Materials Testing, Montanunivesität Leoben, Franz-Josef-Strasse 18, A-8700 Leoben, Austria
[3]Institute of Materials Science, Silesian University, Katowice, Poland
[4]Technical University of Hamburg-Harburg, Dept. of Materials Science and Technology, Eisendorferstrasse 42, D-21071 Hamburg, Germany
[5]Institut for Materials Research, GKSS Research Centre, Max-Planck-Strasse 1, D-21502 Geesthacht, Germany

ABSTRACT

An intermetallic Ti-46Al-9Nb (at%) alloy with different microstructures (near gamma, duplex, and fully lamellar) was studied by internal friction measurements at 300 K to 1280 K using different frequency ranges: (I) 0.01 Hz to 10 Hz and (II) around 2 kHz. The loss spectra in range I show (i) a loss peak of Debye type at $T \approx 1000$ K which is only present in duplex and fully lamellar samples; (ii) a high-temperature damping background above ≈ 1100 K. The activation enthalpies determined from the frequency shift are H = 2.9 eV for the loss peak and H = 4.1 - 4.3 eV for the high-temperature damping background. The activation enthalpies for the visco-elastic high-temperature damping background agree well with values obtained from creep experiments and are in the range of those determined for self-diffusion of Al in TiAl. These results indicate that both properties (high-temperature damping background and creep) are controlled by volume diffusion-assisted climb of dislocations. The loss peak is assigned to diffusion-controlled local glide of dislocation segments which, as indicated by transmission electron microscopy observations, are pinned at lamella interfaces.

INTRODUCTION

Intermetallic γ-TiAl based alloys exhibit increasing technical importance for high-temperature applications in the automotive and aerospace industries (see, for example [1]. These are based on their properties at elevated temperatures, such as high yield strength, advanced creep properties, high stiffness, and good oxidation/corrosion resistance. The mechanical properties at high temperatures are strongly influenced by the creep deformation behavior, which is largely controlled by diffusion mechanisms occurring in the different phases of the TiAl-system (self and solute diffusion). As an alternative to creep tests – which are time consuming and require considerable technical effort – mechanical loss (internal friction) experiments can also give access to the high-temperature mechanical properties. This was demonstrated for various materials such as intermetallic compounds [2], quasicrystalline materials [3], and ceramics [4]. The basic source of information in mechanical loss studies is the high-temperature damping background, which exhibits viscoelastic behavior and requires measurements far below 1 Hz. It was found that the activation enthalpy, determined from the frequency shift of the high-temperature background corresponds closely with those of creep tests and self-diffusion studies [2-4].

The aim of the present paper was to determine the activation enthalpy of the underlying mechanism in a Ti-46Al-9Nb (at%) alloy by applying mechanical spectroscopy. Sheet samples with fine-grained near gamma/duplex microstructures and a coarse-grained fully lamellar microstructure were investigated.

MATERIAL

The starting material used in this study had a nominal chemical composition of Ti–46Al–9Nb (at %) and was fabricated by a powder metallurgical process. Sheets were hot-rolled from hot-isostatically pressed pre-alloyed powder compacts as described in ref. [5]. Three microstructures were obtained by appropriate heat treatments. The fine-grained equiaxed near-gamma (NG) microstructure was prepared by annealing for 3 hours at 1000 °C and subsequent furnace cooling (Fig. 1a). The microstructure consists of areas of equiaxed γ-TiAl grains with an average size in the range of 10 – 15 μm. The grains are surrounded by recrystallized regions consisting of small γ-TiAl and α_2-Ti$_3$Al grains.

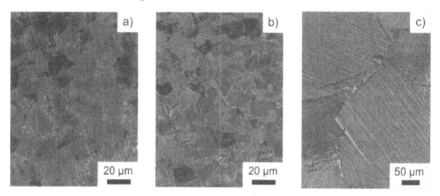

Figure 1. Ti-46at%Al-9at%Nb sheet with (a) near-gamma, (b) duplex and (c) fully lamellar microstructure (SEM images taken in back-scattered electron mode).

The duplex (D) microstructure as shown in Fig. 1b was adjusted by annealing within the (α + γ) phase field (1300 °C / 30 min) and subsequent cooling at ~ 100 K min^{-1}. The microstructure consists of a similar amount of γ grains and lamellar α_2/γ grains showing a grain size in the range of 10 – 15 μm. In the lamellar grains a mean interface spacing of ~ 110 nm, including α_2/γ as well as all variants of γ/γ interfaces, was determined from TEM images with the interface oriented "edge on" along <011]. Additionally, TEM micrographs were recorded in order to resolve the dislocation structure within lamellar grains.

The coarse-grained fully lamellar (L) microstructure was obtained by short-duration annealing within the α-phase field (1360 °C / 10 min) followed by controlled cooling (20 K min^{-1}) to room temperature. The L microstructure is shown in Fig. 1c. The colonies consist of alternating γ/α_2 or γ/γ laths. The average colony size in the lamellar microstructure is about 300 μm and the mean interface spacing is ~ 170 nm. Fig. 2. shows dislocation segments constrained between the interfaces of a lamellar grain within the duplex microstructure.

Specimens with dimensions of 50 x 5 x 1mm³ for mechanical spectroscopy were prepared from Ti–46Al–9Nb sheets with NG, D and L microstructures. For all samples the longer axis was oriented parallel to the rolling direction.

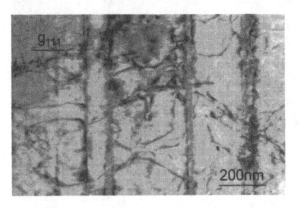

Figure 2. TEM bright field image of a lamellar grain within the duplex micro-structure recorded with a g_{111} two-beam condition. Note the presence of dislocations constrained by γ/γ and γ/α₂ interfaces.

EXPERIMENTAL RESULTS

Low-frequency internal friction measurements were carried out with a subresonance torsion apparatus using forced vibrations in the frequency range of 10^{-3} Hz to 50 Hz [6]. The mechanical loss angle φ was determined from the phase shift between applied stress and strain in the temperature range from 300 K to 1280 K.

Figure 3a shows mechanical loss measurements obtained on 3 specimens with NG, D and L microstructure. Depicted is the internal friction Q^{-1} = tan φ versus temperature T for a frequency of f = 0.1 Hz. The loss spectra Q^{-1}(T) show two phenomena: (i) a loss peak in the temperature range of 1000 K and (ii) a high-temperature background above 1100 K. The 1000 K loss peak is only present in specimens with D and L microstructure. Fig. 3b shows, as an example, measurements conducted on a specimen with D microstructure for various measuring frequencies (0.01 Hz – 10 Hz). The inset shows the loss peak in a more expanded scale. Both the loss peak and the background are shifted to higher temperature with increasing measuring frequency. This indicates that both phenomena are due to thermally activated relaxation processes. The relaxation time τ is then determined by an Arrhenius equation

$$\tau^{-1} = \tau_\infty^{-1}\exp(-H / kT),\qquad(1)$$

where H is the activation enthalpy, k is the Boltzmann's constant and τ_∞^{-1} represents an atomic attempt frequency. The loss peaks in figure 3 are of Debye-type. The activation enthalpy can be determined from the temperature shift of the peak with frequency. The results obtained for specimens with D and L microstructure are 3.1±0.1 eV and H = 2.9±0.1 eV, respectively. The high-temperature background, as shown e.g. in Fig. 3, increases at constant T proportiona to 1/f which is characteristic for viscoelastic relaxation. Analysis of the data by applying a Maxwell rheological model, for which $Q^{-1} = 1 / (2\pi f \tau)$, gives for the activation enthalpies of NG, D, and L samples the values listed in Table 1 (for details of the evaluations see [7] and [8]).

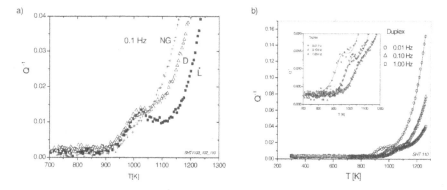

Fig. 3. Internal friction versus temperature of Ti-46Al-9Nb with **(a)** NG, D and L microstructure for f = 0.1 Hz and **(b)** D microstructure for 0.01, 0.1 and 1 Hz. Inset: loss peak at ~ 1000 K.

Table 1. Activation enthalpies (in eV) for Ti-46Al-9Nb obtained by mechanical spectroscopy (internal friction). For comparison the data from creep tests [9,13] and temperature/strain rate cycling tests [10] conducted on alloys with similar compositions are included.

Material/microstructure	High-temperature background	Creep tests	Temperature/strain rate cycling tests
Ti-46Al-9Nb / NG	4.3±0.2	4.1 – 4.2	4.2 – 4.5 Ti-45Al-(5-10)Nb /L
Ti-46Al-9Nb / D	4.1±0.2	4.2*	
Ti-46Al-9Nb / L	4.2±0.2	3.9 – 4.1	

DISCUSSION

The low-frequency mechanical-loss spectra of Ti-46Al-9Nb show two basic phenomena: (i) In all specimens (NG, D, L) a high-temperature damping background is observed, which exhibits a viscoelastic behavior. In the following it will be shown that the damping background in these materials is closely related to the creep behavior corresponding to diffusion-controlled deformation. (ii) In the specimens containing lamellar structure, an additional loss peak appears around 1000 K which is not present in fine-grained NG specimens. The peak is assigned to local motion of dislocation segments. Both phenomena, the high-temperature damping background and the loss peak, are thermally activated. Their interpretation is mainly based on the experimentally determined values of the activation enthalpies.

The activation enthalpy of the high-temperature damping background is $H = 4.3\pm0.2$ eV for NG, H= 4.1 ± 0.2 for D, and $H = 4.2 \pm0.2$ eV for L material, i.e. about equal within error limits. The activation enthalpies are slightly higher than those observed for a low-Nb-bearing Ti-46.5-4(Nb,Cr,Ta,B) alloy (≈ 3.9 eV) [7]. The H values correlate very well with the activation enthalpy obtained from creep tests (4 – 4.2 eV) conducted on specimens with identical composition and microstructure [9]. In addition, the results obtained from mechanical spectroscopy agree with data derived from an activation parameter study which was recently conducted on Ti-45Al-(5-10Nb)) alloys [10]. In this study, temperature and strain rate cycling tests were used to determine activation enthalpies (see Table 1). Accompanying TEM investigations indicate that climb of ordinary dislocations is the predominant deformation mechanism at elevated temperatures and slow strain rates.

The activation enthalpies of the high-temperature damping background determined in this study seem to be rather high compared with activation energies from self-diffusion measurements for binary γ-TiAl based alloys. Herzig et al. [11] obtained $H^{*}_{Ti} = 2.6$ eV and H^{*}_{Al} = 3.7 eV by means of tracer diffusion experiments. The activation energy for Nb tracer solute diffusion in TiAl was determined as $H^{*}_{Nb} = 3.4$ eV which is distinctly higher than that reported for Ti self-diffusion. On that account, the higher activation enthalpies of the high-temperature background in the present study might be explained by the higher activation energy necessary to promote diffusion in high-Nb-bearing γ-TiAl based alloys. This implies that diffusion-assisted climb of dislocations, controlled by diffusion of Nb, is related to the high-temperature damping background.

The anelastic damping peak around 1000 K, which is only present in Ti-46Al-9Nb material with D and L microstructures (see Fig. 3a), is attributed to dislocation relaxation. The activation enthalpies of 2.9 to 3.1 eV are significantly lower than that of the high-temperature background but approximately the same as found for the peak present in Ti-46.5Al-4(Nb, Cr, Ta, B) material with L microstructure (3 eV) [7]. It is assumed that this peak originates from stress-induced (reversible) movement of dislocation segments anchored at their ends either at lamellar interfaces or at obstacles present within the γ-lamellae (Fig. 2). The existence of such segments or loops in fully lamellar TiAl materials is well established and described in detail in [12]. This peak is smaller in D specimens, obviously due to the smaller lamellar volume fraction. This peak does not occur in fine-grained NG material which can be explained with the absence of the dislocation segments present in F and D specimens. The activation enthalpy for the peak of about 3.0 eV indicates that the local movement of the dislocations occurs with lower activation enthalpy than diffusion-controlled climb of dislocations which is related to the high temperature damping background and the creep properties.

SUMMARY AND CONCLUSIONS

(i) Activation enthalpies of 4.1 – 4.3 eV were determined for the viscoelastic, high-temperature, damping background in two-phase Ti-46Al-9Nb alloys with different microstructures (fine-grained near gamma, duplex and coarse-grained fully lamellar). These values agree well with those obtained from creep experiments and are in the range of those determined for self-diffusion of Al in TiAl. These results indicate that both properties (high-temperature damping background and creep) are controlled by volume diffusion assisted climb of dislocations.

(ii) A mechanical loss peak at about 1000 K is only observed in specimens with duplex and fully lamellar microstructure. This is assigned to the local movement of dislocation segments which are pinned at γ/α_2 and γ/γ interfaces as well as within γ lamellae. The bowing out of the dislocation segments occurs with a lower activation enthalpy (H $\approx$ 3.0 eV) than diffusion-controlled dislocation climb.

ACKNOWLEDGEMENTS.

We thank Plansee Aktiengesellschaft (Reutte, Austria) for providing sample material.

REFERENCES

1. Y.-W. Kim, H. Clemens, H. Rosenberger, eds, Gamma Titanium Aluminides 2003, TMS (The Minerals, Metals & Materials Society), Warrendale, PA, 2003.
2. M. Weller, M. Hirscher, E. Schweizer, and H. Kronmüller, *J. Phys. (Paris)* IV **6**, C8, 231 (1996).
3. M. Weller, and B. Damson, *Quasicrystals - Structure and Physical Properties*, edited by H.-R. Trebin, Wiley-VCH, Weinheim, p. 539.
4. A. Lakki, R. Herzog, M. Weller, H. Schubert, C. Reetz, O. Görke, M. Kilo, and G. Borchardt, *J. European Ceramic Society* **20**, 285 (2000).
5. R. Gerling, A. Bartels, H. Clemens, H. Kestler, F.-P. Schimansky, *Intermetallics* **12**, 275 (2004).
6. M. Weller: J. *Phys. (Paris)* IV **5**, C7, 199 (1995).
7. M. Weller, A. Chatterjee, G. Haneczok, E. Arzt, F. Appel, and H. Clemens, *Z. Metallkde.* **92**, 1019 (2001).
8. M. Weller, H. Clemens, G. Haneczok, G. Dehm, A. Bartels, S. Bystrzanowski, R. Gerling, and E. Arzt, *Phil. Mag. Lett* **84**, 383 (2004).
9. S. Bystrzanowski, A. Bartels, H. Clemens, R. Gerling, F.-P. Schimansky, F.P. Kestler, G. Dehm, G. Haneczok, M. Weller, *Gamma Titanium Aluminides 2003*, Y.-W. Kim, H. Clemens and H. Rosenberger eds., TMS (The Minerals, Metals & Materials Society Warrendale, PA), 2003, 465.
10. F. Appel, M. Oehring, R. Wagner, *Intermetallics* **8**, 1283 (2000).
11. Ch. Herzig, T. Przeorski, and Y. Mishin, *Intermetallics* **7**, 389 (1999).
12. F. Appel and R. Wagner, *Mater. Sci. Eng.* **R22**, 187 (1998).
13. Bystrzanowski, A. Bartels, H.Clemens, R. Gerling, F.-P. Schimansky, G. Dehm, these proceedings.

Mater. Res. Soc. Symp. Proc. Vol. 842 © 2005 Materials Research Society   S4.3

Investigation of the structure and properties of hypereutectic Ti-based bulk alloys

Dmitri V. Louzguine-Luzgin[1,*], Larissa V. Louzguina-Luzgina[2,&] and Akihisa Inoue[1]

[1] Institute for Materials Research, Tohoku University, Katahira 2-1-1, Aoba-Ku, Sendai 980-8577, Japan
[2] Research and Development Project, CREST, Japan Science and Technology Agency, Sendai 985-8577, Japan

Keywords: Ti-based alloys, mechanical properties, intermetallic compound, high strength, ductility.

ABSTRACT

Structure and mechanical properties of binary Ti-TM (TM-other transition metals) and ternary Ti-Fe-(TM, B or Si) alloys produced in the shape of the arc-melted ingots of about 25 mm diameter and 10 mm height are studied. The formation of high-strength and ductile hypereutectic alloys was achieved in the Ti-Fe, Ti-Fe-Cu and Ti-Fe-B systems. The structures of the high-strength and ductile hypereutectic alloys studied by X-ray diffractometry and scanning electron microscopy were found to consist of the primary cubic $Pm\bar{3}m$ intermetallic compound (TiFe-phase or a solid solution on its base) and a dispersed eutectic consisting of this $Pm\bar{3}m$ intermetallic compound + BCC $Im\bar{3}m$ β-Ti supersaturated solid solution phase. The hypereutectic Ti-Fe alloy showed excellent compressive mechanical properties. The addition of Cu improves its ductility. B addition increased mechanical strength. Ni, Cr and Mn additions caused embrittlement owing to the formation of alternative intermetallic compounds. The deformation behaviour and the fractography of the Ti-based alloys were studied in details. The reasons for the high strength and good ductility of the hypereutectic alloys are discussed.

INTRODUCTION

The ultimate tensile strength of typical structural Ti-based alloys somewhat exceeds 1000 MPa [1]. At the same time, a special squeeze casting technique [2] allows to produce bulk glassy alloys [3,4] with high ultimate tensile strength which attains 2200 MPa for the $Ti_{50}Ni_{20}Cu_{23}Sn_7$ bulk glassy alloy [5] (alloy's compositions are given in nominal atomic percents) and 2480 MPa for the $Ti_{45}Cu_{25}Ni_{15}Sn_3Be_7Zr_5$ one [6]. Ti alloys have relatively high corrosion resistance at room temperature [7]. The relatively low density of the main alloying element Ti (4.5 Mg/m^3) implies a higher strength/density ratio compared to Fe- or Zr-based bulk glassy alloys. However, small critical diameter of 5-8 mm for the Ti-based bulk glassy alloys [4,6,8,9] attained so far and low ductility restrict their applications. Only addition of a toxic element Be helps to improve slightly their compressive ductility [6].

Recently, it also has been shown that a 3 mm diameter cylindrical rod of cast $Ti_{60}Cu_{14}Ni_{12}Sn_4Nb_{10}$ alloy consisting of a micron–size β-Ti dendrites exhibit high ultimate compressive strength of 2.4 GPa and 14.5 % plastic strain [10]. The deformation behavior of the

---

* Louzguine is official French spelling. English spelling is Luzgin.
& Louzguina is official French spelling. English spelling is Luzgina.

nanostructured Ti-Cu-Ni-Sn-(Ta,Nb) alloys has been studied recently [11]. These alloys can be used for biomedical applications [12]. One can also remind that BCC $Im\bar{3}m$ β-Zr phase can be used for ductilization of bulk glassy alloys [13].

Recently, bulk Ti-Fe alloys exhibiting high mechanical properties exceeding 2000 MPa and good ductility of 4-7 % were obtained at a low cooling rate of about 10 K/s after pre-melting in an arc-furnace [14,15]. It was also shown that the hypereutectic alloy has better mechanical properties than the eutectic and hypoeutectic ones. The high strength Ti-Fe alloys obtained by arc-melting do not require additional rapid solidification procedure (for example, Cu-mould casting). Arc-melting has already proven to be a suitable technique for direct production of a high-strength Ti-Ni-Cu-Nb alloy without additional treatment [16]. In the present paper, we study the influence of some other late transition metals (LTM) additions on the mechanical properties of the Ti-Fe alloys. A binary Ti-Mn alloy is also studied for comparison. An influence of small B and Si additions of 0.5-2 % is also investigated as B and Si are known to be good grain size modifiers of β-Ti.

EXPERIMENTAL PROCEDURE

The ingots of the $Ti_{65}Fe_{35}$ (base alloy obtained earlier [12]), $Ti_{60}Mn_{40}$, $Ti_{70}Fe_{15}Cr_{15}$, $Ti_{70}Fe_{15}Ni_{15}$, $Ti_{70}Fe_{15}Cu_{15}$, $Ti_{64.7}Fe_{34.8}B_{0.5}$, $Ti_{64.4}Fe_{34.6}B_1$, $Ti_{74}Fe_{24}B_2$, $Ti_{64.4}Fe_{34.6}Si_1$ and $Ti_{74}Fe_{24}Si_2$ alloys of about 25 mm diameter and 10 mm height were prepared by arc-melting the mixture of preliminary degassed sponge Ti (with initial purity of 99.7 mass%) and other metals and metalloids (>99.9 mass% purity) in an argon atmosphere purified with Ti getter. Mass purities of other metals and metalloids exceeded 99.9 %. The ingots were turned and re-melted 4 times to ensure compositional homogeneity. After arc-melting the cooling rate was in the order of 10 K/s. The structure of the central part of the ingots was examined by X-ray diffractometry with monochromatic CuK_{α} radiation and scanning electron microscopy (SEM) carried out using a microscope operating at 20 kV. The samples for SEM were cut from the central part of the as-cast ingots. Room temperature mechanical properties were measured at 298 K with an Instron-type testing machine at a strain rate of 5×10^{-4} s^{-1}. The sample for compressive mechanical testing cut from the central part of the as-cast ingot was a 5 mm long rectangular parallelepiped with 2.5x2.5 mm cross section. A strain-gauge was attached to the sample. The density was measured at 298 K by the Archimedean method using tetrabromoethane ($CHBr_2CHBr_2$).

RESULTS

The $Ti_{70}Fe_{15}Cr_{15}$, $Ti_{74}Fe_{24}B_2$, $Ti_{64.4}Fe_{34.6}Si_1$ and $Ti_{74}Fe_{24}Si_2$ alloy ingots being brittle were crushed to pieces upon cutting. These alloys were not used for further studies. A few typical X-ray diffraction (XRD) patterns are shown in Fig. 1. β-Ti phase is a solid solution of the other 3-d transition metals in β-Ti. According to X-ray diffracted intensities $Ti_{60}Mn_{40}$ and $Ti_{70}Fe_{15}Cu_{15}$ alloys exhibit a textured structure.

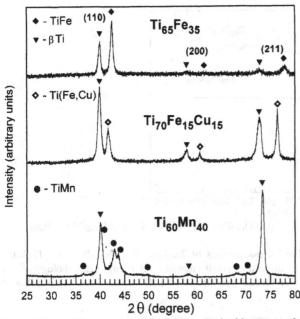

Fig. 1. X-ray diffraction patterns of the arc-melted ingot samples.

The mechanical properties of the samples cut from the arc-melted ingots of the studied alloys are shown in Table 1. $Ti_{70}Fe_{15}Cu_{15}$ alloy shows the best combination of compressive mechanical properties: Young's modulus (E) of 120 GPa, compressive fracture strength (σ_{true}) of 1.61 GPa, 0.2 % yield strength ($\sigma_{0.2}$) of 1.53 GPa, and 8 % ductility (ε).

Table 1. Mechanical properties, density (ρ) and phase composition of the studied alloys.

Alloy	E [GPa]	σ_{true} [MPa]	$\sigma_{0.2}$ [MPa]	ε [%]	ρ [Mg/m^3]	Phase composition
$Ti_{65}Fe_{35}$	149	2220	1800	6.7	5.796	βTi, TiFe
$Ti_{60}Mn_{40}$	127	1630	1630	0.2	5.149	βTi, TiMn
$Ti_{70}Fe_{15}Ni_{15}$	134	1260	1260	0.05	5.120	βTi, Ti$_2$Ni
$Ti_{70}Fe_{15}Cu_{15}$	121	1610	1530	8.0	5.551	βTi, Ti(Fe,Cu)
$Ti_{64.7}Fe_{34.8}B_{0.5}$	150	2470	1840	4.3	4.475	βTi, TiFe
$Ti_{64.4}Fe_{34.6}B_1$	157	2050	1750	4.5	4.448	βTi, TiFe
$Ti_{74}Fe_{24}B_2$	169	1500	1500	0.2	5.377	βTi, TiFe

The strain-stress curves of some selected alloys are shown in Fig. 2. The $Ti_{70}Fe_{15}Cu_{15}$ alloy does not show strain hardening compared to the Ti-Fe and Ti-Fe-B alloys.

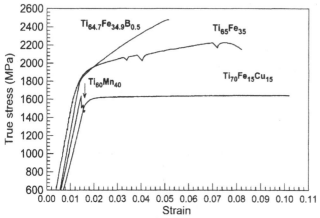

Fig. 2. The strain-stress curves of $Ti_{65}Fe_{35}$, $Ti_{64.7}Fe_{34.8}B_{0.5}$, $Ti_{70}Fe_{15}Cu_{15}$ and $Ti_{60}Mn_{40}$ alloys.

SEM images of the polished cross-sections of $Ti_{65}Fe_{35}$, $Ti_{64.7}Fe_{34.8}B_{0.5}$ and $Ti_{60}Mn_{40}$ alloys are shown in Fig. 3. The addition of a small amount of B changes morphology of the primary TiFe crystals from rounded to faceted (see Fig. 3 (a,b)). TiMn phase has a needle-shaped morphology.

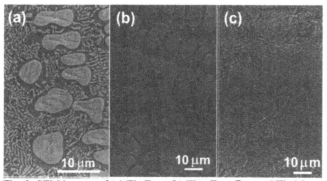

Fig. 3. SEM images of (a) $Ti_{65}Fe_{35}$, (b) $Ti_{64.7}Fe_{34.8}B_{0.5}$ and $Ti_{60}Mn_{40}$ alloys.

DISCUSSION

According to X-ray diffractometry and scanning electron microscopy data the structure of high-strength and ductile hypereutectic alloys, except for $Ti_{60}Mn_{40}$ and $Ti_{70}Fe_{15}Ni_{15}$ is similar. It consists of the primary cubic $Pm\overline{3}m$ intermetallic compound (TiFe-phase or a solid solution on its base) with a dendritic morphology and eutectic consisting of this $Pm\overline{3}m$ intermetallic compound and BCC $Im\overline{3}m$ β-Ti supersaturated solid solution phase (Fig. 1).

The $Ti_{65}Fe_{35}$ hypereutectic alloy showed excellent compressive mechanical properties. The addition of Cu improved its ductility. The composition is slightly changed to $Ti_{70}LTM_{30}$ because the eutectic point shifts to higher Ti concentration with Ni and Cu additions. Ni, Cr and Mn additions caused embrittlement owing to the formation of the alternative intermetallic compounds which have different morphology. For example, the reason for brittle fracture of $Ti_{60}Mn_{40}$ alloy is the needle-shaped morphology of the primary and eutectic TiMn crystals which act as crack initiation sites upon deformation even though the structure is disperse (see Fig. 3).

The $Ti_{70}Fe_{15}Cu_{15}$ alloy shows a good combination of mechanical properties – high strength and reasonable compressive ductility (Fig. 2 and Table 1). An interesting result which is not explained yet is connected with the fact that the $Ti_{70}Fe_{15}Cu_{15}$ alloy does not show strain hardening (Fig. 2) compared to the other alloys (Table 1). B addition in small quantities (0.5 at.%) increases mechanical strength up to 2470 MPa, but reduces ductility. Si is another grain size modifier that, however, caused drastic embrittlement of the samples.

The deformation behaviour and the fractography of the Ti-based alloys were also studied. The location of the crack propagation planes in the $Ti_{65}Fe_{35}$ alloy sample after mechanical testing indicate that the vector normal to the compressive fracture surface is inclined at an angle of about 90 degrees to the load direction which indicates that the fracture occurs along the plane of the maximum normal stress.

SUMMARY

The structure and mechanical properties of a number of binary Ti-TM and ternary Ti-Fe-(TM, B or Si) alloys were studied. High-strength and ductile hypereutectic arc-melted alloys are formed in the Ti-Fe, Ti-Fe-B and Ti-Fe-Cu systems. The structures of the high-strength and ductile hypereutectic alloys studied by X-ray diffractometry and scanning electron microscopy were found to consist of the primary cubic $Pm\bar{3}m$ intermetallic compound (TiFe-phase or a solid solution on its base) and a dispersed eutectic consisting of this $Pm\bar{3}m$ intermetallic compound and BCC $Im\bar{3}m$ β-Ti supersaturated solid solution phase. The Ti-Fe hypereutectic alloy showed good compressive mechanical properties. The addition of Cu improved its ductility. Ni, Cr and Mn additions caused embrittlement owing to the formation of alternative intermetallic compounds with different morphology. B addition in small quantities (0.5 at.%) increases mechanical strength up to 2470 MPa but decreases ductility. Another grain size modifier Si caused drastic embrittlement of the samples.

REFERENCES

1. E. A. Brandes and G. B. Brook:Smithells, Metals Reference Book 7-th Edition Ed.(Butterworth-Heinemann Ltd., Hartnolls Ltd. Bodmin UK, 1992), p. 22-84.
2. A. Inoue, N. Nishiyama, K. Amiya, T. Zhang, and T. Masumoto *Mater. Lett.* **19**, 131 (1994).
3. X. H. Lin, and W. L. Johnson *J. Appl. Phys.* **78**, 6514 (1995).
4. T. Zhang, A. Inoue, and T. Masumoto *Mater Sci. Eng.* **A181/182**, 1423 (1994).
5. T. Zhang, and A. Inoue *Mater. Trans. JIM* **39**, 1001 (1998).

6. Y. C. Kim, W. T. Kim and D. H. Kim *Materials Science and Engineering A* **375-377**, 127 (2004).
7. W.F. Gale and T. C. Totemeier Eds. Smithells, Metals Reference Book 8-th Edition (Elsevier Burlington 2004), p 31-1.
8. W. L. Johnson *MRS Bull*, **24**, 42 (1999).
9. Y.C. Kim, S. Yi, W.T. Kim and D.H. Kim *Mater. Sci. Forum* **360–362**, 67 (2001).
10. G. He, J. Eckert, W. Loser, and L. Schultz *Nature Materials* **2**, 33 (2003).
11. G. He, M. Hagiwara, J. Eckert, W. Loser *Phil. Mag. Lett.* 84, 365 (2004).
12. G. He, J. Eckert, Q. L. Dai, M. L. Sui, W. Löser, M. Hagiwara, E. Ma *Biomaterials* 24, 5115 (2003).
13. C. C. Hays, C. P. Kim, and W. L. Johnson *Phys. Rev. Lett* **84**, 2901 (2000).
14. D. V. Louzguine, H. Kato and A. Inoue *J. Alloys Comp*, 2004., **384**, L1 (2004).
15. D. V. Louzguine, H. Kato and A. Inoue *J. of Met. and Nanocr. Mater.*, 2004, in press.
16. D. V. Louzguine, H. Kato and A. Inoue *J. Alloys Comp.* **375**, 171 (2004).

\

Mater. Res. Soc. Symp. Proc. Vol. 842 © 2005 Materials Research Society   S7.2

TEM characterisation of fatigue tested lamellar Ti-44Al-8Nb-1B

H Jiang, D Hu* and I P Jones
Department of Metallurgy and Materials and *IRC in Materials, The University of Birmingham, Edgbaston, B15 2TT UK

ABSTRACT

Detailed microstructural examination by TEM of fine-grained polycrystalline lamellar Ti-44Al-8Nb-1B after fatigue testing (R=0.1) at room temperature has been carried out. The operative slip systems were identified as 1/2<110> ordinary dislocations and 1/6<112> twinning. The results showed no strong relation between operative slip systems and macroscopic Schmid factor and it is believed the local stress conditions control the operation of the slip systems. Translamellar cracking was observed to be associated with fine transverse twins in the gamma lamellae.

INTRODUCTION

Gamma (TiAl)-based titanium aluminides have received wide attention over the past two decades as possible high temperature aerospace materials due to their low density, excellent high temperature strength and oxidation resistance [1]. The fatigue behavior of TiAl-based alloys has been investigated extensively, mainly focused on the property-microstructure relationship. Many investigations of the fatigue crack propagation of γ-TiAl alloys [2-4] have shown that the fully lamellar microstructure presents superior crack growth resistance as compared to the other possible microstructures. It has been shown that deformation under fatigue conditions at room temperature is through the glide of mainly ordinary dislocations and via twinning in the TiAl phase [7-10]. Some efforts have been directed to relating the defect characteristics to the macro-stress condition in PST TiAl crystals because they offer simplicity in controlling the stress orientation [5]. However, such efforts have not been extended sufficiently into polycrystalline lamellar microstructures, although some work on local deformation behavior near cracks has been reported [6]. The purpose of this paper is to report a detailed TEM study on fatigue tested polycrystalline fine-grained lamellar Ti-44Al-8Nb-1B in an attempt to provide a general picture of the defects' nature, density distribution and relation to the crystal orientations.

EXPERIMENTAL PROCEDURE

The alloy, with a nominal composition of Ti-44Al-8Nb-1B (at.%), was prepared by double plasma cold hearth arc melting and cast into a 100mm diameter ingot weighing about 25kg. The oxygen content was about 700 wtppm. A section of the ingot was isothermally forged at 1150°C at a strain rate of $5x10^{-3}$ s^{-1} to a reduction in height of about 70%. The forged alloy was heat treated at 1310°C for 2 hours followed by furnace cooling to room temperature. The heat treatment gave rise to a near fully lamellar microstructure with about 12% grain boundary gamma grains. As-machined cylindrical samples with a diameter of 4mm and a gauge length of 20mm were subjected to fatigue testing at 500MPa (80% of the 0.2% proof stress) in air with a

stress ratio R=0.1 at 82Hz.
The volume fraction of gamma grains was examined by optical microscope (OM) and the fracture surfaces were analysed by JEOL 6300 scanning electron microscope (SEM). A Philips CM20 transmission electron microscope (TEM) with a double-tilt goniometer stage was used to investigate the dislocation configuration and slip systems. Samples for TEM were prepared from slices of 0.5mm thickness which were cut perpendicular to the specimen axis. The slices were mechanically ground from both sides to a thickness of 200μm and subsequently twin-jet electropolished at 26V and a temperature of -30°C, using a solution of 5% perchloric acid, 35% butan-1-ol and 60% methanol.

RESULTS AND DISCUSSION

Starting microstructure and failure modes

The mainly lamellar microstructure of Ti-44Al-8Nb-1B is shown in Fig. 1. The average lamellar colony size is about 70μm and that of the grain boundary gamma grains is about 10μm. The volume fraction of the gamma grains is about 12%. The gamma grains were formed through coarsening at grain boundaries which is very common in lamellar TiAl alloys when subjected to slow cooling. Figure 1b show the TEM microstructure of a lamellar area before testing. The lamellar grains consist of regular γ and α_2 laths, with the orientation relationship between γ and α_2 being expressed as $\{111\}_\gamma$ // $(0001)_{\alpha2}$ and $<110>_\gamma$ // $<11\bar{2}0>\alpha_2$. The α_2 platelets are long and straight and quite regularly spaced between the γ lamellae.

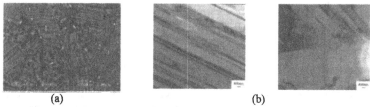

| (a) | (b) |

Figure 1 Scanning electron micrograph of NFL microstructure Ti44Al8Nb1B (a) and many beam transmission electron micrographs taken from undeformed samples of Ti44Al8Nb1B (b)

Figure 2 shows the fracture surface of a testpiece failed at 8.6×10^6 cycles. No obvious crack initiation site on the fracture surface was observed and both translamellar failure and interlamellar failure of lamellar colonies may clearly been seen.

Figure 2 SEM images of 4481 fracture surface with different magnifications a) Translamellar, b) Interlamellar regions.

Slip systems

The Burgers vectors of the dislocations were determined using the (g·b) criterion. It is revealed that the predominant deformation mechanisms are by 1/2<110> dislocations and 1/6<11$\bar{2}$> twinning. Figure 3 shows the dislocation characteristics in a coarse gamma lamella at a grain boundary in a foil taken away from the fracture surface. The majority of the dislocations seen are single dislocations which are visible with g=020, as seen in figure 3 (d). Dislocations marked A are visible with g=11$\bar{1}$ (as seen in fig 3.a) but invisible with g=002 and g=1$\bar{1}$1 (as seen in fig 3. b, c), and can thus be identified to have the Burgers vector 1/2[110], slipping on (1$\bar{1}$1) (as seen in fig 3.c). Dislocations marked B and C can be identified to have Burgers vector 1/2[112] and 1/2[1$\bar{1}$0]. So 1/2<110> ordinary dislocation and 1/2<112> superdislocation were found in the fatigued alloy. However, the 1/2[112] dislocations are not likely to result from fatigue deformation. They might be the dislocations left there by coalescence of two neighbouring gamma lamellae during coarsening as indicated by fig..3d in which their glide plane is edge-on. Twining was also observed in fatigue tested samples and an example is given in fig.3f. Dislocations with slip planes parallel to the lamellar interfaces can glide through the full lamellar length for there are only weak barriers (APDBs) in their paths. However, those gliding across the lamellae will be blocked by the lamellar interfaces as for the case of dislocations A in fig.3a where the dislocation pile-up at the interface is evident.

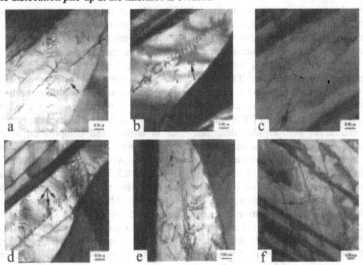

Figure 3 Burgers vector analysis of the dislocations (a,b,c,d,e) and twins (f) in fatigued alloy

Dislocation distribution

Deformation induced dislocations are not evenly distributed throughout the fatigue deformed microstructure. Dislocation density varies significantly between gamma lamellae. There are two factors affecting the dislocation density in gamma lamellae: lamellar width and orientation.

Figure 4a and 4b show two gamma lamellae in the sample foil of which one has a dislocation density of about 10^9 (/cm^2) and the other has a dislocation density of about 10^7(/cm^2) only. Dislocation density varies much less significantly between gamma grains and most of the gamma grains have a high dislocation density. Figure 4c shows a gamma grain with a high density of dislocations.

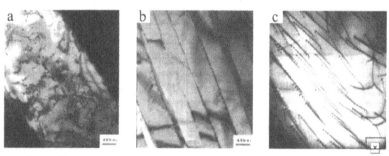

Figure 4 Dislocation distribution in fatigue test sample (a) γ lamellae with high dislocation density; (b) γ lamellae free from dislocations; (c) γ grain with high dislocation density;

The observed active slip systems including 1/2<110> ordinary dislocations and 1/6<112> twins, were correlated to the macroscopic stress condition. The microscopic Schmid factors of the operative slip systems and possible slip systems in the observed gamma lamellae and gamma grains were calculated and are summarised in Table 1. It seems that the macroscopic Schmid factor is not the only controlling factor of the operative slip system. Some observed operative slip systems have very low Schmidt factor whilst other possible slip systems were not observed although they have high macroscopic Schmid factor. This is entirely different from early work on PST crystals. Yashida et al [11] reported a good correlation between operative slip systems and macroscopic Schmid factor in PST crystals. They observed ordinary dislocations, twinning and superdislocations in PST crystals deformed both in tension and compression at different orientations over a wide range of temperatures. For comparison, only ordinary dislocations and twinning from that work are accounted. Out of about 90 observed slip systems, half of them had Schmid factor above 0.3 and 1/3 of them were between 0.2-0.3. Most of the operative slip systems with Schmid factors below 0.2 were from compression along the normal to the lamellar interface, which is the hardest orientation. More importantly, almost no slip systems with high Schmid factor failed to be operative. Thus, it is clear that the macroscopic Schmid factor is the controlling factor for the operation of slip systems in PST crystals during deformation, whilst this is not the case in polycrystalline lamellar Ti-44Al-8Nb-1B subjected to fatigue deformation. Thus in fine-grained polycrystalline lamellar TiAl alloys it is the local stress conditions which determine the local Schmid factor, which are the controlling factor

<u>Twinning around cracks</u>

Crack propagation within lamellar colonies occurs in both interlamellar and translamellar manners. The former is actually a debonding process along the lamellar interfaces and the latter involves cracks advancing across gamma lamellae. Figure 5 shows a crack in a foil taken near

Table 1 Schmid factor analysis in fatigued Ti44Al8Nb1B with near fully lamellar structure.
T in the table indicate twinning

	Slip system	Tensile axis	Macroscopic Schmid factor	Max Schmid factor
Lamellae with high dislocation density	1/2[1T0](11T)	[132]	0.12	0.47
	1/2[1T0](11T)	[314]	0	0.38
	1/2[1T0](11T)	[121]	0.14	0.41
	1/2[1T0](111)	[631]	0.27	0.32
				0.45$_{(T)}$
	1/2[110](1T1)	[112]	0.27	0.31$_{(T)}$
	1/2[1T0](111)	[T34]	0.38	
	1/2[110](T11)	[032]	0.47	
Lamellae with low dislocation density		[134]		0.38
		[635]		0.42
		[122]		0.41
		[110]		0.47$_{(T)}$
γ-grains with high dislocation density	1/2[110](T11)	[133]	0.43	
	1/2[1T0](111)	[133]	0.30	
	1/2[1T0](11T)	[25T]	0.33	
γ-grains without dislocations		[2T5]		0.44$_{(T)}$
Twins	1/6[112](11T)	[131]	0.39	
	1/6[112](11T)	[543]	0.42	
	1/6[112](11T)	[321]	0.47	
	1/6[112](11T)	[332]	0.43	

the fracture surface. This crack propagated along lamellar interfaces at locations P and across gamma lamellae at locations C. TEM examination shows that fine twins are often present at the crossing locations in the vicinity of the crack, but less often twins were observed far away from the crack in the same lamellar area.

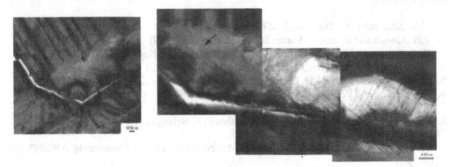

Figure 5 Mechanical twinning close to the crack

The twins in fig. 5 are identified to be of $1/6[\bar{1}12](1\bar{1}1)$ type. The external tensile axis can be identified to be $[\bar{3}\,\bar{1}\,2]$. The macroscopic Schmid factor for the twins is nearly 0, which means it is impossible to generate twins here responding to the macroscopic tensile stress. The generation of twins may have been caused by the stress concentration at the end of the crack, and that is why we only can find twins close to the crack instead of away from the crack. Such a phenomenon was observed in previous work on fatigue tested lamellar Ti-46Al-5Nb-1W [12]. The twin interface is one of the γ/γ interfaces which, like other γ/γ interfaces, acts as a preferential crack advance path. Twins around the cracks can be present long before the arrival of the crack, such as those observed in the foils away from the fracture surface. They can also be induced in front of the advancing cracks. In this case, twins across gamma lamellae are localized to the vicinity of the cracks.

CONCLUSIONS

1. The operative slip systems in fine-grained near lamellar Ti-44Al-8Nb-1B after testing at room temperature with a maximum stress of 500MPa and a stress ratio of 0.1 at 82Hz involve 1/2<110] ordinary dislocations and 1/6<112] twins
2. No obvious relation between macroscopic Schmid factor and operative slip systems has been found.
3. The operation of slip systems is likely to be controlled by local stress conditions.

ACKNOWLEDGEMENTS

This work was supported by EPSRC under GR/R56273/01(p) and by Rolls-Royce; this support is gratefully acknowledged. HJ is thankful to Professor M.H. Loretto, Dr. X. Wu and Dr. H.S. Jiao for supplying the TiAl material and for their helpful discussions.

REFERENCES

1. Y.W. Kim and D.M. Dimiduk, J. of Metals 43, 40 (1991)
2. P.B. Aswath and S. Suresh, Mater. Sci. Eng. A114, L5 (1989)
3. W.O. Soboyejo, J.E. Deffeyes and P.B. Aswath, Mater. Sci. Eng. A138, 95 (1991)
4. A.W. James and P. Bowen, Mater. Sci. Eng. A153, 486 (1992)
5. H. Inui, A. Nakamura, M.H. Oh and M. Yamaguchi, Acta Metal. Mater. 40, 3095 (1992)
6. Z.W. Huang and P. Bowen, Acta Mater 47, 3189 (1999)
7. S.M.L. Sastry and H.A. Lipsitt, Met Trans A, 8A, 299 (1977)
8. A.L. Gloanec, G. Hena, M. Jouiad , D. Bertheau , P. Belaygue , M. Grange, Scripta Met., 52, 107 (2005)
9. Y. Umakoshi, H.Y. Yasuda, T. Nakano, Materials Science and Engineering A192/193 511 (1995)
10. T. Nakano, H.Y. Yasuda, N. Higashitanakat and Y. Umakoshi, Acta Mater. 45 4807 (1997)
11. Yashida, Inui, Yamaguchi, Phil Mag (1998)
12. X.Y. Li, X. Wu, unpublished work

Mater. Res. Soc. Symp. Proc. Vol. 842 © 2005 Materials Research Society   S6.11

On the Effects of Interstitial Elements on Microstructure and Properties of Ternary and Quaternary TiAl based alloys

Jean-Pierre Chevalier[1,2], Mélanie Lamirand[1] and Jean-Louis Bonnentien[1]
[1] Centre d'Etudes de Chimie Métallurgique,
Centre National de la Recherche Scientifique,
15, rue Georges Urbain,
F 94407 Vitry cedex
[2] also at Chaire des Matériaux Industriels Métalliques et Céramiques,
Conservatoire National des Arts et Métiers,
2, rue Conté
F 75003 Paris

ABSTRACT

Ti-Al-Cr ternary and Ti-Al-Cr-Nb quaternary alloys have been studied as a function of initial purity and added interstitial content. Using strict clean processing together with either ultra high purity or commercial purity alloys, the effects of interstitial elements (essentially O, but also C and N) on microstructure and hardness, yield stress and fracture strain have been studied for both fully lamellar microstructures and duplex microstructures. The results are clear and similar trends are observed : as long as they do not precipitate, these stabilise the lamellar microstructure and affect the kinetics of the α-γ phase tranformation, leading to a higher than equilibrium value for the γ_2 phase for continuous cooling. Both the lamellar spacing and the γ_2 phase fraction correlate with increased hardness and yield stress, and also with decreasing fracture strain. The effects are significant.

INTRODUCTION

It is well known that titanium and its alloys can have notable levels of interstitial elements in solid solution [1]. Furthermore, the alloying reaction for the formation of Ti-Al based alloys is strongly exothermic, leading to a further increase in oxygen content of the resulting alloy via contamination of the molten alloy. For example the oxygen level of alloys studied are frequently around 1000 wt ppm. The effects of interstitial elements on phase stability, microstructrure and properties have been studied previously (e.g. [2, 3]), but notably less than the effects of substitutional alloying elements, probably due to the experimental difficulties of very clean processing, and use of high purity starting materials.

Our previous work on the effect of O, N and C content on binary Ti-Al alloys showed clear trends [4]. After continuous cooling leading to fully lamellar microstructures, the volume fraction of γ_2 phase (Fvγ_2) increased as a function of interstitial content, up to interstitial concentrations where precipitation of oxides, nitrides or carbides could be detected. This led to a larger number of γ_2 lamellas, with a concurrent decrease in the interlamellar spacing (i.e. the spacing corresponding to the different variants of γ lamellas). In terms of mechanical properties, hardness increased with Fvγ_2 and the decrease in γ_2 lamellar spacing.

The aim of the study summarised here was to ascertain :

- whether the addition of ternary (Cr) or quaternary (Cr and Nb) substitutional elements alters the trends previously observed ;
- what is the effect O, C and N on fracture strain (taken as an indication of ductility) ;
- whether the effect of interstitial elements on microstructure is due to modifications of phase equilibria or to kinetic effects during the $\alpha - \gamma$ phase transformation.

EXPERIMENTAL PROCEDURES

Alloys were prepared by high frequency induction levitation melting (containerless) under pure helium atmosphere, monitored by gas chromatography, and cast in a cooled copper mould [4, 5]. These ingots were remelted in a floating zone induction furnace under pure helium and cylindrical bars (~15 mm in diameter and ~150 mm long) were obtained. The high purity $Ti_{50}Al_{48}Cr_2$ and $Ti_{48}Al_{48}Cr_2Nb_2$ alloys were prepared from Van Arkel (iodide process) titanium [6], pure aluminium, industrial chromium purified by zone melting and pure commercial niobium. Alloys containing from 150 to 185 wt. ppm oxygen are obtained. The other alloys were obtained from an initial alloy containing ~ 1000 wt. ppm oxygen and provided by Snecma. O, N and C were then introduced prior to remelting as TiO_2, TiN and TiC.

Heat treatments were carried out also using HF induction heating under a controlled gas atmosphere, with self-supporting specimens. Two heat treatments have been studied here, aimed at producing either fully lamellar or duplex microstructures. After heating for a few minutes at $1350 \pm 20°C$, just above the γ transus, the first treatment consisted in cooling to room temperature at a constant rate of 35°C/min, whereas the second, in cooling to 1150°C followed by holding for five hours and then in cooling to room temperature at 35°C/min.

Back scattered electron images were obtained using a LEO 1530 field emission gun scanning electron microscope. These were used to obtain the γ_2 and γ volume fractions, the interlamellar spacing γ and the γ_2 lamellar thickness using images analysis. After etching, the microstructures were also observed with a Reichert metallographical microscope.

Tensile test specimens with a cylindrical shape and a 10 mm gauge length (4 mm in diameter) were machined from the samples. Given the low ductility of these alloys, care was taken (by e.g. surface polishing and self-aligning mountings) to reduce extraneous causes of fracture. The experimental procedures are decribed in full detail elsewhere [5].

RESULTS :

Three aspects have been chosen to summarise the results. The first two concern the effects of interstitial elements on the fully lamellar microstructures for ternary and quaternary alloys and the third is a comparison of volume fraction of γ phase for fully lamellar and duplex microstructure for the quaternary alloy.

The choice of heat treatment, or essentially that of cooling rate from the γ phase, to produce the fully lamellar microstructure, was made on the basis of previous work on microstructures obtained as a function of cooling rate for binary alloys [7]. The value of $35° min^{-1}$ which has been used here allows comparisons to be made, but may no longer be necessarily optimised for ternary and quaternary alloys.

$Ti_{50}Al_{48}Cr_2$ and continuous cooling

For the high purity alloy (figure 1a), the microstructure is essentially duplex. Comparing this observation with those for binary high purity alloys where near fully lamellar microstructures were obtained for the same conditions, Cr addition has had a marked effect. However this effect is rapidly mitigated by added interstitial O, and for 1000 wt ppm a nearly fully lamellar microstructure is obtained (figure 1b). With more O the microstructure becomes fully lamellar. N and C additions lead to similar results, but furthermore reduce grain size [8, 9, 10] : the grain size with respect to an alloy with 1000 wt ppm O is reduced by a factor of 3 for 5000 wt ppm added N and by a factor of 5 for 5000 wt ppm added C.

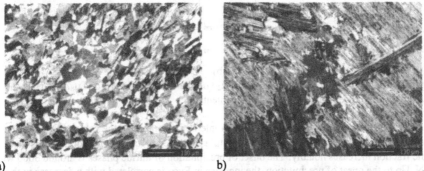

a)   b)

Figure 1. Optical micrographs after treatment at 1350°C and continuous cooling at 35° min^{-1} for $Ti_{50}Al_{48}Cr_2$: a) "pure" , b) 1000 wt. ppm oxygen

$Ti_{48}Al_{48}Cr_2Nb_2$ and continuous cooling

For the pure alloy (figure 2a), the addition of Nb can be seen to have increased the amount of fully lamellar grains, although the microstructure is still duplex. It can be seen

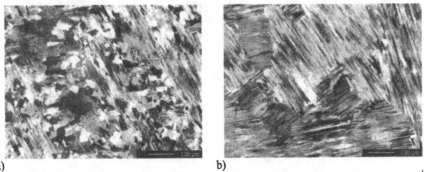

a)   b)

Figure 2. Optical micrographs after treatment at 1350°C and continuous cooling at 35° min^{-1} for $Ti_{48}Al_{48}Cr_2Nb_2$: a) "pure" , b) 1000 wt ppm oxygen

that 1000 wt ppm added O leads to a fully lamellar microstructure (figure 2b), and further additions simply increase the value of Fva_2 from ~9% for 1000 wt ppm O to ~ 18% for 4000 wt ppm O (figure 3a).

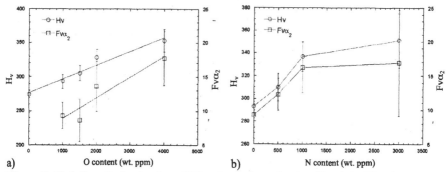

a)   O content (wt. ppm)   b)   N content (wt. ppm)

Figure 3. Variation of microhardness (H_v) and α_2 volume fraction ($Fv\alpha_2$) for $Ti_{48}Al_{48}Cr_2Nb_2$ as a function of : a) oxygen content, b) nitrogen content

Figure 3 shows the effects of the O and N interstitial content on the value of $Fv\alpha_2$ and it can be seen that for the fully lamellar microstructures obtained by continuous cooling, this volume fraction increases notably. The onset of precipitation of nitride phases is around 3000 wt ppm N. Up to the onset of precipitation, the increase in $Fv\alpha_2$ is correlated with a decrease in the spacing between α lamellas, and a corresponding increase in hardness. Yield stress increases with hardness. Similar results are obtained for C additions.

Figure 4 shows tha variation of fracture strain (which is here considered as a means of assessing ductility) and this decreases very markedly with O content. Similar effects are observed for added N and C. It should be noted that N and C are added to a base alloy already containing 1000 wt ppm O. Thus the added interstitial elements do harden the alloys, but appear to be very detrimental to ductility.

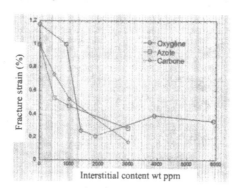

Figure 4. Fracture strain as a function of interstitial content.

Ti₄₈Al₄₈Cr₂Nb₂ heat treated in the α+γ phase field

For heat treatments in the γ +γ phase field, duplex microstructures are obtained. Measuring accurately the volume fractions of the γ and γ phases is difficult because of the large differences in scale between the large γ grains and the fine lamellas. Hence Fvγ is overestimated for duplex microstructures with the most γ grains (i.e. for pure and low O content alloys). The results are summarised in figure 5. The striking effect is that Fvγ is clearly higher, and with a

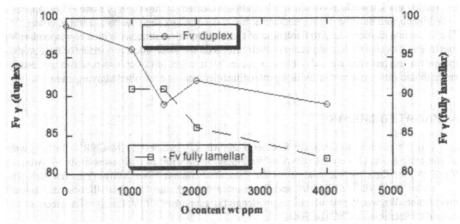

Figure 5. Variation in Fvγ as a function of O content for duplex and fully lamellar microstructures

weaker dependence on O content, for the duplex microstructure, than for the fully lamellar microstructure. It is therefore possible to suggest that phase equilibria has nearly been reached through the heat treatment leading to the duplex microstructure. *A contrario*, this suggests this has not been reached in the case of continuous cooling, and that the variation in Fvγ (and correspondingly Fvγ) for the fully lamellar microstructure is a kinetic effect.

DISCUSSION

The key to understanding why interstitial elements stabilise the lamallar microstructure is the large difference in solubility of these elements in the γ 2 and γ phase as demonstrated, notably for O, by atom probe field ion microscopy (APFIM) studies [11, 12]. The necessary partitioning of the interstitial elements during continuous cooling implies short diffusion lengths, and this when coupled with displacive/diffusive tranformation on well defined crystallographic planes will tend to favour thin lamellas.

The relation between interstitial content and hardness and yield stress for the fully lamellar microstructure can be rationalised in terms of both the increased γ 2 volume fraction (the harder phase, which will moreover be strengthened by interstitial solid solution effects) and the decreased spacing between γ 2 lamellas. To a certain extent the loss of ductility might be due to difficulties in strain accomodation between hardened lamellar grains which deform in an anisotropic manner.

CONCLUSIONS

The effect of interstitial elements on the fully lamellar microstructures of ternary and quaternary Ti-Al alloys is strong and similar to that previously observed in binary alloys. This can readily mask the effects of added substitutional elements. The origin of these effects appear to be kinetic, and thus the duplex microstructure should be less sensitive to interstitial content. The interstitial elements lead to hardening of the lamellar microstructure with a corresponding loss of ductility. No systematic study of their effects on the mechanical properties of the duplex microstructure has been carried out by our group, but it might be possible to reach an interesting strength/ductility compromise by using interstitial elements as controlled alloying elements.

ACKNOWLEDGMENTS

It is a pleasure to thank Dr. S. Guérin and Mr. C. Leroux (CECM-CNRS) for help with tensile testing, Dr. G. Ferrière (CNAM) for guidance with image analysis and Mr M. Armand (ONERA) for the chemical analysis of the interstitial content of the alloys. This research was funded by the CNRS, SNECMA and Turbomeca through the "Intermétalliques base titane" programme. We would also like to acknowledge the support of Dr. V. Pontikis for access to the laboratory facilities at CECM-CNRS.

REFERENCES

1   C. Ouchi, H. Iizumi and S. Mitao, *Mater. Sci. Eng. A*, **A243**, 186 (1998).
2   N. Saunders, *Gamma Titanium Aluminides 1999*, ed. Y.W. Kim, D.M. Dimiduk and M.H. Loretto (TMS, Warrendale, 1999), pp 183-188.
3   D.E. Larsen, United State Patent n° 5,685,924 (Nov. 11, 1997).
4   F. Perdrix, PhD Thesis, Université Paris XI, Orsay (2000).
5   M. Lamirand, PhD Thesis, Université Paris XII, Créteil (2004).
6   J. Bigot, *C.R.Acad.Sc.Paris* , **279**, 6 (1974).
7   F. Perdrix, M. Cornet, J. Bigot and J-P. Chevalier, *Microstructure Design for improved Mechanical Behaviour of Advanced Materials*, ed. J. Petit, H. Takahashi, D. Blavette, N. Igata and O. Dimitrov (J. de Physique IV, Proceeding, Les Ulis, 2000) pp 15 – 20.
8   F. Perdrix, M-F. Trichet, J-L. Bonnentien, M. Cornet and J. Bigot, *Intermetallics*, **9**, 147 (2001).
9   F. Perdrix, M-F. Trichet, J-L. Bonnentien, M. Cornet and J. Bigot, *Intermetallics*, **9**, 807 (2001).
10   H.S.Cho, S.W. Nam, J.H. Yun and D.M. Lee, *Mater. Sci. Eng. A*, **A262**, 129 (1999).
11   A. Menand, A. Huguet, and A. Nerac-Partaix, *Acta mater.*, **12**, 4729 (1996).
12   A. Menand, H. Zapolsky-Tatarenko and A. Nerac-Partaix, *Mat. Sci. Eng. A*, **A250**, 55 (1998).

The Correlation of Slip between Adjacent Lamellae in TiAl.

G. Molénat[1], A. Couret[1], M. Sundararaman[2], J. B. Singh[2], G. Saada and P. Veyssière
LEM, CNRS-ONERA, Chatillon, France;
[1]Materials Science Division, Bhabha Atomic Research Centre, Mumbai, India;
[2]CEMES, CNRS, Toulouse, France.

ABSTRACT

In lamellar TiAl, concerted deformation occurs in adjacent true-twin related variants (O/OT). When deformation takes place in a given variant, O, by activation of the slip system with highest shear stress, slip proceeds in the companion variant, OT, by operation of a slip system of the same family in mirror orientation, including the slip direction. Depending on load orientation, the latter system may not necessarily exhibit the highest resolved shear stress. The correlation does not depend upon Burgers vector orientation with respect to the interface.

This property applies to ordinary dislocations and twinning. The variant that conforms the Schmid law is referred to as the *pilot* variant, the twin-related variant is dubbed *driven*.

INTRODUCTION

Processed under adequate thermal conditions [1] Ti-rich TiAl alloys exhibit a well-documented lamellar structure that consists of an arrangement of layers of γ-TiAl (L1$_0$) and hexagonal α_2-Ti$_3$Al phases such that the basal plane of the latter is in coherency with one {111} plane of the γ phase which is tetragonal (c/a = 1.02). Due to the peculiar atomic ordering in γ-TiAl, this results in 6 different orientation variants and several types of interfaces, of which only those separating true-twin related variants are perfectly coherent [2]. A fully lamellar grain is often dominated by one particular orientation together with its twin-related variant whose density is, however, much less [3].

Slip transmission is hampered by geometrically necessary dislocations at interfaces and by the fact that unit translations do not match between two contiguous variants unless these are true-twin-related. Perfect Burgers vectors in one variant may become partial unit translations in the adjacent variant. As an example, 1/2<110], the Burgers vector of the so-called ordinary dislocations (which are perfect) may become 1/2<011] in an adjacent not true-twin-related variant, defining a superpartial submitted to a back force from an antiphase boundary (APB). By convention the <hkl] notation generates all permutations between ±h and ±k, with ±l ascribed the third position.

Polycrystals are inherently poorly ductile. In single-grained polysynthetically twinned crystals (PST) materials, whereas strains up to 20 % can be achieved before fracture when slip is parallel to the interfaces [4, 5], this is no longer the case when slip planes intersect the variant interfaces in which case samples undergo reduced strain and fail. Difficulties in slip transmission between adjacent variants are commonly regarded as being at the origin of this behavior.

The present experimental investigation address this question. It is shown that correlations between a slip system activated in a given variant and that activated in the neighbour variant should be discussed not only in terms of the Schmid factors applied to the two systems but also as a function of the orientation relationship between the two variants.

EXPERIMENTAL PROCEDURE

Arc melted buttons of Ti-47Al-1Cr-0.2Si (at.%) were prepared in the Materials Science Department at ONERA, Châtillon, France. After homogenization during 3 hours at 1400°C in a platinum furnace under flowing argon, the alloy was furnace-cooled down to room temperature; between 1400 and 900°C the cooling rate was about 15°C/min. The final average grain size was about 3 mm.

Samples (about 5x5x10 mm^3) were compression tested at 25°C and 600°C at a strain rate of 3.33 10^{-4} s^{-1} to about 2% of permanent strain. The 0.2% offset yield stresses were 335 and 260 MPa, respectively. The fact that grain size scales with sample dimension has resulted in significant scatter but the general trend is that the yield stress is the largest at 25°C, that is, we do not find an anomalous flow stress dependence upon temperature. Special care was taken to keep track of the compression axis later in the course of TEM investigations. Several foils have been investigated for each temperature.

The microstructure of the undeformed alloy has been statistically characterized in two samples. As an example, Table 1 accounts for the arrangement of 72 contiguous lamellae comprising 52 γ and 20 α_2 lamellae with average widths of about 340 nm and 115 nm, respectively. γ-lamellae are referred to as O1, O2, O3, OT1, OT2 and OT3 such that O1, O2 and O3 are in twin orientation with OT1, OT2 and OT3, respectively. Consistent with a previous result [3], Table 1 shows that one orientation prevails (here O3) together with its twin (OT3). In fact, it would be pure coincidence that the dominant lamellae and twin-related variant in one sample be the same in another sample, this is since variant name depends on a convention made arbitrarily upon starting a TEM session.

Table 1. Variant arrangement in the two-phase microstructure of an undeformed sample.

Volume Fractions					
α_2	γ Matrix		γ Twins		
0.12	0.76		0.12		
	O1	O2	O3	OT1	OT2	OT3
	0.11	0.07	0.83	0.01	0.30	0.68
The various γ/α_2 and γ/γ interfaces (%)					
γ/α_2 interfaces	γ/γ interfaces				
55	45				
	Twin	Pseudo-twin	Others		
	40	20	40		

MICROSTRUCTURAL ANALYSIS

The microstructural properties of interest here are essentially the same whether the samples were deformed at room temperature or at 600°C. For the sake of clarity within the limited space allowed, we report TEM observations conducted on samples deformed at 600°C (Figure 1) and their analysis in samples strained at room temperature (Table 2). The full analysis of deformation properties is the object of a paper submitted elsewhere.

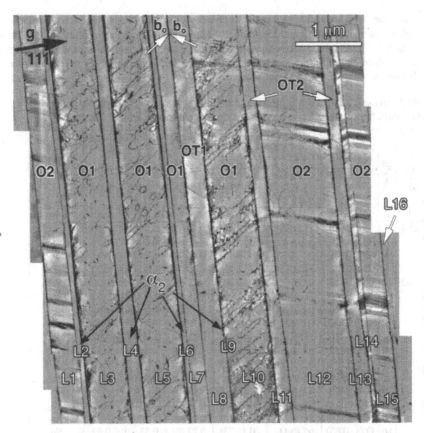

Figure 1. An area about 55 μm^2 showing organization in 16 lamellae (Table 1). The micrographs, taken under weak-beam conditions, are printed in the negative mode for better visibility. b_o, the projection of the Burgers vectors on the plane of observation, is common to every lamella where ordinary slip is activated. In spite of a significantly lesser contrast, screw ordinary dislocations similar to those in L3, L5 and L10 can be seen in the upper part of L7 and L8. A very thin α_2 lamella (L9) separates the twin-related γ-lamellae L8 (OT1) and L10 (O1).

The montage in Figure 1 indicates that two types of deformation microstructures are to be distinguished. The first type, found in lamellae L3, L5, L7 and L10, all with orientation O1, and lamella L8 (orientation OT1), is comprised of ordinary dislocations with one Burgers vector only. In the second type, represented by L1, L11, L12, L14 and L15 (O2 and OT2) twins predominate. They coexist with a still significant amount of ordinary dislocations. As expected, α_2 lamellae (L2, L4, L6, L9, L13, L16) exhibit few dislocations.

In L3, L5, L7, L8 and L10, all dislocations have a 1/2[110] Burgers vector and exhibit analogous properties. Ordinary dislocations whose Burgers vector is inclined to the interface plane was the most frequent situation - some cases of a Burgers vector parallel to the interface were observed in another sample. The frequent observation of loops anchored at γ/α_2 interfaces (marked "loop" in Figure 1) is consistent with the emission of ordinary dislocation taking place at γ/α_2 interfaces. This property holds true whether the Burgers vector of ordinary dislocations is parallel or not to the lamellae interfaces.

Table 2. Analysis of a sequence of variants in terms of the crystallographic orientation relationships between neighbours and of the slip systems activated (SF : Schmid factor).

Sample id	Lamella No.	Lamellae OR	OR of Bounding Lamellae		SF of operating system	Highest SF available		Schmid Law
		OR	L.H.S.	R.H.S.	ASFO	HSFO	HSFT	
1	2	3	4	5	6	7	8	9
25-1	1	O1 (0.70)	OT2 (0.38)	α2 (0.42)	0.49 (0)	0.49 (0)	0.21	Yes
	2	O1 (1.17)	α2 (0.42)	OT1 (0.11)	0.49 (0)	0.49 (0)	0.21	Yes
	3	OT1 (0.11)	O1 (1.17)	O1 (0.88)	0.44 (0)	0.44 (0)	0.21	Yes
	4	O1 (0.88)	OT2 (0.38)	α2 (0.42)	0.49 (0)	0.49 (0)	0.21	Yes
25-3	5	OT1 (0.12)	α2 (0.25)	O1 (0.60)	0.22 (0)	0.33 (0)	0.32	*No*
	6	O1 (0.60)	OT1 (0.12)	α2 (0.05)	0.43 (0)	0.45 (60)	0.32	Yes
	7	O1 (0.85)	α2 (0.05)	OT1 (0.22)	0.43 (0)	0.45 (60)	0.32	Yes
	8	OT1 (0.22)	O1 (0.85)	O1 (1.12)	0.22 (0)	0.33 (0)	0.32	*No*
	9	O1 (1.10)	OT1 (0.22)	α2 (0.15)	0.43 (0)	0.45 (60)	0.32	Yes
	10	OT2 (0.30)	α2 (0.15)	α2 (0.20)	0.30 (60)	0.44 (0)	0.32	*No*
	11	O1 (1.55)	α2 (0.05)	OT1 (0.25)	0.43 (0)	0.45 (60)	0.32	Yes
	12	OT1 (0.25)	O1 (1.55)	α2 (0.18)	0.22 (0)	0.33 (0)	0.32	*No*
	13	O1 (0.38)	α2 (0.18)	OT1 (1.45)	0.43 (0)	0.45 (60)	0.32	Yes
	14	OT1 (1.45)	O1 (0.38)	O1 (0.40)	0.22 (0)	0.33 (0)	0.32	*No*
	15	O1 (0.40)	OT1 (1.45)	α2 (0.15)	0.43 (0)	0.45 (60)	0.32	Yes
	16	OT1 (0.85)	α2 (0.11)	O1 (0.75)	0.22 (0)	0.33 (0)	0.32	*No*
	17	O1 (0.75)	OT1 (0.85)	α2 (0.11)	0.43 (0)	0.45 (60)	0.32	Yes
	18	OT1 (0.65)	α2 (0.11)	O1 (1.30)	0.22 (0)	0.33 (0)	0.32	*No*
	19	O1 (1.30)	OT1 (0.65)	α2 (0.30)	0.43 (0)	0.45 (60)	0.32	Yes

At 600°C twinning is the dominant deformation mode in certain lamellae as L1, L11, L12, L14 and L15 (Figure 1). The twins lie on the ($\bar{1}11$) planes of the lamellae with O2 and OT2 orientations. They are true twins in nature. Twins nucleating at the L13/L12 γ/α_2 interface together with some necessary ordinary dislocations are arranged as if they originated from internal stresses produced by twinning dislocations piled-up in the vicinity of the next L14/L13 interface. On the other hand, though shear has transferred through the α_2 (L13) lamella, hardly any dislocation contrast can be observed within this lamella suggesting that the α_2 phase has elastically transmitted the shear.

Table 2 shows the role of (i) the applied stress, (ii) lamella width, and (iii) lamellae environment at 25°C. For both deformation temperatures, a number of lamellae (column 2) taken from different samples (column 1) have been analysed. Column 3 indicates the orientation of the lamella with its width in parenthesis expressed in μm. Columns 4 and 5 provide the same information for the left-hand side (LHS) and right-hand side (RHS) lamellae. Column 6 the Schmid factor for the active ordinary slip system (ASFO). Columns 7 and 8 indicate the highest Schmid factor for ordinary dislocations (HSFO) and for twins (HSFT), respectively, within each lamella. The angle of the Burgers vector of the operating ordinary dislocations to the interface plane (0°) or (60°) is given in columns 6 and 7. Finally, column 9 indicates whether the Schmid law applies ('Yes', the ASFO is equal to the HSFO) or not ('No'). Experimental uncertainties imply that the answer be also "Yes" when the ASFO is within - 0.05 from the HSFO.

Lamellae with a given orientation all deform by activation of a unique slip system, either of ordinary dislocations or twins (although the presence of ordinary dislocations in this latter case is dictated by reaction occurring at the interface)
In sample 25-1 (Table 2), deformation in all the lamellae listed obeys the Schmid law. This is not true in sample 25-3, where the two twin-related lamellae O1 and OT1 deform by ordinary dislocations on mirror slip systems. The Schmid law is obeyed in O1, however, not in OT1 (Table 2). This observation is important for it eliminates the width and the environment of a lamella as pertinent parameters in controlling the slip system activation. Table 2 thus shows that only one Burgers vector operates in twin-related adjacent variants. We have repeatedly checked that when in a variant Oi the ASFO is equal to the HSFO (answer for Schmid law is "Yes"), then the corresponding ASFO in the conjugate OTi variant may (as in sample 25-1) or may not (sample 25-3) be equal to the HSFO. This property is independent of the orientation of the incident Burgers vector with respect to the interface plane. It is also independent of the volume fraction of the different variants. It is unchanged for lamellae separated by a thin layer of α2.

DISCUSSION AND CONCLUSION

The present observations are consistent with the following scenario. Sources of ordinary dislocations are activated within lamellae Oi whose orientation provides an ASFO (i) equal to the HSFO and (ii) higher than the HSFO of the OTi orientation. In other words, the operation of sources of ordinary dislocations is controlled by the applied stress in Oi, the orientation in which the CRSS is the highest due to a highest Schmid factor. Operating Burgers vector may either be inclined to or parallel to the interface plane. Dislocations generated in Oi subsequently invade lamellae OTi. The Oi orientation that is capable of imposing the operation of an unfavourable ordinary system in the OTi orientation is regarded as the *pilot* orientation whereas the OTi orientation itself is referred to as the *driven* orientation.

Figure 1 shows this scenario taking place, in a sample deformed at 600°C, between lamellae L7 and L10 (orientation O1) on the one hand, and lamella L8 (orientation OT1) on the other hand. L7 and L10 are pilot variants (ASFO = HSFO = 0.49, note that L10 is separated from L8 by an α2 layer). L8 is the driven orientation (ASFO = 0.28 when HSFO = 0.39 is available).

Ordinary dislocations having a Burgers vector parallel to the interface transmit slip across interfaces by a simple cross-slip process. Hence no problem is expected for such a transmission from the pilot orientation to the driven orientation. However, the transmission of an ordinary dislocation whose Burgers vector is at 60° from the interface (e.g. 7 to 9 and 11 to 19 in sample 25-3) is not as easy. Zghal et al. [3] actually reported in situ observations of such dislocations impacting an interface between twin-related variants. No transfer at all occurs in this case. In the vicinity of the interface and within the OTi lamella, these authors observed one case of the activation of a source of ordinary dislocations whose Burgers vector lying at 60° from the interface, which they interpreted as resulting from the built-up of internal stresses by accumulation of incident dislocations.

Situations of a violation of the Schmid law as that exhibited by lamella 10 in sample 25-3 can also be encountered. Here the other set of ordinary dislocations is activated within lamella. Clearly, Burgers vector transmission from the pilot to the driven orientation cannot apply since the OT2 orientation is pseudo-twin related to the O1 pilot orientation. That a slip system with less favoured Schmid factor is activated in this case is likely to be governed by internal stresses.

The notion of pilot and driven variants apply to twinning in the same way. An additional property is that twinning is activated in the driven variant even though the resolved shear stress is smaller than the critical resolved shear stress for ordinary slip. Finally, a property common to ordinary slip and twinning is that the operating systems are in mirror orientations in the pilot and driven variants and that includes the slip directions. In particular, the orientation of the <110] slip directions (parallel or inclined) is conserved so that most of the incident shear is transferred to the adjacent lamella. For twins, the stain component normal to the interface is transmitted.

Two studies differing from one another with regards to load orientation and deformation modes, elaborate on slip transmission in PST TiAl in terms of the continuity of macroscopic strain across the various interfaces [6, 7]. By consideration of various sample shapes resulting from monotonous tension or compression straining, Kishida et al. [7] conclude that the condition that strain be continuous is essential in determining the operative slip systems and the relative activities of these. In fatigued lamellar TiAl, Nakano et al. [6] point out a violation of the Schmid law at Oi/OTj and Oi/Oj (i ≠ j) semicoherent interfaces (thus different from the cases considered here) which they also interpret in terms of strain continuity. It is rather clear that the present pilot/driven effect proceeds from the same type of reasoning on strain continuity.

Finally, it is interesting to note that while the critical shear stress for ordinary dislocations in lamellar TiAl decreases with increasing temperature, ordinary dislocations exhibit exactly the same pinning behaviour (the pinning is increasingly active with increasing temperature) as in Al-rich alloys which show flow stress anomaly.

ACKNOWLEDGEMENTS

The authors are grateful to the Indo-French Centre for the Promotion of Advanced Research India) for sponsoring this project (No. 2308-3) and for funding the visit of one of the authors (J.B.S.) in France. The authors thank Dr. Rajeev Kapoor of the Materials Science Division, B.A.R.C., for his help in carrying out deformation experiments.

REFERENCES

1. Willey, L.A. and H. Margolin. 1973, ASM: Metal Park, OH. p. 264.
2. Inui, H., et al., Philosophical Magazine A, 1992. **66**: p. 539-555.
3. Zghal, S., et al., Philosophical Magazine A, 2001. **81**(8): p. 537-546.
4. Inui, H., et al., Acta Metallurgica Materialia, 1992. **40**: p. 3095-3101.
5. Umakoshi, Y., T. Nakano, and T. Yamane, Materials Science and Engineering, 1992. **A152**: p. 81-88.
6. Nakano, T., et al., Philosophical Magazine A, 2001. **81**(6): p. 1447-1471.
7. Kishida, K., H. Inui, and M. Yamaguchi, Philosophical Magazine A, 1998. **78**(1): p. 1-28.

Mater. Res. Soc. Symp. Proc. Vol. 842 © 2005 Materials Research Society   S6.7

A Study of the Deformation Behavior of Lamellar γ-TiAl by Numeric Modeling

T. Schaden[1], F.D. Fischer[1,2], H. Clemens[3], F. Appel[4] and A. Bartels[5]

[1]Institute of Mechanics, Montanuniversität Leoben, Franz-Josef-Straße 18, 8700 Leoben, Austria
[2]Erich Schmid Institute of Materials Science, Austrian Academy of Sciences, Jahnstraße 12, 8700 Leoben, Austria
[3]Department of Physical Metallurgy and Materials Testing, Montanuniversität Leoben, Franz-Josef-Straße 18, 8700 Leoben, Austria
[4]Institute for Materials Research, GKSS Research Center, Geesthacht, Germany
[5]Department of Materials Science and Technology, TU Hamburg-Harburg, Eißendorfer Str. 42, D-21073, Hamburg, Germany

ABSTRACT

In this paper the deformation behavior of a fully lamellar microstructure, which is usually present in cast γ-TiAl based alloys, is studied by numerical modeling. After large compressive deformation at elevated temperatures the lamellar colonies are often bent or buckled depending on their orientation, which is representative for a deformation instability characterized by a large wave length. Such a deformation behavior is triggered by both, structural defects of the lamellae and their somewhat irregular arrangement. In addition, a shear band-type deformation mode occurs according to an instability mode exhibiting a short wave length. These two deformation modes interact in a rather subtle way, which leads to a very inhomogeneous deformation pattern.

INTRODUCTION

Processing of titanium aluminides often starts with the break-down of the coarse-grained lamellar microstructure of the ingot [1]. Figure 1a shows a representative microstructure of a cast γ-TiAl based alloy [2]. The lamellar colonies consist of alternating layers of γ-TiAl and α_2-Ti_3Al. After large compressive deformation at temperatures above 1000°C the microstructure changes to that shown in figure 1b. Here buckled colonies and shear localization zones dominate the appearance.

Figure 2a shows the flow curves of Ti-47Al-4(Nb,Cr,Mn,Si) derived from compression tests [1]. The flow-stress response reflects the effect of dynamic recrystallization in that the flow curves exhibit a broad peak at low strains (~10%), followed by flow softening to an approximately constant stress level at strains from $\varepsilon = 60\%$ to 90%. Under this condition the evolution of the microstructure occurs by thermally activated deformation and recovery processes [1].

The goal of this study is to investigate which local effects are responsible for the observed global softening of the material.

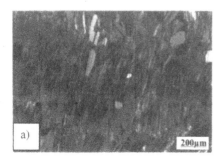
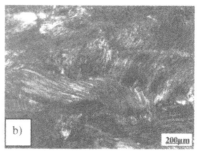

Figure 1. (a) Representative microstructure of a γ-TiAl based alloy ingot. The microstructure consists predominantly of lamellar colonies. (b) The same ingot after hot-deformation of about 80% (compression direction is vertical).

MODELING

In order to simulate the microstructural development during hot-working a finite element unit cell model was established, which is a representative volume element (RVE). The model demonstrates the deformation behavior of the material on a micro-(I), meso-(II) and macroscopic (III) level (Figure 3). Level I describes the center grain, one lamellar colony which shows a very fine structured area with alternating layers of the γ-TiAl and α_2-Ti_3Al lamellae. This colony is surrounded by six neighboring grains, which reflect the stiffness at the mesoscopic level II. Because the number of elements is limited, it is not possible to model the surrounding area with such a fine mesh as the middle grain. As a consequence one has to replace the lamellar structure of the neighboring grains by zones of plastic anisotropy, described by means of Hill-plasticity [3]. The global materials behavior is defined by the embedding (level III), which accounts for the macro deformation and sums up the materials behavior of the six differently oriented grains.

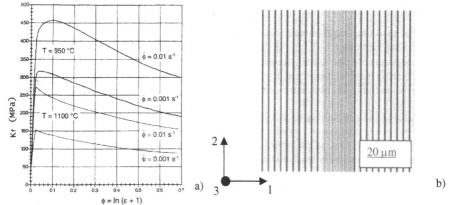

Figure 2. (a) Flow curves of Ti-47Al-4(Nb,Cr,Mn,Si) tested at the conditions indicated [1]. K_f: true stress; $\Phi=\ln(\varepsilon+1)$: true strain. (b) The local model level I represents a lamellar colony with alternating layers of γ-TiAl and α_2-Ti_3Al lamellae.

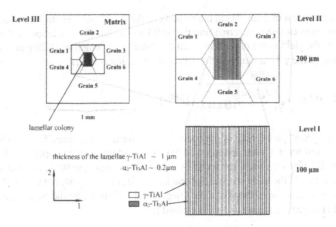

Figure 3. Sketch of the three-level model consisting of a lamellar center grain modeled with a fine mesh. A neighboring assembly of six grains with different orientations is embedded in a matrix with an averaged materials behavior in order to predict the deformation behavior on micro-(I), meso-(II) or macroscopic (III) level. The insert (bottom left) shows the coordinates of the principal axes.

Experimental results are implemented as materials input. The elastic data (Young's moduli and Poisson's ratios) for the γ-TiAl and α_2-Ti$_3$Al phases are linearly dependent on temperature. Data for room temperature are given in Schafrik [4]. The values for the Young's moduli for a temperature of about 1000°C are received by extrapolation of the existing curves resulting in 140GPa for the γ-TiAl and 95GPa for the α_2-Ti$_3$Al phase. The yield stress of the single γ-TiAl and α_2-Ti$_3$Al phase is determined by interpolation from global flow curves (figure 2a) for cylindrical samples which were tested under compression at temperatures in the range of 950°C to 1100°C [1]. Since the γ-TiAl phase is predominant, it is not possible to measure the yield stress of α_2-Ti$_3$Al directly. For this reason the measured global yield stress must be separated into two contributions evoked through the γ-TiAl and the α_2-Ti$_3$Al phase. By varying the phase content of α_2-Ti$_3$Al in a range between 5 to 50%, and the yield stress ratio of γ-TiAl to α_2-Ti$_3$Al in a range of 1:1 to 1:10, however, it is possible to model plausible deformation patterns.

<u>The local model</u>

To define the plastic anisotropy of the neighboring grains it is necessary to find the Hill-plasticity yield locus describing the whole center grain (level I) of the global model (figure 3). The center grain is replaced by a second, local model (figure 2b) in order to calibrate the materials behavior. This local model level I takes the lamellar structure of the colonies into account, described by the geometrical relations taken from realistic microstructures. A colony size of about 100μm is chosen, defining the length of the lamella. The thickness of the alternating layers of γ-TiAl and α_2-Ti$_3$Al lamellae are taken to be ~1μm and ~0.2μm, respectively. The yield stress of the material in lamella direction 1 differs much from the value of the yield stress in direction 2 and 3. This can be easily explained by the different "free" path length for dislocations. In our case a γ-TiAl lamella is seen as a plate with in-plane dimensions of

100µm and a thickness of 1µm. Then, the relation of the "free" path length for dislocations is 1:100. This relation can be transferred into the well known Hall-Petch relation. It describes the dependence of the yield stress of a material on the "free" path length of gliding dislocations. For the yield stress of the γ-TiAl lamellae the following equation is used:

$$\sigma_{y,n} = 30.1 + 0.1161 / \sqrt{\left(d_n (\mu m) \middle/ 1000000 \right)} \text{ [MPa];} \tag{1}$$

d stands for the "free" path length of dislocations. This formula was taken from Marketz et al. [5]. It is only valid for room temperature and has to be a adopted for a temperature level of 1000°C. The Hill-plasticity yield locus can be described according to the Hall-Petch effect by assuming the ratios of the yield stresses for one lamella $\sigma_{y,1}/\sigma_{y,2}$, $\sigma_{y,1}/\sigma_{y,3}$ at elevated temperatures being similar to those at room temperatures. To describe the plastic anisotropy of the α_2-Ti$_3$Al lamellae the same concept is applied [5].

The global model

The materials input of the local model is taken over to the center grain, the microscopic level I of the global model. The results of calibration of the local model level I are load-displacement curves as shown in figure 4a. They help to fix the Hill-plasticity yield locus of the lamellar colony (figure 4b). The load displacement-curves (figure 4a) of the lamellar colonies are compared to the load displacement-curves of a homogeneous material with a sharp defined yield stress. The load level at the beginning of yielding is calibrated exhibiting three different yield stresses of the lamellar colony $\sigma_{y,1}$, $\sigma_{y,2}$ and $\sigma_{y,3}$ depending on the coordinates principal axes 1,2 and 3 (figure 2b). The results of the lamellar package were taken to define the yield loci of the six neighboring grains which allows to describe the materials behavior at the mesoscopic level II (figure 3) over a length scale of 100µm. Here, the topological anisotropic materials behavior of a grain is determined by two contributions, namely by the existence and orientation of the lamellae

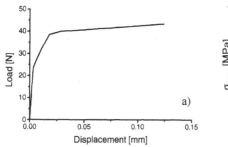
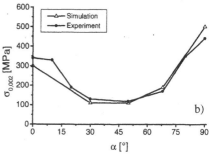

Figure 4. (a) Load-displacement curve of the local model for calibration in direction 2 (characteristics of the curve in direction 1 is identical) helps to define the Hill-plasticity yield locus, describing the plastic anisotropy of the lamellar colony. (b) The yield stress of a lamellar colony as a function of the orientation angle α (angle between the direction of lamellar colony and the loading direction) as obtained from experiments [3] is compared to calculated results [6].

and the crystallographic anisotropy of the γ-TiAl and α_2-Ti$_3$Al lamellae (figure 2b) according to level I. This concept allows to vary the orientation of the six neighboring grains.

An averaging effect is expected by embedding the interior grain into six randomly oriented neighboring grains (figure 3), which are also embedded in a continuum (level III) represented by periodic boundary conditions [7].

RESULTS AND DISCUSSION

Two different types of nonlinear deformation at the local level are distinguished, the "structural" deformation and the "materials" deformation. The materials deformation is simply the deformation under compression. The structural deformation represents the deformation due to - at least - two deformation modes. The structural deformation itself can be split into a buckling type deformation (figure 5a) with a big wave length [8] and shear localization zones (figure 5b), which lead to a deformation pattern with a small wave length [9]. Buckling is a bifurcation mode which appears in the early stage of deformation. Due to the existence of material and geometrical imperfections this type of deformation is followed by shear localization zones.

Three different types of imperfections involving shear localization zones can be found. These are geometrical faults of the layered structure inside the colonies, geometrical discontinuities on the edge of colonies like grain boundaries and finally buckling waves acting as a trigger for shear localization zones.

The numerical model is able to reflect the very inhomogeneous deformed microstructure and demonstrates the concentration of strain in families of shear localization zones with different amounts of localized shear. The experimentally found deformation pattern (figure 1b) can be well predicted by numerical findings.

Due to the high value of imparted deformation energy, recrystallization effects occur (figure 6a), which lead to a softening of the material. The energy concentration in the shear localization zones is calculated (figure 6b) and will be used in further studies to predict the onset of

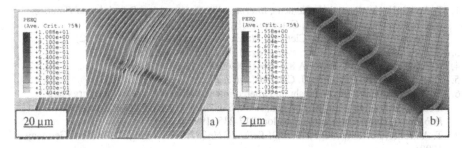

Figure 5. (a) Distribution of PEEQ (Accumulated Equivalent Plastic strain) [10] in the center grain as a result of the three level model. In the early steps of deformation a buckling-type deformation is observed. (b) Imperfections of the material as well as structural defects trigger the deformation in shear localization zones. In this case the discontinuities of the α_2-Ti$_3$Al lamellae are responsible for the localization.

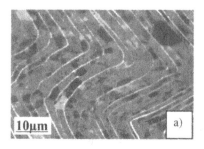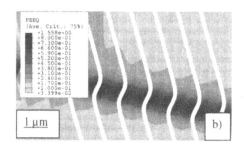

Figure 6. (a) Scanning electron microscope image of a Ti-48Al-2Cr (At.%) alloy which was deformed under compression at 1000°C to a strain of 50% ($\dot{\varepsilon} = 10^{-3}/s$) [11]. Dynamic recrystallization takes place in shear localization zones [1]. (b) Distribution of PEEQ in shear localization zones.

recrystallization and to estimate the size of the newly formed grains. All theoretical details will be published in a forthcoming paper.

SUMMARY

The deformation behavior of fully lamellar γ-TiAl based alloys was studied by numerical modeling. From this investigation it can be concluded that two softening mechanisms occur: one due to the bifurcation-type deformation pattern and the other one due to recrystallization induced softening. In addition, the model allows to verify the interaction of all these mechanisms and to quantify their specific contributions.

REFERENCES

1. F. Appel, H. Kestler and H. Clemens, *Intermetallic Compounds – Principle and Practice,* Vol. 3, John Wiley Publishers, Chicester, UK, 617-642 (2002).
2. A. Bartels, H. Kestler and H. Clemens, *Mater. Sci. Eng. A,* 329-331, 153 (2002).
3. S.M. Schlögl and F.D. Fischer, *Phil. Mag. A,* Vol. 75, No. 3, 621-636 (1997).
4. R.E. Schafrik, *Metall. Trans. A* **8**, 1003-1006 (1977).
5. W.T. Marketz, F.D. Fischer, H. Clemens, *Int. J. Plast., 19,* 281-321 (2003).
6. M. Yamaguchi and Y. Umakoshi, in Ordered intermetallics – Physical metallurgy and mechanical behavior, Liu, C.T., Cahn, R.W., Sauthoff, G., eds., *Kluwer Academic Publishers*, Dordrecht, Boston, London, 217-235 (1992).
7. T. Antretter, *Fortschritt-Berichte VDI Reihe 18 Nr.232*, VDI Verlag, Düsseldorf, Germany, 1998.
8. O.I. Minchev, F.G. Rammerstorfer and F.D. Fischer, *in Material Instabilities: Theory and Applications,* ASME, AMD-Vol. 183, MD-Vol. 50, New York, 357-368 (1994).
9. D. Okumura, N. Ohno and H. Noguchi, *J. Mech. Phys. Solids,* 52, 641-666 (2004).
10. ABAQUS Finite Element Analysis Products, Hibbit, Karlsson & Sorensen Inc., (www.hks.com)
11. R.M. Imayev, V.M. Imayev, M. Oehring and F. Appel, *Met. Mater. Trans.,* in press.

Mater. Res. Soc. Symp. Proc. Vol. 842 © 2005 Materials Research Society   S6.8

Effect of B Addition on Thermal Stability of Lamellar Structure in Ti-47Al-2Cr-2Nb Alloys

Y. Yamamoto, N. D. Evans, P. J. Maziasz and C. T. Liu
Metals and Ceramics Division, Oak Ridge National Laboratory, Oak Ridge, TN37831, U.S.A

ABSTRACT

Thermal stability of fully lamellar microstructures in Ti-47Al-2Cr-2Nb (at.%) alloys with and without B additions has been evaluated in the temperature range of 800 to 1200°C. The alloy with 0.15B exhibits an $\alpha_2 + \gamma$ fully lamellar microstructure containing ribbon-like structures (TiB$_2$+β) inside the lamellar colonies. During aging at 800 and 1000°C, β particles surrounded by γ grains form adjacent to the ribbon-like structures. The formation of these grains is attributed to the precipitation of thin plate-like β phase along $\{1\bar{1}00\}_{TiB_2}$ within the TiB$_2$ ribbon particles, resulting in the loss of α_2 plates around the ribbon-like structure by scavenging Ti. This allows coarsening γ lamellar plates to form new γ grains, and then the ribbon-like structure completely dissolves to leave only β particles. At 1200°C, discontinuous coarsening (DC) of the lamellae becomes the dominant mode of microstructural change for all the alloys. However, the ribbon-like structures in the alloy with B additions play a role in pinning the DC cells, which leads to higher thermal stability of the lamellar microstructure in that alloy.

INTRODUCTION

For improvement of mechanical properties of γ-TiAl alloys at both room temperature and elevated temperatures, it is necessary to refine both the lamellar colony size and the inter-lamellar spacing [1,2]. Liu et al [3,4] have successfully developed a fabrication process for effective microstructural refinement by using powder metallurgy (PM) of Ti-47Al-2Cr-2Nb alloys combined with hot extrusion for direct consolidation. Maziasz et al [5-7] have also reported that even in the cast alloys (ingot metallurgy, IM) followed by hot extrusion, the grain and lamellar refinements can be achieved if small amounts of B and W are added to the alloys. Both PM and IM alloys show a similar colony size and inter-lamellar spacing, and exhibit good mechanical properties. However, when the alloys are exposed to elevated temperatures around 800°C, PM alloys show good thermal stability of the lamellar microstructure even after 5000 h aging, while IM alloys exhibit microstructural instability and coarsening even after a short period of aging due to the formation of coarsened γ grains and β particles [8].

Thermal stability of the microstructure in similar γ-TiAl alloys containing a large amount of B (up to 1.0 at.%) at over 1230°C has been examined by Godfrey et al [9], and they revealed that string-like structures containing TiB$_2$ and β within the lamellae improve microstructure stability at elevated temperatures. However, there seems to be no reported data on the stability of such lamellar structure for those alloys aged at about 800°C.

The objective of this study, therefore, is to identify the effect of B additions on the thermal stability of fully lamellar microstructure in Ti-47Al-2Cr-2Nb alloys during aging in the temperature range of 800 to 1200°C, especially focusing on the effects of the boride phase formed within the lamellar microstructure.

EXPERIMENTAL PROCEDURE

Alloys used in this study were Ti-47Al-2Cr-2Nb-(0, 0.05, 0.15)B (at%), designated as 0B, 0.05B and 0.15B alloys, respectively. They were prepared by arc melting using commercially pure metals, followed by drop casting to a bar shape with 25 mm in diameter and 125 mm in length. The difference in weight before and after melting was less than 0.2%. The ingots were heat-treated in a vacuum furnace at 1400°C for 1h in the α single phase region and then slowly furnace cooled, in order to homogenize the materials and establish a fully lamellar microstructure. Specimens were cut from the ingots by electro-discharge machining, and encapsulated in quartz tubes backfilled with argon, and then aged at 800, 1000 and 1200°C for various periods of time up to 4320 h, followed by air cooling. Microstructures were observed using an optical microscope, and using scanning and transmission electron microscopes. Compositions of the phases were examined using X-ray energy dispersive spectroscopy (XEDS) and electron energy loss spectroscopy (EELS).

RESULTS

Stability of lamellar microstructure with 0.15B

Figure 1 shows backscattered electron images for the 0.15B alloy before and after aging at 800°C. Before aging, the alloy exhibits an $\alpha_2 + \gamma$ fully lamellar microstructure containing a number of ribbon-like structures with about 10 μm in length and less than 0.2 μm in width within the lamellar colonies (Fig.1a). After 240 h aging, α_2 plates adjacent to the ribbons disappear and the contrast of the ribbon becomes slightly brighter than that before aging (Fig.1b). After 720 h aging, some of the ribbons are now surrounded by particles with a white contrast and grains with a gray contrast (Fig.1c). SEM-XEDS analysis revealed that the white and gray regions correspond to β and γ phases, respectively (Table I). After further aging to 4320 h, the ribbon is completely encased by formation of the β particles, and the γ grains are always observed adjacent to the β phase (Fig.1d). It should be noted that the ribbon-like structures are also observed on the lamellar colony boundary, but there is no obvious difference in the microstructural change relative to those observed within the colony. It is also noted that some of the ribbon-like structures show no indication of the formation of β and γ grains, even after aging for 4320 h, instead the α_2 plates around these phase disappear, as shown in Fig.1b. Figure 2 shows a backscattered electron image for the alloy after aging at 1000°C for 168 h. This microstructure exhibits a number of β particles surrounded by γ grains within the lamellae of a given colony, which is similar to the microstructures observed in the sample aged at 800°C for 4320 h. This comparison indicates that the same microstructural

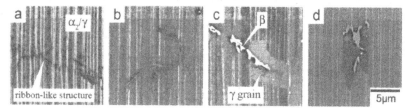

Figure 1. Backscattered electron images near the ribbon-like structure in 0.15B alloy before (a) and after aging at 800°C for 240 h (b), 720 h (c) and 4320 h (d).

change would occur at 1000°C, although such change takes place about 30 times faster than that at 800°C.

Figure 3 shows a TEM bright field image of a typical configuration of a ribbon-like structure within the lamellae after aging at 800°C. Some of the α_2 plates adjacent to the ribbon disappeared, leaving γ/γ interfaces, and a large γ lamellar plate on the left-hand side of the image becomes wide to form a γ grain. This result suggests that Ti in the α_2 plates is absorbed by the ribbon-like structure, resulting in γ/γ interfaces as the α_2 plates dissolve during aging, and the formation of the large γ grains then follows the instability of those remnant γ/γ interfaces.

Figure 4 shows TEM bright field images near the ribbon-like structure in the 0.15B alloy before (a) and after aging at 800°C for 4320 h (b), together with selected area diffraction patterns (SADP) obtained from the ribbon. The ribbon-like structure before aging consists of facets and a thin plate-like phase. SADP analysis reveals that the former is TiB_2 with the C32 (hexagonal) structure, while the latter is β with the B2/bcc structure, with an orientation relationship of $[0001]_{TiB_2}$ // $<001>_\beta$ and $\{1\bar{1}00\}_{TiB_2}$ // $\{100\}_\beta$. The long facet aligned

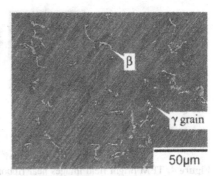

Figure 2. Backscattered electron image of 0.15B alloy aged at 1000°C for 168h.

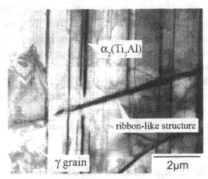

Figure 3. TEM bright field image near ribbon-like strucutre in 0.15B alloy aged at 800°C for 720 h.

parallel to the TiB_2/β inter-phase boundary corresponds to $\{1\bar{1}00\}_{TiB_2}$. These crystallographic configurations are consistent with several reported observations in similar alloy systems [9,10]. After 4320 h aging, however, the ribbon-like structure contains a large number of interfaces parallel to the habit plane with an average spacing of less than 10 nm. The SADP shows the presence of both TiB_2 and β phases, and the diffuse lines of diffracted intensity

Table I. Results of EDS analysis of each phase in 0.15B alloy before and after aging.

Aging	Phase	Composition (at.%)			
		Ti	Al	Cr	Nb
no aging	γ lamellar plate	47.9	49.1	1.2	1.9
	α_2 lamellar plate	58.3	37.9	2.5	1.3
4320 h at 800°C	γ lamellar plate	48.2	49.0	0.9	1.9
	α_2 lamellar plate	62.5	34.9	1.0	1.7
	γ grain (gray phase)	48.7	48.5	0.9	1.9
	β particle (white phase)	65.6	30.5	3.1	0.9

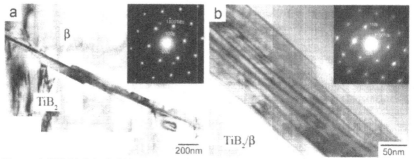

Figure 4. TEM bright field images near ribbon-like structure in 0.15B alloy before (a) and after aging at 800°C for 4320 h (b), together with selected area diffraction patterns obtained from [0001] $_{TiB2}$ // [001]$_\beta$ zone axis.

along <1$\bar{1}$00>$_{TiB_2}$ in the pattern indicate that there are thin plate-defects normal to this crystallographic direction within the TiB$_2$ particle. TEM-XEDS and EELS analyses reveal that the ribbon-like structure before aging is composed of 57% of Ti, 43% of B, and small amounts of other elements. From this result, the volume ratio of TiB$_2$: β can be roughly estimated to be 0.4: 0.6, by assuming that the ribbon-like structure consists of only these two phases, and that the β phase contains no B. On the other hand, the composition and the volume ratio of TiB$_2$: β in the ribbon-like structure after aging change to 65% of Ti, 33% of B with 2% of other elements and 0.3: 0.7, respectively. This result clearly indicates that the precipitation of β phase at the inside of the TiB$_2$ particles takes place as nucleation of a new thin plate-like phase during aging.

Effect of B content on thermal stability

Figure 5 shows optical micrographs of the alloys aged at 1200°C for 168 h. Discontinuous coarsening (DC) of the lamellae can be observed at the lamellar colony boundary in the 0.15B alloy (Fig.5a). More than a half of the area is covered by DC cells in the 0.05B alloy (Fig.5b), and the original lamellar microstructure cannot be detected in the alloy without B (Fig.5c). The ribbon-like structure is often observed on the migrating boundaries of DC cells in the 0.15B alloy, and the amount of the ribbons decreases with decreasing B content. Thus, the ribbon-like structure plays a role in pinning the DC cells at 1200°C. More detailed observations are currently in progress and will be described elsewhere [11].

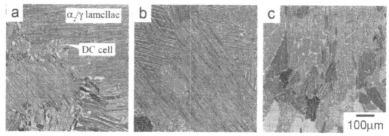

Figure 5. Optical micrographs showing the microstructure of (a) 0.15B, (b) 0.05B and (c) 0B alloys after aging at 1200°C for 168 h.

DISCUSSION

Based on the results observed, it is concluded that the formations of both β and γ grains during aging at 800 and 1000°C are attributed to the precipitation of β phase at the inside of TiB₂ phase particles. Figure 6 schematically illustrates the formation process of the γ grains. At first, the thin plate-like β phase forms at the inside of the TiB₂ by scavenging Ti from α₂ (Ti₃Al) plates adjacent to the ribbon-like structure, leaving only γ/γ interfaces. The loss of α₂ plates allows expansion of the γ lamellar plates across the lamellae, resulting in the formation of larger γ grains with no substructure [12]. Finally, the ribbon-like structure is completely dissolved, leaving the β particles surrounded by γ grains. The B atoms would be dissolved within the β phase as interstitial atomes. It is well known that the addition of Cr and Nb to γ-TiAl alloys enhances the formation of the β phase at elevated temperatures [13], so that the ribbon-like structure would act as a preferential nucleation site for formation of β phase at around 800°C.

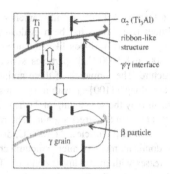

Figure 6. Schematic illustrations of the formation process of γ grain adjacent to the ribbon-like structure.

The formation of the thin plate-like β phase along a {1Ī00} plane within TiB₂ can be explained as follows: Figure 7 is a schematic illustration of superimposed atomic arrangements on (0001)ₜᵢ𝐵₂ // {001}ᵦ of two periodic layers for each phase including the inter-phase boundary on {1Ī00}ₜᵢ𝐵₂ // {100}ᵦ. The lattice parameters of the a- and c-axes in TiB₂ and a-axis in β are reported as 0.305, 0.322 and 0.320 nm, respectively [9], so that the misfit along each axis on the inter-phase boundary is relatively low (//aₜᵢ𝐵₂: 4.3% , //cₜᵢ𝐵₂: -0.6%). In addition, the stacking of {1Ī00}ₜᵢ𝐵₂ planes can be expressed as ..., I., II., II., I., II., II., I.,..., where the stacking I shows 2 Ti-B (attractive) and 2 Ti-Ti (repulsive) bonds across the plane for each Ti atom, whereas the stacking II shows 4 Ti-B and 1 Ti-Ti bonds for each Ti. From the viewpoint of the inter-plane bond energy, TiB₂ would easily transform to β for stacking I as shown in Fig. 7,

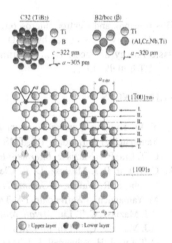

Figure 7. Schematic illustrations of crystal structures of TiB₂ (C32) and β (B2/bcc), together with a superimposed atom arrangements on (0001)ₜᵢ𝐵₂//{001}ᵦ plane of two periodic layers for each phase including an inter-phase boundary on {1Ī00}ₜᵢ𝐵₂ // {100}ᵦ.

because of relatively fewer attractive atom bonds across that particular plane. The experimental evidence of the selective habit plane as well as the transformation of TiB₂ may be obtained by using high resolution TEM observations, and work on this topic is currently in progress.

167

SUMMARY

The thermal stability of fully lamellar microstructures in Ti-47Al-2Cr-2Nb alloys with and without B additions has been evaluated in the temperature range of 800 to 1200°C. In the alloy with 0.15B, ribbon-like structure (TiB$_2$+β) forms inside lamellar colonies. During aging at 800 and 1000°C, β particles surrounded by γ grains form adjacent to the ribbon-like structure. The formation of these grains is attributed to the precipitation of thin plate-like β phase along $\{1\bar{1}00\}_{TiB_2}$ within TiB$_2$, resulting in the loss of α$_2$ plates around the ribbon-like structure by the scavenging of Ti. This reaction allows instability of the remaining γ lamellar plates to form the larger γ grains. The ribbon-like structure is completely dissolved due to the formation of β particles. At 1200°C, discontinuous coarsening (DC) of the lamellae becomes the dominant mode of microstructural change for all the alloys, and the growth rate of DC cell decreases with increasing B contents. The ribbon-like structure in the alloys with B plays a role in pinning the DC cells, leading to the relatively higher thermal stability of the lamellar microstructure.

ACKNOWLEDGMENTS

This research, and electron microscopy conducted at Oak Ridge National Laboratory SHaRE User Center, was sponsored by the Division of Materials Science and Engineering, Office of Basic Energy Sciences, U.S. Department of Energy under contract DE-AC05-00OR22725 with UT-Battelle, LLC.

REFERENCES

1. Y. W. Kim, in *Gamma Titanium Aluminide*, edited by Y. W. Kim, R. Wagnerand M. Yamaguchi, (Warrendale, PA, TMS, 1995) 637.
2. C. T. Liu, P. J. Maziasz, *Intermetallics*, **6**, 653 (1998).
3. C. T. Liu, P. J. Maziasz, D. R. Clemens, J. H. Schneibel, V. K. Sikka, T. G. Nieh, J. L. Wright, L. R. Walker, in *Gamma Titanium Aluminide*, edited by Y. W. Kim, R. Wagnerand M. Yamaguchi, (Warrendale, PA, TMS, 1995), 679.
4. P. J. Maziasz, C. T. Liu, *Metall. Mater. Trans.* **29A**, 105(1998).
5. P. J. Maziasz, C. T. Liu, *Mat. Res. Soc. Symp. Proc.* **460**, 219 (1997)
6. C. T. Liu, J. H. Schneibel, P. J. Maziasz, J. L. Wright , D. S. Easton, *Intermetallics*, **4**, 429.
7. P. J. Maziasz, R. V. Ramanujan, C. T. Liu, *Intermetallics*, **5**, 83 (1997).
8. C. T. Liu, unpublished work (1998).
9. A. B. Godfrey, M. H. Loretto, *Intermetallics*, **4**, 47 (1996).
10. D. S. Schwartz, D. S. Shih, *Mat. Res. Soc. Symp. Proc.* **364**, 787 (1995)
11. Y. Yamamoto, P. J. Maziasz, C. T. Liu, to be submitted (2004).
12. Y. Yamamoto, M. Takeyama, T. Matsuo in *Structural Intermetallics 2001*, edited by K. J. Hemker, D. M. Dimiduk, H. Clemens, R. Darolia, H. Inui, J. M. Larson, V. K. Sikka, M. Thomas and J. D. Wittenberger, (Warrendale, PA, TMS, 2001), 601.
13. S. Kobayashi, Ph. D. Thesis, Tokyo Institute of Technology (2001).

Mater. Res. Soc. Symp. Proc. Vol. 842 © 2005 Materials Research Society   S5.40

Compressive Creep Behavior in Coarse Grained Polycrystals of Ti₃Al
and its Dependence on Binary Alloy Compositions

Tohru Takahashi[1], Yuki Sakaino[2], and Shunzi Song[2]
[1]Tokyo University of Agriculture and Technology,
Department of Mechanical Systems Engineering;
2-24-26 Naka-cho; Koganei, Tokyo 184-8588, JAPAN
[2]Graduate Student, Tokyo University of Agriculture and Technology

ABSTRACT

Compressive creep behavior has been investigated on coarse grained Ti₃Al alloys with aluminum contents ranging from 15mol%Al to 42mol%Al, in order to obtain basic information concerning the chemical composition effect on creep of Ti₃Al. Pure aluminum and titanium of 99.99% purity were arc-melted into small ingots weighing about 10grams under an argon atmosphere. The resulting microstructures after the hot deformation and vacuum annealing contained equiaxed grains whose average diameter ranged from 125 to 192 micrometers except alloys containing 40 and 42 mol% aluminum.

Compressive creep tests were performed in vacuum on parallelepiped specimens with dimensions of 2mm×2mm×3mm. The applied compressive stress was 159MPa, and the test temperature was around 1200K.

A very small primary transient and the minimum creep rate region followed by a gradual creep acceleration were observed in the materials containing aluminum up to 25mol%. In contrast to this, the materials containing more aluminum than 25mol% showed greater primary transient where creep deceleration continued up to about 0.1 true strain. Dual phase materials containing the γ phase showed small primary transient probably due to the constraint from the γ phase.

INTRODUCTION

The ordered intermetallic phases formed in titanium-aluminum system are prospective for application in light-weight and heat-resistant structure [1,2]. Concerning the α₂ phase with its stoichiometric composition of Ti₃Al, the DO₁₉ type ordered structure is considered to be stable for a wide range of aluminum content from about 22mol%Al to about 35mol%Al according to the well-known binary alloy phase diagram. On the aluminum lean side of the α₂ phase stands the α+α₂ dual phase region, and in this region the fractions of both constituent phases would vary as a function of temperature and alloy composition. On the aluminum rich side of the α₂ phase stands α₂+γ dual phase region whose constituent phases would merge into the α solid solution phase at higher temperature than about 1400K. In the present study, the chemical composition effect on the creep of the α₂ phase is to be clarified, especially at its outermost compositions bordering either α+α₂ or α₂+γ region.

EXPERIMENTAL DETAILS

Materials and materials preparation

Aluminum and titanium of 99.99% purity were arc-melted into small ingots weighing about 10grams under an argon atmosphere. The chemical compositions of the alloys were varied from 15mol% to 42mol%. Such a wide composition range covers the whole composition range of α_2 single phase region, and partly covers $\alpha+\alpha_2$ and $\alpha_2+\gamma$ dual phase regions standing next to the α_2 region on the left and right sides, respectively[1]. The prepared alloys contained 15, 18, 20, 22, 23, 24, 25, 26, 27, 28, 30, 32, 34, 35, 40 and 42 mol% aluminum, and, hereafter, the alloys will be denominated by their aluminum content, like 15Al.

Small pieces were hot deformed at 1300K by about 50% reduction of height in air and subsequently vacuum annealed at 1400K for 100ks. The resulting microstructures contained equiaxed grains whose average diameter ranged from 125 to 192 micrometers except 40Al and 42Al alloys that contained small amounts of γ phase.

Disc pieces with about 8mm diameter and 8mm thickness were cut from the ingot, and were compressed at 1200-1300 K to about 50 % reduction in height under a strain rate of $10^{-4}\,s^{-1}$. Vacuum annealing at 1400K for 100ks recrystallized the materials into equiaxed polycrystals containing coarse grains. Small rectangular parallelepiped specimens with dimensions of about 2mm×2mm×3mm were cut and finished by emery paper polishing.

Crystallographic characterization

Crystallographic characterizations were carried out on the diffraction patterns obtained with copper $K\alpha_1$ characteristic radiation; $K\beta$ radiation was filtered by a nickel foil and the $K\alpha_2$ contribution was stripped off by the analyzing software of RIGAKU RAD-IIC diffractometer.

Compressive creep tests

Compressive creep tests were performed in vacuum on parallelepiped specimens with dimensions of 2mm×2mm×3mm. The amount of creep deformation was measured by an LVDT sensor attached straight upon the load-conveying rod above the specimen. Based on the recorded contraction in height of the specimen, the applied load was intermittently adjusted in order to keep the true stress constant within about 1% error. The applied compressive stress was 159MPa, and the test temperature 1200K. The test temperature was monitored with an alumel-chromel thermocouple directly welded to the specimen and was kept constant within about 1K error.

RESULTS AND DISCUSSION

Microstructures

Figure 1 shows the microstructures of the selected materials; (a) 15Al, (b) 25Al, (c) 35Al and (d) 42Al. The apparent microstructures in the materials were coarse grained and equiaxed microstructures except for the 42Al material that contained $\alpha_2+\gamma$ dual phase lamellar microstructure. The 40Al material contained a small amount of γ phase. Except for these aluminum rich materials, the microstructures showed ordinary polycrystalline appearance. The average grain diameters were evaluated using the linear intercept method; the measured grain diameters are summarized in Table I. The average grain diameters were about 150μm in almost all the materials used in the present study.

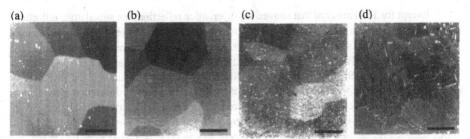

Figure 1. Microstructures of the tested materials; (a) 15Al, (b) 25Al, (c) 35Al, (d) 42Al.
Scale bars indicate 100 micrometers.

Table I. The average grain diameters (in micrometers) of the examined materials.

15Al	154	23Al	167	27Al	179	34Al	128
18Al	134	24Al	125	28Al	===	35Al	158
20Al	146	25Al	192	30Al	136	40Al	153
22Al	165	26Al	182	32Al	132	42Al	136*

* The grain diameter for 42Al shows the average size of its $\alpha_2+\gamma$ lamellar colony.

Crystallography

Figures 2 (a), (b), (c) and (d) show diffraction patterns of 15Al, 25Al, 35Al and 40Al materials, respectively. The numbers attached to the peak show the Miller indices of reflecting planes. The peaks in 15Al material are indexed as the disordered hexagonal structure, while the peaks in others were indexed as the DO_{19} ordered structure. Peaks from the γ phase overlapped in 40Al.

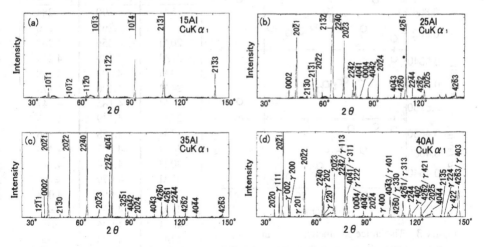

Figure 2. X-ray diffraction patterns in (a) 15Al, (b) 25Al, (c) 35Al and (d) 40Al materials.

171

Except for 15Al material that showed no superlattice reflection, the superlattice reflections from DO$_{19}$ structure can be identified. The diffraction peaks observed in the 35Al material could all be indexed as DO$_{19}$ structure so that the 35Al material was considered to contain no γ. The reflections from the γ-TiAl phase were found only in 40Al and 42Al materials.

Lattice parameters evaluated from the diffraction patterns are summarized in Figure 3. The a in 15Al material was doubled for comparison. From the observed diffraction patterns, it is difficult to elucidate the coexistence the disordered α phase and the ordered α_2 phase for 18Al to 22Al materials. So all the diffraction peaks were assumed to come from α_2 phase for 18Al to 35Al. Analysis on the lattice parameters of the γ phases in 40Al and 42Al were performed, but not included here. The c parameters, and consequently the c/a ratios, in 20Al and 22Al materials showed some peculiarity, but the present authors could not reach the conclusive results.

A monotonous decrement can be found in both a and c from 18Al to 35Al with increasing aluminum content, probably because the atomic radius of aluminum is slightly smaller than that of titanium. The axial ratio, c/a, in the disordered 15Al showed a little smaller value than those in other ordered materials. The c and c/a values in both 40Al and 42Al materials showed somewhat clear difference from the ones in 35Al material. The 35Al composition is considered to be close to the composition in α_2 phase within $\alpha_2+\gamma$ dual phase region. So, the abrupt changes of c and c/a in 40Al and 42Al could be attributed to the coexisting γ phase.

The lattice parameters as listed below were used to estimate the density of the materials by assuming the excess amount of titanium or aluminum can interchangeably substitute into anti sites to form solid solutions. Then the density of the α_2 phases can be calculated from their chemical compositions, atomic weights, Avogadro's constant, and lattice parameters. Then the estimated densities of the materials are compared with their experimentally measured values that were determined from the mass to volume ratios.

Table II Density (ρ) values in 10^3 kg/m^3.

Materials	ρ estimated	ρ measured
15Al	4.290	4.284
18Al	4.260	4.266
20Al	4.256	4.262
22Al	4.189	4.244
23Al	4.200	4.188
24Al	4.190	4.173
25Al	4.185	4.147
26Al	4.149	4.133
27Al	4.158	4.166
28Al	4.147	4.149
30Al	4.114	4.116
32Al	4.079	===
34Al	4.052	4.038
35Al	4.038	===
40Al	4.026	3.919*
42Al	4.023	===

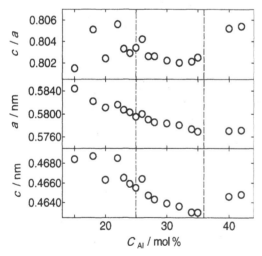

Figure 3 The lattice parameters (a, c and c/a) and their change with aluminum content.

The measured density decreased monotonously with increasing aluminum content. The measured and the estimated densities showed a fairly agreement in each alloy composition from 15Al to 34Al. This suggests that the α_2 phase has little tendency to form compositional vacancies on both sides of the stoichiometric composition of 25mol%Al. The 40Al material containing some amount of lighter γ phase showed about 2.5% less density than the estimated density of the α_2 phase.

Compressive creep behavior

Figure 4 (a), (b), (c) and (d) show the observed creep rates in 15-22Al, 22-25Al, 25-27Al, and 27-35Al materials, respectively, as a function of creep strain. The test temperature was 1200K, and the applied true stress was about 159MPa. Creep rate strongly depended upon aluminum content of the materials.

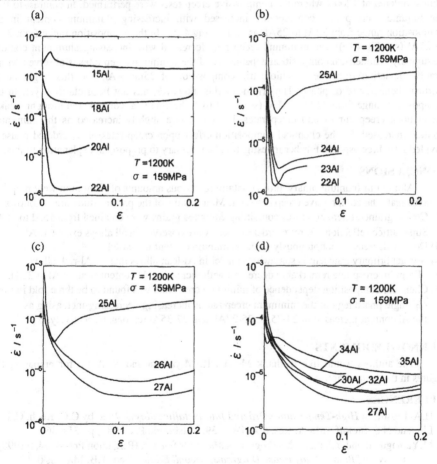

Figure 4. Creep rate vs. creep strain relationship in various aluminum content.

Creep curves were composed of three portions, in general; a normal primary transient creep, minimum creep rate region, and steady creep rate region or accelerating creep region were observed in all the tested materials. The extent of the normal primary transient is rather small (up to a few per cent strain) in 15-25Al, however, it becomes greater (around 10 per cent strain) in 25-35Al. The details are not included in Figure 4, but 40Al and 42Al material that contain γ phase besides α_2 phase again showed very small normal primary transient.

The chemical composition dependence of creep rates were considerably different depending upon the regions of their aluminum content. In the region ranging from 15Al to 22Al (see Fig.4 (a)), the creep rate strikingly decreased as the aluminum content increased; the minimum creep rate decreased by three order of magnitude. According to the binary alloy phase diagram[3], the 15Al material would be totally disordered, and the 22Al material would keep almost ordered at 1200K where the compressive creep tests were performed. In contrast to this, the minimum creep rate considerably increased with increasing aluminum content in the composition range from 22Al to 25Al, as seen in Fig.4 (b). In the composition range from 25Al to 27Al (see Fig.4 (c)), the minimum creep rate decreased with increasing aluminum content. Assumedly symmetrical composition dependence of the minimum creep rates is observed within 3mol% deviation from the stoichiometric composition of 25mol%Al, and this might be an intrinsic behavior of α_2 phase. The origin for this, however, has not been clarified yet. In the composition range from 27Al to 35Al (see Fig.4 (d)), the chemical composition brought weaker effect upon creep curves and creep rates, but creep rate slightly increased as the aluminum content increased. So the chemical composition effect upon creep rates of α_2 ordered phase is divided into three aspects. Further investigation is necessary to propose an explanation to this.

CONCLUSIONS

Aluminum-titanium binary alloys containing various amounts of aluminum were prepared to investigate the compressive creep behavior. Main results of the present study are as follows.
(1) Coarse grained microstructures containing equiaxed grains were obtained from 15Al to 35Al.
(2) Superlattice reflections from α_2 ordered phase were observed in all alloys except 15Al.
(3) Density decreased monotonously as the aluminum content increased.
(4) Normal primary transient is somewhat smaller in Al-lean alloys than in Al-rich alloys.
(5) Minimum creep rate remarkably decreased with increasing Al content from 15Al to 22Al.
(6) Chemical composition dependence of minimum creep rate was found to be three fold in the α_2 single phase region; the minimum creep rate is increasing-decreasing-increasing as the Al content increased in 23-25Al, 25-27Al, and 27-35Al regions, respectively.

ACKNOWLEDGMENTS

The authors would like to thank Messrs. R. Miyashita and N. Asato for preparing the figures in the manuscript.

REFERENCES

1. H.A. Lipsitt, in *High-Temperature Ordered Intermetallic Alloys*, edited by C.C. Koch, C.T. Liu, and N.S. Stoloff, (Mater. Res. Soc. Porc. **39**, Pittsburgh, PA, 1985) pp.351-364.
2. M.Yamaguchi and Y.Umakoshi, *Progress in Materials Science*, (Pergamon Press), **34**,1(1990).
3. J.L. Murray, in *"Binary Alloy Phase Diagrams, Second Edition,"* ed. T.B. Massalski, H. Okamoto, P.R. Subramanian and L. Kacprzak (ASM International, 1990) vol **1**, pp.225-227.

Mater. Res. Soc. Symp. Proc. Vol. 842 © 2005 Materials Research Society

Micro Fracture Toughness Testing of TiAl Based Alloys with a Fully Lamellar Structure

K. Takashima, T. P. Halford, D. Rudinal, Y. Higo and M. Takeyama[1]
Precision and Intelligence Laboratory, Tokyo Institute of Technology,
R2-35 4259 Nagatsuta, Midori-ku, Yokohama, 226-8503, Japan
[1] Department of Metallurgy and Ceramics Science, Tokyo Institute of Technology,
S8-8 2-12-1 Ookayama, Meguro-Ku, Tokyo, 152-8552, Japan

ABSTRACT

A micro-sized testing technique has been applied to investigate the fracture properties of lamellar colonies in a fully lamellar Ti-46Al-5Nb-1W alloy. Micro-sized cantilever specimens with a size $\approx$ 10 x 10 x 50 μm^3 were prepared by focused ion beam machining. Notches with a width of 0.5 μm and a depth of 5 μm were also introduced into the micro-sized specimens by focused ion beam machining. Fracture tests were successfully completed using a mechanical testing machine for micro-sized specimens at room temperature. The fracture toughness (K_Q) values obtained were in the range 1.4 − 7 MPam$^{1/2}$. Fracture surface observations indicate that these variations are attributable to differences in local lamellar orientations ahead of the notch. These fracture toughness values are also lower than those having been previously reported in conventional samples. This may be due the absence of significant extrinsic toughening mechanisms in these micro-sized specimens. Fracture mechanisms of these alloys are also considered on the micrometer scale. The results obtained in this investigation give important and fundamental information on the development of TiAl based alloys with high fracture toughness.

INTRODUCTION

TiAl (γ) based aluminides have been developed as candidate materials for gas turbine engine components to replace existing nickel based superalloys. This is because of their high specific strength and modulus coupled with their relatively good elevated temperature strength, oxidation resistance and creep properties [1, 2]. Their disadvantages, which include poor room temperature ductility, low fracture toughness and difficult fabricability, have, however, restricted industrial applications. Gamma single phase alloys are essentially very brittle. In contrast, mechanical properties are improved for two phase alloys containing small amount of Ti_3Al (α_2) [3]. Their mechanical properties depend strongly upon microstructure, which varies widely with differing heat-treatments [4]. The microstructures of ($\gamma+\alpha_2$) alloys of interest can be roughly divided into lamellar and duplex microstructures. The lamellar microstructure has a good balance of elevated temperature properties and fracture toughness. This increased fracture toughness, relative to duplex structures, is attributed to crack tip shielding by extrinsic toughening mechanisms, including shear ligament bridging [5]. These extrinsic toughening mechanisms are dominated by microscopic lamellar structures including lamellar orientation [6, 7], colony size [8] and plate thickness [9]. It is therefore important to investigate the fracture properties of lamellae materials on the micro meter scale for the design of these alloys.

Fracture properties of the lamellar structure in TiAl alloys have been investigated using PST (Poly-Synthetically Twined) crystals [6, 7]. It is, however, rather difficult to prepare PST crystals, and the processing conditions are usually different from those of ordinary lamellar alloys. The microstructural features, including lamellar spacing, are thus also different from ordinary lamellar structures,making it difficult to compare the mechanical properties of PST crystals with ordinary lamellar structured alloys.

We have developed a testing machine which enables the mechanical testing of micro-sized materials [10]. Micro-sized tensile, bending, fracture toughness and fatigue tests have been carried out for specimens with dimensions of 10 – 50 μm [11-14]. The size of these specimens is smaller than one lamellar colony in the lamellar structure, making it possible to directly measure the mechanical properties of an individual lamellar colony.

In this investigation, a micro fracture toughness testing method was developed for micro-sized specimens prepared from a TiAl based alloy with a fully lamellar structure. Fracture toughness values of lamellar colonies were measured on the micrometer scale. The effect of size on the fracture toughness was also discussed.

EXPERIMENTAL PROCEDURE

The material used was a Ti-46Al-5Nb-1W (at. %) alloy, known as Alloy 7. This alloy was prepared at the University of Birmingham by vacuum arc melting and subsequently canned and hot extruded by Plansee, Austria. The microstructure is fully lamellar with a colony size of 75 μm as shown in Fig. 1. Thin slices with a thickness of 80 μm were cut from the extruded material by electro-discharge machining and ground and polished on both sides to provide foils with a thickness of approximately 20 μm. Disks with a diameter of 3 mm were cut from the foil and micro-sized cantilever beam type specimens introduced into the disks by focused ion beam machining. Figure 2 shows a scanning electron micrograph of a micro cantilever specimen. The length (L), width (B) and thickness (W) of specimens were 50, 10 and ~20 μm, respectively. The loading point was located 40 μm from the fixed end of the specimen. Notches with a depth of ~5 μm were introduced into the specimens by focused ion beam machining. The notch width was 0.5 μm, and a notch root radius is thus deduced to be 0.25 μm. The notch position was set 10 μm from the fixed end of the specimen.

Fracture toughness testing was carried out using a mechanical testing machine for micro-sized specimens, which was developed in our laboratory. The applied loading rate was 0.1 μm/s. The load resolution of this testing machine is 10 μN and the displacement resolution is 5 nm. The loading position can be adjusted by a precise X-Y stage at a translation resolution of 0.1 μm.

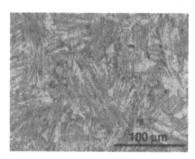

Figure 1. Optical micrograph of Ti-46Al-5Nb-1W alloy with a fully lamellar structure.

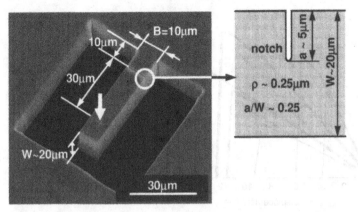

Figure 2. Scanning electron micrograph of a micro-sized specimen prepared from one lamellar colony and schematic view of notch geometry.

The details of the testing machine have been described elsewhere [10]. Fracture surfaces after fracture toughness testing were observed using a scanning electron microscope.

RESULTS AND DISCUSSION

Figure 3 shows load-displacement curves obtained during fracture toughness testing. As fracture in these specimens occurred in a brittle manner, the crack is deduced to start to propagate at the maximum load. Fracture toughness values were calculated from the maximum load using the following equations for stress intensity (K) for a notched cantilever beam [15].

$$K = \frac{6PS}{W^2 B}\sqrt{\pi a}\, F(a/W), \quad (a/W < 0.6) \tag{1}$$

where

$$F(a/W) = 1.22 - 1.40(a/W) + 7.33(a/W)^2 - 13.08(a/W)^3 + 14.0(a/W)^4 \tag{2}$$

In equation (1), a, P and S are total crack length, failure load and distance between the loading point and notch position, respectively. The total crack length was measured from the fracture surface by scanning electron micrographs after testing. Fracture toughness values for each specimen are also shown in Fig. 3. As some of these values do not satisfy small scale yielding conditions, fracture toughness values are indicated by K_Q. K_Q values are scattered from 1.4 to 6.9 MPam$^{1/2}$.

Figure 4 shows a scanning electron micrograph of the fracture surface of a specimen with a low K_Q value. Flat facets dominate the fracture surface and this indicates that fracture occurred by the interlamellar mode. Figure 5 shows a scanning electron micrograph of the fracture surface of a specimen with a higher K_Q value. Greatly increased amounts of translamellar fracture are observed. These fractographic features demonstrate a significant variation in the amount of the different lamellar failure modes present, and a large scatter in K_Q values is thus attributed to variations in fracture mode.

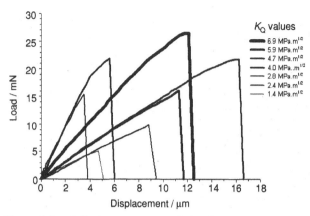

Figure 3. Load-displacement curves during fracture toughness testing of micro-sized TiAl specimens.

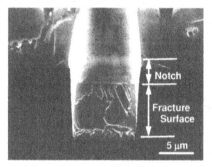

Figure 4. Scanning electron micrograph of fracture surface of a specimen with a low K_Q (2.4 MPam$^{1/2}$).

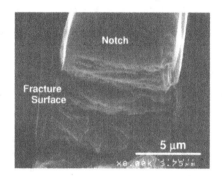

Figure 5. Scanning electron micrograph of fracture surface of a specimen with a high K_Q (5.9 MPam$^{1/2}$).

The trend of K_Q values with lamellar orientation is consistent with the results obtained for ordinary sized PST crystal specimens with a fully lamellar structure. The low K_Q values (1.4 – 2.8 MPam$^{1/2}$) in this investigation are considered to be equivalent to that of the interlamellar mode fracture (i.e., the notch direction is parallel to lamellar interface). K_{IC} values of interlamellar fracture in PST crystals have been reported to be approximately 4 MPam$^{1/2}$ [7]. This value is high compared to K_Q values of our samples. Fracture toughness testing of the interlamellar mode in PST crystals have been carried out using a specimen with a notch only, as it is extremely difficult to introduce a fatigue precrack into the interlamellar direction. In contrast, a notch with a root radius of 0.25 μm was introduced between lamellar plates for our micro-sized specimens. Actually, a K_C value of cleavage fracture for (111) planes in γ-TiAl was calculated to be 1.47 MPm$^{1/2}$ [16] based on the cleavage energy calculation [17]. This K_C value is consistent with the K_Q values obtained in this investigation. This suggests that more accurate K_Q of the interlamellar mode for TiAl can be evaluated by micro-size fracture testing.

The K_Q values increase with increasing the amount of translamellar fracture. The K_Q values for such specimens are, however, much lower than those obtained for macro-sized specimens of the same material (18 MPam$^{1/2}$) [18]. For macro-sized specimens, extrinsic toughening mechanisms, including shear ligament bridging, act in the crack wake increasing the crack growth resistance of fully lamellar structured specimens. The crack growth resistance increases rapidly with increasing length of crack wake for lamellar structured TiAl alloys [19]. In contrast, the crack length in micro-sized specimens is only 2-3 μm. This indicates that extrinsic toughening mechanisms are not activated in micro-sized specimens. This is one reason that K_Q values are believed to be lower compared to those of macro-sized specimens. This also suggests that intrinsic fracture toughness can be evaluated using micro fracture toughness testing developed in this investigation. The results obtained in this investigation give important and fundamental information for designing TiAl based alloys with high fracture toughness.

CONCLUSIONS

A micro-sized testing technique has been applied to investigate the fracture properties of lamellar colonies in a fully lamellar Ti-46Al-5Nb-1W alloy. The following conclusions were obtained.
1) Fracture tests were successfully completed using a mechanical testing machine for micro-sized specimens at room temperature.
2) Fracture toughness (K_Q) values were obtained in the range 1.4 – 7 MPam$^{1/2}$. Fracture surface observations indicate that these variations are attributable to differences in local lamellar orientations ahead of the notch.
3) The fracture toughness values obtained are lower than those of macro-sized specimens. This may be due the absence of significant extrinsic toughening mechanisms in these micro-sized specimens. This suggests that intrinsic fracture toughness can be evaluated using micro fracture toughness testing developed in this investigation.

ACKNOWLEDGEMENTS

This work was supported by a "Grants-in-Aid for Scientific Research (C)" from Japan Society for Promotion of Science (JSPS).

REFERENCES

1. R. E. Schafrik, in *Proceedings of third International Symposium on Structural Intermetallics (ISSI 3)*, edited by K. J. Hemker, D. M. Dimiduk, et al. (TMS, Warrendale, PA, 2002), pp.13-17.
2. R. Pather, A. Wisbey, A. Partridge, T. Halford, D. N. Horspool, P. Bowen and H. Kestler, in *Proceedings of third International Symposium on Structural Intermetallics (ISSI 3)*, edited by K. J. Hemker, D. M. Dimiduk, et al. (TMS, Warrendale, PA, 2002), pp.207-215.
3. S. C. Huang and E. L. Hall, *Metall. Trans. A.*, **22A**, 427 (1991).
4. Y. Kim, *Acta Metall.*, **40**, 1121 (1992).
5. R. O. Ritchie, *International Journal of Fracture*, **100**, 55 (1999).
6. T. Nakano, T. Kawanaka, H. Y. Yasuda and Y. Umakoshi, *Mater. Sci. Eng. A*, **194**, 43 (1995).
7. S. Yokoshima and M. Yamaguchi, *Acta Mater*, **44**, 873 (1996).
8. K. S. Chan, *Metall. Trans. A*, **24A**, 569 (1993).
9. K. S. Chan and Y. W. Kim, *Acta Metall.*, **43**, 439 (1995).
10. Y. Higo, K. Takashima, M. Shimojo, S. Sugiura, B. Pfister and M. V. Swain, in *Materials Science of Microelectromechanical Systems (MEMS) Devices II*, edited by M. P. de Boer, A. H. Heuer, S. J. Jacobs and E. Peeters, (Mater. Res. Soc. Proc. **605**, Pittsburgh, PA, 2000), pp. 241-246.
11. S. Maekawa, K. Takashima, M. Shimojo, Y. Higo, S. Sugiura, B. Pfister and M. V. Swain, *Jpn. J. Appl. Phys.*, **38**, 7194 (1999).
12. K. Takashima, Y. Higo, S. Sugiura and M. Shimojo, *Mat. Trans.*, **42**, 68 (2001).
13. K. Takashima, M. Shimojo, Y. Higo and M. V. Swain, *ASTS STP-1413*, 72 (2001).
14. K. Takashima, S. Koyama, K. Nakai and Y. Higo, in *Nano- and Microelectromechanical Systems (NEMS and MEMS) and Molecular Machines*, edited by D. A. LaVan, A. A. Ayon, T. E. Buchhiet and M. J. Madou, (Mater. Res. Soc. Proc. **741**, Pittsburgh, PA, 2000), pp. 35-40.
15. H. Okamura, *Introduction to Linear Fracture Mechanics*, (Baifukan, Tokyo, 1976) pp.218 (in Japanese).
16. K.S. Chan, P. Wang, N. Bhate and K. S. Kumar, *Acta Mater.*, **52**, 4601 (2004).
17. M. H. Yoo and K. Yoshimi, *Intermetallics*, **8**, 1215 (2000).
18. T.P.Halford, *Fatigue and Fracture of a High Strength, Fully Lamellar γ-TiAl based Alloy*, PhD Thesis, The University of Birmingham, (2003).
19. K. T. V. Rao, Y. W. Kim, C. L. Muhlstein and R. O. Ritchie, *Mat. Sci. Eng. A*, **192/193**, 474 (1995)

Mater. Res. Soc. Symp. Proc. Vol. 842 © 2005 Materials Research Society S5.45

High-Temperature Environmental Embrittlement of Thermomechanically Processed TiAl-Based Intermetallic Alloys with Various Kinds of Microstructures

T. Takasugi, Y. Hotta, S. Shibuya, Y. Kaneno, H. Inoue and T. Tetsui[1]
Department of Metallurgy and Materials Science, Graduate School of Engineering, Osaka Prefecture University, 1-1 Sakai, Osaka 599-8531, Japan
[1]Mitsubishi Heavy Ind Co Ltd, Nagasaki Research & Development Center, 5-717-1 Fukahori-Machi, Nagasaki, 851-0392, Japan

ABSTRACT

Thermomechanically processed TiAl-based intermetallic alloys with various alloy compositions and microstructures were tensile tested in various environmental media including air, water vapor and a mixture gas of 5vol.%H_2+Ar as a function of temperature. All the TiAl-based intermetallic alloys showed reduced tensile fracture stress (or elongation) in air, water vapor and a mixture gas of 5vol.%H_2+Ar not only at ambient temperature (RT~600K) but also at high temperature mostly from 600K to 1000K (sometimes higher temperature than 1000K). The high-temperature environmental embrittlement of TiAl-based intermetallic alloy depended upon the microstructure. The possible species causing the high-temperature environmental embrittlement are hydrogen atoms decomposed from water vapor (H_2O) or hydrogen gas (H_2), similar to those causing the low-temperature environmental embrittlement.

INTRODUCTION

Gamma (γ) TiAl alloys have been considered as a potentially important aerospace and vehicle structural material because of their lightweight, good high-temperature mechanical properties, and oxidation resistance [1]. Various kinds of microstructures could be obtained by compositional modification and microstructural control on intermetallic alloys based on gamma (γ) TiAl. Depending on desired properties and applications, the most appropriate microstructure could be chosen from these microstructures. However, they generally exhibit low ductility and poor fracture toughness until intermediate temperature.

The influence of hydrogen on the ambient mechanical properties of intermetallic alloys based on gamma (γ) TiAl has been focused by some researchers [2-6]. Also, it has been reported that the so-called environmental embrittlement occurs at ambient temperature in intermetallic alloys based on gamma (γ) TiAl [7-11]. In these cases, hydrogen is introduced from test atmospheres such as hydrogen gas or air, the moisture in which reacts with the alloy and to generate atomic hydrogen, resulting in reduced tensile elongation. Using Ti-49at.%Al alloy with a dual-phase microstructure and Ti-48Al-2Cr-2Nb alloy with a γ phase structure, it has been recently reported that the presence of hydrogen resulted in a reduction in the tensile elongation in the temperature range until 973K [12-14]. In this study, thermomechanically processed TiAl-based intermetallic alloys with various kinds of microstructures are tensile tested in various environmental media including air, water vapor and a mixture gas of 5vol.%H_2+Ar from room temperature to 1100K. It will be shown that the reduction of tensile fracture stress and elongation generally occur in atmosphere such air, water vapor and a mixture gas of 5vol.%H_2+Ar at high temperature between 600K and 1000K (or higher temperature).

EXPERIMENTAL PROCEDURES

Three different kinds of TiAl-based intermetallic alloys, which have γ grain microstructure, fully lamellar microstructure and duplex microstructure containing β phase, were thermomechanically fabricated in this study. Metallographic and structural investigations of

TiAl-based intermetallic alloys prior to or after tensile test were carried out by optical microscopy (OM), X-ray diffraction (XRD), scanning electron microscopy (SEM) attached with electron back scattering pattern (EBSP).

Tensile specimens with a gauge dimension of approximately 10x2x1 mm^3 were prepared by an electro discharge machine (EDM) and abraded on sufficiently fine SiC paper. Tensile tests were conducted within a metal tube surrounded by an electric furnace, at a temperature from room temperature to 1100K, mostly at a fixed strain rate of 1.66×10^{-4} s^{-1}. As test atmospheres, seven kinds of environmental media were used: vacuum, laboratory air, water vapor, a mixture gas of 5vol.%H$_2$+Ar, oxygen (O$_2$) gas, nitrogen (N$_2$) gas and Ar gas. Repeated evacuation of and subsequent gas purges into a metal tube were done before the tensile test. The degree of vacuum was $<1.2 \times 10^{-3}$ Pa. The purities of H$_2$ gas, Ar gas, oxygen gas and nitrogen gas were >99.998%, >99.999%, >99.995% and >99.999%, respectively. The humidity in laboratory air was kept between 50% and 75% through day and season. As the environmental media of water vapor, water vapor boiled under 1 atmospheric pressure at 373K was introduced to a metal tube in which tensile test was performed. Tensile fracture stress and tensile elongation were evaluated from the recorded charts. Also, the data taken from test in each atmosphere were normalized by those taken from test in vacuum.

RESULTS

Figure 1 shows the optical microstructures of Ti-46Al-7Nb-1.5Cr, Ti-44Al-7Nb-0.2C and Ti-42Al-5Mn alloys. The microstructure of Ti-46Al-7Nb-1.5Cr alloy exhibits relatively small-grained γ phase. The γ grains are slightly elongated to a direction perpendicular to forging axis. The microstructure of Ti-44Al-7Nb-0.2C alloys exhibits a fully lamellar structure and has a grain size of approximately 45 μm. The microstructure of Ti-42Al-5Mn alloy exhibits a duplex microstructure of γ and lamellar (γ/α$_2$) grains mixed with isolated β phase.

Ti-46Al-7Nb-1.5Cr alloy

Figure 2 shows the changes of tensile fracture stress and tensile elongation, and normalized tensile fracture stress by temperature for Ti-46Al-7Nb-1.5Cr alloy that was deformed in vacuum, air and a mixture gas of 5vol.%H$_2$+Ar. In this case, an apparent plastic deformation after yielding was not observed except for the specimen deformed in vacuum at high temperature beyond 900K. The measured tensile fracture stress tends to increase with increasing temperature and then makes a peak around at 800K, followed by a decrease at further high temperature. It is evident from the normalized curves that the tensile fracture stress in the alloys deformed not only in air but also in a mixture gas of 5vol.%H$_2$+Ar was clearly reduced at ambient temperature, indicating the

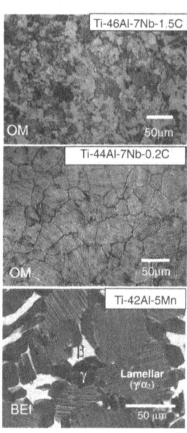

Figure 1 Optical or SEM microstructures of Ti-46Al-7Nb-1.5Cr, Ti-44Al-7Nb-0.2C and Ti-42Al-5Mn alloys.

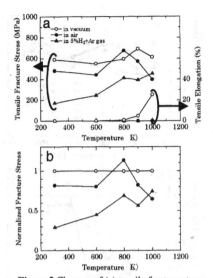

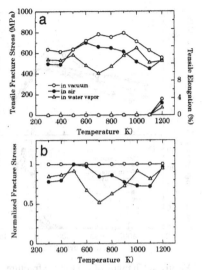

Figure 2 Changes of (a) tensile fracture stress and tensile elongation, and (b) normalized tensile fracture stress by temperature for Ti-46Al-7Nb-1.5Cr alloy, which was deformed in vacuum, air and a mixture gas of 5vol.%H₂+Ar.

Figure 3 Changes of (a) tensile fracture stress and elongation, and (b) normalized tensile fracture stress by temperature for Ti-44Al-7Nb-0.2C alloy, which was deformed in vacuum, air and water vapor.

occurrence of the low-temperature environmental embrittlement. Then, such a low-temperature environmental embrittlement tends to diminish around at 800K. However, the tensile fracture stress is again reduced at high temperature beyond 800K, indicating the existence of the high-temperature environmental embrittlement. Both types of the environmental embrittlement appear to be severer in a mixture gas of 5vol.%H₂+Ar than in air, except for the result at the highest temperature (1000K).

Ti-46Al-7Nb-0.2C alloy

Figure 3 shows the changes of tensile fracture stress and tensile elongation, and also normalized tensile fracture stress by temperature for Ti-44Al-7Nb-0.2C alloy, which was deformed in vacuum, air and water vapor. In this alloy, an apparent plastic deformation after yielding was not observed even except for the specimen deformed at the highest temperature (1200K). The measured tensile fracture stress tends to increase with increasing temperature, and then makes a peak around at 800K, followed by a decrease at further high temperature. In the specimen deformed in air as well as in water vapor, the low-temperature environmental embrittlement tends to almost diminish at 500K, while the high-temperature environmental embrittlement becomes obvious at 600K, and then more substantial with increasing temperature. It is interesting to note that the high-temperature embrittlement maximizes at lower temperature in the specimen deformed in water vapor than in the specimen deformed in air.

Ti-42Al-5Mn alloy

Figure 4 shows the changes of tensile fracture stress and tensile elongation by temperature for Ti-42Al-5Mn alloy, which was deformed in various atmospheres. The measured tensile fracture stress generally tends to increase with increasing temperature and then makes a peak around at 900K, followed by a decrease at further high temperature. Also, the tensile elongation shows the

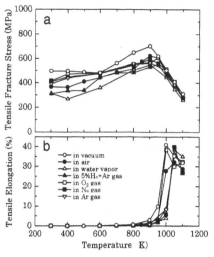

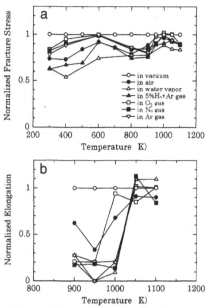

Figure 4 Changes of (a) tensile fracture stress and (b) tensile elongation by temperature for Ti-42Al-5Mn alloy, which was deformed in various atmospheres.

Figure 5 Changes of normalized (a) tensile fracture stress and (b) tensile elongation by temperature for Ti-42Al-5Mn alloy, which was deformed in various atmospheres.

behavior such as brittle-ductile transition (BDT) by which the tensile elongation rapidly increases with increasing temperature, and the brittle-ductile transition temperature BDTT is dependent on atmosphere. It is assumed that at temperature beyond BDTT, considerable glide deformation and climb motion by dislocations begin to take place owing to the reduction of Peierls stress and also the increase of atomic diffusion, resulting in high elongation values.

It is evident from the normalized curves (Fig. 5) that the tensile fracture stress in the specimens deformed in water vapor, air and a mixture gas of 5vol.%H_2+Ar was substantially reduced at ambient temperature, indicating the occurrence of the low-temperature environmental embrittlement. Then, the low-temperature environmental embrittlement tends to reduce with increasing temperature. However, the high-temperature environmental embrittlement alternatively begins to take place from 600K before the low-temperature environmental embrittlement disappears, and then maximizes at 800-1000K, followed by a recovery at sufficiently high temperatures beyond 1000K. On the other hand, the reduction of the tensile fracture stress in the specimens deformed in oxygen, nitrogen and Ar gas were smaller than that of the specimens deformed in the former group of atmospheres. Consequently, the degree of the environmental embrittlement may be ranked in sequence, water vapor > a mixture gas of 5vol.%H_2+Ar > air > oxygen gas ≅ nitrogen gas ≅ Ar gas. This rank is commonly applicable to both of the low-temperature and high-temperature environmental embrittlement. Here, it is noted that the environmental embrittlement observed in the specimens deformed in oxygen gas, nitrogen gas and Ar gas is caused by a trace amount of water vapor contained in these gases. From the normalized tensile elongation evaluated at temperature above 900K, it is clearly found that the high-temperature environmental embrittlement maximizes at 950K, and then almost fully recovers at temperature beyond 1100K. It is again noted that the BDT (recovery) of tensile elongation (or fracture stress) with increasing temperature takes place at a lower temperature in the specimens deformed in air and oxygen gas than in the specimens deformed in the other atmospheres. This result may indicate that air or oxygen gas is less effective in causing the high-temperature environmental embrittlement, rather, beneficial to reducing the high-temperature environmental embrittlement.

DISCUSSION

In the previous study using isothermally forged TiAl-based intermetallic alloys with various microstructures [11] and other TiAl-based alloys [7], the reduction of fracture stress or tensile elongation has been observed at ambient temperature when they are deformed in air and hydrogen gas. It has been suggested that the moisture-induced (or hydrogen-induced) embrittlement of many intermetallic alloys including TiAl occurs by the decomposition of moisture (or hydrogen gas) on alloy surface (or freshly exposed grain boundaries or cleavage planes), and subsequent micro-processes and condensation of atomic hydrogen to grain boundaries (or lattice planes) in front of a propagating micro crack [15-16]. The grain boundary cohesion (or lattice cohesion) and the associated plastic work can be reduced by hydrogen condensation. It has been suggested that the decomposition of hydrogen atom from moisture and hydrogen gas is caused by the reaction of $M + H_2O \rightarrow 2H^+ + MO + 2e^-$ (where M is active component element, e.g. Al in the present alloys) and $H_2 \rightarrow 2H^+ + 2e^-$ (on alloy surface with the help of catalysis such as Ti in the present alloys), respectively. The severest low-temperature environmental embrittlement observed in water vapor and a mixture gas of $5vol.\%H_2+Ar$ is attributed to high contents of H_2O and H_2 molecule in atmosphere. When temperature increases, the amount of absorbed hydrogen becomes small due to the reduction in the capacity of hydrogen condensation into relevant place. Consequently, hydrogen cannot reach the critical content facilitating the subsequent propagation of a micro crack.

For the high-temperature environmental embrittlement observed in the present study, it is suggested that the hydrogen atoms decomposed not only from water vapor (H_2O) but also from hydrogen gas (H_2) are the species causing the high-temperature environmental embrittlement, as they behaved in the low-temperature environmental embrittlement. Again, the severest high-temperature environmental embrittlement observed in water vapor and a mixture of $5vol.\%H_2+Ar$ may be attributed to high contents of H_2O and H_2 molecule in atmosphere. Also, it is likely that the oxidized scale formed on alloy surface reduces the decomposition to hydrogen atoms and/or the penetration of hydrogen atoms into alloy interior, resulting in the reduction of the high-temperature environmental embrittlement in air and oxygen gas.

In the recent study using Ti-49at.%Al and Ti-48Al-2Cr-2Nb alloys, it has been reported that the presence of hydrogen resulted in a reduction in the tensile elongation in the range of temperature until 573K [12,13] or 973K [14]. It has been suggested that the internal hydrogen embrittlement may be attributed to the reduced cohesive strength of the lattice, whereas the external hydrogen embrittlement may be attributed to hydrogen-enhanced localized plastic deformation [12-14]. The temperature range in which the high-temperature environmental embrittlement was operating in this study similarly extended to 1000K. The mechanism responsible for the high-temperature environmental embrittlement may be different from the low-temperature environmental embrittlement reported in many metals and alloys, and also from the low-temperature environmental embrittlement observed in the present TiAl-based intermetallic alloys. Certainly, more detailed observation for hydrogen behavior and kinetics is required in association with the plastic deformation and the microstructure of TiAl-based alloys.

Last, it was found that the high-temperature environmental embrittlement of TiAl-based intermetallic alloy depends upon the microstructure, as the low-temperature environmental embrittlement has depended on the microstructure [11]. For examples, the reduced tensile fracture stress observed in Ti-44Al-7Nb-0.2C (with a fully lamellar microstructure) did not recover up to the highest temperature tested, while that observed in Ti-42Al-5Mn (with a duplex microstructure mixed with isolated β phase) fully recovered at 1050K. It is speculated that the high-temperature environmental embrittlement cannot occur when extensive plastic deformation promoted by climb and glide motion of dislocations takes place and thereby the flow strength become lower.

CONCLUSIONS

Some thermomechanically processed TiAl-based intermetallic alloys with various kinds of alloy compositions and microstructures were tensile tested in various atmospheres including air, water vapor and a mixture gas of 5vol.%H_2+Ar from room temperature to 1100K. The following results were obtained from the present study.

(1) All the TiAl-based alloys showed reduced tensile fracture stress or elongation in air, water vapor and a mixture gas of 5vol.%H_2+Ar in the low temperature range mostly reaching 600K, and also in the high temperature range from 600K to 1000K (sometimes higher temperature).

(2) It was found that the high-temperature environmental embrittlement of TiAl-based intermetallic alloy depends upon the microstructure as well as the low-temperature environmental embrittlement.

(3) The possible species causing the high-temperature environmental embrittlement are hydrogen atoms decomposed from water vapor (H_2O) or hydrogen gas (H_2), similar to those causing the low-temperature environmental embrittlement.

(4) It was demonstrated that the oxidized scale is effective in reducing the high-temperature environmental embrittlement of TiAl-based intermetallic alloys

ACKNOWLEDGEMENTS

This work was supported in part by the New Energy and Industrial Technology Development Organization (NEDO) through Ministry of Economy, Trade and Industry (METI). One of the authors (T.T.) gratefully acknowledges the financial support by the Light Metals Educational Foundation of Japan.

REFERENCES

1. S. C. Huang and J. C. Chesnutt, *Intermetallic Compounds, Volume 2, Practice,* ed. J. H. Westbrook and R. L. Fleischer, (John Wiley and Sons, 1995), pp. 73-90.
2. D. E. Matejczyk and C. G. Rhode, *Scripta Metall.* **24**, 1369 (1990).
3. W. Y. Chu and A. W. Thompson, *Scripta Metall,* **25**, 2133 (1991).
4. K. S. Chan and Y. –W Kim, *Metall. Trans. A,* **23**, 1663 (1992).
5. W. Y. Chu and A. W. Thompson, *Metall. Trans. A,* **23**, 1299 (1992).
6. K. W. Gao, W. Y. Chu, Y. B. Wang and C. M. Hsiao, *Scripta Metall,* **27**, 555 (1992).
7. T. Takasugi and S. Hanada, *J. Mater. Research,* **17**, 1739 (1992).
8. C. T. Liu and Y. –W. Kim, *Scripta Metall.* **27**, 599 (1992).
9. M. Nakamura, K. Hashimoto, T. Tsujimoto and T. Suzuki, *J. Mater. Res.* **8**, 68 (1993).
10. M. Nakamura, N. Itoh, K. Hashimoto, T. Tsujimoto and T. Suzuki, *Metall. Trans. A* **25**, 321 (1994).
11. T. Tsuyumu, Y. Kaneno, H. Inoue and T. Takasugi, *Metall. Mater. Trans. A* **34**, 645 (2003).
12. K. –W. Gao and M. Nakamura, *Mater. Sci. Eng. A,* **325**, 66 (2002).
13. K. –W. Gao and M. Nakamura, *Interemetallics,* **10**, 233 (2002).
14. M. Nakamura, E. Abe, K. Gao, L. Qiao and W. Chu, *ISIJ International,* **43**, 489 (2003).
15. T. Takasugi and O. Izumi, *Acta Metallurgica,* **34**, 607 (1986).
16. C. T. Liu, *6th Int. Symp. Intermetallic Compounds - Structure and Mechanical Properties,* ed. O. Izumi, (JIM, 1991) pp. 703-712.

Mater. Res. Soc. Symp. Proc. Vol. 842 © 2005 Materials Research Society
S5.47

Experimental Studies and Thermodynamic Simulation of Phase Transformations in γ-TiAl Based Alloys

Harald F. Chladil[1], Helmut Clemens[1], Harald Leitner[1], Arno Bartels[2], Rainer Gerling[3], Wilfried T. Marketz[4], and Ernst Kozeschnik[5]

[1] Dept. of Physical Metallurgy and Materials Testing, Montanuniversitaet, A-8700 Leoben, Austria
[2] Materials Science and Technology, TU-Hamburg-Harburg, D-21071 Hamburg, Germany
[3] Institute for Materials Research, GKSS-Research Centre, D-21502 Geesthacht, Germany
[4] Boehler-Schmiedetechnik GmbH&CoKG, A-8605 Kapfenberg, Austria
[5] Institute for Materials Science, Welding and Forming, Graz University of Technology, A-8010 Graz, Austria

ABSTRACT

Phase transformations and phase transition temperatures in several Nb-rich γ-TiAl based alloys were investigated experimentally and compared to thermodynamic simulations. The present study combines light-optical and scanning electron microscopy, X-ray diffraction and differential-scanning-calorimetry for the characterization of the prevailing phases and phase transformations. Thermodynamic simulation based on the CALPHAD method was used for predict phase stabilities. The results from experiments on a variety of γ-TiAl based alloys are compared to thermodynamic calculations. Finally, the influence of carbon on the transition temperatures is presented.

INTRODUCTION

Research and development on γ-TiAl-based alloys have progressed significantly within the last 15 years [1,2]. Current γ-TiAl based alloys are complex multi-phase materials. Effective alloy development, hot working, and subsequent heat treatments require the knowledge of the constituent phases and their transformation kinetics. For casting, the solidification path has to be known, whereas for thermo-mechanical processing, α and β transus temperatures, which sensitively depend on the alloy composition, are of particular importance. Knowledge of the influence of alloying elements on the amount and the thermal stability of the stable and metastable phases is the basis for heat treatments, which are applied in order to optimize mechanical properties. In particular, these high Nb bearing alloys with the baseline composition Ti-(42-45)Al-(5-10)Nb+X (at%) have attracted a lot of attention due to their combination of high creep strength, good ductility at room temperature and excellent oxidation resistance [3,4]. The term X stands for small amounts of metallic and non-metallic alloying elements. Here, carbon is of particular interest because of its applicability to precipitation hardening [3].

In the present paper the phase transformations in five different variants of high Nb bearing γ-TiAl based alloys were investigated: Ti-Al-7.5Nb-1(Si, B, C) with 42 and 45 at% Al and a Ti-45Al-7.5Nb base alloy with varying amounts of C (0, 0.25 and 0.5 at%).

EXPERIMENTAL AND SIMULATION

The Ti-Al-7.5Nb-1(Si, B, C) alloys with 42 and 45at% Al were prepared by means of vacuum arc remelting (VAR). After double remelting, the levels of O, N and H were analyzed to be 700ppm, 70ppm and 30ppm, respectively. In order to reduce chemical inhomogeneities caused by segregational effects during solidification, the ingots were hot-extruded at 1250°C.

For the manufacture of the Ti-45Al-7.5Nb base alloys with 0, 0.25 and 0.5 at% C, a powder metallurgical approach was used which guarantees a more homogeneous chemical distribution of the constituting elements [5]. Pre-alloyed powders were produced by means of gas atomization and subsequently densified by hot-isostatic pressing (HIPing) at 200MPa and 1280°C for 2 hours. The content of O, N and H analyzed in the HIPed materials was comparable to that obtained for ingot materials.

The sequence of phase transformations was studied by differential-scanning-calorimetry (DSC) using a Netzsch STA 409C. The measurements were performed in a dynamic Ar atmosphere with a flow rate of 50 ml/min. Dynamical DSC experiments were carried out from room temperature up to 1450°C at heating rates of 10, 20 and 40 Kmin^{-1} using platinum crucibles with Al_2O_3 inserts.

The phase constitution of the different alloys was determined using X-ray diffraction (XRD), scanning electron microscopy (SEM) in back-scattered electron (BSE) mode and electron back-scattered diffraction (EBSD). XRD patterns were recorded with a Siemens D500 diffractometer in the Bragg–Brentano mode using Cu K$_\alpha$ radiation. A Cambridge Instruments Stereoscan 360 equipped with an Oxford Instruments Inca Crystal 300 was used for SEM and EBSD investigations.

Thermodynamic equilibrium calculations were performed using the software ThermoCalc [6] and MatCalc [7] and a commercially available TiAl database [8]. Details of modeling can be found in the reviews of Ansara [9] and Saunders and Miodownik [10]. For multi-component γ-TiAl based alloys with alloying element concentrations < 5 at%, good agreement between calculated and experimentally determined α-transus temperatures was found [8,11,12]. However, it should be noted that C, which is an essential alloying element in advanced γ-TiAl based alloys, is currently not available in the commercial TiAl database.

RESULTS AND DISCUSSION

Figure 1a shows the microstructure of HIPed Ti-45Al-7.5Nb. Due to HIPing in the middle of the (α+β) phase field, a duplex microstructure is formed upon furnace cooling to room temperature. The microstructure is very homogeneous with no visible segregation. In contrast, Figure 1b displays the rather inhomogeneous microstructure of a Ti-45Al-7.5Nb-1(Si,B,C) ingot after extrusion at 1250°C which corresponds to a temperature in the (α+γ) phase field. Macro-segregations which are present in the as-cast state can also be observed after extrusion leading to the appearance of a banded structure. EDX analyses exhibited a strong variation of the Al and Nb content especially transverse to the extrusion direction. The former interdendritic regions are rich in Al, whereas in the former dendrite cores an increase of Nb is observed [13]. As a consequence, the extruded microstructure consists of a banded duplex structure of small μm-sized equiaxed γ-TiAl grains with strings of sub-μm sized $α_2$-Ti$_3$Al grains in between. It should be noted that these chemical fluctuations lead to locally different phase transition temperatures which, for example, make the evaluation of DSC data more difficult. Additionally, titanium borides enriched in Nb can be detected in Figure 1b. These rod-shaped particles are aligned in extrusion direction.

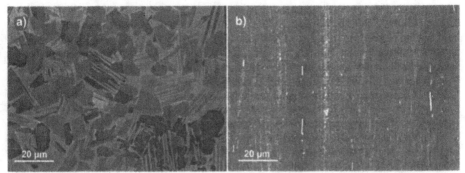

Figure 1. Microstructure of **(a)** HIPed Ti-45Al-7.5Nb and **(b)** Ti-45Al-7.5Nb-1(Si,B,C) ingot after hot extrusion (extrusion direction is vertical).

In previous studies, the α-transus temperature (T_α) of Ti-45Al-7.5Nb and other γ-TiAl based alloys was determined by metallographic examination of samples which were annealed little above and below T_α followed by fast cooling to room temperature. Since the annealing time was kept sufficient long, these T_α temperatures are considered to be determined under "thermodynamic equilibrium" conditions. Simultaneously, the T_α temperatures were obtained from DSC measurements ("dynamic T_α") using a heating rate of 10 Kmin^{-1}. A comparison of the differently determined T_α temperatures shows good agreement.

Figure 2a presents the results of DSC measurements conducted on Ti-45Al-7.5Nb with 0, 0.25 and 0.5 at% C. The evaluation of the DSC curves was carried out according to Cagran [14]. From Figure 2a it is evident that Nb and C shift the eutectoid temperature (T_{eu}), whereas the α-transus temperature T_α of all three alloys is almost identical.

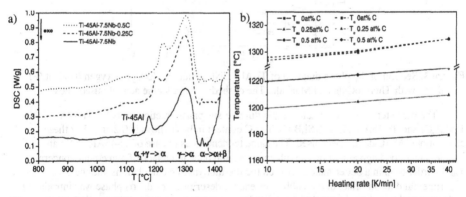

Figure 2. (a) Effects of C additions on phase transformations. The DSC curves were obtained from a ternary and two C-containing alloys (see inset) on heating the specimens with 20 Kmin^{-1}. The eutectoid (T_{eu}), α-transus (T_α) and $\alpha/(\alpha+\beta)$ transition temperatures were determined by extrapolation of the peaks in heat flow. The arrow indicates the eutectoid temperature of binary Ti-45Al [3]. For better readability the curves belonging to C-containing alloys were shifted along the ordinate; (b) T_{eu} and T_α versus DSC heating rate.

The variation of T_{eu} and T_α as a function of the DSC heating rate is illustrated in Figure 2b. A Nb content of 7.5 at% shifts T_{eu} to about 1175°C (heating rate: 10 Kmin^{-1}). A similar effect has been observed for Ti-(44-49)Al alloys with 8 at% Nb [15]. Alloying with C has an even more pronounced impact on the position of T_{eu} (Figure 2b). From the present results it is tempting to speculate that C stabilizes the ordered α_2-Ti$_3$Al phase and thus raises the eutectoid temperature. The influence of Nb and C on T_α is rather small. For a heating rate of 10 Kmin^{-1}, a T_α of about 1280°C was determined, independent of the alloy composition. The influence of the different heating rates on the phase transformation temperatures is comparable which indicates that C raises the eutectoid temperature, but has no significant effect on the kinetics

Figure 3 shows a calculated section of the Ti-Al-Nb phase diagram. The thermodynamic simulations were carried out by keeping the Nb content constant (7.5 at%). In addition, an oxygen content of 0.15 at% was taken into account. The calculated temperatures are: T_{eu}=1080°C and T_α=1330°C. Comparison with the experimental data from Ti-45Al-7.5Nb demonstrates that the calculation gives a good prediction of the α-transus temperature. However, a discrepancy with regard to the eutectoid temperature was observed. The experimental data (Figure 2) as well as data from literature [3,15] indicate an increase of T_{eu}, whereas the calculated value is below that determined for binary TiAl ($T_{eu} \sim$ 1125°C) [16]. Obviously, the current TiAl database cannot describe the eutectoid reaction in high Nb-containing γ-TiAl based alloys correctly.

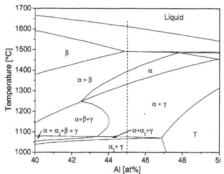

Figure 3. Section of the phase diagram of Ti-Al-7.5Nb including 700ppm oxygen (0.15 at.%) calculated with ThermoCalc and MatCalc. The dashed line marks the actually used alloy

The next step was to calculate temperature versus phase fraction plots for Ti-42Al-7.5Nb-1(Si,B,C) and Ti-45Al-7.5Nb-1(Si,B,C). The C content in both alloys is 0.2 at%. For these calculations, MatCalc has been used. The same tendencies as in case of Ti-45Al-7.5Nb are visible, i.e. a good agreement of T_α with experimental data and a lack of agreement concerning T_{eu}. In order to gain a better match between the thermodynamic calculations and the experimental data an offset in the Gibbs free energy description of the α_2 phase was introduced and optimized based on the results of the DSC measurements. Details of the applied procedure will be published in a forthcoming paper. With this approach, the effect of C as an α_2-Ti$_3$Al stabilizer has been considered indirectly. The results of these modified calculations are shown in Figure 4. As a result the calculated eutectoid temperature increases up to 1160°C, accompanied by a small change of the alpha-transus temperature. A comparison of the predicted data with the experimental ones shows that the experimental value of T_{eu} is about 30°C higher and T_α about 20°C lower than the predicted values.

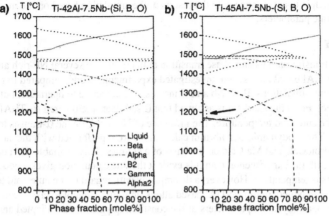

Figure 4. Calculated phase fraction (mole%) versus temperature for (a) Ti-42Al-7.5Nb-1(Si,B,C) and (b) Ti-45Al-7.5Nb-1(Si,B,C). The C content in both alloys is 0.2 at%. An oxygen content of 0.15 at% was taken into account. The arrow in (b) denotes the B2 phase field.

In order to obtain additional verification of the thermodynamic calculations, annealing treatments were performed on the Ti-45Al-7.5Nb-1(Si,B,C) alloy. The arrow in Figure 4b indicates a small phase field where the B2 phase is stable. An annealing treatment was conducted at 1180°C where the maximum phase fraction of B2 is expected. Additionally, a sample was annealed at 1100°C. At this temperature MatCalc predicts only γ and α2 as stable phases. After annealing for 3 hours, the samples were quenched in water. Figure 5 displays a SEM image in BSE mode of the sample that was annealed at 1180°C.

Figure 5. Microstructure of a Ti-45Al-7.5Nb-1(Si, B,C) alloy after annealing at 1180°C for 3 hours and subsequent water quenching. Bright contrast: B2 particles; dark contrast: γ grains; medium contrast: α2 phase.

The micrograph shows bright particles at colony boundaries and triple junctions of γ grains. These particles were identified as B2 phase by means of XRD and EBSD. B2 and β phase can be distinguished due to the different atomic configuration within their unit cells. A quantitative analysis of the EBSD data yields a volume fraction of about 5 percent which agrees well with the predicted fraction (Figure 4a). However, in the sample annealed at 1100°C no B2 phase could be detected, which is in accordance with the results of the simulation based on the modified thermodynamic parameters.

SUMMARY

Phase transformations and phase transition temperatures in several Nb-rich and C-containing γ-TiAl based alloys were investigated experimentally and compared to results of thermodynamic simulations. Both Nb and C increase the eutectoid temperature, but C shows a stronger effect. From the results it is assumed that C stabilizes the ordered α_2-Ti$_3$Al phase and thus raises the eutectoid temperature. The influence of Nb and C on the α-transus temperature is rather small. Thermodynamic equilibrium calculations were performed with the aid of the software ThermoCalc and MatCalc on the basis of a commercially available TiAl database. From comparison with the experimental data it is evident that the calculation gives a good prediction of the α-transus temperature. However, a discrepancy with regard to the eutectoid temperature is observed in high Nb-containing γ-TiAl based alloys. In a first attempt to take the effect of C into account, the thermodynamic parameters of the constituting phases were adapted and reasonable agreement between experiment and thermodynamic simulation could be achieved.

REFERENCES

[1] F. Appel and M. Oehring, Titanium and Titanium Alloys, ed. M. Peters and C. Leyens (Wiley-Vch, Weinheim), pp. 89-152

[2] H. Kestler and H. Clemens, Titanium and Titanium Alloys, ed. M. Peters and C. Leyens (Wiley-Vch, Weinheim), pp. 351-392

[3] F. Appel, M. Oehring and R. Wagner, Intermetallics 8, 1283 (2000)

[4] V. A. C. Haanappel, H. Clemens and M. F. Stroosnijder, Intermetallics 10, 293 (2002)

[5] R. Gerling, H. Clemens and F.P. Schimansky, Advanced Engineering Materials 6, 23 (2004)

[6] B. Sundman, B. Jansson and J.-O. Andersson, CALPHAD 9, 153 (1985)

[7] E. Kozeschnik and B. Buchmayr, Mathematical Modelling of Weld Phenomena 5, eds. H. Cerjak and H.K.D.H. Bhadeshia (IOM Communications, London, book 738, 2001), p. 349.

[8] N. Saunders, Gamma Titanium Aluminides 1999, ed. Y-W. Kim, D.M. Dimiduk and M.H. Loretto (Minerals, TMS, Warrendale, PA, 1999), p. 183

[9] I. Ansara, Int. Met. Reviews 22, 20 (1979)

[10] N. Saunders and A.P. Miodownik, CALPHAD-A Comprehensive Guide (New York: Elsevier Science, 1998)

[11] M. Beschliesser, H. Clemens, H. Kestler and F. Jeglitsch, Scripta Materialia 49, 279 (2003)

[12] S. Malinov, T. Novoselova and W. Sha, Mat. Science and Engineering A 386, 344 (2004)

[13] U. Brossmann, M. Oehring and F. Appel, Structural Intermetallics 2001m ed. K.J. Hemker, D.M. Dimiduk, H. Clemens, R. Darolia, H. Inui, J.M. Larsen, V.K, Sikka, M. Thomas, J.D. Whittenberger (TMS, Warrendale, PA, 2001), p. 191

[14] C. Cagran, B. Wilthan, G. Pottlacher, B Roebuck, M. Wickins, R.A. Harding, Intermetallics 11, 1327 (2003)

[15] G.L. Chen, W.J. Zhang, Z.C,. Liu, S.J. Li and Y.W. Kim, Titanium Aluminides 1999, ed Y-W. Kim, D.M. Dimiduk and M.H. Loretto (TMS, Warrendale, PA, 1999), p. 371

[16] H. Okamoto, Journal of Phase Equilibria 14, 120 (1993)

Mater. Res. Soc. Symp. Proc. Vol. 842 © 2005 Materials Research Society   S5.48

Massive Transformation in High Niobium Containing TiAl-Alloys

A. Bartels[a], S. Bystrzanowski[a], H. Chladil[b], H. Leitner[b], H. Clemens[b], R. Gerling[c], and F.-P. Schimansky[c]

[a]Materials Science and Technology, TU Hamburg-Harburg, D-21073 Hamburg, Germany
[b]Department of Physical Metallurgy and Materials Testing, Montanuniversität Leoben, A-8700 Leoben, Austria
[c]Institute for Materials Research, GKSS-Research Centre, D-21502 Geesthacht, Germany

ABSTRACT

Massive transformation in high Nb bearing γ-TiAl-based alloys, Ti-45Al-7.5Nb and Ti-46Al-9Nb (at.%), and the thermal stability of the resulting microstructure were investigated. Using a quenching dilatometer, a nearly complete massive transformation in Ti-45Al-7.5Nb was found at about 1050°C after annealing at 1305°C for 10min and subsequent cooling with a rate of 55K/s. Higher starting temperatures and higher cooling rates lead to incomplete massive transformation and small transformed areas situated at the grain-boundary triple points of the parent α-grains are observed. By means of EBSD only in one case the same orientation of the close-packed planes of parent α-grains and of massively transformed γ_M-areas was observed.

The thermal stability of the microstructure of massively transformed Ti-46Al-9Nb sheet material was tested by annealing samples for 1 hour between 400 and 1200°C. Above 800°C a drop of hardness was measured and X-ray diffraction patterns show an increasing separation of $(200)_\gamma$ and $(002)_\gamma$ reflections as expected from a tetragonal γ-TiAl lattice. After annealing at 1100°C α_2-phase segregates at grain boundaries and after 1200°C α_2-lamellae appear inside the γ_M-grains parallel to all four $\{111\}_\gamma$-planes.

INTRODUCTION

Over the last years new concepts in alloy design of γ-TiAl based alloys were employed and a new alloy class exhibiting a high Nb content (5-10at.%) was developed [1,2]. These so-called TNB alloys combine high tensile strength at elevated temperatures with good ductility at room temperature. The high Nb content reduces the stacking fault energy and leads to widely dissociated superdislocations. As a consequence cross-slip and climb processes are effectively hampered [1]. Additionally, a high Nb-content diminishes diffusion processes in these alloys and thus decreasing the climb rate of dislocations [1,3]. Normally, in these new alloys a low Al-content around 45at.% is adjusted. The resulting lower α-transus temperature in combination with decelerated diffusion due to niobium facilitates massive transformation during cooling from the α-phase field. The massive transformation in γ-TiAl based alloys is subject of intensive research and of special interest because technical alloys seldom show massive transformation behavior [4]. In TiAl alloys these microstructural features can be used to increase the strength or to establish special microstructures by subsequent heat treatments in the $(\alpha+\gamma)$-phase field, e.g. fine γ-grains with crossed α_2-lathes [5]. Therefore, the aim of this study was to investigate massive transformation in two high Nb-containing γ-TiAl based alloys as well as the influence of a subsequent heat treatment on the thermal stability of this microstructure.

EXPERIMENTAL

Two high Nb containing γ-TiAl based alloys with nominal chemical compositions of Ti-45Al-7.5Nb and Ti-46Al-9Nb (at.%) were produced via the powder metallurgical route at GKSS (PIGA-facility) [6]. After densification by hot-isostatic pressing at 200MPa and 1280°C the powder compacts of Ti-46Al-9Nb were hot-rolled to sheets at Plansee AG [7]. After a final heat treatment for 3 hours at 1000°C the sheets show a fine-grained near-gamma microstructure.

For the quenching experiments cylindrical specimens with the dimension Ø4x10mm were cut from a compact of Ti-45Al-7.5Nb. The experiments were performed in a quenching dilatometer DIL805A/D supplied by Bähr-Thermoanalyse GmbH. The samples were heated under vacuum up to 1305°C, which is 5°C above the α-transus temperature, held for 10min and subsequently quenched with cooling rates of 55K/s, 12K/s and 2.2K/s.

Additionally, Ti-45Al-7.5Nb samples were quenched in oil from the α-phase field. After quenching the orientation image of small massively transformed areas was determined by means of EBSD using a Cambridge Instruments Stereoscan 360 equipped with an Oxford Instruments Inca Crystal 300 system.

Specimens of Ti-46Al-9Nb of 21x20x1mm^3 were annealed in a furnace for 7min at 1330°C under atmospheric conditions and subsequently quenched in oil. Due to the lost of Al from the surface during exposure to this high temperature a layer of about 100μm of α-phase has formed on both sides. However, the inner part is massively transformed. The massively transformed samples were then cut into pieces for subsequent 1h annealing treatments. After annealing the α-layer was removed and the surface was polished to examine the microstructure with light microscopy and/or SEM in back-scattered electron (BSE) mode. Vickers hardness measurements were conducted applying a load of 20N (HV2). The X-ray diffraction measurements were performed with a Bruker AXS D8 using Cu-K$_\alpha$ radiation.

RESULTS AND DISCUSSION

Figure 1 presents the rapid cooling experiments conducted on Ti-45Al-7.5Nb in terms of length change ΔL as well as the differentiated length change dL/dT versus temperature. Figure 2 summarizes the resulting microstructures. The transformation of α→γ is attended with a volume reduction appearing as peak in the dL/dT vs. T curves. Before the peak temperature is reached α-phase is present and the observed slope of dL/dT = 0.2 μm/K leads to a thermal expansion coefficient of ~ 2×10^{-5}K^{-1}. Applying a cooling rate of 55K/s an almost 100% massively transformed microstructure can be obtained (Figure 2a). The transformation starts at ~1080°C and ends at ~1030°C. The total length change during transformation is 17μm which corresponds to ΔL/L$_0$=1.7x10^{-3}. A cooling rate of 12K/s leads to a mixture of massively transformed grains and fine lamellar colonies (figure 2b). The transformation temperature is 20K higher than in the case of almost complete massive transformation and the change of length is 18μm (ΔL/L$_0$=1.8x10^{-3}). Cooling with 2.2K/s leads to a fully lamellar microstructure. The formation of the two-phase lamellar structure requires long-range diffusion and thus takes place above 1100°C. The broad peak in figure 1b indicates a duration of the transformation longer than 30s which obviously is enough for long-range diffusion of some 100nm to form a fully lamellar microstructure. The total change of length with 26μm (ΔL/L$_0$=2.6×10^{-3}) is 50% higher, because in the lamellar microstructure the transformed γ-phase has a higher Al-content compared to massively transformed γ$_M$. At temperatures below the transformation peaks (figure 1a) the thermal expansion coefficient is about 1.5x10^{-5}K^{-1}.

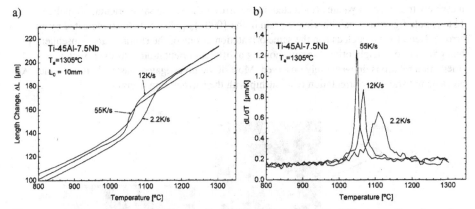

Figure 1. a) Change of length during cooling of Ti-45Al-7.5Nb with different cooling rates after soaking for 10min at 1305°C in He-atmosphere. The initial specimen length was 10mm. b) Differentiated shrinkage dL/dT versus temperature.

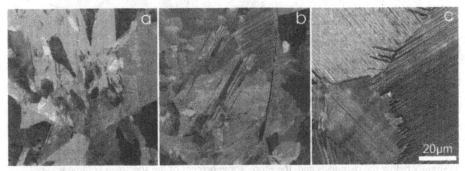

Figure 2. Microstructure of the 3 rapidly cooled Ti-45Al-7.5Nb specimens from Figure 1. a) cooling rate 55K/s, b) 12K/s and c) 2.2K/s. SEM images taken in BSE mode.

Ti-45Al-7.5Nb samples with a small fraction of massive transformed areas were prepared by oil quenching from the α–phase field. The α-grain size was about 500μm. After quenching only 10% of the specimen was massively transformed and the remaining part consisted of α_2 as the ordered form of the α-phase. The massively transformed areas occurred preferred at triple points of grain boundaries. Orientation images were performed by EBSD. As an example figure 3 shows the orientation maps of a very small massively transformed area and of the surrounding four α_2-grains (see the 'quality'-image in figure 3). The orientation image has 98 × 116 analysis points (about 53×63μm^2) with a resolution below 1μm. The pattern analysis for γ-TiAl was performed assuming a cubic structure, because the c/a-ratio is near 1 and thus no analysis of the tetragonal structure is possible. The hexagonal phase was analyzed as α instead of ordered α_2.

In figure 3 the orientation images are shown separately for the α- and γ-phase together with the pole figures of the close-packed planes, i.e. $\{0001\}_\alpha$ and $\{111\}_\gamma$. Only a small fraction of the massively transformed γ_M area (indicated by the arrow) shows the same orientation in the $\{111\}_\gamma$-pole figure as a neighboring α-grain in the $\{0001\}_\alpha$-pole figure (also indicated by an arrow). Here, the γ_M nucleus had the same orientation as one of the α-grains and subsequently had grown into a neighboring α-grain showing a different orientation. However, such a orientation relationship was rarely observed. Most of the investigated massively transformed γ_M-areas do not show any orientation relationship with the surrounding α-grains.

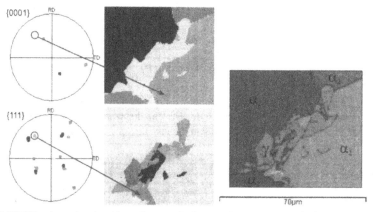

Figure 3. EBSD orientation imaging of a massively transformed area and of the surrounding α_2-grains. The analysis assumes the α_2 as α-Ti and γ_M as a cubic phase. The arrows indicate areas of γ_M and α where the $\{111\}_\gamma$ planes show the same orientation than the $\{0001\}_\alpha$-planes.

Ti-46Al-9Nb sheet specimens were oil-quenched from 1330°C. This annealing temperature is slightly below the α-transus and, therefore, some percent of the specimen volume was not transformed to α-phase. The small remaining γ-grains hinder very effectively the grain growth of α during annealing (4min at 1330°C). The α-grains had a grain size below 100μm. The high density of grain boundaries causes a high density of nuclei which leads to an almost complete massive transformation during quenching. Figure 4a shows the micrograph of the transformed γ_M after removing the α-layer from the surface of the specimens. Small γ-grains, which did not transform to α during annealing, appear in the micrograph as small white or black dots.

Massively transformed Ti-46Al-9Nb sheet specimens were annealed for 1h in 100°C steps in the temperature range of 400-1200°C. For each annealing treatment a new sample was used. After annealing the Vickers hardness was measured (figure 5) and a XRD 2θ-scan was recorded (figure 6a). Due to the high density of internal boundaries after massive transformation a hardness of 408HV2 was measured, which is about 25% higher than that of the initial near-gamma microstructure. In the X-ray diffraction spectrum of the massively transformed samples the $(200)_\gamma$ and $(002)_\gamma$-reflections could not be resolved which might be a consequence of high internal stresses and due to partially wrong site occupation of the constitutes atoms in the γ_M-lattice.

Figure 4. a) Microstructure of Ti-46Al-9Nb after massive transformation, (b) after 1h at 1100°C and (c) after 1h at 1200°C.

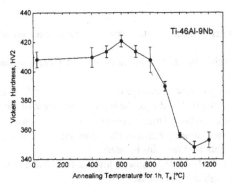

Figure 5. Vickers hardness HV2 of massively transformed Ti-46Al-9Nb after 1h annealing treatments.

a)   b)

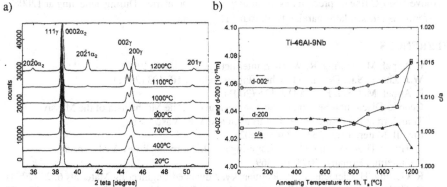

Figure 6. a) XRD scans of massively transformed Ti-46Al-9Nb after annealing for 1h. The $(20\bar{2}1)_{\alpha 2}$-reflection occurs in this spectrum, which is caused by a not completely removed α-layer. b) Lattice parameter and c/a-ratio evaluated from $(002)_\gamma/(200)_\gamma$-reflections in figure 6a.

Above 700°C the hardness decreases and an increasing separation of $(200)_\gamma$ and $(002)_\gamma$ reflections is observed. At this temperature recovery processes start and diminish internal stresses. From the X-ray diffraction data the lattice parameter and the c/a-ratio were calculated. After massive transformation the c/a-ratio is very low (c/a=1.006). Above 700°C the c/a-ratio increases during the 1h annealing treatment. Here, the precipitation of α_2 starts leading to an increase of the Al-content in the γ_M-phase. After annealing at 1100°C the α_2-precipitations become visible at the grain boundaries of the γ_M-phase (figure 4b). At 1100°C the hardness shows a minimum, indicating that recovery processes are finished. In addition, the $(200)_\gamma$ and $(002)_\gamma$-reflections are clearly separated (figure 6a). During annealing at 1200°C α_2-precipitations occur as laths inside the γ_M-grains (figure 4c) with their $\{0001\}_\alpha$-planes oriented parallel to all four $\{111\}_\gamma$-planes. These crossed α_2-laths cause an increase in hardness. The high α_2-fraction after annealing at 1200°C leads to a decrease of the Al-content in the γ-phase and consequently to an increase of the c/a-ratio up to 1.015.

SUMMARY

- A quenching dilatometer was used to study massive transformation in high Nb-containing γ–TiAl based alloys.
- In Ti-45Al-7.5Nb massive transformation takes place at 1050°C during cooling with 55K/s from 1305°C.
- The transformation $\alpha \rightarrow \gamma_M$ causes a change of length of 1.7×10^{-3}, whereas a fully lamellar transformation ends with 2.6×10^{-3} due to the higher Al-content of the γ-phase.
- Small areas of γ_M at α-grain boundaries were used for orientation imagining by means of EBSD. An orientation relationship between the close-packed planes $\{111\}_\gamma$ of γ_M and the $\{0001\}_\alpha$ planes of neighboring α-grains is rarely observed.
- Massively transformed Ti-46Al-9Nb shows a low c/a-ratio and a high hardness (410HV2) which is attributed to high internal stresses.
- Annealing for 1h above 700°C leads to a decreases in hardness and an increase in the c/a-ratio.
- Above 1000°C fine α_2-precipitates occur at γ_M-grain boundaries. During annealing at 1200° α_2 forms as crossed laths parallel to all four $\{111\}_\gamma$-planes.

REFERENCES

1. F. Appel, M. Oehring, R. Wagner, Intermetallics, **8**, 1283 (2000)
2. W.J. Zhang, S.C. Deevi, G.L. Chen, Intermetallics, **10**, 403 (2002)
3. F. Appel, Mat. Sci. Eng. **A317**, 115 (2001)
4. General Discussion Session of the Symposium on „The Mechanism of the Massive Transformation", edited by H.I. Aaronson and V.K. Vasudevan, Metall. Mater. Trans., **33A**, 2445 (2002)
5. D. Hu, in Ti-2003, edited by G. Lütjering and J. Albrecht (Wiley-VCH Verlag, Weinheim, Germany, 2004) p. 2369
6. R. Gerling, H. Clemens, F.-P. Schimansky, Advanced Engineering Materials 6, 23 (2004)
7. R. Gerling, A. Bartels, H. Clemens, H. Kestler, F.-P. Schimansky, Intermetallics **12**, 275 (2004)

Mater. Res. Soc. Symp. Proc. Vol. 842 © 2005 Materials Research Society   S5.50

The Creep Behaviors of Two Fine-grained XD TiAl Alloys Produced by Similar Heat Treatments

Hanliang Zhu[1],   Dongyi Seo[2],   Kouichi Maruyama[1],   Peter Au[2]
[1] Graduate School of Environmental Studies, Tohoku University, Sendai 980-8579,Japan
[2] Structures, Materials and Propulsion Laboratory, Institute for Aerospace Research, National Research Council of Canada, Ottawa, Ont., Canada K1A 0R6

ABSTRACT

The microstructural characteristics and creep behavior of two fine-grained XD TiAl alloys, Ti-45Al and 47Al–2Nb–2Mn+0.8vol%TiB$_2$ (at%), were investigated. A nearly lamellar structure (NL) and two kinds of fully lamellar (FL) structures in both alloys were prepared by selected heat treatments. The results of microstructural examination and tensile creep tests indicate that the 45XD alloy with a NL structure possesses an inferior creep resistance due to its coarse lamellar spacing and larger amount of equiaxed γ grains at the grain boundaries, whereas the same alloy in a FL condition with fine lamellar spacing lowers the minimum creep rates. Contrary to 45XD, the 47XD alloy with a NL structure exhibits the best creep resistance. However, 47XD with a FL structure with finer lamellar spacing shows inferior creep resistance. On the basis of microstructural deformation characteristics, it is suggested that the well-interlocked grain boundary and relatively coarse colony size in FL and NL 47XD inhibit sliding and microstructural degradation at the grain boundaries during creep deformation, resulting in better creep resistance. Therefore, good microstructural stability is essential for improving the creep resistance of these alloys.

INTRODUCTION

TiAl-based alloys are potential candidate structural materials for high temperature applications in turbine and automotive engines due to their low density and good elevated temperature properties. However, the poor ductility at ambient temperature has limited their applications. To overcome this obstacle, development efforts have been focused on refining the lamellar structure [1]. It has been shown that investment cast XD TiAl alloys have uniform cast microstructures with relatively fine grain size by adding boron [2]. Also, the lamellar structure can be further refined by appropriate heat treatment. However, because of different Al contents in 45XD and 47XD alloys, the microstructural features resulting from similar heat treatment conditions are different [2], and this microstructure dependence leads to different creep resistance [3]. So far, limited work has been carried out to investigate the effects of heat treatment as well as the resulted microstructure on the creep behaviors of 45XD and 47XD. In this study, a NL structure and two kinds of FL structures in both alloys were produced by selected heat treatments. The effects of microstructure on creep behavior and microstructural evolution during creep deformation were investigated.

EXPERIMENTAL PROCEDURE

In order to produce different microstructures, three heat treatment conditions were selected for both alloys. Annealing treatments consisting of 1350°C/2 hours was applied to 45XD, and 1400°C/2 hours was applied to 47XD; both were followed by oil quench (OQ). A second group of 45XD and 47XD samples was annealed at 1350°C and 1400°C for 0.5 hours, respectively, and then air-cooled (AC). A third group of the same alloys was heat treated at 1010°C for 50 hours and then air-cooled. The details of creep test and scanning electron microscopy-back scattered electron (SEM-BSE) examination procedures are explained in Ref.[4].

RESULTS

Heat treated microstructures

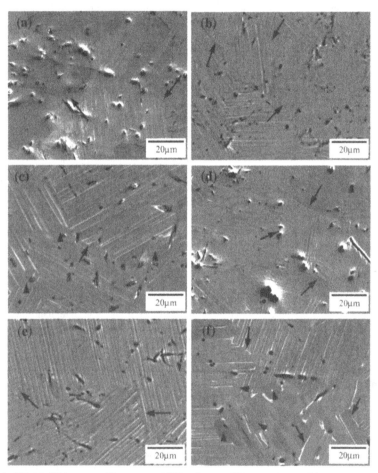

Figure 1. SEM-BSE images of heat treated XD alloys. 45XD: (a) 1350°C/2h/OQ, (b) 1350°C/0.5h/AC, (c) 1010°C/50h/AC, 47XD: (d) 1400°C/2h/OQ, (e) 1400°C/0.5h/AC, (f) 1010°C/50h/AC. Long arrows show grain boundary morphologies and short arrows indicate γ grains.

The results of microstructures and associated parameters after various heat treatments are present in Fig.1 and Table 1, respectively. In 45XD and 47XD alloys, air cooling and oil quenching at 1350°C and 1400°C, respectively, from the single α phase region, produced FL structures, whereas 1010°C/50h/AC produced NL structures, see Fig.1c and Fig.1f. The NL structures consist of a mixture of predominantly lamellar colonies and less than 15vol% of equiaxed γ grains. It can be observed that OQ produced a fine-grained fully lamellar structure (FGFL) in both alloys. Moreover, the lamellar spacing in the OQ FL structures is narrower than that in the AC FL structures due to a higher cooling rate associated with oil quench, and the lamellar spacing in the NL structures are much coarser for both alloys (Table 1). In addition, the lamellar structures for 47XD are coarser in comparison with those for 45XD subjected to similar heat treatments. Another noticeable difference among all the structures is the morphology of the grain boundaries. The grain boundaries in the three microstructures of 45XD and OQ FL of 47XD are relatively smooth, consisting of a mixture of interlocked and planar boundaries (Fig.1a-1d), whereas those in the AC FL structure of 47XD are well interlocked (Fig.1e). The grain boundaries in the NL structure of 47XD are also well interlocked, though there are some equiaxed γ grains at the grain boundaries (Fig.1f).

Table 1 Summary of microstructural parameters of 45XD and 47XD alloys.

Alloy	Heat treatment	Grain size, µm	Lamellar fraction, %	α_2 fraction, %	Lamellar spacing, nm	Grain boundary morphology
45XD	1350°C/2h/OQ	10-50	>99	34.7	40	Planar
	1350°C/0.5h/AC	20-100	>99	30.7	60	Mixture
	1010°C/50h/AC	20-100	86	<10	400	Mixture
47XD	1400°C/2h/OQ	50-200	>99	34.5	80	Mixture
	1400°C/0.5h/AC	100-250	>99	16.6	110	interlocked
	1010°C/50h/AC	100-200	85	<5.0	420	interlocked

Creep behavior

The creep test results for various microstructures are present in Fig.2. All the creep curves have the same basic shape, exhibiting a primary creep stage, followed by a steady-state creep regime, and an extended tertiary creep regime until rupture occurs. It is evident that the FL structures exhibit greater creep life and lower minimum creep rate compared to the NL structure for 45XD alloys, and the FGFL structure has the best creep resistance due to its fine lamellar spacing. In contrary, the 47XD alloy with the NL structure has the best creep resistance, and the AC FL structure shows longer creep life and lower minimum creep rate than the FGFL structure. The creep results of 47XD also contradict most of the results of previous studies on relatively coarse-grained materials, which suggest that the presence of equiaxed γ grains and wide lamellar spacing are detrimental to creep resistance [5]. In addition, by comparing the creep behavior of different microstructures of both alloys, the FGFL structure of 47XD has the worst creep resistance, whereas the AC FL and NL structures of 47XD have better creep resistance than the three microstructures of 45XD alloy.

Deformation microstructures

The deformed microstructures in the longitudinal sections of creep ruptured specimens for various heat treated conditions are shown in Fig.3. It can be observed that, as a result of spheroidization of the lamellar structure, fine globular structures are formed in the colony boundaries. For both alloys with a FGFL structure, the area fraction of these regions constitutes approximately 50% whereas in the AC FL and NL structures of 47XD, these regions constitute less than 15% of area fraction, suggesting that less microstructural degradation has occurred. It is also noted that the lamellae are severely distorted in the AC FL and NL structures of 47XD, whereas the colonies exhibit minimal deformation in other microstructures of both alloys. Furthermore, voids have formed and propagated along the grain boundaries, particularly along the fine globular regions, leading to intergranular cracking.

DISCUSSION

It is well known that grain boundary sliding (GBS) is one of the major high temperature creep deformation mechanisms, and a fine grain microstructure is more vulnerable to GBS, leading to an increase in creep rate when GBS occurs. As demonstrated in [6], GBS raises the stress in the colony and causes stress concentration at the colony boundaries. The latter is thought to accelerate microstructural degradation of TiAl alloys. On the other hand, microstructural degradation at the grain boundaries involving spheroidization of the lamellae further enhances GBS. It is suggested that during creep deformation of fine-grained TiAl alloys, microstructure degradation and GBS promote each other mutually, eventually forming a fine globular structure along the grain boundaries. Extensive local deformation is believed to take place in these regions, and further increases the strain rate.

For the three microstructures of 45XD and the FGFL structure of 47XD, the grain boundaries are relatively smooth, and the size of some grains are very fine with a size less than 100 μm (Table 1). For these microstructures, it is postulated that GBS operates during creep deformation. This conjecture is supported by the presence of extremely broad regions containing fine globular structures in the creep-ruptured microstructures (Fig.3). From previous studies, it is known that the presence of equiaxed γ grains at the grain boundaries and wide

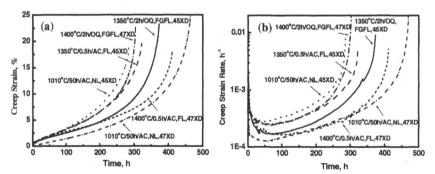

Figure 2. Creep curves of XD TiAl alloys after various heat treatments

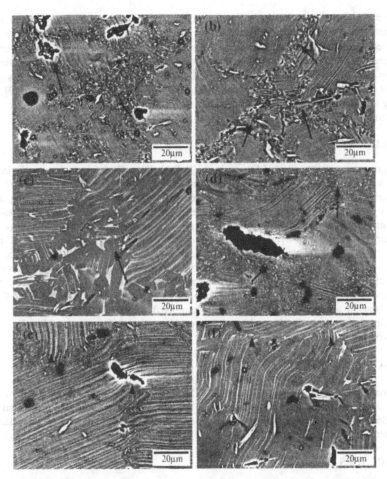

Figure 3. SEM-BSE images of XD alloys after creep failure. 45XD alloy: (a) 1350°C/2h/OQ, (b) 1350°C/0.5h/AC, (c) 1010°C/50h/AC: 47XD alloy: (d) 1400°C/2h/OQ, (e) 1400°C/0.5h/AC, (f) 1010°C/50h/AC . Long arrows show fine globular regions and short arrows indicate bending or shear bands.

lamellar spacing are detrimental to creep resistance [5]. Since the NL structure of 45XD alloy has these characteristics, the minimum creep rate of this structure is higher and has inferior overall creep resistance.

On the other hand, well-interlocked grain boundaries in the AC FL structure of 47XD inhibit GBS effectively by lamellar incursions, reducing the amount of microstructural degradation at the grain boundaries greatly (Fig.3e). This is consistent with the severely distorted lamellae near the grain boundaries. Without the contributions from GBS, the minimum creep rate can be significantly lower. Moreover, the well-interlocked grain boundaries resist intergranular crack propagation [7], resulting in an extended tertiary stage of creep. Although there are many equiaxied γ grains (15vol%) at the grain boundaries in the NL structure of 47XD, deformation of these grains is reduced by the bridging ligaments of the interlocked grain boundaries during primary creep and inhibit GBS effectively. Finally, a low volume fraction of α_2 provides good thermal stability [8] and minimize microstructural degradation at the grain boundaries [9], leading to lower creep rate and an extended tertiary stage of creep compared to the other microstructures.

CONCLUSIONS

During creep deformation of fine-grained XD TiAl alloys, GBS is thought to occur in local grains with smooth grain boundaries and fine grain size. Microstructural degradation at the grain boundaries causes GBS, which further accelerates microstructural degradation, resulting in an increase in creep rate and a decrease in creep life. In the cases of the three microstructures of 45XD and OQ FL of 47XD, a fine lamellar spacing provides a major contribution in lowering the minimum creep rates. However, if the grain boundaries are well interlocked, as in the AC FL and NL structures of 47XD, incursions of the lamellae can inhibit GBS effectively and reduce microstructural degradation at the grain boundaries greatly, resulting in lower creep rates and longer creep life. By combining good thermal stability, the NL structure of 47XD has the best creep resistance, even though it has many equiaxied γ grains at the grain boundaries and the widest lamellar spacing. Therefore, in order to improve the creep resistance of fine-grained TiAl alloys, it is important to develop a microstructure with good grain boundary stability by selecting an appropriate heat treatment.

ACKNOWLEDGEMENTS

Financial supports from the 21st century COE program in Tohoku University Materials Research Center and the Ministry of Education, Science, Sports and Culture, Japan (Grant No. 15360361) are gratefully acknowledged. This research was also supported by the Canadian Department of National Defence.

REFERENCES

[1] L. M. Hsiung, T. G. Nieh: Mat. Sci. Eng. A364 (2004) 1-10
[2] D.Y. Seo, L. Zhao, J. Beddoes: Mat. Sci. Eng. A329–331 (2002) 130-140
[3] H. Y. Kim, K. Maruyama: Acta Mater. 51(2003) 2191-2204.
[4] H.Zhu, D.Y.Seo, K.Maruyama, Mat. Trans. 45(2004) 2618-2621
[5] K. Maruyama, R. Yamamoto, H. Nakakuki, N. Fujitsuna: Mat. Sci. Eng. A239–240 (1997) 419-428.
[6] A. Chakraborty, J. C. Earthman. Met. Mat. Trans. 1997; 28A: 979.
[7] W. R. Chen, J. Triantafillou, J. Beddoes, L. Zhao: Intermetallics 7(1999) 171-178.
[8] H.Zhu, D.Y.Seo, K.Maruyama, P.Au, Scripta Mat. 52(2005): 45-50.
[9] R. Yamamoto, K. Mizoguchi, G. Wegmann, K. Maruyama: Intermetallics 6 (1998) 699-702

Mater. Res. Soc. Symp. Proc. Vol. 842 © 2005 Materials Research Society   S7.8

Creep of TiAl alloys at 750°C under moderate stress

Alain Couret and Joel Malaplate
CEMES/CNRS, 29 rue J. Marvig, BP 4347, 31055 Toulouse Cedex 4, France

ABSTRACT

The creep properties at 750°C of TiAl alloys produced by cast and powder metallurgy processes are investigated by creep tests and subsequent TEM investigations. It is demonstrated that the same dislocation mechanism is controlling the primary and secondary creep stages in these two alloys. This mechanism is the mixed climb of ordinary dislocations. Explanations to the creep behaviour during primary and secondary stages are then proposed.

INTRODUCTION

Creep of TiAl-based alloys with different microstructures has been extensively studied during the last two decades (for reviews see [1,2]). Lamellar or nearly lamellar alloys exhibit a higher creep resistance than duplex or single γ-phased alloys. The aim of the present work is to investigate the creep behavior of TiAl alloys under moderate stress, close to the operating conditions in turbine engines. Experiments have been thus conducted on two γ TiAl alloys with the so-called GE composition ($Ti_{48}Al_{48}Cr_2Nb_2$) produced by cast (C) and powder metallurgy (PM) routes. Attention is mainly focused on the temperature of 750°C (0.55 T_m) and on the stress of 150 MPa for C alloy and 80 MPa for PM alloy.

Powder metallurgy and as-cast alloys were provided by Turboméca (Groupe Snecma) and Snecma moteurs (Groupe Snecma). PM alloy was hipped at 1200°C under 140 MPa during 4h under argon and then thermally treated at 1300°C during 1h before air cooling. C alloy was hipped under the following conditions: 1260°C/172MPa/4H under argon, during one cycle and followed by a furnace cooling. It is then heat-treated as follows : 1300°C/20h under vacuum with gas fan cool. According to the usual classification, powder metallurgy and as-cast alloy exhibit duplex and near lamellar microstructures, respectively. In the case of the PM alloy, the microstructure is fine and homogeneous with 10 μm as an average grain size. 20% of these grains have a lamellar microstructure, in which α_2 lamellae represent surprisingly 60% of the volume fraction. The as-cast alloy is mainly lamellar and strongly textured with a columnar area at the periphery and an equiaxed area at the centre. The columnar area contains elongated fully α_2/γ lamellar grains, and single γ-phased grains situated at lamellar grain boundaries, whose sizes are respectively 500 to 800 μm and 10 to 50 μm. The interface planes are perpendicular to the radial direction of solidification. The equiaxed area, which extends over 1 mm consists of lamellar and single γ-phased grains in the same proportion. Their size varies between 20 and 50 μm. The lamellar structure of the columnar grains exhibits mainly large γ lamellae (2 μm) separated by twins and ordered domain interfaces.

EXPERIMENTAL RESULTS

Creep behaviour

Figure 1a shows a creep curve for the PM alloy (80 MPa ; 750°C) interrupted at about 2% strain. It exhibits primary and secondary stages, the partition of which being determined from the minimum of derivative curve. For this example, this minimum is reached at 0.32% strain after 61 hours and the minimum creep rate is 9.6 10^{-9} s^{-1}. Creep tests under real constant stress have confirmed that secondary stage is formed by a short steady stage followed by a long period during which the creep rate increases slowly [3]. It is followed by a ternary stage yielding to the rupture of the sample. Table 1 gives the features of primary and secondary stages at different stress levels.

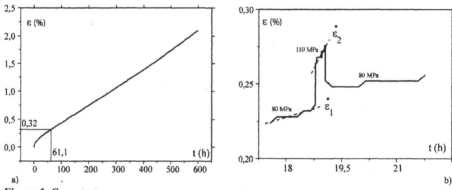

a)

b)

Figure 1. Creep tests
a) Creep curve interrupted at 2% strain in the PM alloy at 750°C under 80 MPa.
b) Stress jump performed during primary stage in the PM alloy at 750°C under 80 MPa.

Table 1. Features of primary and secondary stages at different stress levels.

	Duration of primary stage	Strain during primary stage	Minimum creep rate (10^{-9} s^{-1})
PM 80 MPa	51	0.3	9
PM 150 MPa	33	0.8	57
C 150 MPa	113	0.9	15

The stress dependence of creep can be directly analysed by performing stress jumps during creep tests, which allow measuring the creep activation parameters, as the stress exponent and the activation volume. Figure 1b is an example of a stress jump performed during primary stage in the PM alloy. At 0.24% of deformation, the stress is suddenly increased from 80 MPa to 110 MPa. After a short time at 110 MPa, the stress is decreased down to 110 MPa in order to perform the following jump. The stress exponent n and the apparent activation volume V_a are calculated from the instantaneous variation of the creep rate according the following relations:

$$n = \frac{\log\dot{\epsilon}_2 - \log\dot{\epsilon}_1}{\log\sigma_2 - \log\sigma_1} \qquad V_a = kT\frac{\ln\dot{\epsilon}_2 - \ln\dot{\epsilon}_1}{\sigma_2 - \sigma_1} = \frac{nkT}{\sigma}$$

where $\dot{\epsilon}$ is the creep rate, σ the applied stress, k the Boltzmann constant and T the temperature. Subscripts 1 and 2 refer to values before and after the jumps.

Stress jumps were performed all along the creep curves in PM and C alloys. The most important result is that these activation parameters are constant all along the curve, namely in primary and secondary stages. Table 2 gives the average values that were obtained. In both PM and C alloys, the same order of magnitude have been obtained.

Table 2. Activation parameters calculated from stress jumps.

Alloy	n	V_a (b_{or}^3)
Cast	11	36
Powder metallurgy	7	46

Considering that the applied stress is the sum of an internal stress and of the effective stress, strain change dip tests allow measuring this internal stress, which represents the main force opposing to the dislocation motion. This method consists to reducing step by step the applied stress and to recording the variation of the sample length. If the stress reduction is higher than the effective stress, it is followed by zero creep or to reverse movements. If recovery occurs, creep will start again after a so called "incubation period". Figure 2 shows the results for the C and PM alloys initially crept at 150 MPa and 80 MPa, respectively. First stress reductions of $\sigma_a/20$ amplitude were applied at 1.5% strain and the following ones after new creep rate stabilizations. The cumulative stress reductions $\sigma_a - \Sigma\sigma_a$ are plotted as a function of the cumulative incubation periods, $\Sigma\Delta t_i$. Exponential fits have been superimposed to the experimental results. Such fits describes fairly the PM variation whereas the C curve can be divided into two ranges, with a transition around 112.5 MPa. In both alloys, the curve approaches an infinite value, which is close to 45 MPa for the C alloy and to 12 MPa for the PM alloy. This indicates the presence of an internal stress that is difficult and even probably impossible to recover.

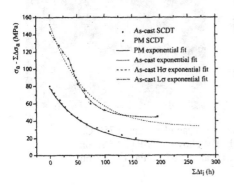

Figure 2. Plot of the reduced stresses as a function of the incubation time during strain change dip tests. Experimental results are fitted with exponential variations.

Deformation microstructure

Figure 3 shows that the deformation microstructure is dominated by ordinary dislocations in both C and PM alloys whatever the strain amount. As illustrated by Figures 3 a & b for the case of PM alloy, at 2% strain, two types of dislocation have been identified. Dislocations of first type are elongated along their screw directions and are anchored at many pinning points (Figure 3a). These dislocations are moving by glide. In the high temperature range, their movement occurs by the activation of bursts of dislocations and is controlled by dynamic strain ageing processes [4]. Dislocations of the second type are curved and not elongated along any preferential direction (Figure 3b). It has been recently demonstrated that the dislocations are moving by mixed climb, occurring in planes close but distinct from the $\{111\}$ glide plane [6]. The corresponding elementary process is the nucleation and the lateral propagation of jog pairs. In the PM and C alloys under 80 MPa and 150 MPa, respectively, glide dislocations and mixed climb dislocations are evenly distributed in samples crept in secondary stage.

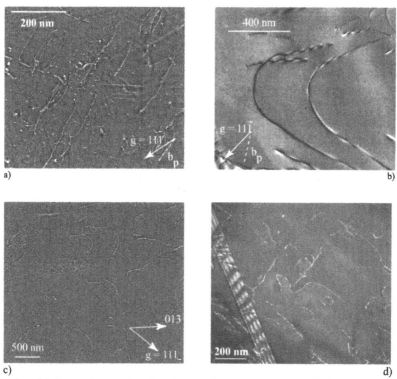

a)   b)

c)   d)

Figure 3. TEM observations of the microstructure in samples crept at 750°C.
a) PM alloy : 80 MPa ; 2% strain,   b) PM alloy : 80 MPa ; 2% strain
c) C alloy ; 150 MPa ; 0.5% strain,   d) C alloy ; 250 MPa ; 0.5% strain

As illustrated by Figure 3c for the C alloy, microstructure in samples crept up to 0.5% strain, namely in the primary stage, is dominated by dislocations moving by mixed climb. On the contrary, an increase of the stress up to 250 MPa leads to the activation of glide dislocations and to twinning (Figure 3d).

DISCUSSION

TEM observations have shown that, at the conditions investigated in the present study, the microstructure of crept samples is largely dominated by ordinary dislocations. The present section will be restricted to the behavior and the role of this kind of dislocations.

On the one hand, the stress exponent and the activation volume have been found to be constant all along the creep curve and to have close values in the two alloys. This indicates that the same mechanism is controlling primary and secondary stages. On the other hand, deformation of samples crept in primary stage is dominated by ordinary dislocations moving by mixed climb. Moreover, a fair consistency has been evidenced between mixed climb mechanism occurring by the nucleation and propagation of jog pairs and the values of the activation parameters that have been measured in the present study [6]. These observations tend thus to prove that primary and secondary creep rate are governed by the mixed climb mechanism at 750°C under 150 MPa for the C alloy and under 80 MPa for the PM alloy.

As the stress and/or the strain are increased, the density of ordinary dislocations moving by glide increases. In particular, under 250 MPa in the C alloy, screw, elongated dislocations are observed in majority. Under moderate stress, glide dislocations are confined in bands containing a high number of dislocations [3, 5], which reveals that this movement is probably controlled by dynamic strain ageing. The glide in C and PM samples crept at 150 MPa and 80 MPa, respectively, is thus probably activated under the effect of high local stresses. A similar combination of glide and climb has been evidenced in single phased γ TiAl alloys deformed at 900°C [7]. Glide is observed at a high strain rate 10^{-2} s^{-1} whereas climb is activated at smaller strain rates (10^{-3} s^{-1} and 10^{-5} s^{-1}).

Strain change dip tests have shown that internal stresses of 45 MPa in the C alloy and 12 MPa in the PM alloy are present after creep in secondary stage. For tests leading to the same minimum creep rate (80 MPa in the PM alloy and 150 MPa in the C alloy), a higher primary creep is measured for the C alloy. It is thus proposed a correlation between a wide primary creep and a high fraction of internal stress difficult to recover. Besides, a huge density of dislocations has been observed in the large γ grains of the C alloy [3,5]. The primary stage would thus be a stage during which the strength of the microstructure is homogenised by the hardening of soft areas (γ grains in C alloy) through the building of an internal stress. The curve resulting from the strain change dip test performed in the as-cast alloy exhibits two stress ranges. The high stress range is probably correlated to the deformation of the large γ grains, which would lead to an internal stress of the order of 45 MPa according to the corresponding exponential fit. The low stress range leading to an internal stress of 30 MPa could be attributed to the lamellar grains in which recovery is assisted by interfaces. This last point is in consistency with the measurement of moderate primary creep in pure lamellar alloys oriented in hard mode [8]. Note that in the same as-cast alloy, a scatter has been measured on the primary stage [3]. Primary duration and strain vary strongly from one specimen to another whereas the minimum creep rates are close for the same group of samples. This scatter has been ascribed to the variation of the density of large γ grains, which is correlated to the initial casting defects.

During secondary stage, dislocations are moving in both alloys apparently by the same mechanism since any difference was detected from measurements of the activation parameters and from microstructural analyses. Applied stresses of 80 MPa in the PM alloy and of 150 MPa in C alloy lead to similar creep rate. The instantaneous internal stresses are not the infinite values deduced from strain change dip tests since these value are the result of a long recovery process. In fact, the instantaneous internal stress should be deduced from the smallest stress decrease leading to an incubation period. Due to experimental uncertainties, it is difficult to unambiguously identify an incubation period after the first stress reduction (σ_a /20). In both alloys, the instantaneous internal stress is thus superior to $0.9\sigma_a$. Consequently, the effective stresses are smaller than 10% of the applied stress and so apparently not different enough to explain the higher creep resistance of the C alloy. In this as-cast alloy, at the end of primary stage, the soft areas have been hardened so that the global resistance is uniform. Therefore, in consistency with TEM observations showing dislocations in lamellar areas [3], ordinary dislocations are moving in these areas in which their mean free path is the lamella width (2 μm), since they are oriented in hard mode. In the powder metallurgy alloy, this main free path is equal to the grain size (10 μm). The lower creep rate of the as-cast alloy is thus probably correlated to the lower free path of ordinary dislocations moving by mixed climb.

ACKNOWLEDGEMENTS

This work is part of the research contract "CPR : Intermétalliques base Titane" involving the CNRS, Snecma moteurs - Groupe Snecma, Turboméca - Groupe Snecma and the DGA. The authors want to thank all these partners for financial support and scientific discussions.

REFERENCES

1. J. Beddoes, W. Wallace and Zhao L., *International Materials Reviews*. **40**, 197 (1995).
2. F. Appel and R. Wagner, *Materials Science & Engineering R: Reports*. **R22**, 1 (1998).
3. J. Malaplate, *PhD Thesis*, Université Paul Sabatier (2003).
4. A. Couret, A., *Intermetallics* **9**, 899 (2001).
5. J. Malaplate and A. Couret, *International Conference on Processing and Manufacturing of Advanced Materials*, Vol. 426-432, edited by T. Chandra, J. Torralba and T. Sakai (Leganes - Madrid, Spain, Trans Tech, Publications LTD, Switzerland) (2003).
6. J. Malaplate, D. Caillard and A. Couret, Accepted in *Philosophical Magazine* (2005).
7. B.M. Kad and H.L. Fraser, *Philosophical Magazine A* **69**, 689 (1994).
8. M. Thomas and S. Naka, *Matériaux and techniques*, **1-2**, 13 (2004).

Mater. Res. Soc. Symp. Proc. Vol. 842 © 2005 Materials Research Society   S5.46

Abnormal deformation behavior in Polysynthetically-twinned TiAl crystals with A and N orientations—— an AFM study

Yali Chen and David P. Pope
Department of Materials Science and Engineering, University of Pennsylvania
Philadelphia, PA 19104-6272, U.S.A

Abstract

Polysynthetically-twinned TiAl crystals were deformed by compression with loading axis parallel and perpendicular to the lamellar interfaces. The deformation structures on the free surfaces were scanned using a dimension AFM with scan directions parallel and perpendicular to the lamellar interfaces. Abnormal deformation behaviors were observed to occur in both orientations. When the compression axis is parallel to the lamellar interfaces, the gamma and alpha lamellae deform primarily by shear in planes inclined with the lamellar interface, while the shear vectors lie in the interface. However, in-plane shear, shear in slip planes parallel to the lamellar interfaces, also occurs along the lamellar interfaces. When the loading axis is perpendicular to the lamellar interface, in-plane shear was found to be dominant at the beginning stage of plastic deformation and contributes more to the macroscopic strain. These behaviors are controversial to the Schmid's Law since the applied resolved shear stress for these deformation systems is zero. The abnormal phenomenon was explained by the large coherency stresses along the lamellar interfaces.

Introduction

It has long been accepted that the lamellar interfaces play an important role in the deformation process of γ-based TiAl materials with lamellar structures. [1-3] However, it is a difficult job to investigate the interaction between the lamellar boundaries and the deformation systems, due to the complexity of boundaries in the polycrystals. In early 90s, Dr. Yamaguchi's group [4] produced the so-called polysynthetically-twinned (PST) TiAl crystals, which are big lamellar grains, and used them to study the deformation process with the existence of lamellar interfaces. Since then, a lot of achievements have been made in discovering the deformation behavior of PST crystals. It was found that the yield of PST samples changes with their orientations, [2-6] and the plastic deformation is anisotropic at some orientations. [3,6] TEM results show that the special deformation behavior comes from the selective activation of deformation systems in the gamma phase and alpha 2 phase, which results from the confinement of the lamellar interfaces [3].

In our research, Atomic Force Microscopy (AFM) was chosen to study the deformation behavior of PST crystals. By scanning the free surfaces of the deformed samples with different orientations, deformation structure at different size scale can be investigated so that a more comprehensive view can be obtained. The AFM results showed [7] that there are three deformation modes with the change of sample orientation. When the angle between the loading axis and the lamellar interfaces is below 20 degree, the lamellae deform by shear on slip planes inclined with the lamellar interfaces, but the shear vectors are parallel to the lamellar boundaries. In-plane shear (shear in planes parallel to the lamellar interfaces) dominates when the angle is between 20 and 80 degrees. When the loading axis is close to perpendicular to the lamellar interfaces, cross-lamellar shear dominates at high strain levels. The results suggest that in-plane shear is the easiest deformation mode in PST crystals, because it brings lest disturb to the lamellar interfacial structure. Consequently, in-plane shear was even observed in A orientation (loading axis parallel to the lamellar interface) and N orientation (loading axis perpendicular to the lamellar interface) even though the applied shear stress on these slip planes is zero. In this paper, the abnormal deformation behavior occurred in these two orientations will be discussed.

Experimental details

PST crystals with a nominal composition of Ti-48at%Al were grown using optical floating zone furnace. Crystals were oriented by Laue back diffraction technique and samples were cut using electrical discharge machining (EDM). The free surfaces of the samples were mechanically polished and electro-polished. Compression tests were carried out under room temperature at a speed of 1×10^{-6}/s using an Instron machine. The geometries for the compression tests at A and N orientations are schematically shown in figure 1.

The deformation structures emerged on the free surfaces were studied using a DI 3000 Dimension AFM. In AFM observation, the feature parallel to the scan direction is weakened due to the signal process. In order to obtain the true topography, scan directions parallel and perpendicular to the lamellar interfaces were both chosen.

Results and discussions

1. In-plane shear in A-oriented sample

In-plane shear typically occur as interfacial sliding at the lamellar interfaces. As we discussed elsewhere, [8] in A orientation samples, the neighboring lamellae tend to deform in such a way that the deformation bands have the same offset height and offset angle, so that no strain incompatibility will be produced at the lamellar interface. However, this cannot be always satisfied since some of the gamma domains deform by combination of twinning and ordinary dislocation slip with vectors inclined with lamellar interfaces. Strain incompatibility and complex stress state will be inevitably induced at the lamellar interfaces, although the offset across the lamellar interfaces is only tens of nanometers. The built-up stress along with the coherency stress might be high enough to stimulate the in-plane shear at these places, since the C.R.S.S for the in-plane shear is much lower. [3,6,9]

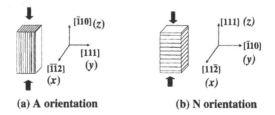

(a) A orientation   (b) N orientation

Figure 1. Geometry of the PST samples during compression test

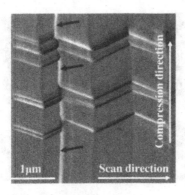

Figure 2. Interfacial sliding at the γ/γ lamellar interface (Deflection image)

As shown in figure 2, there is an intense band at the γ/γ interface pointed by the black arrows. The lamella at the left side has deformation bands going across the lamellar interface, as a result, stresses were built up and in-plane shear occurs to dissipate the energy. The interfacial sliding also occurs at the α_2/γ interfaces when there is a strain incompatibility.

There are some fine γ lamellae separated by fine α_2 lamellae in the PST crystals. These lamellae have the width of several tens of nanometers and look like a pure α_2 lamella in the SEM. [10] Because of the high density of lamellar interfaces inside this kind of lamellar bundle, the deformation process is strongly restricted. Figure 3 (a) shows the deformation structure of a sample deformed by about 3%. Fewer deformation bands exist inside the lamella between the two dashed lines. The deformation bands look like vertical traces which might be formed by prismatic slips in the α_2 phase. [8] However, a closer look in figure (b) reveals that these bands are actually not vertical, but composed of many zigzagged bands in different lamellae. The width of the bundle is about 2μm, and it contains about 20 lamellae. There are not only fewer inclined bands inside the bundle, but also fewer and smaller in-plane bands. Only at the two boundaries of the bundle are there large in-plane shear steps, which have the height of tens of nanometers and are comparable to that of the inclined bands. The possible reason for this deformation behavior is that it is difficult for the fine lamellae to deform while maintaining compatible strain with each other in such a small scale. In other words, the energy increase associated with the deformation might be huge, so the PST crystal tends not to deform inside these lamellae bundles. Rather, it dissipates the energy by interfacial sliding at the boundaries of the bundles.

It is not clear whether the lamellar sliding is a superficial effect or it occurs throughout the whole sample. No direct TEM observations and discussions are available on this issue, although it can be seen from the TEM images that fringes of tens of nanometers in width usually present at the lamellar interfaces. [11,12] The details of the structure inside these fringes need further investigation. In fact, based on the AFM observations, interfacial sliding does not occur everywhere. Resultantly, additional strain incompatibility will be produced at the domain boundaries, especially at the intersections of the domain boundaries and the lamellar interfaces. However, since the domain boundaries are weak barriers to the shear transmission, and the lamellar shear is just on the nanometer scale, the strain incompatibility will not cause large stress concentration at the domain boundaries.

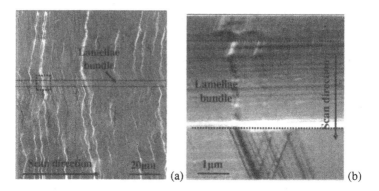

(a)   (b)

Figure 3. The deformation structure inside a 'lamellar bundle'.
Picture (b) is a magnified view of the dashed square in picture (a).

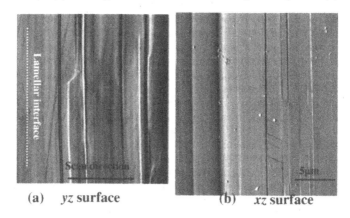

(a)   *yz* surface   (b)   *xz* surface

Figure 4. The deformation structure in the lightly deformed area
on the two free surfaces. (a) Deflection image; (b) Amplitude image.

2. In-plane shear in N orientation

AFM results show that in-plane bands are the dominant species at low strain level as shown in figure 4. Detail analyses [13] indicate that only in-plane shear can produce bands parallel to the lamellar interfaces on both free surfaces. But how can this happen since the Schmid factors for the in-plane shear systems are all zero when the loading axis is perpendicular to the lamellar interface?

One possible reason for the appearance of these unexpected slip traces might be that the crystal is incorrectly oriented so that the loading axis is not perfectly perpendicular to the lamellar interfaces. This would induce deformation mode similar to those seen in the B-oriented samples. [14] Since the possible error in sample orientation using the Laue back diffraction technique is about 2 degrees, the maximum Schmid factor for the deformation systems on the (111) plane is about 0.001. The CRSS of deformation systems on (111) plane parallel to the lamellar interfaces in the γ phases is around 100MPa, [2,3,6,9] so if the in-plane shear is caused by this misorientation, the axial yield stress should be around 10^5MPa, which is 100 times higher than the experimental results, about 1000MPa.

Clearly, misorientation is not the primary reason for the occurring of in-plane shear. A reasonable explanation might be to include the coherency stress and complex stress state at the lamellar interfaces. Previous calculations and measurements have shown that coherency stresses with amplitude of 100MPa exist in the PST crystals. Hazzledine's calculation and CBED results [15] suggest that the residual stress is biaxial in the lamellae, but pure shear in the interfaces. Results from Appel et al. [1] also suggest that the direction of the coherency stress is randomly oriented and the amplitude of the stress ranges from several MPa to over 400MPa. This shear stress, along with the induced shear stress by the pileups might be high enough to activate the dislocations in the lamellar interfaces and produce parallel traces on the free surface. The interfacial shear will then lead to shear stress inside the lamellae and cause the broadening of the parallel bands.

Similar to the A orientation, it is unclear whether the formation of these parallel bands are just a superficial effect or not, since no twinning or dislocation slip on planes parallel to the lamellar interfaces in N-oriented samples was ever observed in TEM. And since many misfit dislocations are present in the lamellar interfaces, it might be very difficult for the dislocations to move throughout the entanglements. However, this question still needs further investigation since the TEM samples are typically cut parallel to the lamellar interfaces [1-3] it is difficult to observe in-plane shear using this kind of thin films.

With the increase of strain level, cross-lamellar shear becomes to take over since only these bands can produce strain in the compression direction. However, in-plane shear still contributes a lot to the overall strain as shown in figure 5.

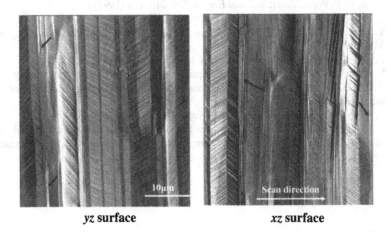

yz surface   *xz* surface

Figure 5. The deformation structure in the heavily deformed area. Black arrows indicate the in-plane shear bands. (Deflection images)

Conclusions

In-plane shear is the easiest deformation mode for the PST crystals due to the confinement of the lamellar interfaces. As a result, in-plane shear occurs whenever possible even when the applied shear stress for this deformation system is zero. In A-oriented samples, in-plane shear occurs along lamellar interfaces when there is a strain incompatibility and stress concentration. In N-oriented samples, in-plane shear is the dominant deformation mode at low strain level and contribute a lot to the overall strain. This abnormal deformation behavior might result from the complex coherency stress state at the lamellar interfaces.

References

1. F. Appel, R. Wagner, *Mater. Sci. & Eng. R*, **22**, 187 (1998)
2. H. Inui, A. Nakamura, M. H. Oh and M. Yamaguchi, *Acta. Metall. Mater.*, **40** (11), 3095 (1992)
3. K. Kishida, H. Inui, M. Yamaguchi, *Philos. Mag. A*, **78** (1), 1 (1998)
4. T. Fujiwara, A. Nakamura, M. Hosomi, S. R. Nishitani, Y. Shirai, and M. Yamaguchi, *Philos. Mag. A*, **61** (4), 591(1990)
5. Y. Umakoshi, T. Nakano, T. Yamane, *Mater. Sci. & Eng. A*, **152**, 81 (1992)
6. M. Kim, M. Nomura, V. Vitek and D. Pope, in in High-Temperature-Ordered Intermetallic Alloys VIII, edited by E. George, M. Mills, M. Yamaguchi, (Mater. Res. Soc. Symp. Proc. **552**, Boston, MA, 1998) pp KK3.1.1
7. Y. Chen, D. Pope, in *Materials Research Society Symposium S*, Boston, MA, 2004 submitted
8. Y. Chen, D. Pope, AFM study of the plastic deformation of Polysynthetically twinned (PST) TiAl crystals in hard orientations —— I: A orientation, *Acta Mater.*, to be submitted
9. L. Lu, D. P. Pope, *Mater. Sci. and Eng. A*, **239-240**, 126 (1997)
10. L. Pan, *Ph.D Thesis*, University of Pennsylvania, 2002
11. Y. Umakoshi, H. Y. Yasuda, T. Nakano, K. Ikeda, *Metal. & Mater. Trans. A*, **29A** (3), 943 (1998)
12. T. L. Lin, M. Chen, in *Gamma Titanium Aluminides, TMS*, pp. 323, 1995
13. Y. Chen, D. Pope, AFM study of the plastic deformation of Polysynthetically twinned (PST) TiAl crystals in hard orientations —— II: N orientation, *Acta Mater.*, to be submitted
14. Y. Chen, D. Pope, *Microscopy Research and Techniques*, submitted
15. P. M. Hazzledine, B. K. Kad, et al. , in Intermetallic Matrix Composites II, edited by D.B. Miracle, D.L. Anton, J.A. Graves, (Mater. Res. Soc. Symp. Proc. **273**, San Francisco, Ca, 1992) pp. 81

Rationalization of the plastic flow behavior of Polysynthetically-twinned (PST) TiAl crystals based on slip mode observation using AFM and Schmid's law

Yali Chen and David P. Pope
Department of Materials Science and Engineering, University of Pennsylvania
Philadelphia, PA 19104-6272, U.S.A

Abstract

PST TiAl samples of different orientations were prepared and deformed by compression at room temperature. The deformation structures on the free surfaces were scanned using an AFM. It was found that when the angle between the lamellar interfaces and the loading axis is between 20 degree and 80 degree, PST samples deform primarily by shear in slip planes parallel to the lamellar interfaces. When the angle is below 20 degree, both the gamma phase and the alpha 2 phase deform by shear in slip planes inclined with the lamellar interfaces, but the shear vectors lie in the interface. When the angle is close to 90 degree, complex deformation behavior occurs. Shear in planes parallel to the lamellar interfaces contributes more to the overall strain in the directions perpendicular to the loading axis and the out-of-plane shear contributes to the strain in the compression direction. The characteristic U-shape curve of the yield stress versus the angle between the loading axis and the lamellar interfaces can be explained quite well using different C.R.S.S. for the three different deformation modes.

Introduction

It is well known that the deformation behavior of polysynthetically twinned (PST) TiAl crystals changes with their orientations. When the loading axis is parallel or perpendicular with the lamellar interface, the yield stress is quite high. While when the loading axis is inclined with the lamellar interfaces, the crystal is relatively easy to be deformed. [1-4] This property was demonstrated by Kim with an irregular U-shaped yield stress-orientation curve. [3] Clearly, the change of the curve is not only related to the change of the applied shear stress on the deformation systems. In fact, TEM results [1,4] have shown that at different orientation, different deformation systems are activated owing to the confinement of the lamellar interfaces. However, due to the limits of the TEM observation, it is hard to get a comprehensive picture of the plastic deformation behavior.

In our previous studies, the deformation behaviors at A, B and N orientations were investigated using AFM. [5-7]The results can be summarized as a simple cartoon shown in figure 1. When the compression axis is parallel to the lamellar interface, i.e., in the so called A orientation, the gamma phase deforms primarily by shear in slip planes inclined with the lamellar interface.[5] In each gamma domain, the shear vectors lie in the lamellar interfacial plane, or the combination of shear vectors produce a zero strain across the lamellar interfaces in a scale of tens of nanometers. The alpha 2 phase deforms by prismatic slip, with shear vector also parallel to the lamellar boundaries. As a result, the microscopic and macroscopic strain in the direction perpendicular to the lamellar interfaces is zero when the PST crystals are in the A orientation.

When the compression axis is inclined by the lamellar interfaces by 45 degree (B orientation), the gamma phase deforms primarily by shear in slip planes parallel to the lamellar interfaces. [6] Shear in planes inclined with the lamellar interfaces also happens at high strain level, but contribute very little to the overall strain. The alpha 2 phase does not deform at this orientation because of the high C.R.S.S for the basal slip system. [8] The alpha 2 layers act as rigid plates and move along with the shear of the gamma lamellae.

When the loading axis is perpendicular to the lamellar interfaces, the deformation behavior is much more complicated. Shear parallel to the lamellar interfaces was found to be the primary deformation process at the low strain level, although the applied shear stress for this deformation system is zero. With the increase of strain level, cross lamellar shear becomes dominant. Prismatic slip in the alpha 2 phase was also observed at high strain level. These abnormal deformation behaviors were attributed to the coherency stress and stress concentration at the lamellar interfaces. [7]

For convenience, the three deformation modes are named A, B and N deformation modes respectively, corresponding to the orientations of the PST samples. It can be seen from figure 1 that because of the selective activation of the deformation systems, the deformation of A and B oriented samples is quite anisotropic, as discovered by Kishida et al., [3,4] which might cause large strain incompatibility among lamellar grains in full-lamellar TiAl polycrystals if the grains are in these orientations. However, in the polycrystals, the lamellar grains are randomly oriented, so it is worthwhile to study the deformation behavior of PST crystals at some intermediate orientations. In this paper, the results from AFM observations will be presented and discussed.

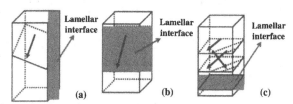

Figure 1. Schematic of the three deformation modes: A type (a), B type (b), N type (c)
(The bold arrows indicate the microscopic shear direction in each layer)

Experimental details

PST crystals with a nominal composition of Ti-48at%Al were grown using optical floating zone furnace. Samples with different orientations were cut by Electrical Discharge Machining. The angle between the loading axis and the lamellar interfaces, φ, was set to be 10°, 20°, 30°, 70° and 80° as shown in figure 2.

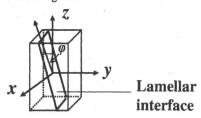

Figure 2 The geometry of PST samples. (φ is the angle between the
compression axis and the lamellar interfaces.)

The free surfaces of the samples were mechanically polished and electro-polished. Compression tests were carried out under room temperature at a speed of 1×10^{-6}/s using an Instron machine. The deformation structures emerged on the free surfaces were studied using a DI 3000 Dimension AFM. For detail discussion about the formation of deformation bands on the free surfaces and the convolution of the tip with the bands, please refer to the references 5-7.

Results

1. φ=10°

Figure 3(a) is a deflection image showing the typical deformation structure when the loading axis inclines with the lamellar interfaces by 10 degree. The scan direction is parallel to the loading axis. It can be seen that the structure is similar to that of the A-oriented samples. [6] The lamellar interfaces are straight in the areas scanned. Some of the uphill deformation bands go across all the lamellae inside the scan area. The quantity of the downhill bands seems to be larger than that in the A-oriented samples. This might be because that the two slip planes which are symmetric in the A orientation are no longer symmetric. However, because of the requirements of strain compatibility and the condition of the local stress state, still only one type of band can freely go through the lamellar boundaries.

As with the A orientation, two types of gamma domains deform by super dislocation slip with Burger vectors parallel to the lamellar interfaces. The other four domains deform either by ordinary dislocation slip with Burger vector parallel to the lamellar interfaces, or by combination of ordinary dislocation slip and twinning with shear vectors inclined to the lamellar interfaces. The net result is still that there is no strain in the direction perpendicular to the lamellar boundaries.

The α_2 lamellae can also deform freely in samples of this orientation as shown in figure 3(b). At this orientation, the Schmid factor for the prismatic slip is slightly lower than that for the A orientation, changing from 0.433 to 0.420, so the deformation of the α_2 lamellae is not strongly affected.

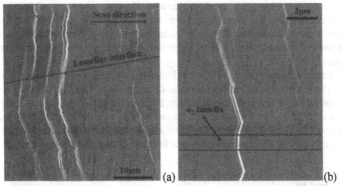

Figure 3 (a) The deformation structure on the yz face of a sample deformed by 5% when φ=10°. (Deflection image with scan direction parallel to the loading axis); (b) The deformation structure in the α_2 lamellae. (Deflection image with scan direction parallel to the lamellar interfaces)

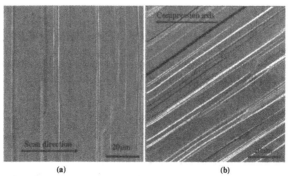

Figure 4 The deformation structure of a φ=30° sample deformed by 3.5%
(a) xz face; (b) yz face. (Deflection images)

2. φ =20° and 30°

Figure 4 shows the deformation structure of a sample deformed by 3% when the loading axis is inclined to the lamellar interfaces by 30°. Figure 4(a) is taken on the xz face and figure 4(b) is taken on the yz face. This deformation structure is similar to that of the B-oriented sample deformed by the same amount. [5] On the front xz surface, only uphill parallel bands emerge. On the yz side surface, both uphill and downhill bands appear which were formed by deformation systems with shear vectors having projections in the opposite directions. As with the B oriented samples, the macroscopic strain in the x direction is zero. The samples with φ =20° have similar deformation structure.

The difference from the B orientation is that, at these orientations, there are more cross lamellar bands. This is reasonable since the resolved shear stresses on the slip planes other than (111) are larger than in the B orientation, especially for those deformation systems with shear vectors parallel to the lamellar interfaces. This feature is similar to type A orientation when the scan direction is parallel to the lamellar interfaces. However, these cross-lamellar bands contribute much less to the overall strain than the parallel bands. The height of the parallel bands varies from tens of nanometers to over 500 nanometers, while the step height of the cross lamellar bands is typically below 10nm, with only a very few having height of tens of nanometers.

Vertical bands perpendicular to the lamellar interfaces were not found in the areas scanned. The Schmid factor for the prismatic deformation systems is 0.318 for φ=30°, about 75% of that for the A oriented samples. Also, because the yield stress at this orientation is much lower that that of the A oriented sample, the resolved shear stress might not be high enough for the prismatic slip systems to move. The basal slip systems cannot be activated either, since the Schmid factor is only 0.22, smaller than that for the prismatic slip systems, while their C.R.S.S. is higher.

2. φ=70° and 80°

When the loading axis is inclined with the lamellar interfaces by 80°, the deformation structures are still similar to that of the B-oriented samples as shown in figure 5. On xz face, only uphill parallel bands appear and on the yz face, both uphill and downhill bands appear. The deformation structure of samples with φ=70° is similar to that shown in these figures.

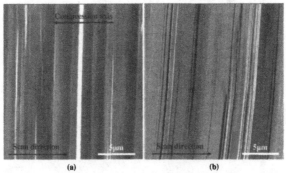

(a)   (b)

Figure 5 The deformation structure of a sample deformed by 2.8% when φ=80° (a) *xz* face; (b) *yz* face (Deflection images)

Cross lamellar bands also exist, especially for φ=80°, but parallel bands still contribute more to the overall strain. As those in the N oriented, [7] the bands usually reside inside only one or several lamellae and do not extend for long distances. The lamellar interfaces no longer stay straight because these bands produce offsets of tens of nanometers perpendicular to the lamellar boundaries.

Discussions and conclusions

From the AFM results, it can be seen that the deformation behavior of the PST crystals at any orientation can be described by one or two of the three deformation modes discussed in the introduction part. When the angle between the loading axis and the lamellar interfaces is below 20 degrees, PST crystals deform mainly by the A deformation mode. When the angle is between 20 and 80 degrees, the B deformation mode is the primary one, with in-plane shear dominates. When the orientation is more close to A and N orientations, cross lamellar shear also occurs, but contribute much less to the overall strain. Only at orientations where the loading axis is very close to perpendicular to the lamellar interface, cross lamellar shear becomes dominate.

Although some theoretical calculations and experimental results have shown that twinning and ordinary dislocation slip might be the easiest deformation modes in the two-phase TiAl materials, [1,4,9] it is still not clear which deformation mode controls the yield point of the PST specimens. According to the AFM results discussed above, there should be three C.R.S.S. values for the three deformation modes separately. Table 1 lists the calculations of the C.R.S.S of the deformation systems for different deformation modes, assuming that these deformation systems are the first ones to be activated and determines the yield stress. The values of the yield stresses are taken from Kim's results [3] for A, B, and N orientations.

Table 1 C.R.S.S in different deformation modes

Deformation mode		A	B	N
Yield stress (MPa) [64]		690	200	800
C.R.S.S (MPa)	Twinning	162.84	86.6	125.6
	Ordinary dislocation	281.52	100	217.6
	Superdislocation	281.52	100	217.6

221

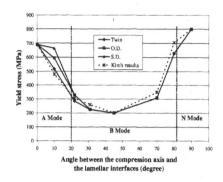

Figure 6 Comparison of the yield stress calculated based on different C.R.S.S

Using the C.R.S.S values from table 1 to calculate the yield stresses of the other orientations, based on the Schmid factors, gives the results shown in figure 6, accompanied by the results from Kim's result. The points with the angle below 20° are calculated using A-type deformation C.R.S.S and those between 20° and 80° are calculated based on the B-type deformation C.R.S.S. The results calculated based on the twinning system is closest to the experimental results, especially in the A type deformation regime. So, it might be reasonable to conclude that twinning controls the onset of plastic deformation in PST crystals. The big change of yield stress from the A and N type to B type deformation mode is due to the difference of C.R.S.S. of twinning on slip planes parallel and those inclined to the lamellar interfaces.

The transition from one mode to another comes from the competition of the resolved shear stresses on different slip planes. When the resolved shear stress is above the C.R.S.S on the planes parallel to the lamellar interfaces, the B type mode will definitely take over because the confinement of the lamellar interfaces is avoided. Since most lamellar grains in the full-lamellar polycrystals should deform by A and B type deformation modes, which lead to anisotropic deformation, strain incompatibility will inevitably be produced at the grain boundaries. As a result, the deformability of the two-phase TiAl crystals with full-lamellar structure is limited.

References
1. H. Inui, A. Nakamura, M. H. Oh and M. Yamaguchi, *Acta. Metall. Mater.*, **40** (11), 3095 (1992)
2. Y. Umakoshi, T. Nakano, *Acta Metall. Mater.*, **41**, 1155 (1993)
3. M. Kim, M. Nomura, V. Vitek and D. Pope, in High-Temperature-Ordered Intermetallic Alloys VIII, edited by E. George, M. Mills, M. Yamaguchi, (Mater. Res. Soc. Symp. Proc. **552**, Boston, MA, 1998) pp KK3.1.1
4. K. Kishida, H. Inui, M. Yamaguchi, *Philos. Mag. A*, **78** (1) 1 (1998)
5. Y. Chen, D. Pope, *Microscopy Research and Techniques*, submitted
6. Y. Chen, D. Pope, AFM study of the plastic deformation of Polysynthetically twinned (PST) TiAl crystals in hard orientations —— I: A orientation, *Acta Mater.*, to be submitted
7. Y. Chen, D. Pope, AFM study of the plastic deformation of Polysynthetically twinned (PST) TiAl crystals in hard orientations —— II: N orientation, *Acta Mater.*, to be submitted
8. H. Inui, M. H. Oh, A. Nakamura, M. Yamaguchi, *Philos. Mag. A*, **66** (4), 539 (1992)
9. Lu L, Pope DP, *Mater. Sci. and Eng. A*, **239-240**, 126 (1997)

Mater. Res. Soc. Symp. Proc. Vol. 842 © 2005 Materials Research Society   S6.10

Texture development during hot extrusion of a gamma titanium aluminide alloy

M. Oehring, F. Appel, H.-G. Brokmeier and U. Lorenz
GKSS Research Centre, Institute for Materials Research,
Max-Planck-Str. 1, D-21502 Geesthacht, Germany

ABSTRACT

The evolution of preferred orientations during processing appears to be of significant importance for the use of γ titanium aluminide alloys, since the desired lamellar microstructures exhibit a strong anisotropy of mechanical properties. As texture evolution certainly is dependent on several factors, involving deformation properties, recrystallization kinetics and particularly the phase constitution, different processing temperatures were investigated. By comparing the results it is indicated that the determined textures can be understood by the deformation modes of the dominating phase at hot-working temperature and the subsequent phase transformations.

INTRODUCTION

Gamma titanium aluminide alloys are an emerging class of high-temperature light-weight alloys for structural applications up to around 750 °C [1 – 3]. As with other structural materials preferred orientations might significantly improve the mechanical properties. For titanium aluminides control of texture appears even to be of particular importance, since the materials show anisotropic elastic and plastic properties and, furthermore, the desired lamellar microstructures of the materials exhibit a strong anisotropy of mechanical properties.

The texture development in γ-TiAl alloys during hot working might be characterized by some peculiarities in comparison to the behavior of solid solution phases [4]. These include the occurrence of perfect and superdislocations, which not only have significantly different critical resolved shear stresses but also differing core structures influencing the deformation, recovery and recrystallization behavior [3]. The complexity of processes during high-temperature deformation is increased by twinning, which is a prominent deformation mode of γ-TiAl alloys [3]. Moreover, recrystallization of intermetallic phases is impeded by the low mobility of grain boundaries and the need that the ordered state has to be restored on primary recrystallization [5]. As γ-TiAl alloys usually are worked in temperature ranges where predominantly the γ, the α and γ or only the disordered hexagonal solid solution phase α are present, not only the texture evolution of different phases but also the phase transformations on cooling and their effect on texture have to be considered.

In this work the texture evolution on hot extrusion of a conventional B containing γ(TiAl) alloy has been investigated. The study included specimens with a lamellar microstructure. Due to the relatively coarse microstructure texture analyses by neutron diffraction have advantages over X-rays in this case, since the absorption for thermal neutrons is extremely low, and thus, the whole body of a sample can be examined and not merely its surface. It further is noted here that the coherent neutron scattering lengths of Ti and Al have nearly the same magnitude but are positive for Al and negative for Ti, which leads to an enhancement of some diffraction peaks and to a cancellation of others compared to X-ray diffraction.

EXPERIMENTAL DETAILS

For the investigations two ingots (diameter 290 mm, length 730 mm) of composition Ti-47 Al- 3.7 (Nb, Cr, Mn, Si)- 0.5 B (at.%) were used as starting material, which were supplied by Howmet Corp. The ingots were produced by triple vacuum arc re-melting (VAR). Part of the material had been hot-isostatically pressed (HIP) and heat-treated prior to extrusion. From the ingots cylindrical billets were taken parallel to the ingot axis at constant radius. These billets were canned and extruded at temperatures in the $(\alpha + \gamma)$ (1250 °C, 1300 °C) or in the α phase field (1380 °C) into round dies. For all extrusions an extrusion ratio of 7:1 was applied. A more detailed description of the extrusion conditions can be found elsewhere [6]. After extrusion, some specimens were subjected to a heat treatment at 1030 °C, which corresponds to about 0.75 T_m (absolute melting temperature). For comparison, also hot extruded powder material was investigated which had been produced by atomizing the same ingot material and compacting it by hot-isostatic pressing. The O_2 and N_2 contents were determined in all material conditions and did not exceed 800 µg/g and 100 µg/g, respectively.

Texture analysis was performed in the TEX-2 neutron diffractometer [7] utilizing the spherical sample method first described by Tobisch and Bunge [8]. Cylindrical samples of 15 mm diameter and 13.5 mm length were cut at about 10 mm from the rim of the ingot or from the center of the extruded bars. The axis of the texture samples was parallel to the ingot axis or to the extrusion axis, respectively. The diffraction experiments were performed under the following conditions: wavelength 0.134 nm, Cu (111) monochromator, beam width 22 mm, distance between sample and detector 80 cm. For neutron detection a ^{3}He single detector was used.

The orientation distribution functions were calculated from the measured pole figures of γ(TiAl): (001), (110), (201), and (112). All measurements were performed with an equal area counting grid of 679 individual sample orientations. These data were transformed into an equal angular grid with a 5x5 matrix that is more suitable for texture analysis. Quantitative textures were calculated using the iterative series expansion method proposed by Dahms and Bunge [9], with a degree of series expansion of $l_{max} = 22$. The advantage of this method is that fewer pole figures are required, when compared with standard series expansion methods. Thus, in the present study four pole figures were sufficient for a quantitative analysis.

RESULTS

The microstructure of the ingot material consisted predominantly of large, elongated lamellar colonies, which had a size of some hundred µm. In addition, small fractions of borides and γ grains were observed. The γ and α_2 lamellae were primarily aligned nearly parallel to the rim of the ingot, i.e. perpendicular to the direction of heat extraction during solidification. Such a preferred orientation of lamellae is often observed for γ alloys and can be easily understood by the growth of α phase dendrites in the [0001] direction on solidification [10]. From the 4 pole figures measured on the ingot material it could be concluded that the observed orientation distribution can be considered as a fiber texture. The fiber axis exhibited a <111> orientation, i.e. the {111} planes of the γ phase are roughly parallel to the ingot axis in agreement to the solidification texture mentioned above.

Extrusion of the ingot material was carried out below and above the α transus temperature which was determined to be 1360 °C for this alloy [6]. After extrusion at 1250 °C and 1300 °C very fine-grained equiaxed microstructures with mean grain sizes between 2 and 3 µm were obtained. Apparently, the material was fully recrystallized because only rarely remnants of lamellar colonies from the cast microstructure could be observed. Since further also no newly formed lamellar colonies were found, it is concluded that up to 1300 °C only minor fractions of the α phase were present in this alloy and thus, the hot-working behavior was dominated by the γ phase. By raising the extrusion temperature up to 1380 °C the microstructural condition changed to a lamellar one which was characterized by a considerably fine mean colony size of 58 μm. As the lamellar interfaces were straight and did not show any bending, the lamellae must have been formed on cooling after extrusion. Also this material is concluded to be completely recrystallized as the colony size is about one magnitude smaller than in the ingot material. Fig. 1 displays pole figures obtained from the extruded material conditions. As expected due to the rotational symmetry of deformation by extrusion, fiber textures with a fiber axis near the extrusion direction were found in all cases. The texture sharpness was in general low or even close to a random orientation distribution. This result means that the relatively strong casting texture was almost completely destroyed by hot extrusion, irrespective of the applied working conditions. Further, it can be concluded that the extrusion temperature, i.e. the dominating phase during extrusion has a significant influence on the texture evolution. The inverse pole figures in Fig. 2 show that specimens extruded below the α transus temperature had an <111> and a <001] orientation component in extrusion direction, whereas the specimen from material extruded above the α transus temperature exhibited a <112] and a very weak <101] component. Fig. 2b and c show inverse pole figures of powder material (PM) and of heat-treated ingot material extruded at 1250 °C under identical conditions as described above (see Fig. 2a). Since the powder material was produced by hot-isostatic pressing of powder, the starting material is believed in this case to be nearly free of texture. Thus, it may be concluded that the texture of the starting material only plays a minor role for texture evolution during extrusion under the applied conditions.

Ingot material extruded at 1250 °C and 1380 °C was subjected to an annealing treatment in the (α₂ + γ) phase field (2 h at 1030 °C) after extrusion. For material extruded at 1250 °C this heat treatment resulted in a decrease of the <001] texture component whereas the <111> component remained constant (Fig. 2). These slight changes in texture were accompanied by a small increase of the grain size by 14 %. For material extruded at 1380 °C, however, a significant reduction of the texture sharpness was observed and in particular, the <112] texture component

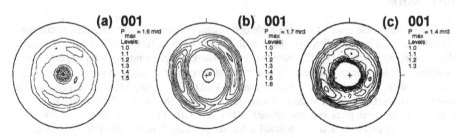

Figure 1: Measured (001) pole figures of extruded specimens. The projection plane is perpendicular to the extrusion direction. Extrusion was carried out on as-cast ingot material at (a) 1250 °C, (b) 1300 °C, and (c) 1380 °C. No heat treatment was applied after extrusion.

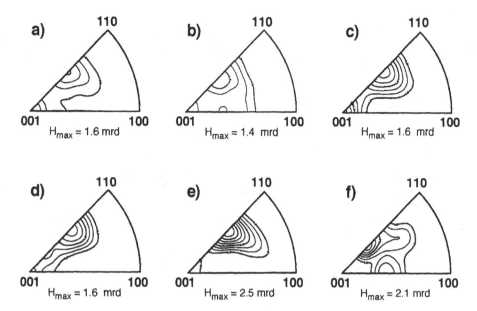

Figure 2: Inverse pole figures in extrusion direction of extruded materials. (a) As-cast ingot material extruded at 1250 °C, (b) powder material extruded at 1250 °C, (c) HIP'ed and heat-treated ingot material extruded at 1250 °C, (d) as-cast ingot material extruded at 1250 °C and subsequently heat-treated at 1030 °C, (e) as-cast ingot material extruded at 1300 °C, and (f) as-cast ingot material extruded at 1380 °C. Counter levels: 1.0, 1.2, 1.4, 1.6, 1.8, 2.0, 2.2, 2.4.

diminished. This decrease of the texture sharpness can easily be explained by a non-equilibrium of the constituent phases, which results in the degradation of the lamellar colonies at the expense of growing γ grains as also was observed by microstructure investigations.

DISCUSSION

A commonly held concept of large-scale deformation of polycrystalline material is that the grains are rotated in a direction that allows multiple-slip by the activation of the most favourable slip or twinning systems [11]. This results in orientations of high symmetry with respect to the main deformation axis and suggests for hot-working temperatures falling in the ($\alpha+\gamma$) field the following states of deformation. Extrusions should lead to an overall preference of γ grains with an <111> orientation parallel to the extrusion direction. This ensures that the imposed radial reduction of the work piece is accomplished by two 1/2<110]{111} slip systems that provide shear processes perpendicular and inclined to the sample axis. A substantial population is also expected for γ grains with an <001] orientation parallel to the extrusion axis, because in this orientation radial shear can be provided by two 1/2<110]{111} slip systems. It is interesting to note that these two grain orientations predicted are the major components of the observed double fiber texture of the materials extruded below the α-transus temperature.

After hot-working the material is fully recrystallized, which indicates that recrystallization and phase transformation can be equally important for texture evolution. As with many other materials, in TiAl alloys static and dynamic recrystallization is certainly triggered by heterogeneities in the deformed state [12], which are associated with blocked slip and twinning and which are manifested in work-hardening. There is good evidence from electron microscopy that dynamic recrystallization starts at immobilized twins or twin intersections. Work-hardening of γ(TiAl)-base alloys at low and intermediate temperatures is ascribed to long-range elastic interactions, which in principle lead to the formation of junctions and sessile multi-poles. These processes certainly increase the amount of stored energy and are beneficial to dynamic recrystallization. However, for the evolution of the deformation structure of the γ phase, propagation of ordinary dislocations with orthogonal Burgers vectors is probably most important. Such dislocations have only a weak elastic interaction and little tendency to form sessile junctions [13]. Furthermore, under these conditions a substantial contribution to work hardening is provided by the formation of dislocation dipoles and debris, which can easily be annealed out. Thus, as recognized in recent studies [14], rapid static and dynamic recovery occurs. Furthermore, due to the non-octahedral dislocation climb, the pile- up structures are probably blunted. For these reasons, the driving force resulting from stored dislocation structures is expected to be relatively small.

On the other hand, recrystallization of ordered structures is considered difficult, when compared with disordered metals, because the ordered state has to be restored and the grain boundary mobility is low [5]. Of perhaps greater significance for texture development is that the new grains grow by bulging; details of this mechanism were investigated in an earlier study on a Ti-45Al-10Nb (at.%) alloy [15]. The characteristic feature of grain boundary bulging is that the new grains have a similar orientation to the old grains from which they have grown. Thus, the operation of the mechanism is expected to result in a recrystallization texture that is closely related to the deformation texture [12]. In view of these considerations it appears plausible that the operated deformation modes remain manifested after recrystallization.

In addition to deformation and recrystallization changes of alloy constitution with temperature also have to be considered. This is apparently reflected in the textures observed after extrusion in the single-phase α field. According to the slip geometry of α(Ti), the radial reduction of the work piece imposed by the funnel shaped die can be realized by symmetrical prismatic glide with a c axis orientation perpendicular to the extrusion direction, for which two cases have to be considered. One of the prism planes can be either perpendicular to the extrusion direction or inclined at an angle of 60 °. This would give rise to $<2\bar{1}\bar{1}0>$ and $<10\bar{1}0>$ deformation textures that would have the extrusion direction alignment. During subsequent cooling of the work piece from α(Ti), the α_2 and γ phases are formed and aligned according to the Blackburn relationship [3], i.e., the {111} plane of the new γ lamellae is parallel to the basal plane of the prior α(Ti) grains. In the final texture, these new γ lamellae should be manifested by significant populations with <112] and <101] orientations parallel to the extrusion direction, which indeed were observed (Fig. 2f).

CONCLUSIONS

The evolution of microstructure and texture during hot working of two-phase titanium aluminide alloys is governed by the operating deformation modes, recrystallization and phase transformation. The relative contribution of these factors depends on alloy composition and processing temperature. The structural features observed after hot-working in the ($\alpha + \gamma$) phase field essentially reflect the kinematics of the propagation of ordinary dislocations and mechanical order twinning that is required to accomplish the imposed shape change. Hot-working in the single phase α(Ti) field leads to textures that largely are governed by the deformation modes of the hexagonal lattice of α(Ti) and the subsequent $\alpha \rightarrow \gamma$ phase transformation occurring upon cooling.

ACKNOWLEDGMENTS

We are grateful to Dr. K. Müller (TU Berlin), Dr. N. Eberhardt and Dr. H. Kestler (Plansee AG, Reutte, Austria) for conducting extrusions. We also gratefully acknowledge the provision of powder material by Dr. R. Gerling and Dr. F.P. Schimansky and stimulating discussions with Dr. J.D.H. Paul and Prof. H. Clemens (University of Leoben, Austria).

REFERENCES

1. Y-W. Kim and D.M. Dimiduk, *Structural Intermetallics 1997*, ed. M.V. Nathal, R. Darolia, C.T. Liu, P.L. Martin, D.B. Miracle, R. Wagner and M. Yamaguchi (TMS, Warrendale, PA, 1997), pp. 531 - 543.
2. D.M. Dimiduk, *Mater. Sci. Eng. A* **263**, 281 (1999).
3. F. Appel and R. Wagner: *Mater. Sci. Eng. R* **22**, 187 (1998).
4. A. Bartels, W. Schillinger, G. Graßl and H. Clemens, *Gamma Titanium Aluminides 2003*, ed Y-W. Kim, H. Clemens, A.H. Rosenberger (TMS, Warrendale, PA, 2003), pp. 275 – 296.
5. R.W. Cahn, *High Temperature Aluminides and Intermetallics*, ed. S.H. Whang, C.T. Liu, D.P. Pope and J.O. Stiegler (TMS, Warrendale, PA, 1990), pp. 245 - 270.
6. M. Oehring, U. Lorenz, F. Appel and D. Roth-Fagaraseanu, *Structural Intermetallics 2001*, ed. K.J. Hemker, D.M. Dimiduk, H. Clemens, R. Darolia, H. Inui, J.M. Larsen. V.K. Sikka, M. Thomas and J.D. Whittenberger (TMS, Warrendale, PA, 2001), pp. 157 - 166.
7. H.-G. Brokmeier, U. Zink, R. Schnieber, and B. Witassek, *Texture and Anisotropy of Polycrystals*, ed. R. Schwarzer (TTP Transtech Publications, Zürich), Mater. Sci. Forum **273–275**, 277 (1998).
8. J. Tobisch and H.J. Bunge, *Textures* **1**, 125 (1972).
9. M. Dahms and H.J. Bunge, *J. Appl. Cryst.* **22**, 439 (1989).
10. D.R. Johnson, H. Inui, and M. Yamaguchi, *Acta mater.* **44**, 2523 (1996).
11. E.A. Calnan and C.J.B. Clews, *Phil. Mag.* **41**, 1085 (1950).
12. F.J. Humphreys and M. Hatherly, *Recrystallization and Related Phenomena* (Pergamon, Oxford, 1995).
13. F. Appel, *Phys. stat. sol. (a)* 116, 153 (1989).
14. F. Appel, U. Sparka and R. Wagner, *Intermetallics* **7**, 325 (1999).
15. W.J. Zhang, U. Lorenz, and F. Appel, Acta mater. **48**, 2803 (2000).

Mater. Res. Soc. Symp. Proc. Vol. 842 © 2005 Materials Research Society   S6.9

Fatigue Testing of microsized samples of γ-TiAl based material

Timothy P. Halford, Kazuki Takashima and Yakichi Higo
Precision & Intelligence Laboratory, Tokyo Institute of Technology,
4259 Nagatsuta, Midori-ku, Yokohama 226-8503, Japan.

ABSTRACT

High strength γ-TiAl based alloys, such as Ti-46Al-5Nb-1W (Alloy 7), which were originally developed for gas turbine and automotive applications are now being considered for application in Micro Electro Mechanical Systems (MEMS). This requires the evaluation of these materials upon the microscale. As international standards do not currently exist for the evaluation of the mechanical properties of samples with dimensions equivalent to those required by MEMS devices, the development of new methods was required. The method developed here is intended for the fatigue testing of samples measuring ≈ 10μm (B) x 20μm (W) x 40μm (L). This is completed using a machine recently developed at Tokyo Institute of Technology to load samples of lamellar γ-TiAl based material to failure in compressive bending. This method is intended to work alongside methods previously developed for the fracture toughness testing of similar microsized cantilever bend specimens.

In this work sample cantilevers of Alloy 7 are Focussed Ion Beam (FIB) machined from foil ≈ 20μm thick and their stress – life (S-N) fatigue behaviour evaluated. The dependence of fatigue life upon lamellar orientation for a given peak stress / stress range is considered. The effect of the reduced scale of these samples upon the mean and scatter of these sample lifetimes is also considered through comparison with previous data obtained from the S-N testing of macrosized samples of the same material.

INTRODUCTION

Micro Electro Mechanical Systems (MEMS) face continuing development in order to facilitate their viable implementation in a wide range of application areas. The success of MEMS as accelerometers controlling the deployment of automotive airbags as well as in ink jet printing have effectively demonstrated their potential for use in innovative functions. In addition to these concepts there is also a market for the implementation of MEMS to complete functions currently requiring larger systems, in order to save cost, space, weight or to improve functionality. These benefits are of particular interest to developers of aerospace programs, where significant reductions in size and weight can be expected to lead to cascading cost benefits, through reduced production expense and greater fuel efficiency. In pursuit of these benefits the research and development of MEMS has received significant investment, with the US Department of Defence allocating $75 million in 1998 and global investment into the field having been estimated at $370 million in 1995 [1]. The elevated temperatures required of devices in many aerospace applications; however, prohibits the use of existing, low melting point, MEMS materials. The development of a material with the capability to operate at temperatures in excess of 400°C for MEMS applications is therefore required. One such material class under consideration for this role are the γ-TiAl based aluminides. These compositions have been under development for macroscale implementation in the aerospace industry largely due to their combination of elevated temperature strength, stiffness, creep and burn resistance in a low density material. These properties also make them attractive possibilities for application in MEMS devices, where their low fracture toughness (which has undermined confidence in their potential for macroscale applications) may be desirable in cases where component failure is preferable to the generation of a false signal [2]. As MEMS devices require microsized material, plane strain criteria [3] are unachievable in many cases [4-6]. This means that the mechanical properties of a potential new MEMS material must be evaluated by micro-scale testing. This issue is particularly relevant in the case of lamellar γ-TiAl based materials, as the directional nature of the lamellar structure provides the potential for distinctly differing properties, which are not demonstrated on the macro-scale. For these reasons fatigue testing of microsized cantilevers of a γ-TiAl based composition (known as Alloy 7) is the focus of this work. This testing was completed upon a new machine developed for the mechanical testing of microsized materials and involves an examination of the effect of colony orientation upon the fatigue life of samples under a given loading regime.

EXPERIMENTAL PROCEDURE

Alloy 7 (Ti-46Al-5Nb-1W, at.%), having been extruded from a vacuum arc melted ingot [7], was Electro Discharge Machined (EDM'ed) into slices of thickness ≈ 80μm. These slices were then mechanically ground and polished until they had a surface finish ≈ 0.05μm on both sides and a thickness ≈ 20μm. Microsized cantilevers were introduced into this material using a Focussed Ion Beam (FIB). Two specimen types were produced, see figure 1, in order to allow the allocation of the film thickness as W ("A" Type) or B ("B" Type) within the tested samples. Four "A" type samples (~ 10 x 20 x 40μm) were tested, followed by one "B" type sample (~15 x 30 x 80μm). Although the directionality in this material is a result of a local microstructure whose orientation dependence can be selected through the control of cantilever location, the "B" sample type is required in cases where properties parallel and perpendicular to the film surface differ. This is often the case in MEMS materials having been produced by deposition. Fabrication was completed while the material was fixed in the test holders. This allows the completion of fabrication, measurement and testing of samples without the individual handling

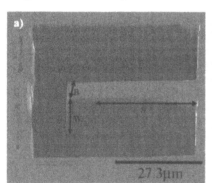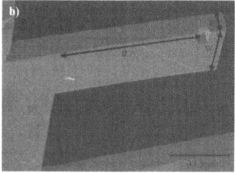

Figure 1: Scanning Electron Microscope (S.E.M) Images of microsized cantilever samples having been fabricated using a Focussed Ion Beam (F.I.B). a) "A" Type sample and b) "B" Type sample.

of microsized cantilevers leading to damage. Upon the completion of machining, location markers, see figure 1, were introduced into cantilevers in order to allow accurate identification of the intended loading point. Samples were then measured using a Field Emission S.E.M, which was also used for the subsequent examination of fracture surfaces.

In order to test microsized cantilevers of this type, a new testing machine has recently ᵤ ᵣen developed at the Tokyo Institute of Technology. This equipment follows the MFT-ᵤ000 machine previously developed for microsized material testing [4, 6, 8]. In this case the MFT-2004 machine utilises a horizontally aligned nanoindentor to apply loads with a resolution of 10µN, using either load or displacement control. The sample position is located using a microscope and two high magnification cameras with an X-Y-Z stage having a resolution of 0.1µm, see figure 2. This equipment can be adapted to different sample types by the repositioning of the microscope and cameras to provide the required visual angles. Testing in this research was completed using a sinusoidal waveform at a frequency of 10Hz, an R ratio (minimum / maximum stress in the fatigue cycle) of 0.1 and a nominal maximum stress of 550MPa (a stress range of 495MPa). These criteria were selected in order to allow comparison with data which has previously identified these conditions (mainly at frequencies of 78-90Hz) as the fatigue limit of this material in macrosized samples [7].

RESULTS

Microsized cantilever samples were successfully tested, under load control, using the MFT-2004 machine. Throughout these tests the development of methods was required in order to allow the correct setting of load and displacement alarms to provide an exact count of the number of cycles to failure. For this reason the lifetimes of "A" type samples were only recorded within inspection intervals. In this case, where the scatter in lifetimes is considered in relation to orders of magnitude, these inaccuracies are not believed to be significant. Upon the completion of this development the successful determination of an exact number of cycles to failure was achieved, with the final cycle(s) being recorded, see figure 3. This is achieved

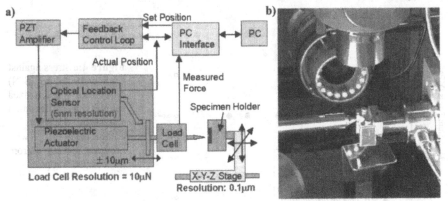

Figure 2: MFT-2004 machine newly developed at The Tokyo Institute of Technology for the mechanical testing of microsized samples, showing a) a block diagram of the machines components and b) a photograph of a sample being loaded.

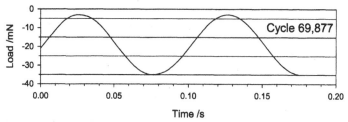

Figure 3: Load – Time trace for final cycles in the life of "B" type sample.

due to the collection of time, load, displacement data for pre-defined cycles within the fatigue lifetime, as well as, the final failure cycle. This information can be used to analyse the effective waveform applied to the sample at varying stages throughout exposure and failure.

Comparison of the fatigue lifetimes obtained from these microsized samples, see figure 4, with data for the macrosized behaviour of this material [7] shows the microsized cantilevers to have failed within 10^6 cycles at the fatigue limit for macrosized samples. There is a large degree of scatter in the lifetimes obtained from the "A" type samples tested, with two samples providing lifetimes ~ 10^3 cycles and another two approaching 10^6 cycles prior to failure. Although there is insufficient data for a comparison of the relative lives of "A" and "B" type samples it is noted that the life observed from the "B" type sample tested falls within the scatter observed from "A" type samples.

Examination of S.E.M fractography images taken, see figure 5, shows a distinct lamellar orientation difference between samples with a lifetime of ~ 10^3 cycles and those which approached 10^6 cycles prior to failure. Fracture surfaces of samples with a comparatively long fatigue lifetime show extensive crack growth along lamellar interfaces orientated perpendicular to the fracture surface. This can be seen to have led to an effective lengthening of the crack surface, as well as, increased interlocking of the two surfaces. In comparison, the fracture

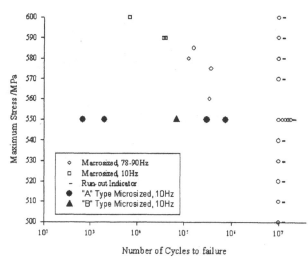

Figure 4: Stress against Cycles to failure (S-N) data for microsized samples, at a maximum stress of 550MPa, in comparison with previous data from macrosized samples.

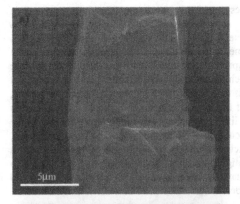

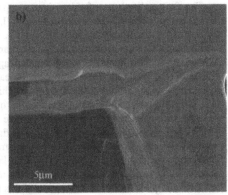

Figure 5: S.E.M fractographs showing cantilever fracture surfaces after a) ~10^6 cycles and b) ~10^3 cycles in "A" type samples, as well as c) after 69,877 cycles in a "B" type sample.

surfaces of samples having survived for a life in the order of 10^3 cycles can be seen to have comparatively flat fracture surfaces, with crack growth along lamellar interfaces orientated approximately perpendicular to the tensile surface.

DISCUSSION AND CONCLUSIONS

A machine has been developed that is capable of the load controlled fatigue testing of microsized materials. This machine uses multiple video images to position the sample, in situ, with the required accuracy prior to accurate application of the desired waveform to the sample. This is in contrast to the previous machine used, which required observation of sample location by a microscope mounted parallel to the indentor, followed by the accurate calibration of the translation distance between the microscope and indentor [8]. Testing has been successfully completed for two different sample types through the utilisation of the capability to re-angle the microscope and reposition the high magnification cameras. This equipment is also believed to be capable of fracture toughness and tensile testing of microsized materials. For these reasons, it is being considered as the basis for future standard mechanical property testing techniques.

The fatigue lifetimes of microsized samples tested here are lower than those expected of macrosized samples of the same material [7]. It is suggested that this may be attributed to reduced extrinsic toughening in such small scale samples. The scatter identifiable in samples tested appears to be the result of the varying lamellar orientations in the different cantilevers. In cases in which the lamellae are orientated approximately parallel to the tensile surface, the growth of a crack requires reinitiation at each lamellar boundary, often at a location not immediately adjacent to the incident crack. This results in growth parallel to the tensile surface, some extrinsic toughening and a lengthening of the overall free surface created. These factors are expected to provide an increased fatigue life. Comparatively, in samples in which the lamellae are orientated approximately perpendicular to the tensile surface and stress directions, crack growth can proceed directly along a lamellar boundary, providing rapid crack growth, a flat fracture surface and comparatively short fatigue lifetimes. Extended testing is intended to identify the effect of these factors upon this material under a range of conditions, in order to provide a complete S-N characterisation of this behaviour.

ACKNOWLEDGEMENTS

This work was supported by a "Grant-In-Aid for Scientific Research" provided by the Japan Society for the Promotion of Science (JSPS).

REFERENCES

1   J. Benoit, *Mat. Sci. Eng B*, **B51**, 254 (1998).
2   M. Elwenspoek and R. Wiegerink, *Mechanical Microsensors*, (2001).
3   ASTM, Standard E 399: Standard test method for plane-strain fracture toughness testing of metallic materials, *Annual Book of ASTM Standards*, (1997) pp.443-473.
4   Y. Ichikawa, S. Maekawa, K. Takashima, M. Shimojo, Y. Higo and M.V. Swain, in *Materials Science of Microelectromechanical Systems (MEMS) Devices II*, edited by M.P. de Boer, A.H. Heuer, S.J. Jacobs and E. Peeters, (Mater. Res. Soc. Proc. **605**, Pittsburgh, PA, 2000), pp.273-278.
5   K. Takashima, A. Ogura, Y. Ichikawa and Y. Higo, in *Materials Science of Microelectromechanical systems (MEMS) Devices III*, edited by H. Kahn, M.P. de Boer, M. Judy and S.M. Spearing, (Mater. Res. Soc. Proc. **657**, Pittsburgh, PA, 2001) pp.EE5.12.1-6.
6   K. Takashima, T.P. Halford, D. Rudinal, Y. Higo and P. Bowen, in *Thin Films- Stressses and Mechanical Properties X*, edited by S.G. Corcoran, Y-C. Juo, N.R. Moody, Z. Suo, (Mater. Res. Soc. Proc. **605**, Pittsburgh, PA, 2004), pp.153-158.
7   R. Pather, A. Wisbey, A. Partridge, T. Halford, D.N. Horspool, P. Bowen and H. Kestler., in *Proceedings of third International Symposium on Structural Intermetallics (ISSI 3), Structural Intermetallics 2001, edited by K.J. Hemker, D.M. Dimiduk, et al. (TMS, Warrendale, PA, 2002) pp 207 - 215.*
8   Y. Higo, K. Takashima, M. Shimojo, S. Sugiura, B. Pfister and M.V. Swain, in *Materials Science of Microelectromechanical Systems (MEMS) Devices II*, edited by M.P. de Boer, A.H. Heuer, S.J. Jacobs and E. Peeters, (Mater. Res. Soc. Proc. **605**, Pittsburgh, PA, 2000), pp.241-246.

Microstructure-Property Relationships of Two Ti$_2$AlNb-based Intermetallic Alloys: Ti-15Al-33Nb(at.%) and Ti-21Al-29Nb(at.%)

Christopher J. Cowen, Dingqiang Li, and Carl J. Boehlert
School of Engineering, Alfred University, Alfred, NY 14802, U.S.A.

ABSTRACT

Two Ti$_2$AlNb intermetallic orthorhombic (O) alloys, Ti-15Al-33Nb and Ti-21Al-29Nb(at.%), were subtransus processed into sheets, using pancake forging and hot-pack rolling, and evaluated in tension (25 and 650°C) and creep (650-710°C) and the properties and deformation behavior were related to microstructure. Some of the microstructural features evaluated were grain boundary character, grain size, phase volume fraction, and morphology. The alloy Al content was important to strength and elongation-to-failure (ε_f), where higher Al contents lead to greater tensile strengths and lower ε_f values and a corresponding brittle fracture response. However, the room temperature (RT) strengths of Ti-15Al-33Nb, which exhibited greater BCC phase volume fractions than Ti-21Al-29Nb and ductile failure ($\varepsilon_f > 2\%$), were always greater than 775 MPa. The creep stress exponents (n) and activation energies (Q$_{app}$) suggested that a transition in the dominant creep deformation mechanism exists and is dependent on stress and microstructure. Supertransus heat treatment, which increased the prior-BCC grain size and resulted in a lath-type O+BCC microstructure, resulted in reduced creep strains and strain rates. In fact, the supertransus heat-treated Ti-15Al-33Nb microstructures exhibited greater creep resistance than subtransus heat-treated Ti-21Al-29Nb microstructures. Combining the creep observations with the tensile response, the supertransus heat treated Ti-15Al-33Nb lath O+BCC microstructures exhibited the most attractive combination of tensile strength, ε_f values, and creep resistance.

INTRODUCTION

For Ti$_2$AlNb intermetallic alloys, which depending on composition and processing treatment may contain any combination of the O, ordered hexagonal (α_2), and ordered or disordered body centered cubic (BCC) phases, microstructure is of critical importance to the mechanical properties. It has been established that the O phase tends to provide creep resistance while the BCC phase tends to provide processability and ductility [1-4]. These phases have shown compatibility through slip transmission across their interface boundaries [3] and their orientation relationship which evolves from their phase transformations [5]. This is one reason why two-phase O+BCC microstructures can exhibit a more attractive combination of RT and elevated-temperature properties (such as tensile, creep, and fatigue strength and tensile ductility), which are important for structural applications within the aerospace industry, than Ti$_3$Al- and TiAl-based intermetallic alloys. In addition they are starting to show promise for biomedical applications as they compare favorably to non-intermetallic Ti alloys such as Ti-6Al-4V(wt.%) [6]. However, shortcomings exist in the understanding of microstructure-property relationships of Ti$_2$AlNb-based alloys, and in particular the microstructure and mechanical behavior of alloys containing between 12-19 at.% Al is lacking. It is unknown what creep strength will result after decreasing the Al content and corresponding O-phase volume fraction. In order to address this uncertainty, Ti-15Al-33Nb and Ti-21Al-29Nb (at.%) alloys were processed into sheets, heat treated and evaluated in tension (RT and 650°C) and creep (650-710°C) and their properties were compared and related to their microstructure.

EXPERIMENTAL PROCEDURE

The two alloys were double vacuum arc remelted and the resulting ingots subsequently underwent thermomechanical processing, which included hot pack forging and hot rolling at subtransus temperatures, at RMI Titanium Company, Niles, Ohio. The hot rolling produced a reduction in area of 90% and the processing resulted in a total effective true shear strain on the order of two. The Ti-15Al-33Nb sheets were processed at 899°C and the Ti-22Al-28Nb sheets were processed at 982°C. After the hot rolling was complete, the sheets were annealed at their respective rolling temperatures for one hour and then air-cooled. Two heat treatment schedules, which could be described as: 1005°C/3h/FC/855°C/8h/FC/650°C/FC or 960°C/3h/FC/855°C/8h/FC/650°C/FC where FC was furnace cooled at 1 or 10°C/minute, were performed on both alloys and the mechanical properties of the heat-treated samples were compared to those for the as-processed condition. In addition, a third heat treatment was performed on Ti-15Al-33Nb where the only variable changed was the solution-treatment temperature, 1105°C. Henceforth, these heat treatments will be referred to by their solution treatment-temperature. All heat treatments were performed in flowing argon and the samples were wrapped in Ta foil.

Phase volume fractions were determined using high-contrast, digitized backscattered electron (BSE) scanning electron microscopy (SEM) images and NIH image analysis software. For the equiaxed-phase grains, average grain size was determined using the ASTM standard E112. In each case, several micrographs were taken at different magnifications from different sheet orientations. Such measurements were in good agreement with those based on electron backscattered diffraction (EBSD) analysis. Spatially resolved EBSD orientation maps were obtained using a Phillips 515 SEM with LaB_6 filament. The EBSD hardware and software were manufactured by EDAX-TSL, Inc. Low-angle boundaries were defined as those boundaries for grain misorientations less than 15° and high-angle boundaries were those with a misorientation greater than 15°. A tolerance of 5° was used to identify a grain boundary. For each map, more than 500 grain boundaries were analyzed using a step size of 0.5μm.

Flat, dogbone-shaped samples were machined from the rolled sheets using either a mill or an electrodischarge machine. The specimens were then ground through a better than 240 grit finish to remove the oxide scale left from processing and any surface layers that may have been damaged during the machining process. Tensile tests were performed at both RT and 650°C using an Instron 8562 testing machine equipped with a 100kN load cell. Strain was measured with a 25.4 mm gage length Instron extensometer attached to the gage section of each dogbone and all tests were conducted at a strain rate of 1.3×10^{-3} s^{-1}. Tests were performed in duplicate and 0.2% yield stress, ultimate tensile stress, modulus (E), and ε_f values were recorded. Constant load, tensile-creep experiments were performed using two Applied Test Systems, Inc. (ATS) creep frames with 20:1 lever arm ratios. The loading direction was parallel to the rolling direction for all the creep and tensile experiments. Creep strain was monitored during the tests using a linear variable differential transformer (LVDT) that was connected to a 25.4 mm gage length ATS high-temperature extensometer attached directly to the gage length of each sample. All creep tests were conducted in air at stresses between 50-275 MPa and temperatures between 650-710°C. For both the tensile and creep experiments, samples were soaked at the testing temperature for at least 30 minutes before loading and the temperature was controlled within ±

3°C of the targeted test temperature. The steady-state creep strain rate ($\dot{\varepsilon}_{ss}$) was determined for each specimen at each of the specific stresses and temperatures. In order to determine secondary creep stress exponent, n, and activation energy, Q_{app}, constant temperature/load jump tests and constant load/temperature jump tests were performed. This was accomplished by either changing the load or temperature after $\dot{\varepsilon}_{ss}$ had been achieved.

RESULTS AND DISCUSSION

The measured compositions of the two alloys were Ti-15.3Al-33.3Nb(at.%) with 1100 ppm oxygen and Ti-20.6Al-28.7Nb with 790 ppm oxygen. Combining the results from DTA, XRD, phase volume fraction analysis, and applying the disappearing phase method, the BCC-transus temperature was estimated for each alloy. For Ti-15Al-33Nb the β-transus was 980°C and for Ti-21Al-29Nb the β-transus was 1040°C. Table I depicts the microstructural parameters measured for the as-processed and heat-treated samples. For identical heat treatments, Ti-21Al-29Nb contained a larger volume fraction (V_f) of O phase than Ti-15Al-33Nb and a corresponding lower V_f of BCC phase. It is noted that the 1005°C and 1105°C heat treatments were supertransus for Ti-15Al-33Nb and a representative microstructure is illustrated in Figure 1a and b. It is noted that majority of the O-phase grain boundaries were high-angle boundaries, see Table I, and in particular these boundaries tended to cluster at near 90° misorientations. This is expected to be a result of the O/BCC orientation relationship and phase transformation [7] and the effect of such boundaries on the mechanical behavior will be the target of forthcoming work.

Table II lists the averaged tensile properties. The as-processed condition displayed the best

Table I The Measured Microstructural Features of the Ti-Al-Nb Alloys

Alloy	Heat Treatment	Vol. Fraction BCC	HABs***	LABs***	Fraction of 90° bndys*	d, μm**
Ti-15Al-33Nb	AP	0.90	0.688	0.237	0.36	3
	HT:960°C	0.58	0.763	0.170	0.25	6
	HT:1005°C	0.55	0.403	0.074	0.24	120
	HT:1105°C	0.36	0.337	0.046	0.32	173
Ti-21Al-29Nb	AP	0.96	0.466	0.387	0.12	3
	HT:960°C	0.28	0.834	0.060	0.30	8
	HT:1005°C	0.17	0.811	0.059	0.28	12

* a 2° tolerance angle used; ** equiaxed phase grain size; *** O-phase boundaries only; LABs: low-angle boundaries; HABs: high-angle boundaries; AP: as-processed

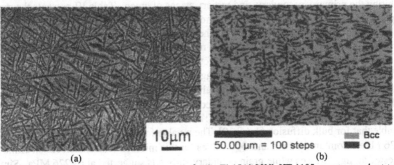

(a) (b)

Figure 1. (a) SEM BSE image and (b) EBSD phase map for the Ti-15Al-33Nb HT:1105 supertransus heat-treated lath O+BCC microstructure.

Table II Tensile Properties of Ti-15Al-33Nb and Ti-21Al-29Nb Microstructures

Alloy	Heat Treatment, °C	RT				650°C			
		E, GPa	YS, MPa	UTS, MPa	ε_f, %	E, GPa	YS, MPa	UTS, MPa	ε_f, %
Ti-15Al-33Nb	AP	94	876	916	12.4	66	592	601	12.8
	HT:960	103	778	867	4.4	66	522	561	8.0
	HT:1005	111	799	852	2.1	76	546	588	5.9
	HT: 1105	113	786	836	2.7	94	531	609	9.5
Ti-21Al-29Nb	AP	103	972	1010	1.7	71	810	825	2
	HT: 960	112	n/a	830	0.8	82	656	664	1.6
	HT:1005	115	803	868	1.6	77	537	657	5.2

YS: yield stress; UTS: ultimate tensile stress; n/a indicates 0.2%YS requirement was not met; AP: as-processed

balance of RT properties, such as strength and ε_f, for both alloys. This is due to the fact that both of these microstructures contained a majority of BCC phase by volume, i.e. 0.90 V_f for Ti-15Al-33Nb and 0.96 V_f for Ti-21Al-29Nb and the finest grain size due to the subtransus processing. Each Ti-15Al-33Nb microstructure exhibited higher RT ε_f values than Ti-21Al-29Nb due to the higher BCC V_f combined with the fact that the BCC phase in Ti-15Al-33Nb was disordered (determined using XRD), while it was ordered in Ti-21Al-29Nb. Due to the O structure's inherently lower number of available slip systems compared to the BCC phase, a larger volume fraction of O phase will in general provide strengthening at the expense of ductility. Comparing both alloys in the subtransus HT:960 condition (0.42 V_f O for Ti-15Al-33Nb versus 0.72 V_f O for Ti-21Al-29Nb), Ti-15Al-33Nb displayed a RT ε_f of 4.4% compared to 0.8% for Ti-21Al-29Nb, and the strength was higher for Ti-15Al-33Nb. For the other microstructures, Ti-21Al-29Nb exhibited slightly higher strengths than Ti-15Al-33Nb. At 650°C, all alloys exhibited lower E and strength values and increased ε_f values which follows the expected trends. The increase in ε_f values for all microstructures can be attributed to the thermally assisted activation of a larger number of active slip systems at 650°C. In Ti-26Al-21Nb, it has been shown that above 450°C, 'c' component dislocations become active along with the 'a' and 'a*' component dislocations that are active in the O phase at RT [8]. The microstructures that exhibited the most dramatic increase in ε_f were the heat-treated microstructures, which contained a larger O V_f. The E values of all the microstructures decreased by ~30% at 650°C compared to RT.

The creep behavior of the two alloys resembled that for most pure metals and alloys, exhibiting the primary, secondary, and tertiary stages of creep. Table III presents the measured creep n and Q_{app} values as a function of σ and T. Figure 2 illustrates log $\dot{\varepsilon}_{ss}$ versus log σ for the heat-treated microstructures at T=650°C. The two Ti-15Al-33Nb supertransus heat-treated microstructures displayed the greatest creep resistance of all the microstructures tested. Over the stress range of 150-226 MPa both microstructures exhibited constant n values (n=2.4 for HT:1005 and n=3.1 for HT:1105) which suggests that a single deformation mechanism was dominant over this stress range. The Q_{app} values determined at σ=150 MPa were 163 kJ/mol and 181 kJ/mol for HT:1005 and HT:1105, respectively. At σ= 275 MPa the Q_{app} values were 288 kJ/mol and 317 kJ/mol for HT:1005 and HT:1105, respectively, and these values closely resemble that for bulk diffusion [1,4,9,10] The n value for the HT:1005 microstructure was 6 at σ>226 MPa. Combining the n and Q_{app} values for this microstructure, dislocation climb is suggested at high stresses while grain boundary sliding is suggested at σ<226 MPa. Similar observations for other Ti$_2$AlNb alloys have been made previously [2,4,10]. Due to the extremely

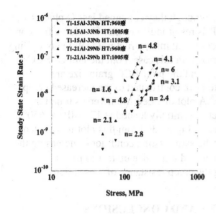

Figure 2. The log ε_{ss} versus log σ for the heat-treated microstructures at T=650°C.

similar creep strain versus life behavior of both supertransus microstructures, it can be concluded that the same deformation mechanisms operating for the HT:1005 microstructure are likely operating for the HT:1105 microstructure. The subtransus microstructures exhibited worse creep resistance than the supertransus microstructures. The Ti-15Al-33Nb HT:960 microstructure had an n value of 1.6 from 49 to 124 MPa, with a Qapp of 115 kJ/mol at 48 MPa, and an n value of 4.8 from 124 to 172 MPa. Such values suggest that the transition to dislocation climb creep occurred at applied stresses greater than or equal to 124 MPa. The subtransus Ti-21Al-29Nb microstructures exhibited similar creep behavior as the HT:960 microstructure had a n value of 2.1 over the stress range of 48-101 MPa and an n value of 4.8 above 101 MPa. At 48 MPa this microstructure had a Qapp of 292 kJ/mol. In the low stress regime, creep deformation is probably occurring by either grain boundary sliding or Nabarro-Herring creep due to its activation energy being closer to that estimated for lattice bulk diffusion (327-346kJ/mol [1,4,9,10]) and the n value of ~ 2. The Ti-21Al-29Nb HT:1005 microstructure had a n value of 2.8 from 48 to 126 MPa, then transitioned to an n value of 4.1 from 126-250 MPa. An activation energy of 215 kJ/mol at 148 MPa was determined, which is in the high stress regime for this microstructure. From 126-250 MPa at 650°C, deformation by dislocation climb is the probable mechanism. Comparing the results, it appears that supertransus heat treatment not only improves the creep resistance but also shifts the onset of the transition in n values to a higher stress.

Table III Measured Creep Exponents and Apparent Activation Energies

Alloy	Heat Treatment, °C	Stress/Temperature, MPa/°C	n	Stress/Temperature, MPa/°C	Qapp, kJ/mol
Ti-15Al-33Nb	HT:960	49-124/650	1.6	48/650-710	115
		124-172/650	4.8		
	HT:1005	148-226/650	2.4	150/650-710	163
		226-273/650	6.0	273/650-690	288
	HT:1105	151-275/650	3.1	151/650-710	181
				275/650-690	317
Ti-21Al-29Nb	HT:960	48-101/650	2.1	48/650-710	292
		101-171/650	4.8		
	HT:1005	48-126,	2.8	148/650-710	215
		126-250	4.1		

The Ti-21Al-29Nb microstructures contained larger O phase volume fractions than each of the identically heat-treated Ti-15Al-33Nb microstructures. This may be the reason for the superior creep resistance of the Ti-21Al-29Nb HT:960 microstructures over the Ti-15Al-33Nb HT:960 microstructure. However, both the supertransus heat-treated Ti-15Al-33Nb micro-structures, HT:1005 and HT:1105, had superior creep resistance to the Ti-21Al-29Nb micro-

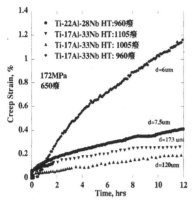

Figure 3. Creep strain versus time for selected microstructures.

structures, although they contained less O phase by volume. This result indicates that O volume fraction is not the dominant microstructural feature controlling the creep rates. In this case, both the lath O+BCC morphology and the prior BCC grain size are expected to have contributed to the increased creep resistance. A plot of creep strain versus time for microstructures initially loaded at 172 MPa at 650°C is presented in Figure 3. From this plot it is evident that under the same testing conditions, increasing the average equiaxed grain size up to 120 μm is increasing the creep resistance of these microstructures.

SUMMARY AND CONCLUSIONS

This work evaluated microstructure-property relationships of two Ti_2AlNb-based intermetallic alloys, Ti-15Al-33Nb and Ti-21Al-29Nb. The focus was to probe the potential for Ti-15Al-33Nb for structural applications. This alloy proved to offer enhanced processability due to its greater BCC volume fraction, which enabled significantly higher tensile ε_f values than those for Ti-21Al-29Nb, which exhibited brittle fracture. The Ti-21Al-29Nb alloy tended to exhibit slightly greater tensile strength than Ti-15Al-33Nb, however, Ti-15Al-33Nb exhibited a better balance of RT and elevated-temperature tensile properties. In addition, the best creep resistance was exhibited by the supertransus heat-treated Ti-15Al-33Nb microstructures. This indicated that O volume fraction was not the dominant microstructural feature for creep resistance, and the lath O+BCC microstructure along with the large prior BCC grain size are preferred for enhanced creep strength. Thus this study showed that lower Al-containing Ti_2AlNb based alloys offer merit for structural applications and supertransus lath O+BCC microstructures offered the best combination of tensile strength, ductility, and creep resistance.

ACKNOWLEDGMENTS

This work was supported by NSF (DMR-0134789) and NYSTAR (No. C020080).

REFERENCES

1. T.K. Nandy, R.S. Mishra, and D. Banerjee, *Scripta Metall. Mater.* **28**, 569 (1993).
2. R.G. Rowe and M. Larsen, *Titanium 1995*, Vol. 1, pp. 364-371, Ed P.A. Blenkinsop, W.J. Evans, and H.M. Flowers, The University Press, Cambridge, UK, 1996.
3. C.J. Boehlert, *Metall. Mater. Trans.* **32A,** 1977 (2001).
4. C.J. Boehlert and D.B. Miracle, *Metall. Mater. Trans.* **30A**, 2349 (1999).
5. K. Muraleedharan and D. Banerjee, *Philos. Mag. A*, **71**, 1011 (1995).
6. C.J. Boehlert, C.J. Cowen, R. Jaeger, M. Niinomi, T. Akahori, *Mater. Sci Eng., C*, (in press).
7. D. Li and C.J. Boehlert, *TMS Letters*, (in press).
8. F. Popille and J. Douin, *Philos. Mag. A*, **73**, 1401 (1996).
9. C.J. Boehlert and J.F. Bingert, *J. Mater. Process. Technol.* **117**, 401 (2001).
10. R.W. Hayes, *Scripta Mater.* **34**, 1005 (1996).

Mater. Res. Soc. Symp. Proc. Vol. 842 © 2005 Materials Research Society   S5.51

Phase Transformation in Orthorhombic Ti$_2$AlNb Alloys Under Severe Deformation

B.A.Greenberg, N.V.Kazantseva, V.P.Pilugin
Institute of Metal Physics,
Ural Division of Russian Academy of Sciences,
Ekaterinburg, GSP-170, Russia

ABSTRACT

It was found that severe plastic deformation of orthorhombic alloys caused phase transformations of the displacement type and those associated with a change in the degree of long-range order, namely B2→ω(B8$_2$), B2→B19 and B2→β (BCC) (in the case of alloy with initial B2-phase structure) and O→B19→A20 (initial O-phase structure, Ti$_2$AlNb). Unlike to ordinary metals, severe plastic deformation of the titanium aluminum intermetallics leads to decreasing of the strength of the material. The B19 and A20 phases are metastable. They are absent in the equilibrium phase diagrams of the compounds under investigation. The formation of the disordered phase states with extensive sliding and having great plasticity under severe deformation makes possible to consider severe deformation as the way for increasing of plasticity of the titanium aluminides.

INTRODUCTION

Severe plastic deformation causes substantial changes in the structure, phase composition, and, therefore, in the mechanical properties of materials. Methods of severe plastic deformation, such as shear under pressure and equal channel angular pressing, are used successfully to obtain fine-grained structure of materials, which substantially improves their mechanical properties [1-2]. Phase transformations of shear type are difficult to analyze in cold-rolled alloys or alloys after shear under pressure, because these treatments always lead to appearance of a texture. From this viewpoint, severe shock-wave deformation represents a more informative process. A texture is not formed under shock wave loading and deformation takes place in the whole volume of the sample [3]. In the present study, we perform a comparative analysis of the phase transformations occurring in the orthorhombic alloys with different initial phase state upon severe deformation by shear under pressure and shock wave loading.

EXPERIMENTAL DETAILS

The samples were loaded by means of the impact by a steel plate, maximum pressure on the surface of the samples was equal to 100 GPa, and deformation by the shear under pressure (10GPa) The setup for studying the shear deformation under pressure consisted of a KM-50-1 standard torsion-testing machine combined with a hydraulic press. The anvils were made of a VK6 hard alloy, which withstands pressures of 10-15 GPa. The rate of rotation of the mobile anvil can be varied in steps of 1 and 0,3 rpm.

Table 1. Chemical composition of the samples (at %)

№	Ti	Al	Nb	Zr	Mo
1	51,4	22	26,6		
2	54,4	24,6	21		
3	59,84	25,6	13,9	0,34	0,32

RESULTS AND DISCUSSION

Before deformation the structure of the 1-2 alloys contained grains with the Widmanstatten structure comprising wide (primary) and small (secondary) plates of the O-phase and grains without plates. The X-ray diffraction analysis showed that the sample included only the orthorhombic O-phase of two types, namely, niobium-enriched and niobium-depleted ones. The lattice parameters of the phases are given in Table 2. The 3-d alloy was done by the pack rolling. Before deformation the samples of 3-d alloy were aged at 1200^0C-1h and quenched in the ice water. The initial phase state of the samples of 3-d alloy before severe deformation was a single-phase one - B2.

Table 2. Phase composition of the samples before deformation

№	Phase composition	Volume fraction, %	Lattice parameters
1	$O^{depl}+O^{enr}$	100%	O^{depl}: $a = 0.6034$ nm, $b = 0.9616$ nm, $c = 0.4666$ nm O^{enr}: $a = 0.6056$ nm, $b = 0.9752$ nm, $c = 0.4676$ nm
2	$O^{depl}+O^{enr}$	100%	O^{depl}: $a = 0.6032$ nm, $b = 0.9752$ nm, $c = 0.4628$ nm O^{enr}: $a = 0.6066$ nm, $b = 0.9558$ nm, $c = 0.4667$ nm
3	B2	100%	B2: $a = 0.324$ nm

The sample 1 and 2.

Lines, which did not belong to the orthorhombic O-phase, appeared in diffraction patterns of the alloy 1 just after it was deformed to $\varepsilon = 0.9$ (hydrostatic compression). After deformation ε=4.7, the intensity of the lines, belonging to the unknown phases, increased. After deformation to ε=6.3, the O-phase lines were still observed in the X-ray diffraction patterns, however their intensity was substantially lower than that of unknown-phase lines. After deformation to ε=7.6, no lines of orthorhombic phases were observed; the X-ray diffraction pattern exhibited reflection of the unknown phase (or phases) only (fig.1). After the shock-wave loading of 100 GPa the diffraction pattern of alloy 2 also showed the lines of a phase, which differed from the orthorhombic O-phase. Interplanar spacings of the new phase were similar to interplanar spacings in the unknown phase, which was found under compression shear (fig.2). We performed the subsequent analysis of X-ray diffraction data using a special code for X-ray diffraction phase analysis and statistical and crystal-chemical investigation and DMPLOT programs, which showed that the new lines could belong to B19 and A20 phases.

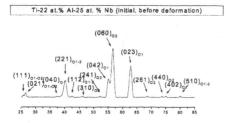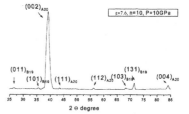

Figure.1. X-ray diffraction patterns of the alloy 1.

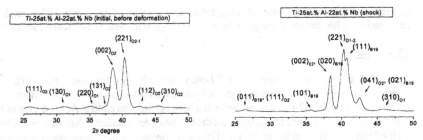

Figure.2. X-ray diffraction patterns of the alloy 2.

TEM study the alloy 2 after deformation ε=4.7 showed that a dispersed structure consisting of the O, B19, and A20 phases was observed. The rings were on the electron diffraction patterns taken from such regions. The average size of fragments determined from dark-field images was about 30-40 nm. As the degree of deformation increased to ε=6.3, the structure of the alloy became very fine and contained only the B19 and A20 phases. The average size of fragments determined from dark-field images was equal to 20-30 nm. As the degree of deformation increased to ε=7.6, the structural constituent refinement continued. The average size of fragments was about 20 nm as determined from dark-field images that showed the presence of only the A20 phase. TEM study of the alloy 2 after shock loading 100 GPA showed the regions consisting of grains with dislocations and plates with small precipitates of B19-phase.

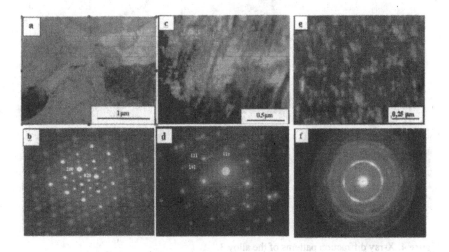

Figure 3. The samples 1-2: a-b the structure of the sample 2 before deformation; c-the microstructure of the sample 1 after shock-wave loading, dark-field image in $(111)_{B19}$; d)-

diffraction pattern to c), zone axis $[111]_{B19}$; e-the microstructure of the sample 2 after the shear deformation $\varepsilon=7.6$, dark-field image taken in the $(110)_{A20}$ reflection; f- diffraction pattern to e).

The sample 3

There are no superstructure lines of B2-phase on the X-ray diffraction patterns of the alloy 3 before deformation. As the degree of the deformation increases up to $\varepsilon=4,3$ the additional lines, which differ from the B2-phase lines, are on the X-ray diffraction pattern. At the degree of the deformation $\varepsilon=4,3$ the line $(110)_{B2}$ splits. At the degree of deformation $\varepsilon=5,6$ we cannot observe such splitting, but the position of the line $(110)B2$ is displaced to high 2θ degree. The parameter of the B2 lattice is decreased and is close to the parameter of disordered BCC phase. The doublet of the $(110)_{B2}$ one can explain as the ω-phase formation, which has the strongest line at those 2θ degree. X-ray diffraction patterns with the small 2θ degree of the alloy after degree of deformation $\varepsilon=4,3$ and $\varepsilon=5,6$ are on the figure 4. The lines in the small 2θ degree can indicate that the new phase is ordered. The lines of the new phase differ from the lines of $\overset{.}{\omega}$- and α_2 - phases. These lines can belong to the orthorhombic phase B19. The parameters of the new phase are close to the ones of the changed B2 phase and the difference increases as the degree of deformation increases. (a=0,293nm, b=0,37nm, c=0,412nm - at $\varepsilon=4,3$; a=0,2799 nm, b=0,414 nm, c= 0.428 nm - at $\varepsilon=5,6$).

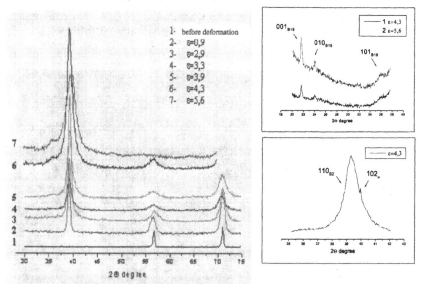

Figure 4. X-ray diffraction patterns of the alloy 3.

TEM study shows that the low ordered B2 in the orthorhombic alloy is stable phase under severe deformation and retains up to high degree of deformation. Under severe deformation the phase transformations $B2\rightarrow\omega$, $B2\rightarrow B19$ and $B2\rightarrow\beta$ occur in the orthorhombic alloy with B2 structure. It was found that severe deformation suppresses transformation B2→α₂ in the alloy and the phase transformations B2→ω, B2→B19 occur simultaneously. As the degree of deformation increases, the quantities of B19 phase increases and the regions of "defect" ω-phase become the places of nano scale disordered BCC phase formation.

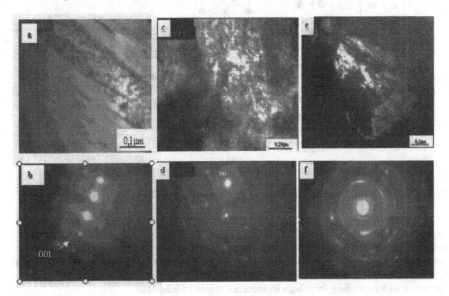

Figure 5. The sample 3, TEM: a-b - the structure before deformation, superstructural reflection $(001)_{B2}$ is pointed by the arrow; c- the structure of the sample 3 after deformation by the shear under pressure, dark-field image in $(211)_{B19}$ reflection (ε=4,3); d-diffraction pattern to c; e-dark-field image in β-reflection (ε=5,6), f-diffraction pattern to d.

CONCLUSION

Study of the phase transformation under severe deformation suggests a link between processes of deformation of the material under usual and severe deformation. Severe deformation made for an evolution of new structure regions with new mechanical properties. Such regions may be form as nano scale fluctuation under usual deformation process and not to be found by the usual methods, but the mechanical properties of material will be changed. In the orthorhombic alloys, severe deformation causes the different phase transformations, which occur at a fixed number. The resulting phase transformation under severe deformation in these alloys is order-disorder phase transformation with retention of crystal singony. Not only do the dislocations can take part in such transformation, but the different defects, for example stacking faults: structural and superstructural (APB) as well. The formation of the disordered phase states with extensive sliding and having great plasticity under severe deformation makes possible to consider severe deformation as the way for increasing of plasticity of the titanium aluminides. On the figure 5, one can follow all phase transformation in the orthorhombic alloys under severe deformation. Unlike titanium alloys, ω-phase in the orthorhombic alloys has ordered crystal lattice and is not stable under severe deformation. The formation of the α_2 crystal

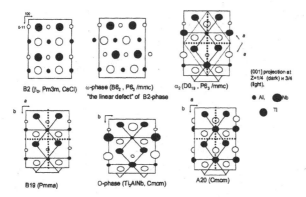

Figure.6. The crystal lattices of the phases in the orthorhombic alloys.

lattice involves the repositions of definite kind of atom, that become impossible under severe deformation. The B19 and A20 phases are metastable. They are absent in the equilibrium phase diagrams of the compounds under investigation. It is evident that the formation of the B19-phase is easier as compared to that for the disordered A20-phase. One can also reach the B19-phase formation with quenching from high temperature region of B2-phase.

ACKNOWLEDGMENTS

Research supported by the Program "National technological basis" №16/03/670-2003 and the Russian Fund for Basic Research Ural № 04-03-96008.

REFERENCES

1. R.Z.Valiev in *Nanostructured Materials Produced by Severe Plastic Deformation* (Moscow, Logos, 2000).
2. A.I.Gusev in *Nanocristalline Materials: Preparation and Properties* (Ekaterinburg, Ural Division of RAS, 1998)
3. N. V. Kazantseva, B. A. Greenberg, A. A.Popov and E. V. Shorokhov, *J.Phys.IV France, 2003, v 110, pp.923-928*

Impact Properties of Hot-Worked Gamma Alloys with BCC β-Ti Phase

Kentaro Shindo, Toshimitsu Tetsui[1], Toshiro Kobayashi, Shigeki Morita[2],
Satoru Kobayashi[3], and Masao Takeyama[4]
[1]Nagasaki R&D Center, Mitsubishi Heavy Industries Ltd.,
5-717-1 Fukahori-Machi, Nagasaki, 851-0392, JAPAN
[2]Dept. of Production Systems Engineering Toyohashi University of Technology,
1-1 Tempaku-cho, Toyohashi, Aichi, 441-8580, JAPAN
[3]Dept. of Microtructure Physics and Metal Forming, MAX PLANK INSTITUTE FOR IRON
RESERCH, Max-Plank-Strasse 1, D-40237 Dusseldorf, GERMANY P
Tokyo Institute of Technology
[4]Dept. of Metallurgy and Ceramics Science, Tokyo Institute of Technology,
2-12-1 Ookayama, Meguro-ku, Tokyo 152-8552, JAPAN

ABSTRACT

Impact resistance of a hot-forged TiAl alloy with a composition of Ti-42~44Al-5~10M:
M=V, Mn(at%), consisting of lamellar, γ and β grain of which the hot-workability was improved
by introducing β phase, has been investigated using an instrumented Charpy impact test, tensile
test at high strain rate and foreign object attack test. In instrumented Charpy impact test the
absorbed energies for crack initiation and propagation were measured, and the effect of
microstructure on the absorbed energies has been analyzed, by paying attention to the grain size,
interlamellar spacing and lamellar area fraction. In tensile test at high strain rate, the dependence
of strain rate of hot-extruded TiAl alloy is obtained and compared with that of Nickel based
superalloy, Inconel713C. In foreign object attack test with taper plate specimen modified turbine
blade and brass ball as foreign object, the relation between impact damage and thickness of
specimen at attack point is investigated. A limit of impact energy, at which there isn't a crack on
buck of attack point, is obtained with each thickness about hot-forged TiAl alloy, Ti-42Al-5Mn.
Therefore, the method of improvement of toughness and assessment of impact resistance of TiAl
alloy is shown in this study.

INTRODUCTION

TiAl intermetallic compounds are characterized by excellent high-temperature specific
strength. This property would be useful for turbine materials in turbo-machinery, and various
research efforts have been conducted over a long period to realize practical development of TiAl
alloys for turbines. Recent years have seen application in the turbine wheel of a mini-turbo for
passenger vehicles[1], where the primary advantage was improved performance in terms of
response. However, expectations are also high for TiAl alloys in turbine blades, since they would
contribute greatly to increasing volume of turbo-machinery by larger blades and to reduced load
stress on the disk.

The biggest issue in terms of practical application for turbine blades, made of TiAl alloy, is
the improvement of impact resistance properties. Since microstructure has a large influence on
the mechanical properties of TiAl alloy, microstructural control is a promising avenue for
countermeasures. This in turn has led to a focus on hot-working, enabling wide-ranging
microstructural control, and hot-worked TiAl alloys were developed with Ti-Al-V and Ti-Al-Mn
ternary systems[2, 3]. These alloys can be hot-extruded or hot-forged, with various structural
control possible in conjunction with subsequent heat treatment. This report presents the results of
consideration of the impact resistance properties of these alloys, developed with the aim of
application for turbine blades. Hot-worked TiAl alloys feature the introduction of β phase with a
high deformability at high- temperature, achieved through the addition of β-stabilizing elements
such as V, Mn, Cr, and Nb serving to improve hot-workability. Figure 1 shows the vertical
section including the alloy composition in Ti-Al-V ternary system[4, 5]. The β phase region
increases with the amount of V, along with hot-workability. Using the hot-worked process

presented in Figure 2, V-added and Mn-added alloys can be produced as hot-extruded or hot-forged materials. Figure 3 indicates the temperature dependence of specific strength (tensile strength divided by specific gravity) in the direction of extrusion for hot-extruded materials. The specific strength of hot-extruded materials is extremely high, i.e, 2~3 times that of Ni-based superalloys such as Inconel 713C or Inconel 718 in the range from room temperature to 700°C.

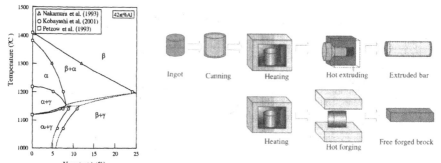

Figure 1. The vertical section of phase diagram including the alloy composition in Ti-Al-V ternary system

Figure 2. Schematic illustration of manufacturing process of hot-worked TiAl alloys.

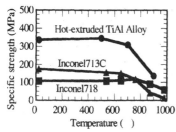

Figure 3. Temperature dependence of specific strength in the direction of extrusion for hot-extruded materials, compared with Nickel based superalloys.

EXPERIMENT

Table 1 shows the microstructure, compositions, manufacturing process, heat treatments, and structural factors for the test alloys. These are classified into 4 types according to microstructure: the $\gamma + \beta$ + lamellar of the hot-worked materials, and, for comparative materials, fully lamellar, duplex, and equiaxed-grain. Of the hot-worked materials, BL1, BL2, and BL3 were V-added alloys, having fine lamellar grain size and interlamellar spacing achieved by hot-extrusion using steel sheath. BL4 was a Mn-added hot-forged alloy, formed into square bar material by means of free forging without steel sheath. Fully lamellar FL1 and FL2 were reaction-sintered materials, while FL3 was isothermal- forged, and these featured grain size and interlamellar spacing that were greater than for the hot-forged materials. The duplex DL material was precision-cast, with a slightly lamellar collonies. The equiaxed EG1 and EG2 materials were both isothermal-forged, and, although the grain size was small, there was no lamellar collonies. In the research reported here, these materials were used in impact resistance evaluation testing.

Table 1. The microstructure, compositions, manufacturing process, heat treatments, and microstructural factors for the test alloys. D is grain size, A_f is lamellar area fraction, and d is interlamellar spacing.

Microstructure	No.	Composition (at%)	Process	Heat treatment	D (mm)	A_f (%)	d (mm)
γ+β+Lamellar	BL1	42Al-10V	hot-extrude	as-extruded	9.0	71.6	0.040
	BL2	45Al-10V	hot-extrude	as-extruded	10.5	84	0.035
	BL3	45Al-10V	hot-extrude	1250°C ~30min	26	41.2	0.092
	BL4	42Al-5Mn	hot-forge	1250°C ~30min	-	-	-
Fully-Lamellar	FL1	-	reaction sinter	1390°C (HIP)	35	100	0.588
	FL2	-	reaction sinter	1390°C (HIP)	132	100	0.357
	FL3	-	isothermal forge	1350°C ~1h	48	100	0.500
Duplex	DL	-	presision cast	1260°C (HIP)	-	4.2	-
Equilaxed grain	EG1	-	isothermal forge	1000°C (HIP)	36	0	-
	EG2	-	isothermal forge	as-forged	3.5	0	-

Impact resistance evaluation was performed using the instrumented Charpy impact test, the impact tensile test, and the ballistic impact test. The influence of microstructure on impact resistance was investigated in the instrumented Charpy impact test, and methods for improved impact resistance were considered based on these results. The strain rate dependence of tensile strength was investigated in the impact tensile test, and the relationship was thereby considered between impact resistance at low load velocity and that at high load velocity (such as in a collision with a foreign object by an actual turbine). Impact resistance in the case of collision with a foreign object was obtained under simulated actual conditions in the ballistic impact test, and the fracture threshold plate thickness for TiAl alloy was considered with respect to impact fracture by collision.

Instrumented Charpy impact test

In this test, a strain gauge attached in the vicinity of the hammer tip was used for the load, while a potentiometer on the revolution axis of the hammer was used to measure the distance, thus obtaining the load-distance curve, and in turn determining the various absorbed energies. With the maximum load point as the boundary, crack initiation energy (Ei, the energy absorbed until crack initiation), crack propagation energy (Ep, the energy required from crack initiation until propagation resulted in fracture), and the total energy absorbed in the process (Et, the sum of Ei and Ep) were calculated. Because the absorbed energy for TiAl alloy is small compared to steel, a compact Charpy impact tester (capacity = 14.9J) was used. Also, as it was difficult to evaluate the differential in absorbed energy until crack initiation using notched specimens, non-notched impact test pieces (3x4x40mm) were used as specified in JIS R 1607. The test was conducted at room temperature, and the hammer angle was 98.5°.

Impact tensile test

A hydraulic servo type high-speed tensile tester was used at room temperature with a crosshead speed of 8.3×10^{-6}m/s~4m/s (initial strain rate range of 4.0×10^{-3}~1.7×10^{4}s^{-1}). The test pieces were tensile test specimens having a double parallel portion as indicated in Figure 4. Strain during the test was measured using a strain gauge attached to the small diameter smooth section. A strain gauge was also placed on the large diameter smooth section, and stress corresponding to the strain was obtained in advance by means of static testing, such that stress could then be determined from the stress-strain curve.

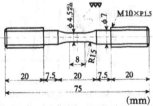

(mm)

Figure 4. Schematic of specimen of the impact tensile test.

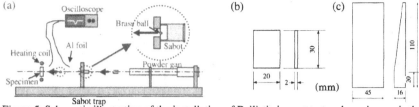

Figure 5. Schematic illustration of the installation of Ballistic impact test and specimens, is the installation of Ballistic impact test, (b) is small test pieces, and (c) is taper model test pieces.

Ballistic impact test

Figure 5 presents a schematic illustration of the Ballistic impact tester and specimens. The foreign object was fired from a powder gun attached to a sabot; the foreign object was detached by means of a sabot trap, such that only the foreign object collided with the Specimen in the collision velocity range of 300~600m/s. The collision velocity was determined from the puncture time differential (measured using an oscilloscope) and the distance between the aluminum foil sheets. Brass balls, 2~9.2mm in diameter, was used for the foreign objects, considering strength reduction of a steel object under the high temperature conditions in an actual installation. Specimens consisted of small test pieces and taper model test pieces (simulating the shape of a turbine blade). The specimens were heated to 450~600°C using a high frequency heating coil, and the temperature at collision was measured by means of thermocouples placed on specimen.

RESULTS

Figure 6 presents representative load-distance curves for the test alloys. The maximum load was recorded for BL1 and BL2; large values for distance until fracture are shown for FL1 and EG2. In each case, Ei is greater than Ep, indicating brittleness characterization. The magnitude of total absorbed energy Et is dependent on Ei, such that Et was the greatest for EG2 (having large Ei), followed by BL1 and then FL1 and BL2. The reason why Ep is low for the duplex and equilaxed grain structures (EG1 and EG2) is considered to be the lesser lamellar structure, given that the lamellar structure is responsible for the comparatively high fracture toughness of TiAl. However, The value of Ep was also low for fully lamellar FL2 and FL3, demonstrating that impact resistance cannot be improved by simply increasing the lamellar structure. Ei and Ep were both comparatively high for the hot-extruded material BL1, and the impact resistance can be said to be superior.

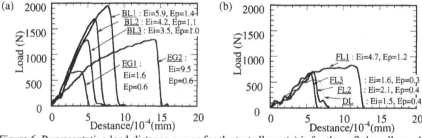

Figure 6. Representative load-distance curves for the test alloys, (a) is for the γ+β+lamellar and equilaxed grain structures, while (b) is for the fully lamellar and duplex structures. The Denominations of Ei and Ep are J/cm^2.

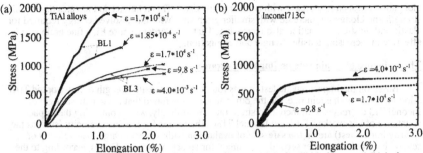

Figure 7. The stress-strain curve for the impact tensile test, (a) is for TiAl alloys and (b) is for Inconel713C.

The stress-strain curve for the impact tensile test is shown in Figure 7, with test results for the common turbine material Inconel 713C also presented for the purpose of comparison. For BL1, tensile strength increased with strain rate, reaching 1904MPa at a high strain rate of $1.7 \times 10^4 s^{-1}$, and the same trend was also confirmed for BL3. In contrast, the tensile strength of Inconel 713C tended to decline slightly with greater strain rate.

For the ballistic impact test, BL4, Mn-added alloy, was used. Because BL4 has superior hot-workability, it can be formed into square bar from a round cylindrical ingot by means of free forging without a sheath, using general purpose hot-forging equipment. In the present context, foreign object collision test pieces were fabricated from hot- free-forged square bar measuring 83T x 53W x 460L (mm). Figure 8 illustrates the relationship between impact energy and the plate thickness of the collision point. The impact energy indicated in the figure is the kinetic energy of foreign object, calculated from the weight of the foreign object and the collision velocity. Fracture state can be categorized into 3 types with reference to the back side of the specimen: uncracked, crack-initiated, and fractured.

DISCUSSION
The influence of structure on impact resistance

Here, the influence that structure has on impact resistance is considered based on the results of the instrumented Charpy impact test. Figure 9 expresses the relationship between absorbed energy and grain size. A clear correlation cannot be discerned between grain diameter and Ep, but Ei becomes greater as grain diameter decreases. This grain size dependence of Ei can be explained by the relationship between grain size and tensile properties. Tensile strength and

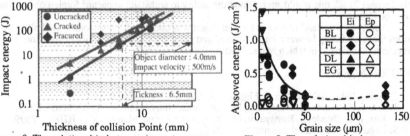

Figure 8. The relationship between impact energy and the specimen thickness of the collision point.

Figure 9. The relationship between absorbed energy and grain size.

251

elongation are seen to increase with smaller grain size[6]. This relationship indicates that tensile strength and elongation increase with smaller grain size, with the absolved energy required for elastic and plastic deformation also increasing. That is, in order to raise Ei, a fine grain size is effective in increasing tensile strength and elongation.

The influence of strain rate on impact resistance

Based on the impact tensile test, it was found that the tensile strength of hot-worked TiAl alloy increases with strain rate. On the other hand, it was found that the tensile strength of Inconel 713 decreases with increased strain rate. Accordingly, when comparing the impact resistance of hot-worked TiAl and Inconel 713C, the results for low load velocity (such as the Charpy impact test) are on the safe side of evaluation with respect to high load velocity of several hundreds meter per second as assumed for an actual turbine. That is, according to the quantitative evaluation of the effect of strain rate on tensile strength in this study, if Charpy impact values for TiAl alloy are found to be higher than for Inconel 713C, this would suggest that the ability to withstand impact of foreign object would also be higher.

Foreign object collision resistance of TiAl alloy

Fig. 8 illustrates the relationship between foreign object collision energy and the plate thickness of the collision point. The line for the uncracked back side is the limit of the region in which collision does not result in single impact fracture. Above the crack initiation line is the region in which single impact fracture definitely occurs. Thus, design guidelines with respect to foreign object collision can be obtained for hot-worked TiAl alloy, 42Al-5Mn, in the case of utilization for actual turbine blades. For example, assuming collision with an object of steel material with a diameter of 4.0mm at a speed of 500m/s, plate thickness of 6.5mm would be required to prevent single impact fracture.

CONCLUSION

In this study, instrumented Charpy impact test, impact tensile test, and ballistic impact test were employed for hot-worked TiAl alloys, and the influences of microstructure and strain rate on impact resistance were considered.
The majority of total absorbed energy Et is accounted for by Ei in the case of TiAl alloy, and the magnitude of Et is dependent upon Ei. The hot-extruded TiAl alloy, 42Al-10V, in particular exhibited high crack initiation energy Ei and crack propagation energy Ep, indicating superior impact resistance.
The tensile strength of TiAl alloys increases with strain rate. In contrast, that of Inconel 713C decreases with greater strain rate. According to the quantitative evaluation of the effect of strain rate on tensile strength, if Charpy impact values for TiAl alloy are found to be higher than for Inconel 713C, this would suggest that the ability to withstand impact of foreign object would also be higher.
Ability to withstand impact of foreign object was obtained for hot-worked TiAl alloy, and from the relationship between impact energy and plate thickness at the point of collision, a safe region was indicated in which impact fracture was not reached.

REFERENCES

1) T. Tetsui, "Materials Science & Engineering", A329-331(2002), P582.
2) T. Tetusi, K. Shindo, S. Kobayashi, M. Takeyama, "Intermetallics", 11[4],(2003), P299.
3) T. Tetsui, K. Shindo, S. Kobayashi, M. Takeyama, "Scripta Materiaria", 47(2002), P399.
4) S. Kobayashi, "Doctoral dissertation", (2002).
5) S. Kobayashi, M. Takeyama, et. al. "Gamma Titanium Aluminides 2003", TMS (2003), P165.
6) C. T. Liu and P. J. Maziasz, "Intermetallics", 6(1998), P 653.

Effects of long-period superstructures on plastic properties in Al-rich TiAl single crystals

Takayoshi Nakano[*], Koutaro Hayashi, Yukichi Umakoshi,
Yu-Lung Chiu[1] and Patrick Veyssière[1]
Department of Materials Science and Engineering, Graduate School of Engineering,
Osaka University,
2-1, Yamada-oka, Suita, Osaka 565-0871, Japan
[1]Laboratoire d'Etude des Microstructures, CNRS-ONERA, BP72, 92322 Châtillon cedex, France

[*] E-mail: nakano@mat.eng.osaka-u.ac.jp

ABSTRACT

In Al-rich TiAl crystals, several long-period superstructures may appear depending on Al composition, annealing temperature and annealing time. Amongst these, Al_5Ti_3 and h-Al_2Ti contain pure Al (002) layers, as in the $L1_0$ structure of the matrix, alternating with Ti (002) layers that exhibit an ordered arrangement of the Al atoms in excess. In single crystals with compositions ranging from Ti-54.7at.%Al to Ti-62.5at.%Al annealed at 1200°C, the Al_5Ti_3 long-period superstructure embedded in the $L1_0$ matrix develops with increasing Al concentration to finally transform fully into h-Al_2Ti for Ti-62.5at.%Al. On the other hand, Al_5Ti_3 precipitates grow with annealing time at 500°C in Ti-58.0at.%Al.

The effects of the Al_5Ti_3 and h-Al_2Ti superstructures on slip properties of $1/2<110]$ ordinary dislocations are examined both at a macroscopic and a microscopic level. The CRSS for $1/2<110]$ ordinary slip increases with Al_5Ti_3 ordering depending on Al composition, or of annealing time in the case of Ti-58.0at.%Al. Dislocations with $1/2<110]$ Burgers vector group into fourfold configurations to avoid the trailing of extended APBs in Al_5Ti_3. The CRSS for slip in the $<110]$ direction further increases with the formation of h-Al_2Ti particles within the $L1_0$ matrix in Ti-62.5at.%Al. By contrast, Ti-62.5at.%Al fully transformed into Al_5Ti_3 exhibits a CRSS significantly lower than that of the two-phase alloy.

INTRODUCTION

In Al-rich off-stoichiometric γ-TiAl, further ordering under long-period superstructures may occur depending on Al composition and heat treatment [1]. The ordering process is controlled by changes in free energy reflecting crystal symmetry [2]. Among all possible superstructures, Al_5Ti_3 and h-Al_2Ti precipitate coherently with the Al (002) layers exhibiting the same periodicity as in the host $L1_0$ structure [2,3] (See fig.1). Although Al_5Ti_3 and h-Al_2Ti coherent particles are not conspicuous, with their identification necessitating thorough structural transmission electron microscope analysis, they strongly influence plastic properties and deformation microstructures. Nakano et al. [4] were the first to report the chemical effect of Al_5Ti_3 on the various deformation modes available in this family of alloys. Subsequently, other effects of the long period superstructures have been identified in Al-rich TiAl [2,5-10]. The microstructure, stability, ordering process, phase diagram of these long-period superstructures is rather well documented [11-19].

The present study is part of more extensive investigation of effects of the Al_5Ti_3 and h-Al_2Ti superstructures. Here we focus on properties of slip along $<110]$ on {111} planes in a wide compositional range.

EXPERIMENTAL PROCEDURE

Single crystals were grown from master ingots with compositions of Ti-54.7at.%Al, Ti-56.0at.%Al, Ti-58.0at.%Al, Ti-60.0at.%Al and Ti-62.5at.%Al at a growth rate of 5.0mm/h using the floating zone method (NEC SC-35HD single crystal furnace). They were annealed at 1200°C for 1h and quenched into ice water. Ti-62.5at.%Al further annealed at 750°C for 48h formed stoichiometric Al_5Ti_3 single-phase single crystals. The detailed procedure can be found in [10,18,20]. To encourage the growth of Al_5Ti_3 precipitates, a single crystal containing 58.0at.%Al was quenched from 1100°C and then annealed at 500°C for various times, 10^2s, 10^3s, 10^4s and 10^5s.

Specimens oriented for compression along [201] were prepared from the above-mentioned single crystals. This orientation favors 1/2<110]{111} ordinary slip which was actually found to dominate deformation microstructures independent of Al composition and annealing condition. Compression tests were performed at a nominal strain rate of 1.7×10^{-4}1/s in vacuum at room temperature. The microstructures before and after deformation were investigated in a Hitachi H-800 transmission electron microscope operated at 200kV.

RESULTS

Figure 1 shows the crystal structures of Al_5Ti_3 and of the h-Al_2Ti V1 variant (for a definition of the two h-Al_2Ti variants, see [2, 14]) together with the $L1_0$ structure. The a- and b-axes of the Al_5Ti_3 structure are defined based on an fcc lattice, this is at variance from a current alternative definition where these axes are rotated by 45° around the c-axis. Al_5Ti_3 and h-Al_2Ti exhibit a tetragonal and orthorhombic lattice, respectively. With regards to the $L1_0$ structure, the periodicity in Al_5Ti_3 along the a- and b- axes is fourfold. It is threefold along the b-axis in h-Al_2Ti. The fact that in all three crystals (i.e. $L1_0$, Al_5Ti_3 and h-Al_2Ti), the stacking sequence

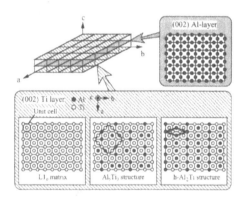

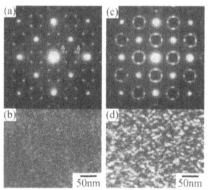

Figure 1 Crystal structures and atomic arrangement of $L1_0$ and of the Al_5Ti_3 and h-Al_2Ti (variant V1) superstructures in Al-rich TiAl.

Figure 2 [001] diffraction patterns (a,c) and the corresponding dark field images (b,d) typical of Al_5Ti_3 (a, b) and h-Al_2Ti (c,d).

254

along the c-axis comprises an Al layer every other plane, enables us to distinguish between crystal structures by transmission electron diffraction analysis near the [001] beam.

Crystal structures and distribution of Al₅Ti₃ and h-Al₂Ti precipitates in Al-rich TiAl single crystals

Crystal structures and distribution of Al_5Ti_3 and h-Al_2Ti precipitates in Al-rich TiAl single crystals

Figure 2 shows representative [001] diffraction patterns and the corresponding dark field images of Al_5Ti_3 and h-Al_2Ti. For Al_5Ti_3, figs. 2(a) and 2(b) are taken from Ti-58.0at.%Al annealed at 1200°C, while for h-Al_2Ti, figs. 2(c) and 2(d) are taken from Ti-62.5at.%Al annealed at 1200°C. The fourfold periodicity showing the existence of Al_5Ti_3 was identified in almost every sample investigated except for three cases (diffuse scattering in Ti-54.7at.%Al, coexisting Al_5Ti_3 and h-Al_2Ti phases in Ti-60.0at.%Al, and h-Al_2Ti only in Ti-62.5at.%Al quenched from 1200°C and). Al_5Ti_3 precipitates grow remarkably in size with increasing Al concentration from Ti-54.7at.%Al to Ti-58.0at.%Al; and, in Ti-58.0at.%Al; with increasing the annealing time at 500°C up to 10^5s.

Al-concentration dependence of plastic behavior and dislocation motion

Al-concentration dependence of plastic behavior and dislocation motion

Figure 3 shows the CRSS for 1/2<110>{111} ordinary slip as a function of Al concentration. The CRSS increases quickly from 54.7at.%Al to 56.0at.%Al, and it subsequently exhibits a plateau-like behavior up to 60.0at.%Al. Beyond 60.0at.%Al, the CRSS again increases reflecting the growth of h-Al_2Ti precipitates. The plateau indicates that the CRSS for ordinary slip is not strongly influenced by the growth of Al_5Ti_3 particles.

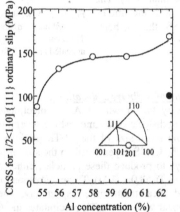

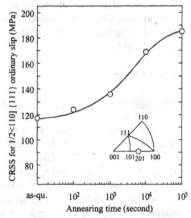

Figure 3 Al concentration dependence of the CRSS for ordinary slip in Al-rich TiAl deformed at RT. The filled circle shows the CRSS of the Al_5Ti_3 single phase for comparison.

Figure 4 Annealing time dependence of the CRSS for ordinary slip in Al-58.0at.%Al crystals annealed at 500°C and then deformed at RT.

We have observed that ordinary dislocations move individually in Ti-54.7at.%Al, whereas, in the 56.0at.%Al and 58.0at.%Al samples, a fraction of the ordinary dislocations start to assemble into fourfold groups. The overall Burgers vector of these grouped configurations is the shortest unit translation in Al_5Ti_3, that is, four times larger than that in the Burgers vector in the L1₀ matrix in the corresponding direction [4]. In pure Al_5Ti_3 (obtained from Ti-62.5at.%Al heat-treated at 750°C), such fourfold configurations are commonly observed. The mean separation between dislocations in the group is narrower in the Al_5Ti_3 single-phase than that in the two-phase Ti-58.0at.%Al single crystal. Interestingly, the CRSS of the former (filled circle in fig. 3) is significantly less than that of the latter (open circle in fig. 3 up to 60.0at.%Al).

By contrast, the growth of h-Al_2Ti particles (62.5.%Al) enhances the CRSS for ordinary slip. In the microstructure, dislocation motion is essentially individual, however, with a slight tendency for threefold grouping in accordance with the periodicity of h-Al_2Ti.

Effect of Al_5Ti_3 precipitate growth

Upon annealing at 500°C, the Al_5Ti_3 particles grow markedly as the annealing time is increased up to 10^5s. Their mean diameter increases from 2.6nm to 10.0nm and their volume fraction from 26% to 62%. Figure 4 illustrates the dependence of the CRSS for ordinary slip at RT on the growth of the Al_5Ti_3 particles.

Figure 5 shows a weak-beam image of ordinary dislocations in Ti-58.0at.%Al annealed at 500°C for 10^3s and subsequently deformed at RT. Almost all the dislocations aligned with the <110] screw orientation are either single fourfold dislocations or dipoles of fourfold dislocations. Part of the ordinary dislocations stabilize along <101] directions, which may reflect a second ordinary slip system. The tendency towards fourfold grouping is conspicuous after short annealing times (i.e. less than 10^3s). The fourfold grouping of the dislocations coincides with the initial stage of moderate time dependence of the CRSS, that is, heat treatment up to 10^3s.

Figure 5 A weak-beam image of ordinary dislocations in Ti-58.0at.%Al single crystal annealed at 500°C for 10^3s.

DISCUSSION

The relationship between CRSS and the grouping of ordinary dislocations.

The CRSS for ordinary slip is influenced by the growth of Al_5Ti_3 precipitates and by resulting changes in configuration of the operating dislocations. Amongst the several parameters indicative of Al_5Ti_3 growth, we make use of the mean size of the Al_5Ti_3 particles measured parallel to the {111} slip plane. Figure 6 shows the CRSS dependence on the size of particles of Al_5Ti_3. As mentioned above, there are two ways to produce these particles, either by way of composition, or as a function of annealing time (in Ti-58.0at.%Al). These correspond in fig. 6 to the two curves dubbed "Al concentration" and "heat treatment", respectively. The fourfold grouping is indicated by asterisks. It starts to appear with the Al_5Ti_3 precipitates in 56.0at.%Al, and it is continuously observed with increasing Al content. This is accompanied by a gradual increase of the CRSS. Beyond a diameter of about 7nm (in Ti-58.0at.%Al heat treated at 500°C for 10^4s and 10^5s), the dislocations moved individually again.

The deformed Al_5Ti_3 single-phase is dominated by fourfold ordinary dislocations bound by relatively narrow APB strips, and the corresponding CRSS is low relative to the two-phase alloys. The relationship between the macroscopic and microscopic properties of Al_5Ti_3-containing alloys may be understood as a function of particle size and of the inter-particle distance following the ideas developed by Gleiter et al. [21]. The size of the particle is not the only controlling parameter since a quasi-isolated particle may be better avoided by an Orowan

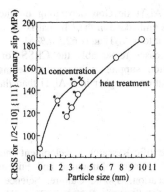

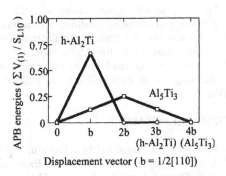

Figure 6 The CRSS for ordinary slip as a function of Al_5Ti_3 particle size. The asterisks indicate sample exhibiting fourfold dislocations

Figure 7 APB energies in the Al_5Ti_3 and h-Al_2Ti superstructures as a function of displacement vector. The perfect dislocation in Al_5Ti_3 and h-Al_2Ti show the Burgers vector of 4b and 3b, respectively.

looping process than being sheared. In the latter case, some grouping is required when the trailed APBs are prohibitively energetic. In the former case, this is all a question of line tension, hence of inter-particle distance. We should however mention that we have observed no residual loops around Al_5Ti_3 particles. The mechanism for changing in the dislocation configuration from grouped to individual with the growth of Al_5Ti_3 particles is therefore not clear yet. Cross-slip in the $L1_0$ matrix may contribute to the effects observed since the frequency of cross-slip should be encouraged by coherency stresses at the interface between the $L1_0$ matrix and Al_5Ti_3 particle.

The energies of APBs created in Al_5Ti_3 and h-Al_2Ti.

h-Al_2Ti seems more efficient in increasing the CRSS for ordinary slip than Al_5Ti_3. We believe that one of the dominant strengthening mechanisms in off-stoichiometric Al-rich TiAl arises from the APBs trailed behind the ordinary dislocations in the Al_5Ti_3 or h-Al_2Ti phases. They can be estimated based on the interaction energy between the first nearest neighbor atoms through the ordering energy, $V^{(1)}_{<1/2\ 1/2\ 0>} = [\phi_{Ti\text{-}Ti\ (1)} + \phi_{Al\text{-}Al\ (1)}]/2 - \phi_{Ti\text{-}Al\ (1)}$, where $\phi_{\alpha\text{-}\beta(1)}$ is the pairwise interaction energy between the first nearest-neighbor atoms α and β along the $<1/2\ 1/2\ 0>$ direction. S_{L1_0} is a unit area in the $\{111\}$ plane of the $L1_0$ unit cell and amounts to $1/2a^2(1+c/a)^{1/2}$, where a and c are the lattice parameters based on the f.c.c. unit in both superstructures. APB energies are shown in fig. 7. The maximum APB energy in h-Al_2Ti is larger than that in Al_5Ti_3. This maximum scales with the maximum back-stress experienced by individual dislocations as these shear a particle, one after the other, on the same plane. This crude approximation reflects rather simply the observed differences in alloy strengthening from Al_5Ti_3 and h-Al_2Ti.

CONCLUSIONS

The CRSS in the $<110]$ direction and dislocation configurations in off-stoichiometric Al-rich TiAl depend on Al composition and heat treatments. The main results of this investigation may be summarized as follows.

(1) The long period superstructures of Al_5Ti_3 and h-Al_2Ti develop depending on the Al concentration and annealing condition. Al_5Ti_3 precipitates embedded in the $L1_0$ matrix develops with increasing Al concentration to finally transform fully into h-Al_2Ti at Ti-62.5at.%Al.
(2) The development of the Al_5Ti_3 superstructure increases remarkably the CRSS for the $1/2<110]$ ordinary slip, but the motion of grouped dislocations suppresses the rapid hardening.
(3) The energy of an APB trailed by an ordinary dislocation is higher in h-Al_2Ti than that in Al_5Ti_3.

ACKNOWLEDGEMENT

This work was supported by a Grant-in-Aid for Scientific Research Development and the 21st COE Program (Project: Center of Excellence for Advanced Structural and Functional Materials Design) from the Japanese Ministry of Education, Sports, Culture, Science and Technology. This work has partly been carried out at the Strategic Research Base "Handai Frontier Research Center" supported by the Japanese Government's Special Coordination Fund for Promoting Science and Technology. T. Nakano would like to thank Iketani Science and Technology Foundation for a financial support. P. Veyssière wishes to express his thanks to the COE program for support, to the Osaka group for providing excellent working facilities. The work was part of a PICS program sponsored by CNRS (France).

REFERENCES

[1] A. Loiseau, A. Lasalmonie, G. Van. Tendeloo, J. Van. Landuyt and S. Amelinckz, Acta Crystallogr. **B41**, 411 (1985).
[2] T. Nakano, A. Negishi, K. Hayashi and Y. Umakoshi, Acta Mater. **47**, 1193 (1999).
[3] M. Palm, L. C. Zhang, F. Stein and G. Sauthoff, Intermetallics **10**, 523 (2002).
[4] T. Nakano, K. Matsumoto, T. Seno, K. Oma and Y. Umakoshi, Phil. Mag. A **74**, 251 (1996).
[5] T. Nakano, K. Hagihara, T. Seno, N. Sumida, M. Yamamoto and Y. Umakoshi, Phil. Mag. Lett. **78**, 385 (1998).
[6] F. Grégori and P. Veyssière, Phil. Mag. A **79**, 403 (1999).
[7] K. Hayashi, T. Nakano and Y. Umakoshi, Sci. Technology Advanced Mater. **2**, 433 (2001).
[8] S. Jiao, N. Bird, P. B. Hirsch and G. Taylor, Phil. Mag. A **81**, 213 (2001).
[9] H. Inui, K. Chikugo, K. Nomura and M. Yamaguchi, Mat. Sci. Engng. A **329**, 377 (2002).
[10] T. Nakano, K. Hayashi and Y. Umakoshi, MRS Proc., Defect Properties and Phenomena in Intermetallic Alloys **753**, pp.261 (2003).
[11] U. D. Kulkarni, Acta Mater. **46**, 1193 (1998).
[12] F. Stein, L. C. Zhang, G. Sauthoff and M. Palm, Acta Mater. **49**, 2919 (2001).
[13] L. C. Zhang, M. Palm and F. Stein, Intermetallics **9**, 229 (2001).
[14] T. Nakano, K. Hayashi and Y. Umakoshi, Phil. Mag. A **82**, 763 (2002).
[15] U. D. Kulkarni, Phil. Mag. A **82**, 1017 (2002).
[16] S. Hata, K. Higuchi, M. Itakura, N. Kuwano, T. Nakano, K. Hayashi and Y. Umakoshi, Phil. Mag. Lett. **82**, 363 (2002).
[17] M. Doi, T. Koyama , T. Taniguchi and S. Naito, Mat. Sci. Engng. A **329**, 891 (2002).
[18] K. Hayashi, T. Nakano and Y. Umakoshi, Intermetallics **10**, 771 (2002).
[19] S. Hata, K. Higuchi, T. Mitate, M. Itakura, Y. Tomokiyo N. Kuwano, T. Nakano, Y. Nagasawa and Y. Umakoshi, Journal of Electron Microscopy **53**, 1 (2004).
[20] T. Nakano, K. Hayashi, Y. Umakoshi, Y.-L. Chiu and P. Veyssière, Phil. Mag. submitted (2004).
[21] H. Gleiter and E. Hornbogen, Mater. Sci. Engng., **2**, 285 (1968).

TEM Analysis of Long-Period Superstructures in TiAl Single Crystal with Composition Gradient

S. Hata, K. Shiraishi, N. Kuwano[1], M. Itakura, Y. Tomokiyo, T. Nakano[2] and Y. Umakoshi[2]
Department of Applied Science for Electronics and Materials, Interdisciplinary Graduate School of Engineering Sciences, Kyushu University, Kasuga, Fukuoka 816-8580, Japan
[1]Art, Science and Technology Center for Cooperative Research, Kyushu University, Kasuga, Fukuoka 816-8580, Japan
[2]Department of Materials Science and Engineering & Handai Frontier Research Center, Graduate School of Engineering, Osaka University, Suita, Osaka 565-0871, Japan

ABSTRACT

The ordering mechanism of long-period superstructures (LPSs) in Al-rich Ti-Al alloys was studied using a TiAl single crystal with a composition gradient. A TiAl single crystal with gradient compositions from 55 to 75 at.% Al was prepared by annealing in a molten Al at 1234°C. The single crystal exhibits long-period ordering into different LPSs depending on the Al concentration as follows: an Al_5Ti_3 type short-range order, h-Al_2Ti and one-dimensional antiphase domain structures. These LPSs show an orientation relationship in which Al (002) layers of the LPSs are parallel to those of the TiAl matrix. The atomic arrangements of the LPSs are characterized in common as the alternate stacking of the Al (002) layers and Ti-Al (002) layers. It is thus concluded that the ordering of this type of LPSs and the phase transition between these LPSs are explained as structural changes in Ti-Al (002) layers of the Al-rich $L1_0$-TiAl crystal.

INTRODUCTION

Non-stoichiometric TiAl alloys that are rich in Al exhibit ordering into long-period superstructures (LPSs), such as Al_5Ti_3 [1], h-Al_2Ti [2] and so on. This ordering phenomenon is explained as ordering of excess Al atoms in Ti (002) layers of the $L1_0$-TiAl matrix [3-10]. According to the Ti-Al phase diagram in figure 1 [10], there are some other LPS phases near the TiAl phase region: r-Al_2Ti, a one-dimensional antiphase domain structure (1d-APS), Al_3Ti, and so on. They are $L1_2$-based long-period structures and often formed with the Al_5Ti_3 and/or h-Al_2Ti LPSs in Al-rich Ti-Al alloys [6-14]. Many research groups studied phase equilibria in the Al-rich part of the Ti-Al system. However, ordering processes into such a mixed state of the different LPSs are not understood well. This may be due to the fact that the long-period ordering is quite sensitive to the Al concentration and annealing conditions [1-14].

The purpose of the present study is to reach a general understanding of the ordering mechanism of LPSs in Al-rich Ti-Al alloys. For this purpose, we prepared a TiAl single crystal with gradient compositions and examined the formation of LPSs systematically as a function of Al concentration. Such a specimen preparation is known as a diffusion couple method to study a phase formation in a wide composition range under a fixed annealing condition [15,16]. Our original point is using a single crystal. The composition-gradient TiAl single crystal is effective in suppressing inhomogeneous ordering at grain boundaries [6,9,11] and in taking account of anisotropic diffusion in the tetragonal TiAl crystal [17,18]. Fundamental features of the

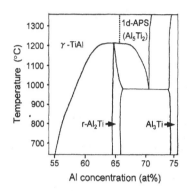

Figure 1. Al-rich part of the Ti-Al phase diagram reported by Palm *et al.* [10].

long-period ordering in Al-rich Ti-Al alloys will be discussed from obtained results.

EXPERIMENTAL DETAILS

A Ti-54.7 at.% Al single crystal was made by a floating zone method [7]. The single crystal was cut and polished on {100}$_{L1o}$ planes and then annealed with an Al plate at 1234°C for 5 min under an Ar atmosphere. At this temperature, the single crystal was dipped into molten Al. The annealing temperature was determined by referring to a diffusion couple experiment by Palm *et al.* [10]. They made a diffusion couple from TiAl (56 at.% Al) and Al$_3$Ti alloys at 1234°C and confirmed a continuous composition gradient from 56 to 71 at.% Al in the diffusion couple. In order to observe all the composition-gradient area in one TEM specimen, the short annealing time, 5 min, was determined from diffusion profile calculations along the [100]$_{L1o}$ direction in the TiAl crystal using diffusion coefficients reported by Ikeda *et al.* [17,18]. After identifying a composition gradient in the single crystal using an electron probe microanalyzer (Shimadzu EPMA-1500), thin foil specimens including the composition-gradient area were fabricated with a focused ion beam (FIB) microsampling technique (Hitachi FB-2000K). Microstructures in the composition-gradient area were observed as a function of Al concentration by using a 200 kV analytical electron microscope (FEI Tecnai-F20) that is equipped with a field emission gun, an energy dispersive X-ray spectroscopy system and a beam-scanning function.

RESULTS AND DISCUSSION

Figure 2(a) shows a (010)$_{L1o}$ cross-sectional view of the TiAl single crystal after the heat treatment described above. A reacted layer about 100-150 μm in thickness is formed at the surface of the single crystal. An enlarged view and concentration profiles across the reacted layer are shown in figures 2(b) and 2(c), respectively. The reacted layer consists of Al$_3$Ti grains and a composition-gradient area from 54.7 to 75 at.% Al.

Figure 3(a) shows a bright field TEM image of the composition-gradient area. The incident

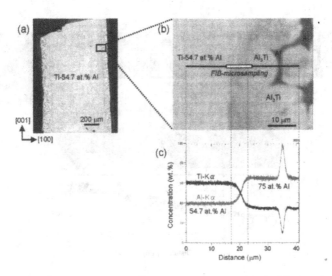

Figure 2. (a) Scanning electron microscope image of a $(010)_{L1_0}$ cross-section of the TiAl single crystal after annealing at 1234°C in the molten Al. (b) Enlarged view at the surface of the single crystal. (c) Concentration profiles across the reacted layer determined by the EPMA experiment.

beam direction is $[001]_{L1_0}$. A grain boundary (G. B.) between the TiAl crystal and the Al₃Ti grain is recognized in the image. Dark areas in both ends of the image are thick areas that remained after the FIB fabrication. Figure 3(b) shows an Al concentration profile recorded along the horizontal line marked in (a). To acquire the concentration profile, the electron beam about 1 nm in diameter was positioned on the specimen and measured an EDX spectrum at each position for 1 sec, and the Al concentration was estimated from the intensity ratio of Ti-Kα and Al-Kα X-rays without absorption corrections. Although experimental errors in the measured Al concentrations are not small due to the short acquisition time, the profile shows a monotonous increase in Al concentration along the $[100]_{L1_0}$ direction, except for a rapid damping near the thick areas due to the X-ray absorption effect. There seems to be a small gap in Al concentration at the grain boundary. This concentration gap may correspond to the phase boundary, TiAl/(TiAl+Al₃Ti), as shown in figure 1.

Figure 3(c) shows selected area electron diffraction patterns taken from areas A to G denoted in (a). Weak diffraction intensities are recognized in addition to the fundamental L1₀ lattice reflections, as indicated by arrows in (c). The diffraction intensity distribution exhibits changes in the crystal structure and degree of order of LPSs with the increase in Al concentration: Al₅Ti₃ type short-range order (SRO) states [19,20] in areas A and B, transitional states between the Al₅Ti₃ type SRO and an h-Al₂Ti type long-range order (LRO) [7,19,20] in areas C and D, and the h-Al₂Ti type LRO state [12] in area E. In area F, the long-period superlattice reflections mostly disappear. It is expected that a 1d-APS [2,10,12-14] is formed along the $[001]_{L1_0}$ direction in area F, as shown later. The diffraction pattern in area G is of the Al₃Ti grain with D0₂₂ structure. The Al₃Ti grain has no special orientation relationship with the other LPSs. This may be due to different formation kinetics: the Al₃Ti grain may be formed by nucleation and growth at the

261

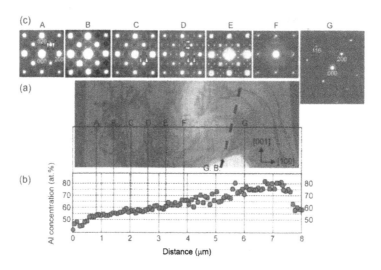

Figure 3. (a) Bright-field TEM image of the composition-gradient area. The incident beam direction is $[001]_{L1_0}$. (b) Al concentration profile along the horizontal line marked in (a). (c) Selected area electron diffraction patterns taken from areas A to G denoted in (a).

surface of the TiAl single crystal.

A TEM analysis in the $[010]_{L1_0}$ incident beam direction is demonstrated in figure 4. The formation of the 1d-APS is clearly identified from the diffraction patterns, as indicated by arrows in figure 4(c). The 1d-APS phase is formed with the h-Al$_2$Ti phase in areas C, D, and E and exhibits an orientation relationship with each other. Dark-field observation revealed that the 1d-APS and h-Al$_2$Ti phases form a lamellar microstructure along the $[101]_{L1_0}$ or $[\bar{1}01]_{L1_0}$ direction, as reported previously [10,12]. These results suggest that the 1d-APS type ordering occurred in the L1$_0$-TiAl crystal. It is again observed that the Al$_3$Ti grain in area G exhibits no special orientation relationship with the other LPSs because of the different formation kinetics.

The present experimental results revealed the orientation relationship among the L1$_0$-TiAl, h-Al$_2$Ti and 1d-APS phases, as illustrated in figure 5. Here, the 1d-APS is considered as a mixture of D0$_{23}$ and D0$_{22}$ structures for simplicity [13,14]. The h-Al$_2$Ti structure is described as a periodic arrangement of hypothetical L1$_2$-Al$_3$Ti unit cells in the L1$_0$-TiAl matrix. The same description can be adopted for the Al$_5$Ti$_3$ structure. The 1d-APS is briefly described by insertion of periodic antiphase boundaries along the [001] direction in the L1$_2$-Al$_3$Ti lattice, although the L1$_2$ lattice actually contains statistical disorder and displacement of atoms [13,14]. It should be noted in figure 5 that all of these LPSs are composed of Al (002) and Ti-Al (002) layers stacked alternately along the [001] direction. In the present composition-gradient TiAl single crystal, the Al (002) and Ti-Al (002) layers of the LPSs are parallel to those of the TiAl matrix. This fact suggests that the ordering of this type of LPSs and the phase transition between these LPSs can be interpreted as structural changes in Ti-Al (002) layers of the TiAl matrix, as predicted previously [3-10]. From this point of view, the r-Al$_2$Ti phase, for example, belongs to the different category [6,8,11], since the r-Al$_2$Ti structure does not consist of the alternate stacking

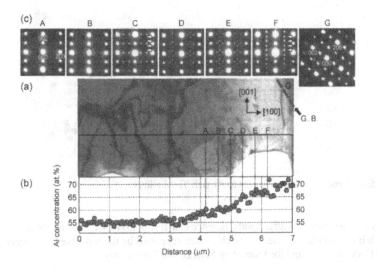

Figure 4. (a) Bright-field TEM image of the composition-gradient area. The incident beam direction is $[010]_{L10}$. (b) Al concentration profile along the horizontal line marked in (a). (c) Selected area electron diffraction patterns taken from areas A to G indicated in (a).

of Al (002) and Ti-Al (002) layers and thus some other ordering mechanisms should be taken into account.

CONCLUSIONS

By using an Al-rich TiAl single crystal with gradient compositions and a TEM-FIB microsampling technique, the composition dependence of the long-period ordering was studied. The observation results give a clear insight into the phase formation in the Al-rich Ti-Al system. A composition-gradient TiAl single crystal was prepared by annealing in a molten Al at 1234°C. The composition-gradient layer forms different LPSs as a function of Al concentration: Al_5Ti_3 type SRO, h-Al_2Ti and 1d-APS. These LPSs that are composed of the alternate stacking of Al (002) and Ti-Al (002) layers exhibit the same orientation relationship with the TiAl matrix: the Al (002) layers of the LPSs are parallel to those of the TiAl matrix. Thus, the ordering of this type of LPSs and the phase transition between these LPSs can be interpreted as structural changes in Ti-Al (002) layers in the Al-rich $L1_0$-TiAl crystal.

ACKNOWLEDGMENTS

The authors thank Dr. R. Kainuma (Tohoku University) and Dr. T. Ikeda (Osaka University) for their valuable comments on preparing TiAl single crystals with a composition gradient. This work was partly supported by a Grant-in-Aid for Young Scientists (B) (15710092) from the

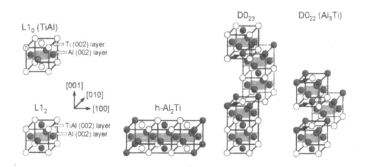

Figure 5. Ordered structures relevant to Al-rich Ti-Al alloys.

Ministry of Education, Culture, Sports, Science and Technology (MEXT) and the Strategic Research Base 'Handai Frontier Research Center' supported by the Japanese Government's Special Coordination Fund for Promoting Science and Technology.

REFERENCES

1. R. Miida, S. Hashimoto and D. Watanabe, *Jpn. J. Appl. Phys.* **21**, L59 (1982).
2. R. Miida, S. Hashimoto and D. Watanabe, *Jpn. J. Appl. Phys.* **18**, L707 (1980).
3 A. Loiseau, A. Lasalmonie, G. Van Tendeloo, J. Van Landuyt and S. Amelinckx, *Acta Cryst.* **B41**, 441 (1985).
4. U. D. Kulkarni, *Acta Mater.* **46**, 1193 (1998).
5. U. D. Kulkarni, *Phil. Mag. A* **82**, 1017 (2002).
6 T. Nakano, A. Negishi, K. Hayashi and Y. Umakoshi, *Acta Mater.* **47**, 1091 (1999).
7. T. Nakano, K. Hayashi and Y. Umakoshi, *Phil. Mag. A* **82**, 763 (2002).
8 C. Lei, Q. Xu and Y.-Q. Sun, *Mater. Sci. Eng.* **A313**, 227 (2001).
9. F. Stein, L. C. Zhang, G. Sauthoff and M. Palm, *Acta Mater.* **49**, 2919 (2001).
10. M. Palm, L. C. Zhang, F. Stein and G. Sauthoff, *Intermetallics* **10**, 523 (2002).
11. L. C. Zhang, M. Palm, F. Stein and G. Sauthoff, *Intermetallics* **9**, 229 (2001).
12 A. Loiseau and C. Vannuffel, *Phys. Stat. Sol. (a)* **107**, 655 (1988).
13. A. Loiseau, G. Van Tendeloo, R. Portier and F. Ducastelle, *J. Physique* **46**, 595 (1985).
14. R. Miida, *Jpn. J. Appl. Phys.* **25**, 1815 (1986).
15. R. Kainuma, M. Palm and G. Inden, *Intermetallics* **2**, 321 (1994).
16. T. Miyazaki, T. Koyama and S. Kobayashi, *Metall. Mater. Trans.* **27A**, 945 (1996).
17.T. Ikeda, H. Kadowaki, H. Nakajima, H. Inui, M. Yamaguchi and M. Koiwa, *Mater. Sci. Eng.* **A312**, 155 (2001).
18. T. Ikeda, H. Kadowaki and H. Nakajima, *Acta Mater.* **49**, 3475 (2001).
19. S. Hata, K. Higuchi, M. Itakura, N. Kuwano, T. Nakano, K. Hayashi and Y. Umakoshi, *Phil. Mag. Lett.* **82**, 363 (2002).
20. S. Hata, K. Higuchi, T. Mitate, M. Itakura, Y. Tomokiyo, N. Kuwano, T. Nakano, Y. Nagasawa and Y. Umakoshi, *J. Electron Microsc.* **53**, 1 (2004).

Influence of Micro-alloying on Oxidation Behavior of TiAl

Michiko Yoshihara[1] and Shigeji Taniguchi[2]

1. Department of Mechanical Engineering and Materials Science, Yokohama National University, 79-5 Tokiwadai, Hodogaya-ku, Yokohama, Kanagawa 240-8501, Japan

2. Department of Materials Science and Processing, Osaka University, 2-1 Yamadaoka, Suita, Osaka 565-0871, Japan

ABSTRACT

The influence of a wide range of elements on oxidation behavior of TiAl was investigated by micro-alloying using ion implantation with ion doses of 10^{19} to $10^{22}\,m^{-2}$ and at acceleration voltages of 40 to 340kV. The oxidation resistance was assessed by a cyclic oxidation test at 1200K in a flow of purified oxygen under atmospheric pressure. The implanted elements can be classified into several groups according to their effect and mechanism. The mechanisms by which the oxidation resistance is improved are as follows: (1) Formation of a protective Al_2O_3 layer through β-phase formation, which was confirmed by TEM observations, in the modified surface layer by the implantation. (2) Reduction of TiO_2 growth rate due to doping effect of the implanted element. (3) Protective Al_2O_3 layer formation through migration of volatile halide. (4) Enrichment of oxide of the implanted element in the scale. On the other hand, the oxidation resistance is decreased by (1) enhanced TiO_2 growth due to doping effect, (2) lattice defects induced by the implantation, and (3) decreased scale strength and enhanced scale spallation.

INTRODUCTION

TiAl-based alloys have attractive properties as light weight heat-resisting material. Numerous engineering alloys have been developed for structural applications in automotive and aerospace components [1]. The alloys thus developed contain at least a few alloying elements that may optimize thermally stable microstructures and/or improved high temperature mechanical properties. Their oxidation resistance is not sufficient at application temperatures higher than 800°C [2, 3], because TiAl alloys do not form a continuous and protective Al_2O_3 layer in spite of their high Al content. The influence of alloying element on their oxidation behavior has not been well clarified yet, and it is an important factor for further alloy development. Since there seems to be no single alloying element which can improve mechanical properties and oxidation resistance simultaneously, the surface treatment is a possible way for their compatibility.

In this sense, the microalloying using ion-implantation technique is not only a useful way for such a surface treatment but also an excellent research tool to investigate the influence of the additional element on oxidation behavior [4-7]. This method can be applied for almost all the elements, including those which cannot be alloyed by the conventional ingot metallurgy. In the present study, 22 elements were implanted, and their influence on the oxidation behavior of TiAl and the mechanisms concerned are discussed. The oxidation was carried out in pure oxygen to avoid nitride formation, which has contradictory effect on oxidation behavior [8], and hence to simplify the consideration.

EXPERIMENTAL PROCEDURES

Two kinds of TiAl specimens were used: cast TiAl in some experiments and forged material in most cases. Their chemical compositions are shown in Table I. These specimens did not form protective Al_2O_3 scales at the temperatures of consideration. Coupon specimens with a size of 15x10x1(or 2) mm were machined out of the ingot or slice of forged pancake. The specimen edges were ground to make 45-degree angle, so as to make a trapezoidal specimen shape. The specimens were polished to a mirror finish and ultrasonically washed in acetone before implantation. Then, they were implanted with elemental ions at acceleration voltages of 40 to 340 kV with doses of 10^{19} to 10^{22} m.$^{-2}$ The implanted elements were B, C, N, F, Mg, Al, Si, P, Cl, Ar, V, Cr, Fe, Cu, Zn, Se, Zr, Nb, Mo, Ag, Ta and W.

A cyclic oxidation test between room temperature and 1200K was carried out in a flow of purified oxygen under atmospheric pressure. The holding time at temperature was 72ks in most cases, but 3.6ks for Si or Al implanted specimen. The oxidation data were not corrected for the slant surface, because its influence is small.

The implanted specimens and oxidized specimens were characterized by conventional methods including Auger electron spectroscopy (AES), glancing angle X-ray diffractometry (GAXRD) with off-set angle of 2 degree, X-ray diffractometry (XRD), scanning electron microscopy (SEM), electron probe microanalysis (EPMA) and transmission electron microscopy (TEM) with energy dispersive spectroscopy (EDS). The TEM specimens were prepared using a focused ion beam (FIB) unit. Secondary ion mass spectroscopy (SIMS) was also used in a few cases.

RESULTS AND DISCUSSION

Figures 1(a) and (b) show the AES depth profiles of the relevant elements after the implantation of W

Table I.   The chemical composition of TiAl specimens (mol %).

	Al	Ti	Fe	Si	C	O	N	H
Cast alloy	50.40	Bal.	-	-	0.026	0.228	0.013	-
Forged alloy	50.41	Bal.	0.11	0.007	0.016	0.079	0.008	0.10

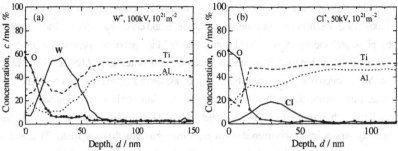

Figure 1. AES depth profiles of the relevant elements for (a) W-implanted specimen (100kV, 10^{21} m^{-2}) and (b) Cl-implanted specimen (50kV, 10^{21} m^{-2}).

(100kV, 10^{21} m^{-2}) and Cl (50kV, 10^{21} m^{-2}), respectively, as examples. The implanted elements follow near-normal distribution. The penetration depth and maximum concentration depend on the implantation conditions, and mass and size of the implanted element. In most cases, no peak was found for a dose of 10^{19} m^{-2} indicating that this dose seems to be below the detection limit of AES.

The GAXRD of the specimens after implantation showed no phases other than γ-TiAl in most cases. The Ti$_2$AlN phase was formed by the implantation of N, while Fe, Nb, Mo, Ta or W resulted in the formation of β-phase. The latter is attributable to the fact that these elements are β-former/stabilizer to TiAl. To confirm the presence of β-phase in a surface zone, detailed TEM observations were performed.

Figure 2 shows a TEM bright field image of a cross section and diffraction patterns of the specimen implanted with W (100kV, 10^{21} m^{-2}). The number in the figure shows positions where diffraction patterns were taken and EDS was performed. The W-modified zone is about 70nm thick and this is consistent with the AES results shown in Figure 1(a). The outermost layer (1) of about 5-10 nm thickness is amorphous.

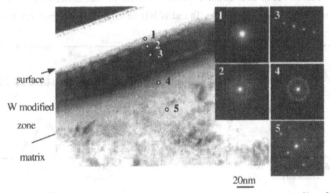

Figure 2. TEM bright field image of the W-implanted specimen (100kV, 10^{21} m^{-2}) and electron diffraction patterns.

The AES and EDS showed that it contains oxygen along with Al, Ti and W, suggesting the layer to be oxide. Position (2) was identified as a mixture of an amorphous and crystalline β-phase. The positions (3) and (4) are β-phase and position (5) is the TiAl matrix. The TEM images of the specimens implanted with Fe, Nb or Mo are similar to Figure 2. These results agree well with the results of AES and GAXRD, and clearly show the formation of β-phase. In all the cases, the modified zones do not show crystalline microstructure. This is attributable to the lattice defects and strain induced by the implantation.

Figures 3(a) and (b) show the cyclic oxidation behavior of Ar-implanted specimen and W-implanted ones, respectively. Ar-implanted specimen shows a larger mass gain than the unimplanted TiAl, and is followed by a mass loss due to scale spallation. On the other hand, W-implanted specimens show excellent behavior except for a dose of 10^{19} m^{-2}, which was found to be too low for the improvement. The result of Ar-implanted specimen suggests that the lattice defects induced by the implantation enhance the diffusion of oxygen or Ti rather than Al. So, the implantation of Ar enhances TiO$_2$ formation and deteriorates the oxidation behavior. Implantation of B, C, N, Mg, V, Cr, Se, Zr or Ag also has no effect. The scale formed on N-implanted specimen has stratified structure and suffered from severe scale spallation. Along with W, the implantation of Fe, Nb, Mo or Ta results in a virtually Al$_2$O$_3$ scale and significant improvement in the oxidation resistance. Since these elements are β-former/stabilizer, β-phase seems to be responsible for this through enhanced Al diffusion in this phase compared to γ-TiAl. The halogen elements, Cl and F, improved the oxidation resistance remarkably. The implantation of Al, Si or P is also effective, though the extent is lower than the β-formers or halogens.

Figure 4 shows AES depth profile of W-implanted specimen oxidized for a very short period. Considering the GAXRD results obtained from the specimen after oxidation, the scale consists of outer and inner Al-rich layers, α-Al$_2$O$_3$, with an intermediate layer of Al$_2$O$_3$ / TiO$_2$ mixture. W is enriched in the middle part of the scale with Ti, suggesting its incorporation in TiO$_2$. The scale formed on specimens implanted with P, Fe, Nb, Mo or Ta is similar to that with W, and the enrichment of P, Nb or Ta in TiO$_2$ was also observed. The valence of P, Nb, Mo, Ta and W is believed to be higher than that of Ti, +4, the

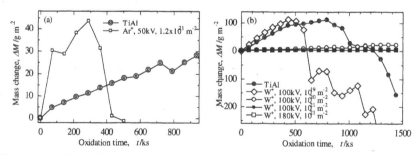

Figure 3. Cyclic oxidation behavior of (a) Ar-implanted specimen and (b) W-implanted specimens, compared with unimplanted TiAl at 1200K in purified oxygen.

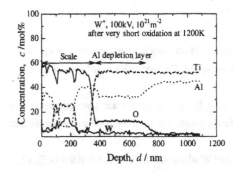

Figure 4. AES depth profile of the W-implanted specimen ($100kV$, 10^{21} m^{-2}) after oxidation at 1200K for 0ks.

incorporation of these elements in TiO_2 is expected to reduce its growth rate due to so-called doping effect. The Al-depletion layer whose composition is consistent with Z-phase, $Ti_5Al_3O_2$, may stabilize the Al_2O_3. The Z-phase was confirmed by GAXRD. Si-implanted specimen showed enrichment of Si at the scale/alloy interface, suggesting the formation of protective Si oxide.

Figures 5 show fractured cross sections of (a) W-implanted specimen ($100kV$, 10^{21} m^{-2}) oxidized for 1440ks and (b) F-implanted specimen ($50kV$, 10^{20} m^{-2}) oxidized for 1800ks. The scale is very thin and consists of virtually Al_2O_3 in both cases. However, the scale on F-implanted specimen is characterized by local convolution and a thick Al-depletion layer of Z-phase beneath the scale. Such convoluted scale was also observed on Cl-implanted specimen. This scale morphology suggests continuous supply of Al from the substrate to grain boundaries in the scale and the growth stress in lateral direction during scale growth. The reaction products between halogens and Al or Ti are volatile. Based on a detailed thermodynamic calculation, Donchev et al. [9] suggested that Al_2O_3 scale can be developed by exclusive transport and subsequent oxidation of a gaseous aluminum-containing species in the inner region of the initially-formed scale. For Cl, the formation of $TiCl_4$ seems to be most probable from the free energy, preferential evaporation of titanium chloride may enhance the formation of Al_2O_3 in the initial stage.

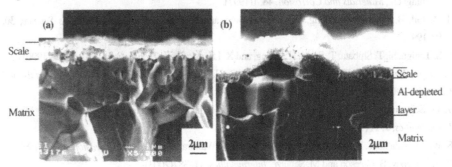

Figure 5. Fractured cross section of (a) W-implanted specimen ($100kV$, 10^{21} m^{-2}) oxidized for 1440ks and (b) F-implanted specimen ($50kV$, 10^{20} m^{-2}) oxidized for 1800ks at 1200K in purified oxygen.

CONCLUSIONS

The micro-alloying of a range of elements has been carried out using ion implantation technique and influence on the oxidation behavior of TiAl has been investigated. The mechanisms can be classified into several groups as follows.

(1) Beta-phase formation: Fe, Nb, Mo, Ta and W form β-phase in the surface layer and result in a protective Al_2O_3 layer formation when the dose is sufficiently high. The oxidation resistance is significantly improved by the implantation of these elements.

(2) Doping effect: The incorporation of P, Nb, Mo, Ta and W whose valence is higher than that of Ti, +4, reduces the growth rate of TiO_2 due to doping effect and thus improves the oxidation resistance. For those with lower valence than Ti, the addition is detrimental due to the enhanced growth of TiO_2.

(3) Halogen effect: Halogen elements react with Al or Ti and form volatile products. The continuous supply of Al from the substrate to scale via volatile Al halide seems to be responsible to protective Al_2O_3 layer formation. Titanium halide may have some influence on the formation of Al_2O_3.

(4) Enrichment of implanted element: Enrichment of Al or Si results in the formation of protective scale rich in these elements and thus improves the oxidation resistance.

(5) Lattice defects: The lattice defects induced by the implantation may enhance the diffusion of oxygen or Ti rather than Al so as to enhance TiO_2 growth.

(6) Decreased scale strength: The decreased scale strength enhances the scale spallation thus deteriorates the oxidation behavior.

REFERENCES

1. Y-W. Kim et al. editors, "*Gamma Titanium Aluminides 2003*", TMS(2003).

2. A. Rahmel, W. J. Quadakkers, and M. Schütze, *Materials and Corrosion*, **46**, 271(1995).

3. S. Taniguchi, *Materials and Corrosion*, **48**, 1(1997).

4. A. Gil, B. Rajchel, N. Zheng, W. J. Quadakkers and H. Nickel, *Journal of Materials Science*, **30**, 5793(1995).

5. S. Taniguchi, T. Shibata, T. Saeki, H. Zhang, and X. Liu, *Materials Transactions, JIM*, **5**, 998(1996).

6. M. F. Stroosnijder, H. J. Schmutzler, V. A. C. Haanappel and J. D. Sunderkötter, *Materials and Corrosion*, **48**, 40(1997).

7. G. Schumacher, F. Dettenwanger, M. Schütze, J. Hornauer, E. Richter, E. Wieser, and W. Möller, *Intermetallics*, **7**, 1113(1999).

8. W. J. Quadakkers, P. Schaaf, N. Zheng, A. Gil and E. Wallura, *Materials and Corrosion*, **48**, 28(1997).

9. A. Donchev, B. Gleeson and M. Schütze, *Intermetallics*, **11**, 387(2003).

Silicides

Mater. Res. Soc. Symp. Proc. Vol. 842 © 2005 Materials Research Society   S2.10

The Effects of Substitutional Additions on Tensile Behavior of Nb-Silicide Based Composites

Laurent Cretegny, Bernard P. Bewlay, Ann M. Ritter, and Melvin R. Jackson
GE Global Research, Schenectady, NY 12301, USA.

ABSTRACT

Nb-silicide based in-situ composites consist of a ductile Nb-based solid solution with high-strength silicides, and they show excellent promise for aircraft engine applications. The Nb-silicide controls the high-temperature tensile behavior of the composite, and the Nb solid solution controls the low and intermediate temperature capability. The aim of the present study was to understand the effects of substitutional elements on the room temperature tensile behavior and identify the principal microstructural features contributing to strengthening mechanisms.

INTRODUCTION

Nb-silicide composites combine a ductile Nb phase with high-strength silicides and Laves phases; they show great promise for high-temperature structural applications [1-4]. These composites consist of Nb_5Si_3 and Nb_3Si type silicides toughened with a Nb solid solution (substitution of Nb by Ti and Hf is abbreviated by (Nb)). The most recent Nb-silicide based in-situ composites, alloyed with elements such as Cr, Ti, Hf, and Al, have demonstrated a promising combination of high-temperature strength, creep resistance, and fracture toughness. The Nb_5Si_3 and Nb_3Si have the tI32 and tP32 ordered tetragonal structures with 32 atoms per unit cell. When Nb_5Si_3 is alloyed with Ti and Hf, the less complex hP16 $(Nb)_5Si_3$ structure can also be stabilized [5, 6]. The effect of substitutional elements on the mechanical behavior of the tI32, tP32, and hP16 silicides has not been investigated previously.

An improved understanding of the mechanisms that control low- and high-temperature deformation of Nb-silicides is required to improve the strength of Nb-silicide based composites. The aim of the present study was to characterize the room temperature tensile behavior of quaternary and more complex engineering Nb-silicide based alloys and to correlate measured strength with microstructure and fracture surface features.

Table 1: Compositions and room-temperature tensile strength data for quaternary and MASC-type alloys.

Ti [at%]	Si [at%]	Cr [at%]	Hf [at%]	Al [at%]	Nb [at%]	Failure stress [MPa]	Failure stress [ksi]	
25.0%	20.0%	-	8.0%	-	47.0%	220.1	31.9	
25.0%	18.0%	-	8.0%	-	49.0%	289.1	41.9	
25.0%	18.0%	-	8.0%	-	49.0%	317.1	46.0	(replicate)
25.0%	16.0%	-	8.0%	-	51.0%	261.6	37.9	
25.0%	14.0%	-	8.0%	-	53.0%	365.6	53.0	
24.7%	16.0%	2.0%	8.2%	1.9%	47.2%	555.1	80.5	
24.7%	16.0%	2.0%	8.2%	1.9%	47.2%	577.1	83.7	(replicate)
24.0%	16.0%	2.0%	<5	1.9%	52.6%	404.8	58.7	
24.0%	13.0%	2.0%	<5	1.9%	55.6%	318.1	46.1	

EXPERIMENTAL PROCEDURE

The samples for tensile testing were prepared using directional solidification [2]. The starting charges were prepared from high purity elements (>99.99%) into a series of Nb-Si-Ti-Hf quaternary alloys and more complex alloys with Hf and Si variations around the composition of the MASC alloy, an established Nb-silicide based engineering alloy (Nb-16Si-24.7Ti-8.2Hf-2Cr-1.9Al), Table 1 [2]. Tensile specimens were manufactured by electro-discharge machining (EDM) according to dimensions adapted from a standard ASTM E8 specimen, Figure 1. The specimen axis was parallel to the crystal growth direction [7]. This geometry was chosen to facilitate machining and make use of the limited available input material. However, the sharp edges and small volume-to-surface ratio in the gage section could potentially affect the measured absolute strengths; a good relative comparison of the various alloys was expected. The gage section surfaces were carefully hand-polished using 800 grit paper, ensuring that only longitudinal marks are left on the surfaces. Room-temperature tensile tests were performed on a 90 kN screw-driven MTS frame and fractography was performed by scanning electron microscopy. Specimen preparation for fractography consisted of grinding back one side of the gage section to about 75% of the original thickness, followed by a fine polish to reveal the microstructure, thus allowing the analysis of the crack interaction with microstructural features.

RESULTS

The room-temperature strength results are summarized in Table 1. For the range of compositions evaluated, specific microstructural features balance the overall strength of these in-situ composite intermetallic alloys. These microstructural elements typically include the primary (Nb) dendrite cores and secondary dendrite arms, reinforcement by wide eutectic cells that experience a low level of constraint, and cleavage fracture of silicides and constrained narrow eutectic cells, Figure 2. Any features that force out-of-plane crack growth create energy consuming obstacles for the crack propagation. For example, when cleavage cracking occurs in a silicide neighboring a metallic dendrite, the crack is immediately arrested at the phase boundary and failure along the interface must be initiated and propagated in an attempt to contour the metallic region, thus raising the energy required to grow a crack. Primary dendrite cores also contribute to the failure mechanisms; however, a significantly higher fraction of the fracture surface consists of secondary arms, suggesting that the phase boundary failure at dendrite arms has a greater effect on the failure mechanisms than the ductile rupture of primary dendrite cores.

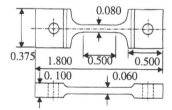

Figure 1: Tensile specimen geometry used for the evaluation of Nb-silicide based alloys. (dimensions in inches)

An in-depth assessment of the relative contribution of each of these microstructure features (metal dendrite, silicides, eutectic cells) requires quantitative analysis of the volume fraction of each phase. This investigation is currently underway and will be reported in a future paper.

Quaternary Nb-xSi-25Ti-8Hf Alloys

The room-temperature tensile test results for quaternary alloys, Figure 3, show a decrease in strength as the Si content increases. A plateau in strength is observed between 16 and 18 at% Si, which is related to the significant changes experienced by the microstructure in this composition range. A qualitative understanding of the influence of the microstructure on the failure mechanisms and the resulting strength was obtained by observing the crack path along microstructural features.

The highest room-temperature strength was measured on the 14 at% Si alloy, whose microstructure consisted mainly of primary (Nb) dendrites and interdendritic silicides. The most prominent features observed on the fracture surface were the out-of-plane excursions in the crack path when intercepting secondary dendrite arms. This phase boundary cracking indicates a weaker interface between the metal and the interdendritic phase than the inherent (Nb) dendrite strength. It also suggests that the surrounding phases do not sufficiently constrain the secondary dendrite arms to cause cleavage fracture, as was reported elsewhere [8].

At 16 at% Si, the tensile behavior of the quaternary alloys is reduced to less than 262 MPa (38 ksi). Observations of the microstructure and fracture surface, Figure 4, clearly illustrate that the formation of large silicide rods leads to an increased level of cleavage fracture. Narrow cells of (Nb)-Nb$_3$Si eutectic are also observed along the crack path, but they are too highly constrained by surrounding (Nb) and Nb$_3$Si phases to be able to contribute to strengthening. Indeed, most of the metallic lamellae in these cells cleave or produce a very fine surface roughening that is not indicative of a strengthening feature of the microstructure.

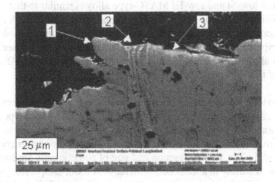

Figure 2: Typical microstructural features observed on fracture surfaces of specimens: (1) crack propagation along secondary (Nb) dendrite arms boundaries, (2) cleavage of constrained eutectic cells and (3) cleavage of large silicides.

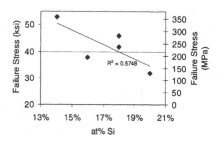

Figure 3: Effect of silicon concentration on room-temperature tensile strength of Nb-xSi-25Ti-8Hf quaternary alloys.

Raising the Si content from 16 to 18 at% does not further reduce the room temperature tensile strength. On the contrary, a slight increase is measured and can be correlated with the formation of wider eutectic cells in the alloy, which allows yielding and delamination of the metallic lamellae due to the lower level of constraint from surrounding phases. Beyond 18 at% Si, the strength drops severely due to a significant increase in volume fraction of silicides, which cleave upon loading at room temperature. The consequence of silicide cleavage is an overall reduction in tensile strength due to an increase in flaw density and size in the material.

MASC-Type Alloys

The three complex alloys investigated showed similar fracture features as were observed in the quaternary alloys. Therefore, one may expect the same trends in room-temperature strength. However, an increase in strength is observed as the Si content is raised from 13 to 16 at% and further improves as the Hf level is raised from less than 5 at% to 8.2 at%.

The microstructure of the low-Si and low-Hf MASC-type alloy is similar to that of the 14 at% Si quaternary alloy, consisting essentially of (Nb) dendrites and Nb$_3$Si silicides. A higher strength is measured in the quaternary alloy, likely from a finer dendritic structure and solution strengthening, both attributed to the higher Hf content. As the Si content is raised from 13 to 16 at% in the MASC-type alloy, the room temperature strength increases to 405 MPa (59 ksi), which is the opposite of that observed in the quaternary alloys.

This difference in behavior correlates with differences in microstructures between the two families of alloys, as highlighted when the 16 at% Si and 8.2 at% Hf MASC-type alloy is included in the evaluation. The respective microstructures show a higher volume of eutectic in the MASC-type alloys than in the quaternary alloys with similar levels of primary elements (Nb, Si, Ti, Hf), Figures 4 and 5. The volume fraction and size of eutectic cells are further increased by raising Hf to 8.2 at%, resulting in a tremendous increase in room-temperature strength. These observations indicate that additional strength in MASC-type alloys with Cr and Al additions can be attributed to the propensity for these compositions to favor eutectic growth compared to the simpler quaternary alloys.

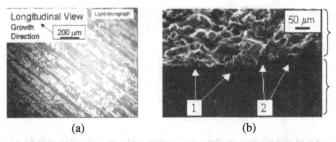

(a)   (b)

Figure 4: Nb-16Si-25Ti-8Hf alloy: (a) Optical micrograph of a polished longitudinal cross-section and (b) SEM micrograph of a cross-section of the fracture surface (1: silicides, 2: (Nb) dendrites).

The effect of unconstrained eutectic cells as a strengthening mechanism on the fracture behavior is illustrated in Figure 6, which clearly shows that the metallic lamellae do not cleave at the location of the incident crack, but force out-of-place crack growth by delamination along the phase boundary and plastic deformation of the metallic phase. This is evidence of an increased energy dissipative capability, which suggests that the fracture behavior in this microstructure is controlled less by nucleation than by propagation [9]. Plastic deformation of the metallic phase is supported both by the dimpled rupture of large (Nb) lamella and by the raised finer lamellae above the incident crack path. No intrusion at metallic lamellae was found in any eutectic cells, indicating that the protruding lamellae are formed through plastic deformation of the (Nb) and not solely by delamination followed by cleavage of both (Nb) and Nb_5Si_3 lamellae, in which case an equal number of intrusions and protrusions would be observed. Further investigation is required to determine the critical size of eutectic cells beyond which unconstrained deformation of the metallic lamellae is possible.

CONCLUSIONS

The room-temperature tensile behavior of quaternary and MASC-type Nb-silicide based alloys was evaluated and correlated with microstructure and fracture surface features. The principal contributors to room-temperature strength were identified as reinforcement by (1) wide eutectic cells that experience a low level of constraint, (2) primary (Nb) dendrite cores and secondary

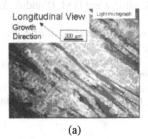
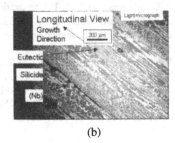

(a)   (b)

Figure 5: Optical micrographs of a longitudinal section of (a) Nb-16Si-24Ti-Hf-2Cr-1.9Al (Hf<5at%) and (b) Nb-16Si-24Ti-8.2Hf-2Cr-1.9Al.

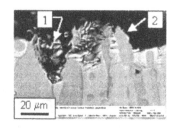

Figure 6: Formation of significant surface roughening at wide eutectic cells in Nb-16Si-24Ti-8.2Hf-2Cr-1.9Al: (1) Plastic deformation and delamination of metallic lamellae and (2) dimple rupture of a lamella.

arms, and (3) cleavage fracture of silicides and constrained narrow eutectic cells. Substitution elements such as Hf, Ti and Cr influenced the strength by direct solid solution strengthening and, more importantly, by significantly increasing the volume fraction and size of eutectic cells, which considerably improved strength. EBSD phase identification and a quantitative analysis of the microstructure is currently underway and will provide the required data for development of physics-based constitutive models for room-temperature strength of these intermetallic in-situ composite alloys.

ACKNOWLEDGEMENTS

The authors would like to thank D.J. Dalpe for preparing the samples, and C. Canestraro for the tensile testing.

REFERENCES

[1]   P.R. Subramanian, M.G. Mendiratta, D.M. Dimiduk and M.A. Stucke, *Mater. Sci. Eng.*, A239-240, 1997, pp. 1-13.
[2]   B.P. Bewlay, M.R. Jackson and H.A. Lipsitt, *Metall. and Mater. Trans.*, 1996, Vol 279, pp. 3801-3808.
[3]   M.G. Mendiratta, J.J. Lewandowski and D.M. Dimiduk, *Metall. Trans.* 22A *(1991)*, pp. 1573-1581.
[4]   P.R. Subramanian, T.A. Parthasarathy, M.G. Mendiratta and D.M. Dimiduk, *Scripta Met. and Mater.*, Vol. 32(8), 1995, pp. 1227-1232.
[5]   B.P. Bewlay, M.R. Jackson and H.A. Lipsitt, *Journal of Phase Equilibria*, Vol 18(3), 1997, pp. 264-278.
[6]   B.P. Bewlay, R.R. Bishop and M.R. Jackson, *Z. Metallkunde*, 1999, Vol 90, pp. 413-422.
[7]   Annual Book of ASTM Standards, ASTM International, Vol. 03-01, 2002.
[8]   J. Kajuch, J. Short, J.J Lewandowski, *Acta Metall. Mater.*, 1995, Vól 43, pp. 1955-1967.
[9]   M.G. Mendiratta, R. Goetz, D.M. Dimiduk, J.J. Lewandoswki, *Metall. and Mater. Trans A*, 1995, Vol 26A, pp.1767-1776.

Mater. Res. Soc. Symp. Proc. Vol. 842 © 2005 Materials Research Society

Effect of Microstructure and Zr Addition on the Crystallographic Orientation Relationships among Phases related to the Eutectoid Decomposition of Nb₃Si in near Eutectic Nb-Si alloy

Seiji Miura[†], Kenji Ohkubo and Tetsuo Mohri

Division of Materials Science and Engineering, Graduate School of Engineering,

Hokkaido University, Kita-13, Nishi-8, Kita-ku, Sapporo 060-8628, Japan

[†] corresponding author. Tel. & Fax : +81-11-706-6347

E-mail address: miura@eng.hokudai.ac.jp (S. Miura)

ABSTRACT

The authors have reported in the previous study that the sluggish decomposition of Nb₃Si phase is effectively accelerated by Zr addition [1]. This is obvious at lower temperature range than the nose temperature of the TTT curve. In the present study a eutectic alloy containing 1.5 % of Zr was investigated. The crystallographic orientation relationships among phases, such as eutectic Nb and product phases formed by eutectoid decomposition of Nb₃Si (eutectoid Nb and Nb₅Si₃ phases) in the Zr-containing sample which was heat treated at 1300°C were investigated by FESEM/EBSD for further understanding of the decomposition process in alloy with a different microstructure.

INTRODUCTION

Recently many efforts have been dedicated to the investigation of alloys based on Nb-silicides because they exhibit superior high temperature strength than the commercial nickel base superalloys [2-4]. Since the lack of room temperature ductility and high temperature oxidation resistance have been the major drawbacks for further development, various additives and controlled microstructures have been introduced to overcome these problems. For further improvement of the toughness of these alloys, advanced microstructure control is needed by which the alloys based on brittle intermetallic compounds are endowed with ductile phase toughening. The Nb/Nb₅Si₃ two-phase structure forms a fine microstructure through the eutectoid decomposition from Nb₃Si, but the kinetics of the reaction in the binary alloy was reported to be so sluggish that it takes more than 100 hr to finish the decomposition even at the nose temperature of about 1500 °C [2]. This slow kinetics may be caused by its high melting point, and is, in turn, an advantage for high temperature applications in a practical viewpoint.

Sekido et al. found that addition of the 10 at.% of Ti increases the decomposition rate of the Nb_3Si [5-9]. Then, the effect of Zr on the eutectoid decomposition reaction was investigated by present authors [1, 10], and it was revealed that the kinetics of the eutectoid decomposition is enhanced by Zr additions. The time-temperature transformation (TTT) diagram for the decomposition was also experimentally determined and the acceleration of the reaction by small Zr addition of 1.5 at.% was confirmed by comparing with the reported TTT curves of binary and ternary alloys containing Ti. Although the role of the ternary elements on the decomposition kinetics is still not fully understood, it was found both crystallographic orientation relationships among phases and Zr distribution in the parent Nb_3Si phase during solidification were remarkably different by Zr addition [1].

In this article, we attempt to understand the eutectoid decomposition process of Nb_3Si in eutectic microstructure because this is one of the promising microstructures for Nb-Si base alloys. The effect of primary Nb phase formed during solidification is focused, and it is examined whether it acts as a preferential nucleation site for the decomposition as was reported on the Nb-25Si alloys in our previous study.

EXPERIMENTAL PROCEDURES

Alloys were arc-melted in Ar atmosphere on a water-cooled copper hearth. The composition of the ternary alloy was determined to be 82.5at.% Nb –16.0at.% Si- 1.5at.% of Zr. A piece of sample was encapsulated in an evacuated silica tube and heat-treated at 1300 °C for 6 hours.

Microstructure observations were conducted on both as-cast and heat-treated specimens which were carefully polished with colloidal SiO_2 (40nm in diameter), and electron probe microanalysis (EPMA) and electron backscatter diffraction (EBSD) analysis are attained by SEM (JEOL-JXA-8900M) and FE-SEM (JEOL-JSM-6500F) with TexSEM Laboratories-OIM software, respectively.

RESULTS AND DISCUSSION

Shown in Figure 1 are the microstructure observed using SEM and the result of EBSD analysis of as-cast alloy. The primary Nb phase (bright) formed from the melt, and surroundings composed of Nb and Nb_3Si phase (dark) formed during the eutectic reaction, liquid -> Nb + Nb_3Si, were observed. Hereafter the former Nb is denoted as a primary Nb and latter is as a eutectic Nb. The phases consisting of the microstructure are identical to those previously reported on Nb-Si binary and Nb-Si-Ti ternary alloys [9], and the volume fractions are almost the same among these alloys. The eutectic Nb is rod-like, and EBSD analysis revealed that all the rod-like Nb within a eutectic colony have almost the same crystallographic orientation as shown in the pole figure in Fig.1 (b). However, no crystallographic orientation relationships between Nb_3Si and primary or eutectic Nb were found. Cockeram et al. and Grylls et al. [11-13] have reported several crystallographic orientation relationships between the eutectic Nb and the Nb_3Si precipitates and matrix. However, none of such crystallographic relationships were identified in the present alloy as was the case in the previous work on 73.5Nb-25Si-1.5Zr alloys [1].

Figure 2 shows the microstructure of the present alloy after the heat treatment. It was confirmed that the bright phase is Nb and the dark phase is the low-temperature Nb_5Si_3 phase (α-Nb_5Si_3) using the inverse pole figure (IPF) maps of the Nb and Nb_5Si_3 phases obtained by EBSD analysis. Near block primary Nb phases, lamellar area is found which was formed by the eutectoid decomposition of the Nb_3Si phase. Also in the eutectic Nb rod area, the eutectic decomposition seems to start as was pointed out previously [1]. Both the primary and eutectic

(a)   (b)

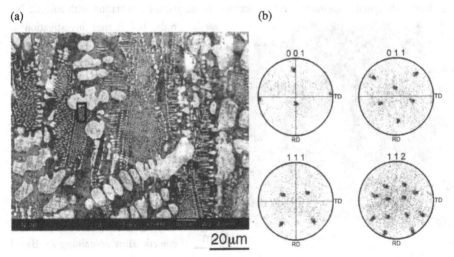

20μm

Figure 1   (a) The microstructure of as-cast alloy and (b) discrete pole figures corresponding to the primary and eutectic (rod-like) Nb phases in the area shown in (a).

Nb seemingly act as nucleation sites. The inter-lamellar spacing in the present alloy is of the order of micron and is almost the same as that of the 75Nb-25Si alloy [1]. The decomposition kinetics also seems to obey the time-temperature transformation (TTT) diagram for Zr-added alloys reported in the previous study [1], and it was confirmed that the fast interface velocity of eutectoid decomposition was not influenced by the overall microstructure. Also confirmed is that Nb lamellae in colony do not have the same crystallographic orientation with the neighboring primary Nb phase using the IPF map. This suggests that the primary Nb is less effective for the nucleation of Nb lamellae for the decomposition reaction.

The crystallographic orientation between Nb and Nb_5Si_3 within a eutectoid lamellar colony was investigated using pole figures shown in Figure 3. Although small deviations are still not negligible, following relationship is proposed.

$$<210>_{Nb} \; // \; <010>_{Nb_5Si_3} \quad \text{and} \quad \{001\}_{Nb} \; // \; \{001\}_{Nb_5Si_3} \qquad (1)$$

This relationship was also found in the previous study when the decomposition of Nb_3Si starts from the primary Nb_5Si_3 in 73.5Nb-25Si-1.5Zr alloys in which the crystallographic orientations of primary and lamellar Nb_5Si_3 are the same [1]. Therefore, the crystallographic orientation relationship expressed by eq.(1) might be corresponding to the case in which Nb_5Si_3 nucleate prior to the lamellar formation. Then, the observed lamellar formation should start with the nucleation of Nb_5Si_3 from Nb_3Si without the preferential nucleation site on primary Nb. On the other hand, Nb lamellae seems to have the same crystallographic orientation with eutectic Nb rods, but further investigation is needed because the lameller spacing was too fine to detect the enough signal from this area with this heat-treatment condition.

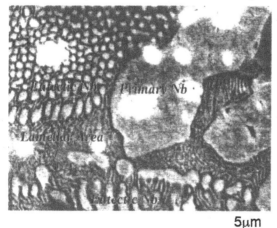

Figure 2   The microstructure of heat-treated alloy at 1300 °C for 6 h.

CONCLUDING REMARKS

An attempt was conducted to understand the eutectoid decomposition process of Nb_3Si in eutectic alloy containing Zr. Based on the microstructure observation and EBSD pattern analysis, one of the crystallographic orientation relationships reported previously [1] was confirmed between Nb and Nb$_5$Si$_3$ in eutectoid lamellar colony. Also confirmed is that the Nb lamellae in lamellar colony has different crystallographic orientation from the primary Nb, which suggests that the primary Nb is less effective as the nucleation site for the decomposition. The decomposition kinetics of the present ternary alloy having a eutectic microstructure is, however, almost the same with that in the previous alloy having the composition of 73.5Nb-25Si-1.5Zr with rather different microstructure.

REFERENCES

1. S. Miura, M. Aoki, Y. Saeki, K. Ohkubo, Y. Mishima and T. Mohri, Met. Mat. Trans., 2005, in press.
2. M. G. Mendiratta and D. M. Dimiduk, Scripta metall. mater., 1991, vol. 25, pp.237-242.
3. B. P. Bewlay, M. R. Jackson, J.-C. Zhao, P. R. Subramanian, M. G. Mendiratta and J. J. Lewandowski, MRS Bulletin., 2003, vol.28, pp. 646-653.
4. E. S. K. Menon and M. G. Mendiratta, "The Fifth Pacific

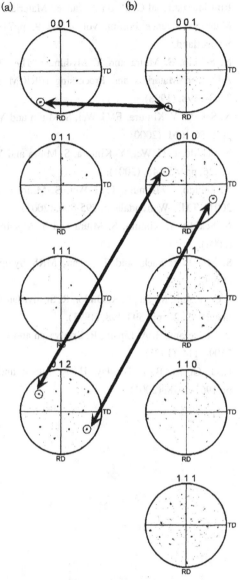

Figure 3 Discrete pole figures of (a) Nb and (b) Nb$_5$Si$_3$ lamellae in a colony near primary Nb phase in heat-treated alloy, respectively.

Rim International Conf. on Advanced Materials and Processing (PRICM5)" Proceedings, Materials Science Forum, vol. 475-479, pp717-720, (2005), Trans. Tech. Publishing, Switzerland.

5. N. Sekido, S. Miura and Y. Mishima, "The Third Pacific Rim International Conf. on Advanced Materials and Processing (PRICM 3)" Proceedings, TMS, Warrendale, pp. 2393-2398, (1998).

6. N. Sekido, Y. Kimura, F.-G. Wei, S. Miura and Y. Mishima, J. Japan Inst. Metals, vol. 64, pp.1056-1061, (2000).

7. N. Sekido, F.-G. Wei, Y. Kimura, S. Miura and Y. Mishima, Trans. Mat. Res. Soc. Japan, vol.26, pp145-148, (2001).

8. N. Sekido, Y. Kimura, F.-G. Wei, S. Miura and Y. Mishima, "Structural Intermetallics 2001", TMS, Warrendale, pp795-800, (2001).

9. N. Sekido, Y. Kimura, S. Miura and Y. Mishima, Mat. Trans., vol. 45, pp. 3264-3271, (2004).

10. S. Miura, Y. Saeki and T. Mohri, MRS Symp. Proc., vol.552, pp. KK6.9.1-KK6.9.5, (1999).

11. B. Cockeram, M. Saqib, R. Omlor, R. Srinivasan, L. E. Matson and I. Weiss, Scripta metall. mater, vol. 25, pp. 393-398, (1991).

12. B. Cockeram, H. A. Lipsitt, R. Srinivasan and I. Weiss, Scripta metall. mater., vol. 25, pp. 2109-2114, (1991).

13. R. J. Grylls, B. P. Bewlay, H. A. Lipsitt and H. L. Fraser, Phil. Mag. A, vol. 81, pp.1967-1978, (2001).

Mater. Res. Soc. Symp. Proc. Vol. 842 © 2005 Materials Research Society   S2.11

Microstructures of LENS™ Deposited Nb-Si Alloys

Ryan R. Dehoff, Peter M. Sarosi, Peter C. Collins, Hamish L. Fraser and Michael J. Mills
Materials Science and Engineering Department, The Ohio State University
Columbus , OH 43201, U.S.A.

ABSTRACT

Nb-Si "in-situ" metal matrix composites consist of Nb_3Si and Nb_5Si_3 intermetallic phases in a body centered cubic Nb solid solution, and show promising potential for elevated temperature structural applications. Cr and Ti have been shown to increase the oxidation resistance and metal loss rate at elevated temperatures compared to the binary Nb-Si system. In this study, the LENS™ (Laser Engineered Net Shaping) process is being implemented to construct the Nb-Ti-Cr-Si alloy system from elemental powder blends. Fast cooling rates associated with LENS™ processing yield a reduction in
· microstructural scale over conventional alloy processes such as directional solidification. Other advantages of LENS™ processing include the ability to produce near net shaped components with graded compositions as well as a more uniform microstructure resulting from the negative enthalpy of mixing associated with the silicide phases. Processing parameters can also be varied, resulting in distinct microstructural differences. Deposits were made with varying compositions of Nb, Ti, Cr and Si. The as-deposited as well as heat treated microstructures were examined using SEM and TEM techniques. The influence of composition and subsequent heat treatment on microstructure will be discussed.

INTRODUCTION

Nb-Ti-Cr-Si Alloy System

The Nb-Si binary systems have the potential to replace Ni-based superalloys as high temperature structural applications. They offer high strength and stiffness at elevated temperatures while maintaining room temperature toughness. The strength in these alloys is achieved by the use of niobium silicides such as Nb_3Si and Nb_5Si_3; however, the silicides in monolithic form are very brittle. When the silicides are incorporated into a matrix phase of Nb to increase the overall toughness, optimal performance at both high and low temperatures may be achieved.

The Nb-Si system has been shown to have creep rates comparable to those of Ni-based superalloys [1].The main drawback to the binary Nb-Si system is catastrophic failure under oxidizing environment. Additions of Ti and Cr have been shown to increase oxidation resistance and decrease the metal loss rate at elevated temperatures. Alloys are typically processed via directional solidification due to the high melting temperature of the constituents. Cooling rates in during direction solidification are relatively slow, resulting in coarse microstructures. Alloying additions have been used to limit grain growth and reduce the size scale of the resulting microstructure.

LENS™ Processing

The starting material for the LENS™ processing technique can be either pre-alloyed powder or an elemental powder blend. Because elemental powder blends are readily available and typically more cost effective than pre-alloyed powder, they were used in this study.

The metal powder flows from a hopper through four different nozzles, all of which converge at a single point. A high-powered laser is focused at the powder convergence point which supplies enough power to melt the powder. This molten powder forms what is termed a melt pool. To build up a single layer the LENS™ rasters the melt pool on the surface of a substrate. Another layer is then subsequently rastered on top of the previous layer which builds the deposit in the z direction. Figure 1 shows a schematic of the LENS™ deposition process including several of the processing parameters.

The processing parameters include the laser power, the travel speed of the laser and resulting melt pool, the hatch width (spacing in the x-y plane), and the layer thickness (distance moved in the z direction between individual layers). These terms are used to calculate the energy density during deposition with more energy supplied during deposition having a higher energy density. Processing parameters must be tailored for a specific alloy composition because too high of an energy density can result in vaporization of lower melting elements resulting in porosity and too low of an energy density can result in all of the powder not being fully melted.

The processing parameters can be varied in order to change the cooling rate which effects the microstructure of the deposit. The substrate material as well as size can also have an effect on cooling rate and it has been shown that a majority of the energy input during the deposition process is conducted away by the substrate [2].

The composition of the alloy can also play a role in the resulting microstructure of the deposition. In particular, Nb-Si and Ti-Si both have large enthalpies of mixing due to the formation of a silicide phase. It has been shown that large negative enthalpies of mixing can add heat to the deposit and results in a more homogeneous microstructure [3].

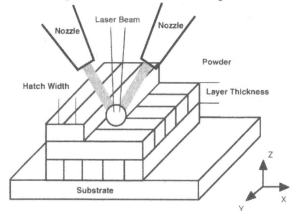

Figure 1: Schematic of the LENS™ processing technique [3].

PROCEDURE

Elemental powder blends were made with nominal compositions of Nb-29Ti-6Cr-5Si and Nb-26Ti-6Cr-15Si (alloy compositions are in atomic percent). The mixtures were blended using a Glen Mills Turbula® Shaker-Mixer for one hour and then loaded into separate powder feeders on the LENS machine. The pillars were cylindrical in shape with a nominal diameter of 0.35 inches. Pillar 1 was made using a laser power of 300 Watts, a hatch width of 0.0075 inches, a layer spacing of 0.005 inches, and a laser on feed rate of 12.5 inches per minute. The resulting energy density for these parameters was 38 MJ/in^3. These processing parameters remained constant through out the deposition. At the start of the deposition, powder feeder 1 (the powder feeder containing the low Si alloy described above) was operating at 100% and powder feeder 2 (the powder feeder containing the high Si alloy described above) was operating at 0%. The pillar was deposited for approximately 0.2 inches, and then the ratio between powder feeders was changed to 80% feeder 1, 20% feeder 2 and allowed to deposit for another 0.2 inches. After every 0.2 inches of deposition, feeder 1 was reduced by 20% while feeder 2 was increased 20%. Once all of the powder was coming from feeder 2, the deposit was continued at constant composition for about 0.5 inches. The Pillar 2 was deposited using a laser power of 300 Watts, a hatch width of 0.0075 inches, a layer spacing of 0.005 inches, and a laser on feed rate of 15 inches per minute. The resulting energy density for Pillar 2 was 32 MJ/in^3. The composition of Pillar 2 was graded in the same manner as Pillar 1.

The pillars were cut from the substrate using a Silicon carbide blade. Each individual pillar was sectioned in half with the cutting direction parallel to the deposition direction. Half of Pillar 1 and Pillar 2 were heat treated in vacuum at 1450° C for 50 hours and then furnace cooled. All of the samples were subsequently polished using Silicon carbide paper down to 1200 grit and final polished in colloidal silica.

TEM foils were extracted using a FEI Strata DB235 dual beam focused ion beam (FIB). Samples were examined using a FEI Sirion field emission gun scanning electron microscope and a Phillips Tecnai TF20 scanning transmission electron microscope.

RESULTS AND DISCUSSION

Standardless EDS analysis showed both Pillars 1 and 2 had the approximate composition of 6-Cr, a Si composition that varied from 5 to 15, and a Nb to Ti ratio of 2:1. Neither pillar exhibited evidence of porosity or unmelted powder. This implies that both energy densities are appropriate for this alloy system. However, because the energy densities affect the microstructure, the resulting microstructures from each energy density were further examined.

Microhardness indents were performed along the length of the graded composition of the non-heat-treated samples. Microhardness values ranged from slightly above 400 Vickers at the 5 % Si composition to a maximum of over 800 Vickers at the 15% Si. There is a direct relationship of hardness to the Si composition. Because of the very low Si solubility in Nb and Ti, almost all of the Si forms a silicide phase. As more Si is added, more silicide forms, resulting in increased hardness. However, there was also variation in hardness in the region with constant Si composition.

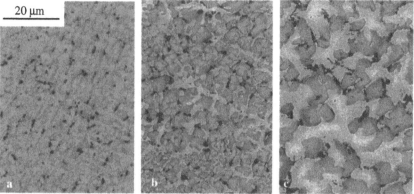

Figure 2: Microstructure of Pillar 1 in the as-deposited condition.

The microstructure of the 15 % Si region is dendritic in nature and shows three distinct phases present. The light phase is the Nb-Ti solid solution matrix, the gray phase is the silicide and a dark phase which will be discussed later. Figure 2a was taken at the very top of the deposit. The microstructure shown in Figure 2b was taken further down the deposit, but still where the composition was 15 % Si. Figure 2c shows the microstructure from one of the first 15% Si layers deposited. These three images indicate that as additional material is deposited, initial layers that were deposited undergo a significant heat treatment and coarsening due to the deposition of additional layers.

The corresponding micrographs of the heat-treated microstructures are shown in Figure 3. The heat treatment was done under vacuum at 1450° C for 50 hours followed by a furnace cool. The microsgraphs in Figure 3 are all at the same magnification. The results of the heattreatments indicate that there is some coarsening of the finer microstructure such as that in Figure 3a, but little coarsening of the microstructure bwetween Figures 2c and 3c. However, there does seem to be a change from a dendritic

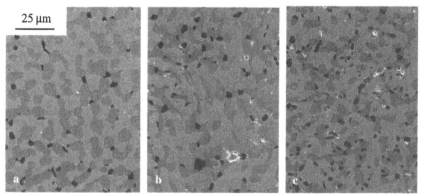

Figure 3: Microstructures of Pillar 1 after a heat treatment of 1450° C for 50 hours. a) is the top of the deposit, b) is the center, and c) is toward the beginning of the deposit.

microstructure to one that is more globular in nature. Cracks were also homogeneously distributed within the silicide phases in Pillar 1, which was attributed to the higher energy density. There was no evidence of cracking in Pillar 2.

Figure 4a is secondary electron image of the microstructure of Pillar 2 after the heat treatment. Figure 4b is a STEM image of a TEM foil extracted using a dual beam FIB. The matrix phase is labeled 1, the silicide phase labeled 2, and a dark phase labeled 3. Standardless EDAX quantification of the dark phase yields a composition of 2at%Si, 8at%Nb and 90at%Ti. Selected area diffraction patterns were analyzed and the crystal structure was determined to be FCC with a lattice parameter of 4.2 Å. This phase appears to be present both before and after the heat treatment. The area fraction of this phase was 5% in the as deposited condition, as well as the heat treated condition, indicating some degree of stability. This phase is not expected based on existing phase analysis for this alloy system [4,5,6], and it is presently under further investigation.

OIM was performed on the heat-treated portion of Pillar 2. Figure 5a is an inverse pole figure map for the Nb-rich phase while Figure 5b is an inverse pole figure map for the silicide phase. Each color indicates crystallographic regions that are within 5° of adjacent regions. It appears that the Nb-rich grains are actually larger than would be suggested by Figure 4a. The Nb grains also appear to be intertwined rather than globular as originally suggested. Both of these features are expected to be beneficial for limiting grain boundary sliding at elevated temperature. Typically, there are several silicide variants within a single Nb grain. This implies that solidification of the Nb phase occurs first, as expected from the phase diagram. The decomposition reactions of Nb_3Si → $Nb + Nb_5Si_3$ is still under investigation.

CONCLUSIONS

It has been shown that LENS™ processing is a viable way to process Nb-Ti-Si-Cr alloys from elemental powder blends. Several advantages of LENS™ processing have

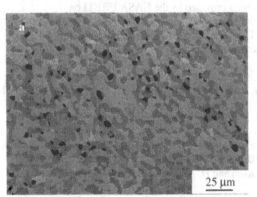 

Figure 4: a) microstructure of Pillar 2 after a heat treatment of 1450° C for 50 hours. b) STEM image of TEM foil cut using a duel beam FIB showing three distinct phases.

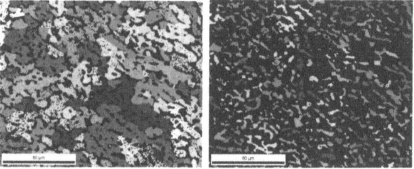

Figure 5: a) inverse pole figure for Nb in Pillar 2 after heat treatment. b) inverse pole figure for the silicide phase in Pillar 2 after heat treatment.

been demonstrated such as the ability to create near net-shaped components with graded compositions. Significant refinement of the microstructure was observed and various microstructures were achieved through variation of processing parameters. Heat treatments at 1450°C for 50 hours show only some coarsening of the finer microstructures. Although it appears that the microstructure changes from dendritic to globular during the heat treatment, OIM results show that the Nb phase consists of relatively large, intertwined grains. The microstructure after heat treatment consisted of 3 phases. A Nb-rich matrix and a Nb_5Si_3 type silicide which coincide with predictions from the phase diagram. A third phase was identified with a composition of Ti-8Nb-2Si having a FCC crystal structure. Additional investigation as to the nature of this phase is presently being conducted.

ACKNOWLEDGEMENTS:
Financial support for this work has been provided by the NASA URETI on Aeropropulsion and Power Technology under subgrant number E-16-V14-G1. The authors would like to thank Dr. R. D. Noebe of NASA Glenn Research Center for performing the high temperature heat treatments.

REFERENCES:
1. M.R. Jackson, B.P. Bewlay, R.G. Rowe, D.W. Skelly, and H.A. Lipsitt, JOM, Vol 48 No 1, 39-44, (1996).
2. W. Hofmeister, M. Griffith, M. Ensz, J. Smugeresky, JOM, Vol 53 No 9,30-34, (2001).
3. K. Schwendner, R. Banerjee, P. Collins, C. Brice and H. Fraser, Scripta Materialia A, Vol. 45 No. 10, 1123-1129 (2001).
4. B. P. Bwelay and M. R. Jackson and H. A. Lipsitt, J. Phase Equilibria, Vol. 18, 264 (1997)
5. P. R. Subramanian, M. G. Mendiratta and D. M. Dimiduk, High Temperature Silicides and Refractory Alloys, Mats. Res. Soc. Sym. Proc., Vol. 322, 491 (1994).
6. J.C. Zhao, M.R. Jackson and L.A. Peluso, Mat Sci and Eng. A, 372, 21-27, (2003)

Microstructures and Mechanical Properties in Ni_3Si-Ni_3Ti-Ni_3Nb-Based Multi-Intermetallic Alloys

T. Takasugi, K. Ohira and Y. Kaneno
Department of Metallurgy and Materials Science, Graduate School of Engineering, Osaka Prefecture University, 1-1 Gakuen-cho, Sakai, Osaka 599-8531, Japan

ABSTRACT

Microstructure, high-temperature tensile deformation and oxidation property of Ni_3Si-Ni_3Ti-Ni_3Nb multi-phase intermetallic alloys with a microstructure consisting of $L1_2$, $D0_{24}$ and $D0_a$ phases were investigated. The tensile stress as well as the tensile elongation of these multi-phase alloys increased with increasing Si content, i.e. the volume fraction of $L1_2$ phase in the wide range of test temperatures. 50-ppm boron addition to these multi-phase intermetallic alloys resulted in increased tensile stress and tensile elongation. The multi-phase intermetallic alloy with a high Si content had good oxidation resistance, and also the boron addition to this alloy resulted in enhanced oxidation resistance. From an overall evaluation of the properties examined, it was shown that the multi-phase intermetallic alloy, which has a high Si content and is composed of $L1_2$ matrix dispersed by $D0_{24}$ and $D0_a$ phases, had the most favorable properties as high-temperature mechanical and chemical materials.

INTRODUCTION

High-temperature structural materials (e.g. super alloys) that are widely utilized today are composed of a two-phase microstructure consisting of γ (Ni solid solution) and γ' ($L1_2$ phase). Also, recent study suggests that multi-phase intermetallic alloys can provide a good balance of room-temperature ductility, high-temperature strength and oxidation resistance, as recognized from recent development of e.g. TiAl-based intermetallic compounds consisting of γ and α_2 phases [1,2]. A group of Ni_3Si, Ni_3Ti and Ni_3Nb which are called as geometrically close packed (GCP) structures (phases) have been of great interest as a constituent phase in advanced intermetallic compounds as well as a strengthener in Ni-based superalloys. They generally show high thermal, chemical and microstructural stabilities, and also attractive mechanical properties such as strength anomaly at high temperatures as well as reasonable deformability at low temperatures owing to their relatively simple crystal structures [3-9]. Ni_3Si, Ni_3Ti and Ni_3Nb have $L1_2$ structure with a lattice parameter of a=0.3497 nm, $D0_{24}$ structure with a tetragonal structure with a lattice parameter of a=0.5101 nm and c=0.8307 nm, and $D0_a$ structure with an ordered structure based on h.c.p. lattice with a parameter of a=0.5106 nm, b=0.4251 nm and c=0.4553 nm, respectively.

An isothermal Ni_3Si-Ni_3Ti-Ni_3Nb pseudo-ternary phase diagram at 1323K has been reported by the present authors [10]. The recent revised pseudo-ternary phase diagram keeping Ni content 79.5at.% (Fig. 1) showed that there was a three-phase region of $Ni_3Si(L1_2)$-$Ni_3Ti(D0_{24})$-$Ni_3Nb(D0_a)$ [11]. The present study focuses on the three-phase microstructure composed of $Ni_3Si(L1_2)$-$Ni_3Ti(D0_{24})$-$Ni_3Nb(D0_a)$. We investigate the relation between microstructure, and mechanical and chemical (oxidation) properties in multi-phase intermetallic alloys based on Ni_3Si-Ni_3Ti-Ni_3Nb pseudo-ternary alloy system. It will be shown that the multi-phase intermetallic alloy with a high Si content, i.e. a high volume fraction of Ni_3Si ($L1_2$ phase) dispersed by the other phases has favorable overall properties among the alloys examined in this study.

EXPERIMENTAL PROCEDURES

The used alloy compositions are plotted in the form of Ni_3Si-Ni_3Ti-Ni_3Nb pseudo-ternary phase diagram [11](Fig.1). An identical 79.5 at.%Ni concentration was kept for all the prepared quaternary alloys. The reason why such an off-stoichiometric composition is adopted in this study is that 79.5 at.%Ni concentration permits stable $L1_2$ Ni_3Si phase [12] and the largest $L1_2$ phase field in Ni_3Si-Ni_3Ti pseudo-binary section [13]. Also, some alloy compositions were doped with 50ppm boron. All the button ingots were homogenized in a vacuum at 1323K for 2 days, followed by a furnace cooling to room temperature.

Metallographic observation was performed by optical microscopy (OM) and scanning electron microscopy (SEM) attached with wavelength- or energy-dispersive spectroscopy (WDS or EDS). The determination of the constituent intermetallic phases was conducted on the basis of SEM electron-probe analysis combined with X-ray diffraction (XRD).

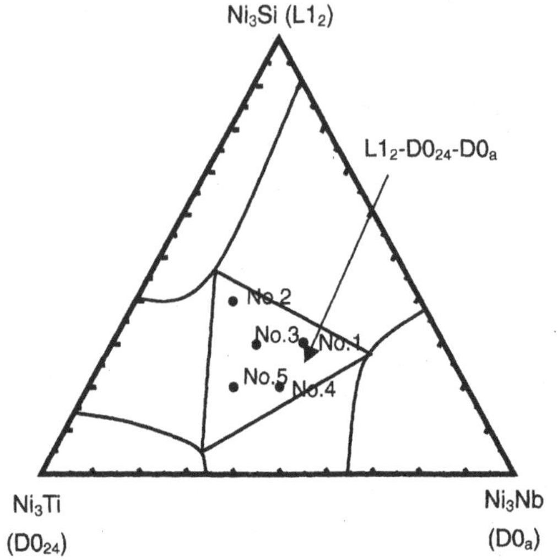

Figure 1. An isothermal Ni_3Si-Ni_3Ti-Ni_3Nb pseudo-ternary phase diagram at 1323K keeping Ni content 79.5at.%. Alloy compositions used in this study are plotted.

Tension tests in the temperature range between room temperature and 1173K were conducted in a vacuum degree of approximately 1.5×10^{-3} Pa within a metal tube. The tensile specimens with a gauge dimension $2 \times 1 \times 10$ mm^3 were cut from the homogenized button ingots. A nominal strain rate used in the tensile tests was 1.65×10^{-4} s^{-1}. Also, the fracture surfaces of the tensile deformed specimens were examined by a scanning electron microscope (SEM).

Oxidation property was evaluated using the specimens with a dimension of $2 \times 2 \times 2$mm^3. Isothermal oxidation tests were performed in a static laboratory air at 1273K until 48h, using TG-DTA. The oxidized surfaces were characterized by XRD using Ni-filtrated CuK_α radiation, and also by SEM-electron probe microscopic analysis (EPMA).

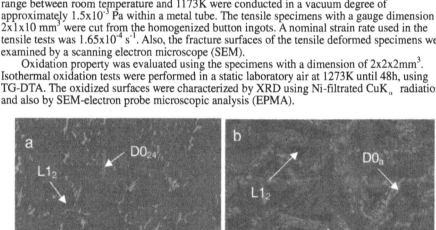

Figure 2. Back scattered (BS)-SEM microstructures of alloy (a) No.2 and (b) No.4 homogenized at 1323K for 2 days. Note that the constituent phases of $L1_2$, DO_{24} and DO_a are imaged as black, gray and white colors, respectively.

292

RESULTS AND DISCUSSION

Microstructure

All the alloys more or less exhibit dendrite microstructure, resulting from the influence of heat flow from bottom to top of the button ingot. All the alloys are also composed of a three-phase microstructure of $L1_2$-DO_{24}-DO_a but show largely different microstructures depending on alloy compositions. Alloy Nos. 1 through 3 with high Si contents exhibit a microstructure composed of $Ni_3Si(L1_2)$ grains dispersed by relatively large $Ni_3Nb(DO_a)$ dispersions and precipitated by plate-like or needle-like $Ni_3Ti(DO_{24})$ phase. As an example, Fig. 2a shows back scattering (BS)-SEM microstructure of alloy No.2. The plate-like or Widmansttaten-like $Ni_3Ti(DO_{24})$ phase was present in $L1_2$ matrix. This result suggests that DO_{24} phase is precipitated by a solubility change by temperature, and therefore could be more finely dispersed by adopting appropriate solution and precipitation heat treatment temperature and time. On the other hand, alloy No. 5 with a highest Ti content exhibits the microstructure composed of large columnar $Ni_3Ti(DO_{24})$ grains and $Ni_3Si(L1_2)$ grains. In the latter grains, a number of the $Ni_3Ti(DO_{24})$ precipitates were observed. Alloy No. 4 exhibits the microstructure composed of lath-type $Ni_3Ti(DO_{24})$ grains dispersed by $Ni_3Nb(DO_a)$ phase, and a small amount of isolated Ni_3Si ($L1_2$) phase (Fig. 2b). The 50ppm boron doping did not change the microstructures of the undoped alloys, regardless of alloy composition.

Mechanical properties

Figure 3 shows tensile fracture stress and tensile elongation as a function of test temperature for all the alloys tested in this study. The behavior is divided to two groups: alloy Nos. 1 through 3 with high Si contents (i.e. high volume fractions of $L1_2$ phase) showed yielding and subsequent plastic elongation at almost whole test temperatures, while alloy Nos. 4 and 5 with low Si contents (i.e. low volume fractions of $L1_2$ phase) did not show yielding and subsequent plastic elongation except at the highest temperature (1173K). Interestingly, the fracture stress observed for the former group of alloys was substantially high at low temperature, and showed a broad maximum at intermediate temperature, followed by a decrease at temperature beyond about 1000K. The tensile elongation ranged between 1 and 3 percent at room temperature, showed a maximum at intermediate temperature (~670K), and then displayed a minimum at (873K~1070K), followed by a rapid increase at temperature beyond about 1070K. Such a marked tensile ductility

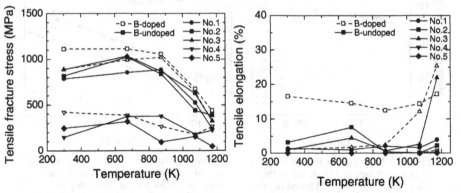

Figure 3. (a) Tensile fracture stress and (b) tensile elongation as a function of test temperature for all the alloys (Nos. 1 through 5) tested in this study.

may be attributed to a specific microstructure (composed of $Ni_3Si(L1_2)$) grains precipitated by plate-like or needle-like DO_{24} phase) shown in Fig. 2a.

The boron doping more or less increased the tensile fracture stress in the wide range of test temperatures as well as alloy composition (i.e. microstructure). In particular, the increase in the tensile fracture stress by boron doping was obvious for alloy No. 2. Regarding the tensile elongation, the effect of boron doping was substantial for alloy Nos. 2 and 3 with high Si contents (i.e. high volume fractions of $L1_2$ phase), but ineffective for alloy No. 4 with a low Si content (i.e. a high volume fraction of DO_{24}phase). The latter result indicates that the boron doping may be not so effective in improving the cohesive strength of the grain boundaries of DO_{24} phase (or DO_a phase) or the interfaces among the constituent phases of $L1_2$, DO_{24} and DO_a. For alloy Nos. 2 and 3 with high volume fractions of $L1_2$ phase, the boron doping generally altered the fracture mode from interfacial or grain boundary fracture to cleavage-like fracture, meaning that the boron enriches on the interfacial planes or the grain boundaries of the constituent $L1_2$ phase. For alloy No. 4 with a high volume fraction of DO_{24} phase, the fracture pattern of the born-doped alloy was cleavage and therefore the same as the boron-undoped alloy. Thus, the observed fractography is consistent with the behavior of the tensile fracture stress and the tensile elongation shown in Fig. 3. However, it is necessary in an extending study to know an optimum boron content resulting in more favorable mechanical properties.

Oxidation property

Figure 4 shows mass gain in air as a function of exposure time at 1273K for all the alloys tested in this study. These alloys drew similar mass gain curves: at an initial stage, a decelerated mass gain with increasing time occurred and at a later stage, a steady state-like mass gain occurred. The growth rate exponents of the oxidation scale except for an initial exposure time were 0.53, 0.44, 0.49, 0.48 and 0.43 for alloy Nos. 1 through 5, respectively. Thus, the growth rate exponents (~0.5) corresponding to a quasi-parabolic oxidation rate were observed for alloy Nos. 1, 3 and 4, while the growth rate exponents lower than 0.5 were observed for alloy Nos. 2 and 5. The highest oxidation resistance was observed for alloy No. 2 with a highest Si content (i.e. a highest volume fraction of $L1_2$ phase) and showed a growth rate exponent of 0.44.

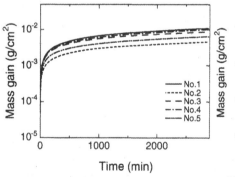

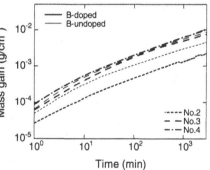

Figure 4. Mass gain in air as a function of exposure time at 1273K for all the alloys (Nos. 1 through 5) tested in this study.

Figure 5. Mass gain in air as a function of exposure time (plotted by a logarithmic scale) at 1273K for the boron-doped alloys (Nos. 2, 3 and 4) tested in this study. Note that the data of the boron-undoped alloys (Nos. 2, 3 and 4) are shown by thin lines, for reference.

Figure 5 shows mass gain in air as a function of exposure time (plotted by a logarithmic scale) at 1273K for the boron-doped alloys (Nos. 2, 3 and 4). Note that the data of the boron-undoped alloys (Nos. 2, 3 and 4) are drawn as thin lines, for reference. The growth rate exponents of the oxidation scale were basically similar to those of the boron-undoped alloys. It is however noted that the boron doping is effective in reducing the oxidation of alloy No. 2 but not so effective in reducing the oxidation of the other alloys. In other words, the boron doping is effective when the alloy shows a non-parabolic growth of the oxidization scale, i.e. is not controlled by bulk diffusion but by diffusion on the grain boundaries or the interfaces. This result suggests that the boron enriched on the free surface, the grain boundaries or the interfaces in the substrate materials (microstructures) has the effect of reducing the diffusion of the necessary atoms to form the oxidized layer.

Figure 6 shows SEM microstructure and elemental maps in the surface layers of alloy No. 2 oxidized in air at 1273K for 48h. It is shown that the scale layer is composed of TiO_2 layer, NiO layer and SiO_2 layer, in sequence from the free surface. Also, these three layers are basically composed of the continuous films. However, NiO layer contains a small amount of isolated TiO_2. Similarly, SiO_2 layer contains a small amount of isolated TiO_2. Also, the substrate beneath the continuous SiO_2 layer contains isolated SiO_2. Therefore, it is demonstrated from the elemental map (Fig. 6) and X-ray result that the superior oxidation resistance observed in alloy No.2 with a high Si content is attributed to the formation of the continuous SiO_2 layer. On the other hand, in alloy (e.g. No. 3) with a moderate Si content, the formed scale was substantially thicker than that in the former alloy, and composed of a mixture zone of NiO, SiO_2 and TiO_2 although the most outward layer was composed of the continuous TiO_2 layer and the NiO layer. Also, the scale formed in this alloy was occasionally peeled out. Consequently, it is concluded from the elemental map and X-ray result that the inferior oxidation resistance observed in alloy No. 3 is attributed to no formation of the continuous SiO_2 layer due to a poor Si content. Thus, it is concluded that the oxidation property does not depend on microstructure but on alloy composition (e.g. Si content).

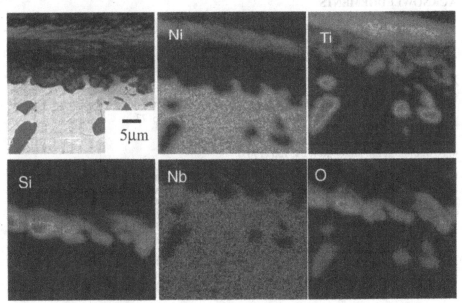

Figure 6. SEM microstructure and elemental maps in the surface layer of alloy No. 2 oxidized in air at 1273K for 48h.

SUMMARY AND CONCLUSIONS

As an overall evaluation of the properties examined, alloy No.2, which has a high Si content and exhibits a microstructure with $L1_2$ matrix dispersed by $D0_{24}$ and $D0_a$ phases, had the most favorable properties among the alloys tested in this study. In terms of density, corrosion property and cost, this alloy is also strongly preferable (or useful) to develop as a new type of high-temperature structural alloys. Therefore, an extending study is planned to confirm whether the precipitation strengthening by the second-phase in solid-state and the thermo-mechanical treatment are applicable or not.

Microstructure, high-temperature tensile deformation and oxidation property of the multi-phase intermetallic alloys, which were based on Ni_3Si-Ni_3Ti-Ni_3Nb pseudo-ternary alloy system and composed of $L1_2$, $D0_{24}$ and $D0_a$, were investigated. The following results were obtained from the present study.

(1) In terms of the volume fraction, the morphology and the size of the constituent phases, the microstructures of these multi-phase intermetallic alloys were largely dependent on alloy composition.

(2) The tensile fracture stress as well as the tensile elongation of these multi-phase intermetallic alloys increased with increasing Si content, i.e. the volume fraction of $L1_2$ phase in the wide range of test temperatures.

(3) 50-ppm boron addition to these multi-phase intermetallic alloys resulted in increased tensile stress and tensile elongation in the wide range of test temperature.

(4) It was found that the multi-phase intermetallic alloy with a high Si content had good oxidation resistance, accompanied with the continuous silicate (SiO_2) layer on the surface scale. Also, the boron doping to this alloy composition resulted in enhanced oxidation resistance.

ACKNOWLEDGEMENTS

This work was supported in part by the Grant-in-aid for Scientific Research (B) from the Ministry of Education, Culture, Sports and Technology.

REFERENCES

1. S. C. Huang and J. C. Chesnutt: in *Intermetallic Compounds, Volume 2, Practice,* ed. J. H. Westbrook and R. L. Fleischer, (John Wiley and Sons, 1995), p. 73.
2. Y. -W. Kim, *J. Metals* **41**, 24 (1989).
3. D. P. Pope and S. S. Ezz, *Int. Mater. Rev.* **29**,136 (1984).
4. N. S. Stoloff, *Int. Mater. Rev.* **34**, 153 (1989).
5. M. Yamaguchi and Y. Umakoshi, *Prog. Mater. Sci.* **34**, 1 (1990).
6. T. Suzuki, Y. Mishima and S. Miura, *ISIJ Int.* **29**, 1 (1989).
7. K. Hagiwara, T. Nakano and Y. Umakoshi, *Mat. Res. Soc. Symp. Proc.* **753**, 357 (2003).
8. K. Hagiwara, T. Nakano and Y. Umakoshi, *Act Mater.* **48**, 1469 (2000).
9. K. Hagiwara, T. Nakano and Y. Umakoshi, *Acta Mater.* **51**, 2623 (2003).
10. K. Tomihisa, Y. Kaneno and T. Takasugi, *Intermetallics* **12**, 317 (2004).
11. K. Ohira, Y. Kaneno, H. Tsuda and T. Takasugi, to be published.
12. Y. Oya and T. Suzuki, *Z Metallkunde* **74**, 21 (1983).
13. T. Takasugi, D. Shindo, O. Izumi and M. Hirabayashi, *Acta Metall. Mater.* **38**, 739 (1990).

Mater. Res. Soc. Symp. Proc. Vol. 842 © 2005 Materials Research Society   S5.30

Experimental Investigation and Thermodynamic Modeling of Phase Equilibria in the Hf-Ti-Si System

Y. Yang[1], B.P. Bewlay[2], M.R. Jackson[2] and Y.A. Chang[1]

[1]University of Wisconsin-Madison,Wisconsin 53706. USA.
[2]General Electric Global Research,Schenectady, New York 12301. USA

ABSTRACT

Phase equilibria in ternary Hf-Ti-Si alloys were studied in the as-solidified and heat treated conditions using scanning electron microscopy, x-ray diffraction, and electron beam microprobe analysis. Selected solid-solid phase equilibria at 1350°C and a partial liquidus projection of the Hf-Ti-Si system at the metal rich end of the phase diagram were established. These data were then used to develop a thermodynamic description of the Hf-Ti-Si system using the CALPHAD (CALculation of PHAse Diagram) approach. The calculated isothermal section at 1350°C and the liquidus projection can satisfactorily account for the available experimental phase equilibria data and solidification paths. Both the calculations and the experimental data suggested that the metal-rich end of the ternary phase diagram possesses one transition reaction: $L + (Hf,Ti)_5Si_3 \rightarrow Hf(Ti)_2Si + \beta(Hf,Ti,Si)$.

INTRODUCTION

Directionally solidified in-situ composites based on niobium and niobium-based silicides are presently being developed for high-temperature structural applications. There has been extensive work on composites generated from binary Nb-Si alloys [1-5], as well as those with additions such as Ti, Hf, Cr and Al. Hf and Ti are important alloying additions because they can improve oxidation resistance and strength [1, 2, 5, 6]. However, there is little previous knowledge of phase equilibria in the Hf-Ti-Si system; this is required for determination of higher order Nb based systems that contain Hf and Ti. This paper first presents the experimental details for the solid-solid phase equilibria at 1350°C and the partial liquidius projection at the metal rich end of the Hf-Ti-Si system, and then describes thermodynamic modeling for the Hf-Ti-Si system using the CALPHAD approach. Finally, comparisons between the calculated results and experimental observations will be presented.

EXPERIMENTAL DETAILS

The alloys for this study were directionally solidified using cold crucible directional solidification [1, 4] after triple melting the starting charges from high purity elements (>99.99%). The directional solidification procedure has been described in more detail previously [1, 6]. Mass losses were measured after preparation of each sample and they were found to be less that 0.1 wt

%. The interstitial levels of the C, H, N and O in the Hf and Ti used were less than 250 weight ppm, respectively.

All of the samples were examined using scanning electron microscopy (back scatter electron (BSE) imaging) and energy dispersive spectrometry (EDS). Electron beam microprobe analysis (EMPA), x-ray diffraction (XRD), and automated electron back scattering diffraction analysis in the SEM (EBSD) were also performed on selected samples. High purity Ti, Hf, and Si were used as standards and conventional matrix corrections (Z, A, and F) were used to calculate the wt. % compositions from measured x-ray intensities. The Hf-rich silicides were identified using both EMPA and XRD data.

The liquidus surface projection of the metal-rich end of the proposed Hf-Ti-Si phase diagram is shown in Figure 1. The substantiating data and microstructural analysis are presented in elsewhere [7]. Three primary solidification phases are shown in the present description of the liquidus surface: a bcc solid solution denoted by β(Hf,Ti,Si), Hf(Ti)$_2$Si, and (Hf,Ti)$_5$Si$_3$. In the present paper, the Hf$_5$Si$_3$ with Ti in solid solution, and the Ti$_5$Si$_3$ with Hf in solid solution, are referred to as (Hf,Ti)$_5$Si$_3$, and the Hf$_2$Si with Ti in solid solution is referred to as Hf(Ti)$_2$Si. The Hf-Si solid solution in the binary alloys is referred to as Hf(Si). The Ti$_5$Si$_3$ and Hf$_5$Si$_3$ both have the hP16 crystal structures, and they were identified to form continuous solid solution at high temperatures. The Hf$_2$Si has the tI12 crystal structure.

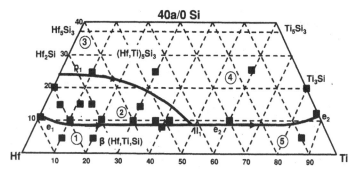

Figure 1. Schematic diagram showing the proposed projection of the Hf-Ti-Si liquidus surface. The projection shows the peritectic ridge, p$_1$, the eutectic valleys, e$_1$, and e$_2$, and the invariant reaction, II$_1$. The actual compositions investigated are also shown.

Four phases were observed in the samples heat treated at 1350°C for 100h. These were β(Hf,Ti,Si), (Hf,Ti)$_5$Si$_3$, Hf(Ti)$_2$Si, and α(Hf,Ti,Si). The EPMA measurements of these samples heat treated at 1350°C are listed in Table I. Again, more detailed data and microstructural analysis are given elsewhere [7]. The following one three-phase equilibria and three two-phase equilibria were observed: β(Hf,Ti,Si)+(Hf,Ti)$_5$Si$_3$+Hf(Ti)$_2$Si, β(Hf,Ti,Si)+(Hf,Ti)$_5$Si$_3$, β(Hf,Ti,Si)+ Hf(Ti)$_2$Si and α(Hf,Ti,Si)+Hf(Ti)$_2$Si.

Table I. EPMA measurements on selected annealed samples at 1350°C

Bulk Compositions	Phases	Phase Compositions		
		Hf	Ti	Si
Hf-60Ti-10Si	β(Hf,Ti,Si)	27.6	1.0	71.4
	(Hf,Ti)₅Si₃	27.7	36.9	35.4
Hf-85Ti-5Si	β(Hf,Ti,Si)	8.9	0.8	90.3
	(Hf,Ti)₅Si₃	14.7	37.2	48.1
Hf-10Ti-5Si	α(Hf,Ti,Si)	87.8	2.7	9.5
	Hf(Ti)₂Si	65.0	33.3	1.7
Hf-10Ti-10Si	α(Hf,Ti,Si)	83.5	2.6	13.9
	Hf(Ti)₂Si	64.9	33.3	1.8
Hf-30Ti-25Si	β(Hf,Ti,Si)	39.0	1.3	59.7
	Hf(Ti)₂Si	57.2	33.0	9.8
	(Hf,Ti)₅Si₃	34.7	36.2	29.1
Hf-30Ti-16Si	β(Hf,Ti,Si)	53.8	1.6	44.6
	Hf(Ti)₂Si	59.8	32.9	7.4
Hf-20Ti-5Si	β(Hf,Ti,Si)	72.8	3.3	23.9
	Hf(Ti)₂Si	64.7	31.9	3.4

THERMODYNAMIC MODELING OF THE HF-TI-SI SYSTEM

In the present modeling, the thermodynamic parameters in the Hf-Si, Si-Ti and Hf-Ti systems are taken from Yan et al.[8], Seifert et al.[9] and Ansara et al.[10], respectively. There are two types of ternary solid solution phases in the Hf-Ti-Si system. The first type is based on the structures of the solid solutions: the β(Hf,Ti,Si) phase is based on bcc Ti and the α(Hf,Ti,Si) is based on hcp Ti. The second type of solid solution is based on intermetallic compounds such as Hf(Ti)₂Si and (Hf,Ti)₅Si₃. The Gibbs energy of the first type of ternary solution is described by,

$$G_m^\varphi = x_{Hf} \, {}^0G_{Hf}^\varphi + x_{Si} \, {}^0G_{Si}^\varphi + x_{Ti} \, {}^0G_{Ti}^\varphi + RT(x_{Hf} \ln x_{Hf} + x_{Si} \ln x_{Si} + x_{Ti} \ln x_{Ti}) + {}^{ex}G_m^\varphi \quad (1)$$

where x_{Hf}, x_{Si} and x_{Ti} are the mole fractions of Hf, Si and Ti, respectively. ${}^0G_{Hf}^\varphi$, ${}^0G_{Si}^\varphi$ and ${}^0G_{Ti}^\varphi$ are the φ phase molar Gibbs energies of the elements Hf, Si and Ti, respectively. ${}^{ex}G_m^\varphi$ is the excess Gibbs energy term of the φ phase. It is expressed by the following equation:

$$^{ex}G_m^\varphi = x_{Hf}x_{Si}L_{Hf,Si}^\varphi + x_{Hf}x_{Ti}L_{Hf,Ti}^\varphi + x_{Si}x_{Ti}L_{Si,Ti}^\varphi + x_{Hf}x_{Si}x_{Ti}L_{Hf,Si,Ti}^\varphi \quad (2)$$

$L_{Hf,Si}^\varphi$, $L_{Si,Ti}^\varphi$ and $L_{Hf,Ti}^\varphi$ are the interaction parameters for Hf-Si, Si-Ti and Hf-Ti, respectively.

The compound energy formalism was used to describe the second type of ternary solid solutions with a form of (Hf,Ti)ₚ(Si)q. The Gibbs energy per mole of atoms of the (Hf,Ti)ₚ(Si)q phase is described by the following equation:

$$G_{(Hf,Ti)_p Si_q}^\varphi = y_{Hf}^l \, {}^0G_{Hf:Si}^\varphi + y_{Ti}^l \, {}^0G_{Ti:Si}^\varphi + \frac{p}{p+q}RT(y_{Hf}^l \ln y_{Hf}^l + y_{Ti}^l \ln y_{Ti}^l) + y_{Hf}^l y_{Ti}^l L_{Hf,Ti:Si}^\varphi \quad (3)$$

in which y'_{Hf} and y'_{Ti} are the site fractions of Hf and Ti on the first sublattice, respectively. $^0G^\varphi_{Hf:Si}$ and $^0G^\varphi_{Ti:Si}$ are the Gibbs energies of the compounds Hf_pSi_q and Ti_pSi_q with the φ phase structure, respectively. $L^\varphi_{Hf,Ti:Si}$ represents the interaction between Hf and Ti in the first sublattice with only Si present in the second sublattice.

In the present modeling, the Gibbs energies of $\beta(Hf,Si,Ti)$, $\alpha(Hf,Si,Ti)$ and liquid were described by Eq. (1). Hf_5Si_3 and Ti_5Si_3 are treated as one phase denoted by $(Hf,Ti)_5Si_3$ due to their similar crystal structures. Similarly, Hf_5Si_4 and Ti_5Si_4 are treated by $(Hf,Ti)_5Si_4$, and $HfSi$ and $TiSi$ by $(Hf,Ti)Si$. The Gibbs energies of $Hf(Ti)_2Si$, $(Hf,Ti)_5Si_3$, $(Hf,Ti)_5Si_4$ are described by Eq. (3). Ti_3Si, $TiSi_2$, $HfSi_2$, and Hf_3Si_2 are treated as binary compounds due to the lack of solubility information. None of the phases described above have experimental thermodynamic data. All parameters were optimized based on the available phase diagram data shown in Fig. 1 and Table 1.

The optimization was performed in the parrot module of Thermo-Calc[11], and Pandat[12] was used to calculate the phase diagrams and solidification paths.

COMPARISONS BETWEEN THE CALCULATION AND EXPERIMENTAL RESULTS

Isothermal Section at 1350°C

The calculated isothermal section at 1350 °C is shown in Figure 2 together with the EPMA measurements of the phases and the bulk compositions of the alloys. The solid squares denote the bulk compositions of alloys and the circles denote the phase compositions. The calculated equilibrium phases presented are generally in good agreement with the microstructural observations and EPMA measurements. The calculation predicts the existence of two three-phase equilibria: $Hf(Ti)_2Si+\alpha(Hf,Ti,Si)+\beta(Hf,Ti,Si)$ and $L+\beta(Hf,Ti,Si)+(Hf,Ti)_5Si_3$. Further experiments are desired to confirm these two three-phase equilibria. The calculated solubility of Ti

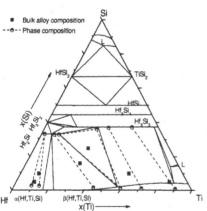

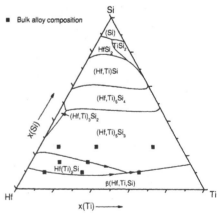

Figure 2. Calculated isothermal section of Hf-Ti-Si at 1350°C compared with experimental data

Figure 3. Calculated liquidus projection along with experimental bulk alloy compositions

in Hf(Ti)$_2$Si is 9.1%, which is in reasonable agreement with the measured value ~10%. This calculation predicts the solubility of Hf in (Hf,Ti)$_5$Si$_3$ is ~48%. This calculation also predicts the Si concentration in β(Hf,Ti,Si) is higher than that in α(Hf,Ti,Si), which contradicts the EPMA measurements listed in Table 1. Further experimental work is needed to clarify this contradiction.

Liquidus projection

The calculated liquidus surface projection is shown in Figure 3 together with the compositions of the investigated alloys. The calculated liquidus surface for Si concentrations from 0 to 40% is made up of 3 primary phase regimes: (Hf,Ti)$_5$Si$_3$, Hf(Ti)$_2$Si, and β(Nb,Hf,Si). There is one type II invariant four-phase reaction at the metal-rich end of the liquid surface, which corresponds to L+Hf(Ti)$_2$Si→(Hf,Ti)$_5$Si$_3$+β(Hf,Ti,Si). The calculated invariant reaction temperature (1509°C) agrees well with the experimental value (~1500°C). Further work is required to establish the liquid composition of the invariant reaction more precisely. The calculated results are considered to be acceptable considering the compromise to fitting both the temperature and the composition data. The calculated primary solidification regions are in general accordance with the experimental observations.

Solidification Simulation

Simulations of the solidification paths were performed for all the experimental alloys; an example is shown in Fig. 4a. These simulations assumed Scheil conditions [13] and were carried out in the solidification module of Pandat. All the simulated results show reasonable agreement with experimental observations. In Figure 4, the Hf-30Ti-25Si alloy is used as an example to show the comparison between the simulated and experimentally observed solidification paths. The solidification sequence of this alloy, as predicted by this calculation, is the following: the first phase to solidify is (Hf,Ti)$_5$Si$_3$, and then the liquid composition reaches the peritectic ridge of Liquid + (Hf,Ti)$_5$Si$_3$→Hf(Ti)$_2$Si, subsequently, the liquid composition traverses this preitectic ridge and moves to the eutectic valley between β(Hf,Ti,Si) and Hf(Ti)$_2$Si, and finally the invariant reaction of Liquid + Hf(Ti)$_2$Si→ (Hf,Ti)$_5$Si$_3$ + β(Hf,Ti,Si) is reached. All the predicted

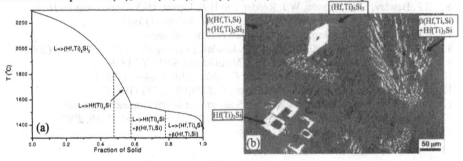

Figure 4. Comparisons between the calculated solidification paths, (a), and the experimentally observed as-cast microstructure, (b), for the Hf-30Ti-25Si alloy.

microstructural features can be found in the as-cast microstructure in Figure 4(b) (BSE image). The Hf-30Ti-25Si alloy contains a small volume fraction of primary (Hf,Ti)$_5$Si$_3$ dendrites (the dark grey phase), peritectic $Hf(Ti)_2Si$ (the lightest phase), interdendritic $\beta(Hf,Ti,Si)$+ $Hf(Ti)_2Si$ eutectic, and the final $(Hf,Ti)_5Si_3+\beta(Hf,Ti,Si)$ eutectic.

SUMMARY

A thermodynamic description of the Hf-Ti-Si system was developed by incorporating the phase equilibria data at 1350°C and the partial liquidus projection of the Hf-Ti-Si system at the metal rich end of the phase diagram. The calculated isothermal section at 1350°C and the liquidus projection can satisfactorily account for the experimental observations. Both the calculated phase diagrams and experimental results suggested that the metal-rich end of the ternary phase diagram possesses one transition reaction: $L + (Hf,Ti)_5Si_3 \rightarrow Hf(Ti)_2Si + \beta(Ti,Hf,Si)$.

ACKNOWLEDGEMENTS

The authors would like to thank D.J. Dalpe for the directional solidification. The authors would also like to thank Dr. D.A. Wark of R.P.I. for the EMPA measurements.

REFERENCES

1. B.P. Bewlay, M.R. Jackson, and P.R. Subramanian, Journal of Metals, 51(4), 32-36 (1999)
2. B.P. Bewlay, M.R. Jackson and H.A. Lipsitt, Metall. and Mater. Trans., 279, 3801-3808 (1996)
3. M.G. Mendiratta, J.J. Lewandowski and D.M. Dimiduk, Metall. Trans. 22A (1573-1581) (1991)
4. M.R. Jackson, B.P. Bewlay, R.G. Rowe, D.W. Skelly, and H.A. Lipsitt, J. of Metals, 48(1), 38-44 (1996)
5. P.R. Subramanian, M.G. Mendiratta and D.M. Dimiduk, Mat. Res. Soc. Symp. Proc., 322, 491-502 (1994)
6. B.P. Bewlay, M.R. Jackson, W.J. Reeder, and H.A. Lipsitt, MRS Proceedings on High Temperature Ordered Intermetallic Alloys VI, 364, 943-948 (1994)
7. B.P. Bewlay, M.R. Jackson, to be submitted to Z. Metallkunde.
8. Y. Yang, Y.A. Chang, J.-C Zhao and B.P. Bewlay, Intermetallics, 11(5), 407-415 (2003)
9. H. Seifert, H.L. Lukas and G. Petzow, Z. Metallkunde, 87(1), 2-13 (1996)
10. H. Bittermann, P. Rogl. Journal of Phase Equilibria, 18(1), 24-47 (1997)
11. B. Sundman, B. Jansson and J.O. Anderson, CALPHAD, 9, 153 (1985)
12. Pandat 4.0m-Phase Diagram Calculation Software for Multicomponent Systems, Computherm LLC, 437S, Yellowstone Dr., Suite 217, Madison, WI 53719, 2003
13. E. Scheil, Z. Metallkunde., 34, 70 (1942)

Mater. Res. Soc. Symp. Proc. Vol. 842 © 2005 Materials Research Society   S2.9

Role of Microstructure in Promoting Fracture and Fatigue Resistance in Mo-Si-B Alloys

J. J. Kruzic*
Department of Mechanical Engineering, Oregon State University, Corvallis, OR 97331, U.S.A.
J. H. Schneibel
Oak Ridge National Laboratory, Metals and Ceramics Division, Oak Ridge, TN 37831, U.S.A.
R. O. Ritchie
Department of Materials Science and Engineering, University of California, and Materials
Sciences Division, Lawrence Berkeley National Laboratory, Berkeley, CA 94720, U.S.A.

ABSTRACT

An investigation of how microstructural features affect the fracture and fatigue properties of a promising class of high temperature Mo-Si-B based alloys is presented. Fracture toughness and fatigue-crack growth properties are measured at 25° and 1300°C for five $Mo-Mo_3Si-Mo_5SiB_2$ containing alloys produced by powder metallurgy with α-Mo matrices. Results are compared with previous studies on intermetallic-matrix microstructures in alloys with similar compositions. It is found that increasing the α-Mo phase volume fraction (17 – 49%) or ductility (by increasing the temperature) benefits the fracture resistance; in addition, α-Mo matrix materials show significant improvements over intermetallic-matrix alloys. Fatigue thresholds were also increased with increasing α-Mo phase content, until a transition to more ductile fatigue behavior occurred with large amounts of α-Mo phase (49%) and ductility (i.e., at 1300°C). The beneficial role of such microstructural variables are attributed to the promotion of the observed toughening mechanisms of crack trapping and bridging by the relatively ductile α-Mo phase.

INTRODUCTION

Intermetallic based Mo-Si-B alloys have been targeted for high temperature turbine engine applications as potential replacements for nickel based superalloys. Two specific Mo-Si-B alloy systems developed by Akinc et al. [1-4] and Berczik [5,6] have received recent attention. While the former is composed entirely of intermetallic compounds, the latter utilizes the relatively ductile α-Mo phase to impart some ductility and fracture resistance to a three phase microstructure also containing Mo_3Si and Mo_5SiB_2 (T2). For any of these alloys to be successful, adequate resistance to oxidation, creep, fracture, and fatigue must be achieved; however, it is recognized that microstructural features which promote improvements in one property are often detrimental to others [7,8]. For example, while a continuous α-Mo matrix with high volume fraction may be beneficial to the fracture and fatigue behavior [9], this tends to compromise the oxidation and creep resistance [7,8,10-12]. Accordingly, a thorough understanding of how microstructure affects each property is needed so that appropriate trade-offs can be made in the optimization of these alloys. Consequently, this present paper seeks to characterize the specific mechanistic role of microstructure in determining the fracture and fatigue resistance of alloys based on the α-Mo, Mo_3Si, and T2 phases, with the objective of

* Corresponding author. Tel: +1-541-737-7027; fax: +1-541-737-2600.
E-mail address: jamie.kruzic@oregonstate.edu (J. J. Kruzic)

providing guidelines for optimizing the properties of this exciting new class of high-temperature structural materials.

EXPERIMENTAL DETAILS

Ground powders of composition Mo-20Si-10B (at.%) containing Mo_3Si (cubic A15 structure) and Mo_5SiB_2 (tetragonal $D8_1$ structure) intermetallic phases were vacuum-annealed to remove silicon from the surface and leave an α-Mo coating on each particle. These were hot-isostatically pressed in evacuated Nb cans for 4 hr at 1600°C and 200 MPa pressure giving final compositions with 7-15 at.% Si and 8-11 at.% B (balance Mo). Five alloys were produced differing in volume fraction of the α-Mo matrix (17 – 49%) and initial coarseness of the intermetallic particles, fine ≤45 μm, medium 45-90 μm, and coarse 90-180 μm. Alloys are designated as F34, M34, C17, C46, and C49, with the letter indicating the microstructural coarseness and the number giving the α-Mo volume percent; microstructures for each may be seen in Figures 1 and 2.

Resistance-curve (R-curve) fracture-toughness experiments were performed on fatigue pre-cracked, disk-shaped compact-tension DC(T) specimens (width 14 mm; thickness 3 mm). Samples were loaded monotonically in displacement control at ~1 μm/min until the onset of cracking. At 25°C, periodic unloads (~10-20% of peak load) were performed to measure the unloading back-face strain compliance, which was used to determine the crack length [13]. 1300°C tests were conducted in gettered argon using (direct-current) electrical potential-drop techniques to monitor crack length [14].

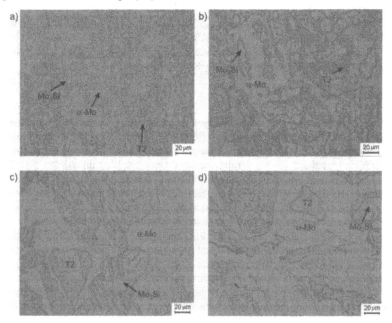

Figure 1. Microstructures of alloys (a) F34, (b) M34, (c) C17, and (d) C46. (C49 is seen in Figure 2b)

Fatigue-crack growth testing (25 Hz, sine waveform) was performed at 25° and 1300°C in identical environments in general accordance with ASTM Standard E647 [15] using computer-controlled servo-hydraulic testing machines at a load ratio R (ratio of minimum to maximum loads) of 0.1. Crack-growth rates, da/dN, were determined as a function of the stress-intensity range, ΔK, using continuous load-shedding to maintain a ΔK-gradient $(=1/\Delta K[d\Delta K/da])$ of ±0.08 mm^{-1} to achieve increasing or decreasing ΔK conditions, respectively. ΔK_{TH} fatigue thresholds, operationally defined at a minimum growth rate of 10^{-10}–10^{-11} m/cycle, were approached under decreasing ΔK conditions. Both fracture and fatigue testing was periodically paused to observe crack profiles using optical and scanning electron microscopy.

RESULTS AND DISCUSSION

R-curves for the five Mo-Mo$_3$Si-T2 alloys are plotted in terms of the stress intensity, K, and clearly indicate rising fracture toughness with crack extension (Figure 2a). Furthermore, there is a trend of increasing toughness with higher α-Mo volume fractions; indeed, alloys C46 and C49 had peak room-temperature toughnesses in excess of 20 MPa√m, i.e., up to seven times higher than that of monolithic molybdenum silicides [16,17]. Crack trapping and bridging by the α-Mo phase were identified as the toughening mechanisms responsible for this behavior (Figure 2b), with the effectiveness of these mechanisms rising with increasing α-Mo content. Experiments at 1300°C on alloys M34, C17, and C46 indicated that the fracture toughness improved at higher temperatures, which was associated with improved α-Mo ductility. The initiation toughness, K_i, which defines the beginning of the R-curve, rose ~65% for alloys M34 and C17, while alloy C46 experienced toughening at 1300°C to such a degree that large-scale crack blunting and deformation occurred (Figure 3a) and linear-elastic fracture mechanics was no longer a valid method for assessing the toughness using the present specimen size. Thus, K_i for alloy C46 at 1300°C is reasoned to be significantly larger than the 12.6 MPa√m measured for alloy M34.

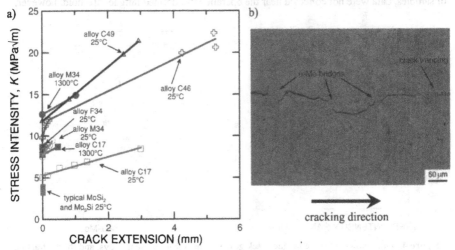

Figure 2. (a) shows R-curve behavior for Mo-Si-B alloys, while (b) shows the active toughening mechanisms, crack trapping and bridging by the α-Mo phase of alloy C49.

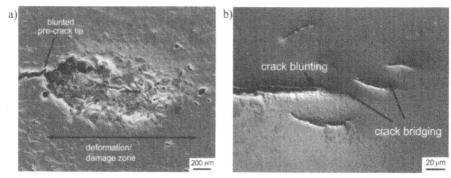

Figure 3. Crack blunting seen in alloys (a) C46 and (b) M34 after R-curve testing at 1300°C. Note the order of magnitude difference in scale between the two micrographs.

Crack blunting was also observed, to a much smaller degree, in the other alloys tested at 1300°C (Figure 3b). Thus, increases in toughness with temperature were attributed to the improved effectiveness of crack trapping due to the enhanced α-Mo ductility at elevated temperatures.

From the fatigue-crack growth results in Figure 4a, it is apparent that at 25°C the Paris-law exponents, m, are extremely high, >78 in all cases; such behavior is characteristic of brittle materials. Fatigue data were also collected for alloys M34 and C49 at 1300°C; although alloy M34 had a similarly high ΔK dependence at 1300°C, alloy C49 displayed a transition to more ductile fatigue behavior, with more than an order of magnitude decrease in the Paris-law exponent from 78 to 4. A Paris exponent of 4 is similar to what is expected for ductile metals, which typically have $m = 2 - 4$ [18]. ΔK_{TH} thresholds ranged between 5 and 9.5 MPa√m for the five alloys at 25°C (Figure 4a). At 1300°C, due to experimental difficulties and limited numbers of samples, data were not collected near the operationally-defined fatigue threshold; however,

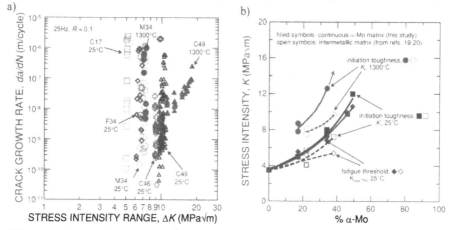

Figure 4. Plots showing (a) the fatigue-crack growth behavior for present alloys, and (b) the fracture toughness and fatigue threshold properties of the present α-Mo matrix alloys compared to that for intermetallic-matrix alloys of similar compositions from refs. [19,20].

based on extrapolation of the data in Figure 4a, the threshold for M34 is expected to be similar to that at 25°C, whereas data for alloy C49 suggests a decrease in the fatigue threshold at 1300°C.

Figure 4b compiles the present fracture and fatigue results along with those for intermetallic matrix Mo-Si-B alloys with similar compositions [19,20]. Here the fatigue thresholds are plotted as the maximum stress intensity, $K_{max,TH}$, along with the initiation toughness values, K_i. Peak toughnesses are not compared since steady-state, or plateau, values were not achieved due to inadvertent failure of specimens and/or limited specimen size. Figure 4b clearly illustrates that the fracture and fatigue properties of all the alloys improve with increasing α-Mo volume fraction. Furthermore, the fracture toughness properties are enhanced with increasing α-Mo ductility, as evidenced by the higher toughness values at 1300°C for all the alloys. Note that a given toughness value may be achieved with lower α-Mo volume fraction if the ductility of the α-Mo can be improved; in the present work, this is accomplished by increasing the temperature. If the room temperature ductility of α-Mo phase can be improved by compositional or microstructural means, lower α-Mo volume fractions will be needed to achieve adequate toughness levels; this is important since the α-Mo phase compromises the oxidation and creep resistance [7,8,10-12] and thus its volume fraction should be minimized if possible.

The fracture and fatigue properties of the present α-Mo matrix materials are also superior to those of the intermetallic-matrix Mo-Si-B alloys [19,20]. This is attributed to higher effectiveness of the crack trapping and bridging mechanisms when there is a continuous α-Mo matrix, since the crack cannot avoid the relatively ductile phase. Furthermore, this effect is enhanced when either the α-Mo volume fraction or ductility (e.g., at 1300°C) is increased, indicating that these factors do not affect the mechanical behavior independently. Finally, alloys with coarser microstructural size-scales demonstrated slightly improved fracture toughness and fatigue properties. This may be seen in the improved crack stability for alloy C17 relative to the tougher alloys F34 and M34. Although F34 and M34 were tougher due to higher α-Mo volume fractions, stable crack growth was more easily accomplished in alloy C17, allowing the collection of R-curve data over several millimeters without catastrophic failure. Similar improved crack stability was also observed during fatigue testing of the coarser microstructures.

CONCLUSIONS

Based on an experimental study of ambient- to high-temperature fracture toughness and fatigue-crack propagation behavior in five Mo-Si-B alloys, containing Mo_3Si and Mo_5SiB_2 intermetallic phases dispersed within a continuous α-Mo matrix, the following conclusions are made:

1. α-Mo matrix Mo-Si-B alloys exhibit far superior fracture and fatigue resistance relative to unreinforced silicides, with fracture toughnesses in excess of 20 MPa√m for α-Mo volume fractions > 45%. Such gains are attributed to crack trapping and crack bridging by the α-Mo phase.

2. Higher α-Mo volume fractions benefited both of these mechanisms, leading to improved fracture and fatigue resistance. Furthermore, the fracture resistance was improved at 1300°C, indicating the role that α-Mo ductility plays in determining mechanical properties. Finally, a given level of fracture resistance may be achieved with lower α-Mo volume fraction by improving α-Mo ductility, a desirable feature since α-Mo compromises the oxidation and creep resistance.

3. Using a continuous α-Mo matrix instead of an intermetallic matrix is also beneficial for the fracture toughness and fatigue-crack growth properties. However, larger beneficial effects are found by increasing the α-Mo volume fraction and ductility. Additionally, coarser microstructures promote fracture and fatigue resistance, specifically by aiding crack bridging and stability.

ACKNOWLEDGEMENTS

This work was supported by the Office of Fossil Energy, Advanced Research Materials Program, WBS Element LBNL-2, and by the Office of Science, Office of Basic Energy Sciences, Division of Materials Sciences and Engineering of the U.S. Department of Energy, under contract DE-AC03-76SF0098 with the Lawrence Berkeley National Laboratory for ROR and JJK. JHS acknowledges funding of this work by the Office of Fossil Energy, Advanced Research Materials (ARM) Program, WBS Element ORNL-2(I), and by the Division of Materials Sciences and Engineering, U.S. Department of Energy, under contract DE-AC05-00OR22725 with Oak Ridge National Laboratory managed by UT-Battelle.

REFERENCES

1. M. K. Meyer, M. J. Kramer, and M. Akinc, Intermetallics **4**, 273 (1996).
2. M. K. Meyer and M. Akinc, J. Am. Ceram. Soc. **79** (10), 2763 (1996).
3. M. K. Meyer, A. J. Thom, and M. Akinc, Intermetallics **7**, 153 (1999).
4. M. Akinc, M. K. Meyer, M. J. Kramer, A. J. Thom, J. J. Heubsch, and B. Cook, Mater. Sci. Eng. **A261** (1-2), 16 (1999).
5. D. M. Berczik, United States Patent No. 5,595,616 (1997).
6. D. M. Berczik, United States Patent No. 5,693,156 (1997).
7. J. H. Schneibel, P. F. Tortorelli, M. J. Kramer, A. J. Thom, J. J. Kruzic, and R. O. Ritchie, in *Defect Properties and Related Phenomena in Intermetallic Alloys*, edited by E. P. George, M. J. Mills, H. Inui, and G. Eggeler (Materials Research Society, Warrendale, PA, 2003), Vol. 753, pp. 53-58.
8. J. H. Schneibel, R. O. Ritchie, J. J. Kruzic, and P. F. Tortorelli, Metall. Mater. Trans. **36A**, in press (2005).
9. J. J. Kruzic, J. H. Schneibel, and R. O. Ritchie, Scripta Mater. **50**, 459 (2004).
10. J. H. Schneibel, C. T. Liu, D. S. Easton, and C. A. Carmichael, Mater. Sci. Eng. **A261** (1-2), 78 (1999).
11. V Supatarawanich, D. R. Johnson, and C. T. Liu, Mater. Sci. Eng. **A344** (1-2), 328 (2003).
12. J. H. Schneibel, Intermetallics **11** (7), 625 (2003).
13. C. J. Gilbert, J. M. McNaney, R. H. Dauskardt, and R. O. Ritchie, J. Test. Eval. **22** (2), 117 (1994).
14. D. Chen, C. J. Gilbert, and R. O. Ritchie, J. Test. Eval. **28** (4), 236 (2000).
15. ASTM E647-00 in *Annual Book of ASTM Standards, Vol. 03.01: Metals- Mechanical Testing; Elevated and Low-temperature Tests; Metallography* (ASTM, West Conshohocken, Pennsylvania, USA, 2002), pp. 595-635.
16. K.T. Venkateswara Rao, W. O. Soboyejo, and R. O. Ritchie, Metall. Trans. **23A**, 2249 (1992).
17. I. Rosales and J. H. Schneibel, Intermetallics **8**, 885 (2000).
18. R. O. Ritchie, Int. J. Fract. **100** (1), 55 (1999).
19. H. Choe, D. Chen, J. H. Schneibel, and R. O. Ritchie, Intermetallics **9**, 319 (2001).
20. H. Choe, J. H. Schneibel, and R. O. Ritchie, Metall. Mater. Trans. **34A**, 25 (2003).

Mater. Res. Soc. Symp. Proc. Vol. 842 © 2005 Materials Research Society S2.8

A Bond-Order Potential Incorporating Analytic Screening Functions for the Molybdenum Silicides

Marc J. Cawkwell[1], Matous Mrovec[2], Duc Nguyen-Manh[3], David G. Pettifor[4] and Vaclav Vitek[1]
[1]Department of Materials Science and Engineering, University of Pennsylvania,
3231 Walnut Street, Philadelphia, PA 19104-6272, U.S.A.
[2]Fraunhofer Institut für Werkstoffmechanik, Wöhlerstr. 11,
79108 Freiburg, GERMANY.
[3]UKAEA Fusion, Culham Science Centre,
Abingdon, OX14 3DB, UNITED KINGDOM.
[4]Department of Materials, University of Oxford, Parks Road,
Oxford, OX1 3PH, UNITED KINGDOM.

ABSTRACT

The intermetallic compound $MoSi_2$, which adopts the $C11_b$ crystal structure, and related alloys exhibit an excellent corrosion resistance at high temperatures but tend to be brittle at room and even relatively high temperatures. The limited ductility of $MoSi_2$ in ambient conditions along with the anomalous temperature dependence of the critical resolved shear stress (CRSS) of the {110}<111>, {011}<100> and {010}<100> slip systems and departure from Schmid law behavior of the {013}<331> slip system can all be attributed to complex dislocation core structures. We have therefore developed a Bond-Order Potential (BOP) for $MoSi_2$ for use in the atomistic simulation of dislocations and other extended defects. BOPs are a real-space, $O(N)$, two-center orthogonal tight-binding formalism that are naturally able to describe systems with mixed metallic and covalent bonding. In this development novel analytic screening functions have been adopted to properly describe the environmental dependence of bond integrals in the open, bcc-based $C11_b$ crystal structure. A many-body repulsive term is included in the model that allows us to fit the elastic constants and negative Cauchy pressures of $MoSi_2$. Due to the internal degree of freedom in the position of the Si atoms in the $C11_b$ structure which is a function of volume, it was necessary to adopt a self-consistent procedure in the fitting of the BOP. The constructed BOP is found to be an excellent description of cohesion in $C11_b$ $MoSi_2$ and we have carefully assessed its transferability to other crystal structures and stoichiometries, notably $C40$, $C49$ and $C54$ $MoSi_2$, $A15$ and DO_3 Mo_3Si and $D8_m$ Mo_5Si_3 by comparing with *ab initio* structural optimizations.

INTRODUCTION

The intermetallic compound $MoSi_2$ and related alloys are promising materials for use at high temperatures. $MoSi_2$ has very good resistance to oxidation and corrosion, high thermal conductivity and a relatively low density [1]. $MoSi_2$ adopts the body-centered tetragonal $C11_b$ crystal structure and this complex crystal structure leads to five active slip systems, the activity of which depends on load orientation with respect to the crystallographic axes [2]. Single crystals of $MoSi_2$ can be plastically deformed in compression along [001] only at temperatures above 1300 °C but plastic deformation can occur even at room temperature for other load orientations.

The {110)<111], {011)<100] and {010)<100] slip systems are observed to exhibit at intermediate temperatures an anomalous increase of their critical resolved shear stresses with increasing temperature which suggests that thermal activation is responsible for transformations of dislocation core structures to some sessile configurations. The critical resolved shear stress for the {013)<331] slip system, which is controlling plastic flow, does not obey the Schmid law. This suggests that <331] dislocations possess a non-planar core structure. The complicated deformation modes found in $MoSi_2$, along with their temperature dependencies, indicate complex dislocation core structures and to perform dependable atomistic simulations, the mixed metallic and covalent bonding must be accurately captured.

In order to generate a reliable and transferable model of interatomic bonding in the molybdenum silicides, in particular $MoSi_2$, that are appropriate for the atomistic simulation of extended defects, it is essential to capture the non-central bonding character arising from the unsaturated $4d$ and $2p$ bands. Therefore, many-body central force schemes such as the Embedded Atom Method [3] or Finnis-Sinclair potentials [4] are not adequate for this purpose. Instead, we have adopted the Bond-Order Potential (BOP) formalism advanced by Pettifor and coworkers, see for example [5, 6]. BOPs are a real-space, $O(N)$ scaling tight-binding based framework that naturally describe materials that exhibit mixed metallic and covalent bonding. BOPs have been successfully developed for both transition metals and intermetallics such as Mo and Ir [7, 8] and TiAl [9]. Recent advances in the implementation of BOPs to the study of materials that exhibit negative Cauchy pressures [6, 9, 10] allow us to fit the negative Cauchy pressures of $C11_b$ $MoSi_2$ through a physically transparent, environmentally dependent many-body repulsive term in the total energy.

Intersite bond integrals in open structures, such as bcc or the bcc-based $C11_b$, show strong environmental dependencies owing to the effects of screening by the local environment. These environmental effects are properly captured in non-orthogonal tight-binding models however such schemes are not an appropriate basis for a semi-empirical interatomic potential due to the large number of adjustable parameters required. The effects of the non-orthogonality of electron orbitals have been captured in the orthogonal BOP scheme through the introduction of analytic screening functions that were derived from non-orthogonal tight-binding theory [7, 11]. These screening functions are described in more detail in the next Section.

There is a degree of freedom in the $C11_b$ structure that permits the Si atoms to move from their ideal lattice sites and the magnitude and direction of the displacement of the Si atoms, Δ, is found to depend on applied strain. Consequently, when fitting elastic constants, lattice parameters and cohesive energy during the construction of the BOP for $MoSi_2$, we have to allow for internal relaxations which requires a self-consistent procedure. The efficacy of the BOP as a description of cohesion in $C11_b$ $MoSi_2$ along with its transferability to other crystal structures and stoichiometries was assessed by comparing the predictions of the BOP to mixed-basis augmented plane-wave plus local orbitals (APW+lo) *ab initio* structural optimizations performed using the WIEN2k package of codes [12].

THEORY

In the BOP formalism, the total energy is written as a sum of three terms, $E_{tot} = E_{bond} + E_{env} + E_{pair}$, where E_{bond} is the bonding contribution arising from the overlap of atomic-like electron orbitals, E_{env} is a many-body repulsive term that describes the overlap repulsion caused by sp-electrons that are being squeezed into the interstitial volumes by strong interatomic bonding and E_{pair} is a pairwise term used to describe any remaining interactions that are not explicitly included in the first two terms [7, 9, 13]. The bond term is constructed using interatomic bond integrals calculated *ab initio*, along with band occupancies and energy differences between different orbitals. The BOP formalism provides for calculation of the bond order, $\Theta_{j\beta,i\alpha}$, between lattice sites i and j and orbitals α and β in real space for a given Hamiltonian, $H_{i\alpha,j\beta}$, using the concept of reconstructing the densities of states through the moments theorem of Ducastelle and Cyrot-Lackman [14] and the Lanczos recursion method of continuous fractions [15].

In order to describe the strong environmental dependence of interatomic bond integrals in the open $C11_b$ structure, we introduce a screened Hamiltonian matrix (see [11] for details). In short, the screened Hamiltonian matrix, $\tilde{H}$, is written in terms of the unscreened Hamiltonian, H, and the nonorthogonality matrix, S, *i.e.*,

$$\tilde{H} = \frac{1}{2}\left(S^{-1}H + HS^{-1}\right) \tag{1}$$

The nonorthogonality matrix can be written as $S = I + O$ where I is the identity operator and O is the overlap matrix. The elements of the inverse matrices $(I+O)^{-1}$ can then be determined using the BOP theory and can be expressed in terms of the unscreened elements of the Hamiltonian and overlap matrices. The screened bond integrals, $\tilde{\beta}$, can then be written as,

$$\tilde{\beta}^{ij}_{ll'\tau} = \beta^{ij}_{ll'\tau}\left(1 - S^{ij}_{ll'\tau}\right) \tag{2}$$

where $l = s, p$ or d and $\tau = \sigma, \pi$ or δ. See references [11] and [7] for the functional form used for the screening function, S.

The BOP for the molybdenum silicides is constructed using a minimal basis of electronic states, *i.e.*, d-orbitals and p-orbitals centered on Mo and Si, respectively. We assume that valence s-orbitals on both the Mo and Si atoms are responsible for the screening of the bond integrals and that the elements of the unscreened Hamiltonian and overlap matrices decay in the same manner and only differ in magnitude. Figure 1 illustrates that our screening formalism accounts very successfully for the environmental dependence of bond integrals in $C11_b$ MoSi$_2$. Not only are the magnitudes of the screened bond integrals in excellent agreement with bond integrals calculated using first principles LMTO, but also their gradients on the isotropic expansion or contraction of the lattice.

The bond term in the BOP correctly predicts the order of structural stability in the MoSi$_2$ system as determined by *ab initio* APW+lo density functional theory calculations: $C11_b \rightarrow C40 \rightarrow C54$ at band fillings of $N_d = 6.0$ and $N_p = 2.6$ electrons with $E_d - E_p = -1.106$ eV. Good convergence to k-space results is found when the bond term is evaluated at five levels of Lanczos recursion, *i.e.*, the first nine moments of the DOS are included in the calculation.

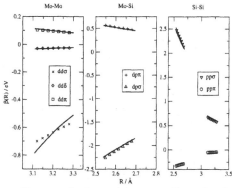

Figure 1. Radial dependencies of bond integrals in $C11_b$ MoSi$_2$. Points represent first principles LMTO calculations and solid lines screened bond integrals in BOP.

The many-body repulsive term, E_{env}, takes the form of a screened Yukawa potential [9, 10] and is fitted to the experimental values of the two Cauchy pressures in $C11_b$ MoSi$_2$, $C_{13} - C_{44}$ and $C_{12} - C_{66}$. The last term in the BOP is the pairwise interaction, E_{pair}, takes the same form as in [9] and is fitted to reproduce as closely as possible the cohesive energy, lattice parameters and remaining elastic constants.

As mentioned above, great care must be taken when fitting the BOP since it is essential to account for the internal degrees of freedom in the $C11_b$ unit cell. Since the effects of the relaxation of the Si atoms can only be evaluated once all three terms in the BOP have been parameterized, the potential must be constructed in a self-consistent manner. This means that in any given step of the potential construction, the effect of the relaxation of silicon atoms on the elastic constants and internal stresses has to be evaluated and the pairwise and many-body repulsive terms then reparameterized in order to improve the fit of experimental elastic constants and lattice parameters.

RESULTS

The BOP developed for the molybdenum silicides is found to be not only a good description of cohesion in $C11_b$ MoSi$_2$ but it also exhibits good transferability to other crystal structures. Table I shows the lattice parameters, internal relaxation parameter, Δ, and cohesive energy per formula unit given by the BOP, compared with experimental or *ab initio* calculated values. It is evident that the BOP provides a good description of the crystal structure of $C11_b$ MoSi$_2$ and predicts the displacement of the Si atoms in the correct direction although the magnitude of this displacement is by a factor of 6 larger than experimental and *ab initio* calculated values. However, it has to be recognized that Δ is very small compared to the lattice parameter and precision less than 1% can not be expected.

In Table II we present the elastic constants for $C11_b$ MoSi$_2$ from both BOP and experiment in units of eV Å^{-1}. Although some elastic constants, in particular C_{11}, C_{12} and C_{33}, are in relatively poor agreement with experiment it is important to note that all elastic constants and shear moduli are positive meaning that the $C11_b$ crystal structure is stable and thus the BOP is suitable for the atomistic simulation of defect structures in MoSi$_2$. The BOP is also able to reproduce the two negative Cauchy pressures in $C11_b$ MoSi$_2$, $C_{13} - C_{44}$ and $C_{12} - C_{66}$, even though their magnitudes are not reproduced accurately.

	BOP	Experiment / *Ab initio*
a / Å	3.0650	3.2056 [16] / 3.21817
c / Å	8.0885	7.8450 [16] / 7.87695
Δ / c	0.0117	0.00195 [16] / 0.00181
E_{coh} / eV f.u.$^{-1}$	-19.307	-17.1406

Table I. Crystallographic parameters for the $C11_b$ structure from BOP compared with experiment or values calculated *ab initio*.

	BOP	Experiment [17]
C_{11}	3.852	2.559
C_{12}	1.234	0.717
C_{13}	0.718	0.546
C_{33}	1.689	3.203
C_{44}	1.023	1.289
C_{66}	1.658	1.246
$C_{13} - C_{44}$	-0.306	-0.743
$C_{12} - C_{66}$	-0.809	-0.529

Table II. Elastic constants and Cauchy pressures calculated using BOP compared with experimental values in units of eV Å^{-1}.

The transferability of the BOP for MoSi$_2$ to crystal structures other than $C11_b$ and also to different stoichiometries versus predictions from *ab initio* calculations was generally found to be very good. A notable success of the BOP for MoSi$_2$ is its transferability to the hexagonal $C40$ and orthorhombic $C54$ structures and also its excellent predictions of the basic structural properties of tetragonal $D8_m$ Mo$_5$Si$_3$. These results are presented in Table III. Structural energy differences predicted by the BOP for MoSi$_2$ are also in good agreement with *ab initio* calculation where $E(C40) - E(C11_b) = 307.4$ meV f.u.$^{-1}$ and $E(C54) - E(C11_b) = 239.0$ meV f.u.$^{-1}$ compared with *ab initio* values of 81.1 and 236.3 meV f.u.$^{-1}$ respectively. It is evident that the BOP overestimates the $C11_b \rightarrow C40$ energy difference although the $C11_b \rightarrow C54$ energy difference is in very good agreement with *ab initio* results. However, the transferability of the BOP to Mo$_3$Si in the $A15$ and $D0_3$ structures is very limited due to the very small interatomic separations in these structures compared with $C11_b$ MoSi$_2$ and the short-range repulsion provided by the pairwise term in the BOP formalism.

	$C40$ MoSi$_2$	$C54$ MoSi$_2$	$D8_m$ Mo$_5$Si$_3$
c / a	1.491 (1.432)	1.119 (1.089)	0.563 (0.506)
b / a	N/A	0.565 (0.572)	N/A
Volume / Å^3 f.u.$^{-1}$	37.218 (40.723)	37.691 (40.760)	117.330 (115.525)
Δ / a	0.00326 (0.00268)	-0.010 (0.00318)	Δ_{Mo}: -0.00193 (0.00176) Δ_{Si}: -0.00663 (0.000658)
E_{coh} / eV f.u.$^{-1}$	-19.000 (-17.059)	-19.068 (-16.904)	-58.732 (-49.224)

Table III. Predictions of the BOP for the crystallographic parameters and cohesive energies of $C40$ and $C54$ MoSi$_2$ and $D8_m$ Mo$_5$Si$_3$. Values calculated *ab initio* are given in parentheses.

CONCLUSIONS

A Bond-Order potential incorporating analytic screening functions to describe the environmental dependence of bond integrals in the open $C11_b$ structure has been successfully developed for the molybdenum silicides. The BOP provides an excellent description of interatomic bonding in $C11_b$ $MoSi_2$ and is suitable for the atomistic simulation of extended defects in this material. The BOP also exhibits good transferability to the $C40$ and $C54$ structures that were not explicitly fitted, as well as to $D8_m$ Mo_5Si_3 that is a promising engineering material in its own right.

ACKNOWLEDGEMENTS

This research was supported by the U.S. Department of Energy, BES Grant no. DE-PG02-98ER45702 (M.J.C. and V.V.) and the United Kingdom EPSRC (D.N.M. and D.G.P.).

REFERENCES

1. M. Yamaguchi, H. Inui and K. Ito, *Acta Mater.* **48**, 307 (2000)
2. K. Ito, H. Inui, Y. Shirai and M. Yamaguchi, *Phil. Mag. A* **72**, 1075 (1995)
3. M. S. Daw and M. I. Baskes, *Phys. Rev. B* **29**, 6443 (1984)
4. M. W. Finnis and J. E. Sinclair, *Phil. Mag. A* **50**, 45 (1984)
5. A. P. Horsfield, A. M. Bratkovsky, M. Fearn, D. G. Pettifor and M. Aoki, *Phys. Rev. B* **53**, 12694 (1996)
6. D. Nguyen-Mạnh, D. G. Pettifor, D. J. H. Cockayne, M. Mrovec, S. Znam and V. Vitek, *Bull. Mater. Sci.* **26**, 43 (2003)
7. M. Mrovec, D. Nguyen-Manh, D. G. Pettifor and V. Vitek, *Phys. Rev. B* **69**, 094115 (2004)
8. M. J. Cawkwell, D. Nguyen-Manh, V. Vitek and D. G. Pettifor, *Mat. Res. Soc. Symp. Proc. Vol.* **779**, W5.5.1 (2003)
9. S. Znam, D. Nguyen-Manh, D. G. Pettifor and V. Vitek, *Phil. Mag.* **83**, 415 (2003)
10. D. Nguyen-Manh, D. G. Pettifor, S. Znam and V. Vitek, *Mat. Res. Soc. Symp. Proc. Vol.* **491**, 353 (1998)
11. D. Nguyen-Manh, D. G. Pettifor and V. Vitek, *Phys. Rev. Lett.* **85**, 4136 (2000)
12. P. Blaha, K. Schwarz, G. K. H. Madsen, D. Kvasnicka and J. Luitz, **WIEN2k**, An Augmented Plane Wave + Local Orbitals Program for Calculating Crystal Properties, (Karlheinz Schwarz, Techn. Universität Wien, Austria), 2001. ISBN 3-9501031-1-2
13. A. P. Sutton, M. W. Finnis, D. G. Pettifor and Y. Ohta, *J. Phys. C: Solid State Phys.* **21**, 35 (1988)
14. F. Ducastelle and F. Cyrot-Lackmann, *J. Phys. Chem. Sol.* **31**, 1295 (1970)
15. C. Lanczos, *J. Res. Nat. Bur. Stand.* **45**, 255 (1950)
16. K. Tanaka, K. Nawata, H. Inui, M. Yamaguchi and M. Koiwa, *Intermetallics* **9**, 603 (2001)
17. K. Tanaka, H. Onome, H. Inui, M. Yamaguchi and M. Koiwa, *Mat. Sci. Eng. A* **239-240**, 188 (1997)

Mater. Res. Soc. Symp. Proc. Vol. 842 © 2005 Materials Research Society   S5.34

High Temperature Oxidation Behavior of Al Added Mo / Mo$_5$SiB$_2$ *in-situ* Composites

Akira Yamauchi, Kyousuke Yoshimi, and Shuji Hanada
Institute for Material Research, Tohoku University,
Sendai 980-8577, Japan

ABSTRACT

Isothermal oxidation behavior of Mo/Mo$_5$SiB$_2$ *in-situ* composites containing small amounts of Al was investigated under an Ar-20%O$_2$ atmosphere in the temperature range of 1073–1673 K. The Mo/Mo$_5$SiB$_2$ *in-situ* composites, (Mo-8.7mol%Si-17.4mol%B)$_{100-x}$Al$_x$ (x=0, 1, 3, and 5mol%), were prepared by Ar arc-melting, and then homogenized at 2073 K for 24 h in an Ar-flow atmosphere. Without addition of Al, Mo/Mo$_5$SiB$_2$ *in-situ* composite exhibits a rapid mass loss at the initial oxidation stage, followed by passive oxidation after the substrate is sealed with borosilicate glass in the temperature range of 1173–1473 K, whereas it exhibits a rapid mass gain around 1073 K. On the other hand, small Al additions, especially of 1 mol%, significantly improve the oxidation resistance of Mo/Mo$_5$SiB$_2$ *in-situ* composites at temperatures from 1073–1573 K. The excellent oxidation resistance is considered to be due to the rapid formation of a continuous, dense scale of Al-Si-O complex oxides. The protective oxide scales contain crystalline oxides, and the amounts of the crystalline oxides obviously increase with Al concentration.

INTRODUCTION

Mo-based composites are attractive materials for high-temperature structural applications to achieve high energy efficiency. Especially, Mo-Si-B alloys have high potential as heat-resistant materials at ultra-high temperature [1-5]. Akinc *et al.* reported that the oxidation resistance of Mo$_5$Si$_3$ is dramatically improved by the addition of boron at temperatures ranging from 1073 to 1773 K in air [6]. However, it would not be possible to use monolithic Mo$_5$Si$_3$ for structural applications because of its fatal brittleness and strong anisotropy of thermal expansion coefficient [7]. Mo$_5$SiB$_2$ is only one ternary compound in the Mo-Si-B system. Ito *et al.* reported that the single crystal of Mo$_5$SiB$_2$ shows excellent strength at high temperature for a few orientations [8-9]. With respect to the oxidation resistance of Mo$_5$SiB$_2$, Yoshimi *et al.* [10] studied the oxidation behavior of Mo$_5$SiB$_2$-based alloy at elevated temperatures, and concluded that the oxidation resistance of Mo$_5$SiB$_2$-based alloy is not as good as that of boron-added Mo$_5$Si$_3$-based alloys. Furthermore, they worked on thermal expansion, strength and oxidation resistance of Mo/Mo$_5$SiB$_2$ *in-situ* composites, and reported that the Mo/Mo$_5$SiB$_2$ *in-situ* composite having a eutectic microstructure shows superior high temperature strength even at 1773 K [11]. However, the oxidation resistance of the Mo/Mo$_5$SiB$_2$ *in-situ* composites is worse than that of Mo$_5$SiB$_2$ [12]. Therefore, the improvement of oxidation resistance is a key to develop Mo-Si-B alloys for ultra-high temperature applications.

Yanagihara *et al.* reported the effect of third elements on the oxidation resistance of MoSi$_2$ [13-14]. The third elements having a larger affinity to oxygen than Si show the advantages not only for the oxidation resistance at high temperature but also for the suppression of the pesting phenomenon in a temperature range from 673 to 973 K. Consequently, their results suggested that the effective fourth elements have the potential to improve the oxidation resistance of Mo/Mo$_5$SiB$_2$ *in-situ* composites. The purpose of this work is to investigate the effect of Al addition on the oxidation behavior of Mo/Mo$_5$SiB$_2$ *in-situ* composites. Isothermal mass change is examined at several elevated temperatures, and cross-section microstructure and morphology of oxide scale are characterized. Based on the obtained results, the oxidation behavior of the Al added Mo/Mo$_5$SiB$_2$ *in-situ* composite is discussed.

EXPERIMENTAL DETAILS

The nominal composition of a ternary Mo/Mo$_5$SiB$_2$ *in-situ* composite used in this study is Mo-8.7mol%Si-17.4mol%B. The Mo-Si-B alloy was produced by an arc-melting method in an Ar atmosphere from 99.9 mass% molybdenum, 99.9999 mass% silicon, and 99.8 mass% boron. Melted button ingots were flipped over and re-melted more than 5 times. And then three Mo/Mo$_5$SiB$_2$ *in-situ* composites having different aluminum concentrations were produced by the arc-melting method from the Mo-Si-B alloy ingots and 99.99mass% aluminum. Homogenization heat treatment was carried out at 2073 K for 86.4 ks in an Ar gas atmosphere. Nominal and chemically analyzed compositions of the present four composites are listed in Table 1.

Specimens for oxidation tests were cut into 5×5×1 mm pieces from the homogenized ingots by electrospark machining. Surfaces were polished with SiC paper up to No.1500, and then polished with a 0.3 µm Al$_2$O$_3$ and fine diamond abrasive for finishing. Prior to oxidation tests, the specimens were ultrasonically cleaned in acetone, and the mass and the surface area were measured. Isothermal oxidation tests were performed at temperatures raging from 1073 to 1673 K for 86.4 ks in a Netzsch STA409A thermogravimeter using Al$_2$O$_3$ crucibles. In the isothermal oxidation tests, the specimen was kept in a high purity Ar gas atmosphere (Ar gas flow rate: 6.67ml s^{-1}) until the furnace temperature reached a desired test temperature. After 900 s passed for the temperature stabilization, high purity oxygen gas began to flow at 0.83ml s^{-1}, and simultaneously the mass change was recorded. After the oxidation test, surface morphology and cross section of the oxidized specimens were examined by SEM, XRD, AES, and EPMA.

RESULTS AND DISCUSSION

Oxidation behavior of Al added Mo/Mo$_5$SiB$_2$ *in-situ* composites

The isothermal mass change curves are shown in figure 1. As shown figure 1 (a), no Al

Table I. Compositions of the Al-doped Mo/Mo$_5$SiB$_2$ *in-situ* composites (mol %).

Composite	Nominal Composition	Chemical Composition
1	Mo-8.7Si-17.4B	Mo-9.0Si-17.2B
2	(Mo-8.7Si-17.4B)-1Al	Mo-9.4Si-16.8B-1Al
3	(Mo-8.7Si-17.4B)-3Al	Mo-9.3Si-16.2B-3Al
4	(Mo-8.7Si-17.4B)-5Al	Mo-8.6Si-16.4B-5Al

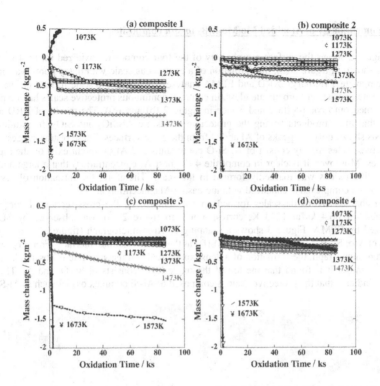

Figure 1. Isothermal mass change curves of the composite 1, 2, 3, and 4 oxidized at temperatures between 1073 and 1673 K in an atmosphere of Ar-20%O_2.

added, ternary Mo/Mo$_5$SiB$_2$ *in-situ* composite (composite 1) exhibits a rapid mass gain at 1073 K. In the higher temperature range, i.e., 1173-1473 K, a rapid mass loss occurs at the initial oxidation stage, and then suddenly changes to the steady state oxidation in which the oxidation rate is slow and almost constant. As reported in the previous studies, this mass loss is interpreted as the volatilization of MoO$_3$ [10-12]. On the other hand, the Al addition considerably improves the oxidation resistance of the Mo/Mo$_5$SiB$_2$ *in-situ* composite in the temperature range between 1073 and 1573 K. The Al addition, especially 1mol% addition, suppresses the initial mass loss more than 50%. The trends of the oxidation behavior of the Al-added Mo/Mo$_5$SiB$_2$ *in-situ* composites appear to be similar irrespective of Al concentration. These results suggest that oxidation mechanism(s) is (are) quite similar for all the Al added Mo/Mo$_5$SiB$_2$ *in-situ* composites. At and above 1673 K, however, the oxidation resistances of all the composites are drastically degraded.

Scale morphology of Al added Mo/Mo₅SiB₂ *in-situ* composites

Figure 2 shows the surface morphology of the four composites oxidized for 86.4 ks at 1273 K. As shown in figure 2 (a), the formation of a continuous scale with many pores was observed for composite 1 (ternary). XRD and EPMA analyses show that the scale formed on the ternary composite consists of borosilicate glass. In contrast, continuous protective scales having no pores are formed onto composite 2 and 3 (see in figure 2 (b) and (c)). The results of XRD analysis show that main products among the protective scales are Al_2SiO_5 and SiO_2. In addition, AES analysis shows only the peaks of Al and O from the scale surfaces. These results suggest that the protective scales mainly consist of Al-Si-O compounds and Al_2O_3 might exist on the top of the surfaces. Moreover, it is clear in composite 4 (highest Al concentration) that a large amount of needle-like oxides was markedly formed in the scale. Therefore, the formation of crystalline Al-Si oxide complexes is enhanced with increasing Al concentration.

To characterize oxide scales formed on the composites, the cross-section of composite 2 oxidized for 86.4 ks at 1273 K, corresponding to figure 2 (b), was observed by SEM and analyzed by EPMA. Figure 3 shows a scanning electron micrograph (figure 3(a)) and element maps of Mo, Si, B, O, and Al (figure 3(b)-(f)) at the scale/substrate interface. From the scanning electron micrograph, the formation of a dense oxide scale on the substrate is observed. From the element maps, it is found that the top of the oxide layer consists of Si, O, and Al. The above results indicate that the protective scale is composed of Ai-Si complex oxides such as Al_2SiO_5. In

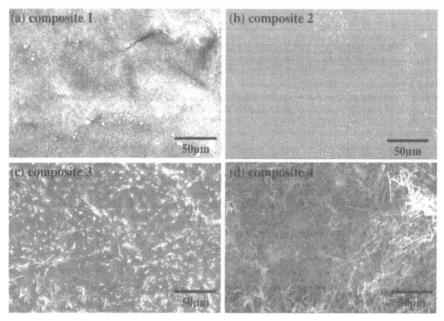

Figure 2. SEM micrographs of the surface morphology of composite 1, 2, 3, and 4 oxidized at 1273 K for 86.4 ks in an atmosphere of Ar-20%O₂.

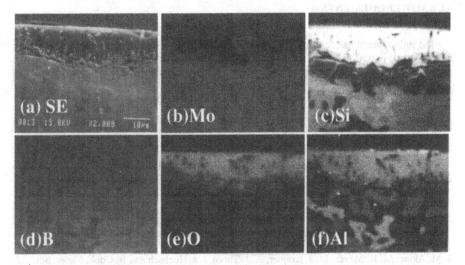

Figure 3. (a) scanning electron micrograph image of cross-section microstructure of composite 2 oxidized at 1273 K for 86.4 ks in an Ar-20%O$_2$ atmosphere. (b)- (f) Element mapping images by EPMA at the same location of (a). (b) Mo, (c) Si, (d) B, (e) O and (f) Al.

the case of 5mol%Al, Al$_6$Si$_2$O$_{13}$ is also detected in the oxide scale. Inclusions of Mo oxides are also observed in the oxide side just around the scale/substrate interface (see in figure 3 (b)-(f)). Furthermore, a Si and B depletion zone is observed under the Al-Si complex oxide scale. These results suggested that the selective oxidation of Si and B caused the formation of the depletion zone.

CONCLUSION

In the present work, the oxidation behavior of ternary Mo/Mo$_5$SiB$_2$ *in-situ* composite and quaternary Al added Mo/Mo$_5$SiB$_2$ *in-situ* composites were investigated. The following results were obtained.
1. Al added Mo/Mo$_5$SiB$_2$ *in-situ* composites show good oxidation resistance in the temperature range from 1073 to 1573 K. Al addition is effective in suppressing the initial mass loss. The oxidation resistances of Al added Mo/Mo$_5$SiB$_2$ *in-situ* composites are almost similar irrespective of Al concentration.
2. In Al added Mo/Mo$_5$SiB$_2$ *in-situ* composites, the crystalline Al-Si oxide complexes are formed on the top of the protective scale. As the Al concentration in the substrate increases, the amount of needle-like Al-Si complex oxides in the scale increases.

ACKNOWLEDGEMENTS

The author would like to thank Y. Murakami for his technical assistance of EPMA and AES measurements in the Laboratory for Advanced Materials, Tohoku University. This work was performed under the inter-university cooperative research program of Laboratory for Advanced Materials, Institute for Materials Research, Tohoku University.

REFERENCES

1. A. K. Vasudevan and J. J. Petrovic, Mater. Sci. Eng. **A155**, 1 (1992).
2. J. J. Petrovic and A. K. Vasudevan in High Tempareture Silicides and Refractory Alloys, edited by C. L. Briant, J. J. Petrovic, B. P. Bewlay, A. K. Vasudevan, and H. A. Lipsitt, (Mater. Res. Soc. Proc. **322**, Boston, MA, 1993) pp. 3-8.
3. C. A. Nunes, R. Sakidja, and J. H. Perepezko in Structural Intermetallics 1997, edited by M. V. Nathal, R. Darolia, C. T. Liu, P. L. Martin, D. B. Miracle, R. Wagner, and M. Yamaguchi, (TMS, Champion, PA, 1997) pp. 831-839.
4. M. Akinc, M. K. Meyer, M. J. Kramer, A. J. Thom, J. J. Huebsch and B. Cook, Mater. Sci. Eng. **A261**, 16 (1999).
5. K. Natesan and S. C. Deevi, Intermetallics **8**, 1147 (2000).
6. M. K. Meyer, and M. Akinc, J. Am. Ceram. Soc. **79**, 938 (1996).
7. F. Chu, D. J. Thoma, K. McClellan, P. Peralta, and Y. He, Intermetallics **7**, 611 (1999).
8. K. Ito, K. Ihara, K. Tanaka, M. Fujikura, and M. Yamaguchi, Intermetallics **9**, 591 (2001).
9. K. Ihara, K. Ito, K. Tanaka, and M. Yamaguchi, Mater. Sci. Eng. **A329-331**, 222 (2002).
10. K. Yoshimi, S. Nakatani, T. Suda, S. Hanada, and H. Habazaki, Intermetallics **10**, 407 (2002).
11. K. Yoshimi, S. Nakatani, N. Nomura, and S. Hanada, Intermetallics **11**, 787 (2003).
12. K. Yoshimi, S. Nakatani, T. Suda, T. Haraguchi, and S. Hanada, Materia Japan **41**, 146 (2002) (in Japanese).
13. K. Yanagihara, T. Maruyama, and K. Nagata, Intermetallics **3**, 243 (1995).
14. K. Yanagihara, T. Maruyama, and K. Nagata, Intermetallics **4**, S133 (1996).

Mater. Res. Soc. Symp. Proc. Vol. 842 © 2005 Materials Research Society   S5.35

Nucleation of (Mo) Precipitates on Dislocations During Annealing of a Mo-rich Mo5SiB2 Phase

Nobuaki Sekido, Ridwan Sakidja, and John H. Perepezko
Department of Materials Science and Engineering, University of Wisconsin-Madison,
1509 University Ave, Madison, WI 53706, USA

ABSTRACT

Upon annealing an as-cast Mo-10Si-20B alloy at high temperatures, a Mo solid solution phase precipitates within a supersaturated Mo_5SiB_2 phase. The precipitation behavior of the Mo solid solution was investigated by means of transmission electron microscopy and X-ray diffractometry. It is found that the Mo_5SiB_2 phase in a Mo-10Si-20B alloy contains a significant amount of structural vacancies in the as-cast state. The excess vacancies are removed to form dislocations during annealing, which provides the heterogeneous nucleation sites for the (Mo) precipitates.

INTRODUCTION

The performance requirements for structural materials in an elevated temperature environment represent some of the most demanding challenges facing contemporary materials design strategies. $MoSi_2$ has been considered as one of the promising candidates because of its high melting point and excellent oxidation resistance at elevated temperatures [1,2]. However, extensive studies have shown there seems to be no feasible way to provide $MoSi_2$ based alloys with the creep, fracture and oxidation resistance required for structural design [3,4]. Recently, the multi-phase alloys based on the Mo-Si-B system have attracted considerable interest for potential high temperature applications, since the oxidation resistance of Mo_5Si_3 (T1) is significantly improved by small boron additions [5-7]. The ternary alloys based upon a Mo solid solution, (Mo), a Mo_3Si (A15), and the Mo_5SiB_2 (T2) are attractive in that the ductile (Mo) phase is in thermodynamic equilibrium with an oxidation resistant borosilicide T2 phase up to, at least, 2200 K [8,9]. Two phase alloys consisting of (Mo) and T2 phases exhibit increased toughness and reasonable oxidation resistance [10,11]. Three phase alloys consisting of (Mo), A15, and T2 offer more favorable oxidation resistance at elevated temperatures [12-14].

Of particular interest in the present study is the (Mo)/T2 two-phase mixture, that develops in a Mo-10Si-20B alloy, where the T2 phase with a non-stoichiometric composition is in equilibrium with (Mo) phase. Previous studies have demonstrated that a Mo-rich T2 phase forms during solidification in a Mo-10Si-20B alloy, which leads to a formation of plate shaped (Mo) precipitates during subsequent annealing at high temperatures [8]. The crystallographic relationship between the two phases has been identified [15], while the nucleation behavior of this precipitation reaction is not clarified. Most of the nucleation events are heterogeneous, and suitable nucleation sites are dislocations, interfaces, and boundaries, all of which increase the free energy of materials. The objective in the present study is to characterize the precipitation behavior of the (Mo) phase in the T2 matrix by means of transmission electron microscopy.

EXPERIMENTAL PROCEDURE

The Mo-10Si-20B (at%) alloys were prepared by arc melting under an argon atmosphere. The alloys were annealed under a flowing Ar atmosphere, followed by furnace cooling. The lattice parameters of T2 phase were determined by X-ray diffractometry (XRD) with powder specimens, and the lattice parameter refinement was done by Cohen's method [16]. The microstructures of the alloys were characterized by means of scanning electron microscopy (SEM), and transmission electron microscopy (TEM). The foils for TEM observation were mechanically polished down to 30 μm, and the final perforation was done by an ion mill with an accelerating voltage of 4 kV. TEM observation was conducted on a JEM200CX-II operated at 200 kV. The Burgers vectors of the dislocations formed in T2 phase were determined by the weak-beam thickness fringe method [17] and the invisibility criterion. The line vectors of the dislocations were determined by a combination of stereo-observation and trace analysis.

RESULTS

In agreement with the previous reports [8,15], the as-cast microstructure of a Mo-10Si-20B alloy shown in Fig. 1a consists of three phases: (Mo), A15, and T2. The primary solidification phase is a faceted T2 phase, and the (Mo)/T2 monovariant eutectic products have formed to enclose the T2 primary, followed by the formation of (Mo)/T2/A15 three-phase microstructure as the final solidification product. After the annealing of the alloy at 1873 K for 150 hours, (Mo) particles have precipitated in the primary T2 matrix (Fig. 1b). The (Mo) precipitates exhibit predominantly a plate-like shape, and have a specific crystallographic relation within the T2 matrix [15]. A precipi-

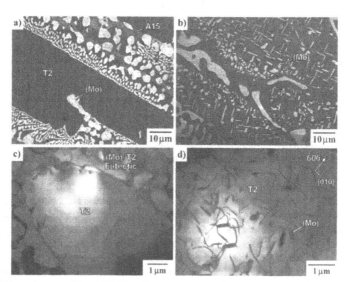

Figure 1. Microstructures of Mo-10Si-20B alloys; a) as-cast, b) annealed at 1873 K for 150 hours, c) annealed at 1827 K for 5 hours, and d) annealed at 1827 K for 20 hours.

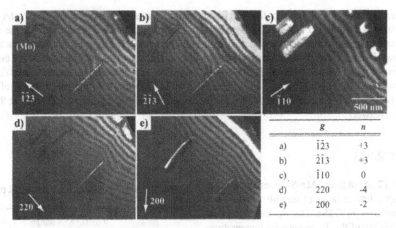

Figure 2. Weak-beam dark field images of the dislocation in T2 phase under a set of operation reflections: a) $g = -1-23$, b) $g = -2-13$, c) $g = -110$, d) $g = 220$, and e) $g = 200$. The numbers of excess thickness fringes terminated at the ends of the dislocation are consistent with the products of each g and $b = [-1,-1,0]$.

	g	n
a)	$\bar{1}23$	+3
b)	$\bar{2}\bar{1}3$	+3
c)	$\bar{1}10$	0
d)	220	-4
e)	200	-2

tate-free-zone (PFZ) has developed near the interface between the primary T2 and the (Mo)/T2 monovariant eutectic product. At the same time, a high density of (Mo) precipitates is observed at the edge of the PFZ. TEM observations have revealed that the dislocation density in the primary T2 phase increases with annealing time. It is confirmed that few dislocations and no precipitates are present in the T2 primary of an as-cast alloy. As shown in Fig. 1c, some dislocations have formed in the vicinity of the boundary between the primary T2 and the eutectic product, while few are formed at the center of the primary T2 phase. After the annealing at 1827 K for 20 hours, many dislocations have developed at the center of the primary T2 phase. The development of the dislocations, which is basically induced only by heat treatment, implies that some amount of vacancies are introduced during solidification, and the vacancies in excess of the equilibrium concentration are annealed out to form dislocations within the T2 matrix. It is also evident that the plate shaped (Mo) particles have mainly precipitated on these dislocations, which indicates that these dislocations act as the preferential nucleation sites for the (Mo) phase precipitation.

The Burgers vectors of the dislocations are determined by the thickness fringe method [17], where the reflection vector, g, the Burgers vector of dislocation, b, and the number of fringes terminated at the end of dislocation line, n, satisfy the following equation:

$$g \cdot b = n \qquad (1).$$

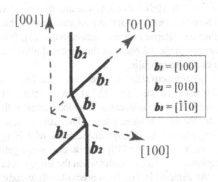

Figure 3. Schematic of a segment of the dislocation network formed in the Mo-10Si-20B alloy annealed at 1823 K for 20 hours. The dislocation network is mainly composed of edge dislocations with Burgers vectors of <100] and <110].

An example is shown in Fig. 2 for the Burgers vector determination, where the Burgers vector of the dislocation is determined as [-1,-1,0]. By a combination of the thickness fringe method, the invisibility criterion, and stereo observation, the dislocation network formed in the alloy is characterized as shown in Fig. 3. The dislocation network is mainly composed of edge dislocations with the Burgers vectors of <100] and <110]. It is also found that many (Mo) particles precipitate on the dislocations with the Burgers vector of <110]. This observation suggests that the precipitation of (Mo) phase preferentially takes place on these dislocations which have the largest strain energy in the dislocation network.

DISCUSSION

The T2 phase in the Mo-Si-B ternary system exhibits a distinct range of solubility at elevated temperatures [8,18,19]. Due to the change in the solubility of the T2 phase with temperature, the (Mo) phase precipitates from a supersaturated T2 matrix during annealing. Generally, there are two ways to provide a solubility range for an ordered intermetallic compound: anti-site substitution and constitutional vacancy development. The Goldschmidt atomic radii of Mo,

Table I. Change in lattice parameters of T2 phase.

	a (nm)	c (nm)	V (nm^3)
As Cast	0.5988	1.1022	0.395
1827 K / 20 h	0.6009	1.1033	0.398
1827 K / 100 h	0.6022	1.1055	0.401

Si and B are 0.140, 0.117 and 0.097 nm, respectively [20]. If a Mo-rich T2 phase were completely derived from the anti-site substitution of Mo atoms for Si, B, or both sites, the lattice volume of the T2 phase would have decreased after the annealing, because larger sized Mo atoms in excess of the equilibrium concentration are eliminated to form the (Mo) precipitates during annealing. However the lattice volume of the T2 phase for an as-cast alloy, as shown in Table I, is found to increase by 1.5% after the annealing at 1823 K for 100 hours. The expansion of T2 lattice volume after the annealing implies that some amount of constitutional vacancies are present in the T2 phase of an as-cast Mo-10Si-20B alloy, and that the excess vacancies are removed during subsequent annealings. There is a tendency for excess vacancies to be attracted together into vacancy clusters, and some clusters collapse into dislocation loops [21]. It is therefore concluded that the development of dislocations in the T2 phase of the annealed alloys stems from the removal process of the excess vacancies during annealing.

Upon precipitation of a second phase from a matrix, it is known that a dislocation acts as a preferential nucleation site [22,23]. As depicted in Fig. 1d, the (Mo) particles have predominantly precipitated on the dislocations, which indicates that these dislocations are the preferential nucleation sites for the (Mo) precipitates. The dislocation network formed in the T2 phase of the present alloy is found to consist of edge dislocations with Burgers vectors of <100] and <110]. Nucleation of the precipitates on the <110] dislocations is energetically favorable, since the activation energy for precipitation becomes smaller when the Burgers vector of a dislocation is larger [22].

As depicted in Fig. 1b, a precipitate-free-zone (PFZ) has developed near the interface between the primary T2 and the (Mo)/T2 eutectic product. The formation of the PFZ can be explained by interaction between the vacancies and the interface as follows. At the early stage of the annealing, as shown in Fig. 1c, formation of the dislocations preferentially takes place in the vicinity of the interface between the primary T2 and (Mo)/T2 eutectic. This fact implies that the interface acts as the

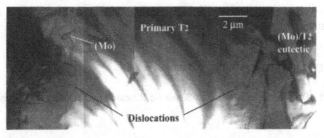

Figure 4. Bright field image of a Mo-10Si-20B alloy annealed at 1823 K for 20h, showing the vicinity of interface between the primary T2 and (Mo)/T2 eutectic product.

heterogeneous site for dislocation nucleation. On the other hand, the interface would also act as a sink for vacancies. Once a dislocation has nucleated at the interface, the dislocation develops further by absorbing a stream of the excess vacancies flowing toward the interface. Consequently the vacancy concentration near the interface locally becomes lower than that at the center part of the primary T2 phase. In such region, nucleation of dislocations is inhibited due to a lower vacancy concentration. This propensity is clearly observed in the alloy annealed at 1823 K for 20 hours as exhibited in Fig. 4. While many dislocations have formed near the interface between the primary T2 and the (Mo)/T2 eutectic, there clearly exists a region where no dislocations and precipitates have formed. Moreover, formation of the dislocations is observed about 10 μm inside the primary T2 phase. Since the dislocations formed during annealing are the preferential nucleation sites for the (Mo) precipitates, no precipitation has occurred in the region without dislocations, which results in the formation of the PFZ near the interface. At the same time, some amount of Mo supersaturation remains in the PFZ. As a result, the (Mo) precipitation takes place more frequently at the edge of the PFZ, which is clearly seen in the SEM microstructure shown in Fig. 1b. It should be noted that few (Mo) particles have precipitated on the dislocations formed in the vicinity of the interface. This fact implies that the excess Mo atoms are easily absorbed into the (Mo) phase of the eutectic product, rather than combining to form the (Mo) precipitates on the dislocations.

Although the interface between the primary T2 and (Mo)/T2 eutectic is found to act as the heterogeneous nucleation site for dislocations, the nucleation mechanism of the dislocations at the center of T2 phase is unclear. More detailed investigations are in progress to elucidate the nucleation mechanism of dislocations, as well as vacancy characteristics in a non-stoichiometric T2 phase. It is worthwhile to note that most of the dislocations formed in the T2 phase of the present alloy must be prismatic loops, since they are derived from an agglomeration of excess vacancies. Moreover motion of the dislocations would be restrained from the (Mo) precipitates and the intersections of dislocations. Therefore these dislocations are not glissile, especially at low temperatures. However the dislocations could act as Frank-Read sources under suitable external stresses, and hence, influence the deformation behavior of the alloy at high temperatures.

CONCLUSIONS

During solidification, a high density of vacancies forms in the T2 phase of a Mo-10Si-20B alloy. Upon high temperature annealing, these vacancies combine to form a network of edge dislocations with Burgers vectors of <100] and <110], which then provides heterogeneous sites for subsequent (Mo) precipitation.

ACKNOWLEDGEMENT

The support of US Air Force Office of Scientific Research (F49620-03-1-0033) is gratefully acknowledged.

REFERENCES

1. J.J. Petrovic, and R.E. Honnell, *Ceram. Eng. Sci. Proc.*, **11**, 734 (1990).
2. A.K. Vasudevan and J.J. Petrovic, *Mater. Sci. Eng.*, **A155**, 1 (1992).
3. J.J. Petrovic and A.K. Vasudevan, *Mater. Sci. Eng.*, **A261**, 1 (1999).
4. D.M. Dimiduk and J.H. Perepezko, *MRS Bulletin*, **28**, 639 (2003).
5. A.J. Thom, M.K. Meyer, M. Akinc, and Y. Kim, *Processing and Fabrication of Advanced Materials for High Temperature Applications III* (TMS, Warrendale, PA, 1993), 413.
6. M.K. Meyer, and M. Akinc, *J. Am. Ceram. Soc.*, **79**, 938 (1996).
7. M.K. Meyer, and M. Akinc, *J. Am. Ceram. Soc.*, **79**, 2763 (1996).
8. C.A. Nunes, R. Sakidja, and J.H. Perepezko, *Structural intermetallics* (TMS, Warrendale, PA, 1997), 831.
9. S. Katrych, A. Grytsiv, A. Bondar, P. Rogl, T. Velikanova, and T. Bohn, *J. Alloys Comp.*, **347**, 94 (2002).
10. M.J. Kramer, O. Unal, and R.N. Wright, *Intermetallics*, **9**, 25 (2001).
11. J.H. Schneibel, M.J. Kramer, and D.S. Easton, *Scripta Mater.*, **46**, 217 (2002).
12. D.M. Berczik, *United States Patent*, 5,693,156 (1997).
13. J.H. Schneibel, C.T. Liu, L. Heatherly, and M.J. Kramer, *Scripta Mater.*, **38**, 1169 (1998).
14. J.H. Schneibel, C.T. Liu, C.A. Carmichael, and D.S. Easton, *Mater. Sci. Eng.*, **A261(1-2)**, 78 (1999).
15. R. Sakidja, H. Sieber, and J.H. Perepezko, *Phil. Mag. Lett.*, **79**, 351 (1999).
16. B.D. Cullity and S.R. Stock, *"Elements of X-ray Diffraction, 3rd Edition"* (Prentice-Hall, 2001).
17. Y. Ishida, H. Ishida, K. Kohra, and H. Ichinose, *Phil. Mag.*, **42A**, 453 (1980).
18. H. Nowotny, E. Dimakopoulou, and H. Kudielka, *Monatsh Chem.*, **88**, 180 (1957).
19. R. Sakidja, J. Myers, S. Kim, and J.H. Perepezko, *Int. J. Refract. Metals Hard Mater.*, **18**, 193 (2000).
20. *"Smithells Metals Reference Book, 6th Edition"*, Ed. Eric A Brandes (Butterworth, London, 1983).
21. D.A. Porter and K.E. Easterling, *"Phase Transformation in Metals and Alloys, 2nd Edition"* (Chapman&Hall, 1992).
22. J.W. Cahn, *Acta Metall.*, **5**, 169 (1957).
23. A. Kelly and R.B. Nicholson, *Prog. Mater. Sci.*, **10**, 151 (1963).

Functional Intermetallics

Mater. Res. Soc. Symp. Proc. Vol. 842 © 2005 Materials Research Society   S3.6

Advanced TEM Investigations on Ni-Ti Shape Memory Material: Strain and Concentration Gradients Surrounding Ni_4Ti_3 Precipitates

Dominique Schryvers, Wim Tirry and Zhiqing Yang
Electron Microscopy for Materials Science (EMAT), University of Antwerp,
Groenenborgerlaan 171, B-2020 Antwerpen, Belgium

ABSTRACT

Lattice deformations and concentration gradients surrounding Ni_4Ti_3 precipitates grown by appropriate annealing in a $Ni_{51}Ti_{49}$ B2 austenite matrix are determined by a combination of TEM techniques. Quantitative Fourier analysis of HRTEM images reveals a deformed nanoscale region with lattice deformations up to 2% while EELS and EDX indicate a Ni depleted zone up to 150 nm away from the matrix-precipitate interface.

INTRODUCTION

NiTi alloys with near-equiatomic composition can exhibit shape memory and superelastic properties resulting from a temperature or stress induced austenite-martensite phase transformation. The behaviour and characteristics of this transformation are strongly influenced by the presence of Ni_4Ti_3 precipitates in the B2 austenite matrix and which can be obtained by appropriate annealing procedures. The atomic structure and morphology of these precipitates have been investigated before [1,2,3]. Due to the anisotropic change of the unit cell dimensions and lattice parameters the precipitates form with a lens shape inside the cubic matrix. Their influence on the transformation temperatures and the occurrence of multiple step transformations was mainly investigated by differential scanning calorimetry (DSC) measurements and conventional transmission electron microscopy (TEM) [4-7]. Small precipitates with a diameter of the central disc up to 300 nm remain coherent or semi-coherent and can act as nucleation centers for the formation of the so-called R-phase [4,6], a rhombohedral distortion preceding the martensitic transformation. Larger precipitates lose their coherency with the matrix and the stress field is partially relaxed by the introduction of interface dislocations [9,10], though they can still act as nucleation centers [4]. This behaviour is explained by the fact that the lattice mismatch between precipitate and matrix induces a stress field in the surrounding matrix favouring particular variants of the product phases. Also the change of Ni concentration in the matrix, due to the higher Ni content in the precipitates, can be expected to have an influence on the local transformation temperatures as is the case for concentration changes at the bulk level [4,8]. However, up till now no quantitative experimental measurements of the strain or concentration gradients exists. In the present work high resolution transmission electron microscopy (HRTEM) is used to measure the actual lattice deformations in the matrix around the Ni_4Ti_3 precipitates. Relative differences in interplanar spacings are determined by Fast Fourier techniques applied to the HRTEM images. To determine the presence of a possible variation in Ni concentration in close proximity of a precipitate nanoprobe electron energy loss spectroscopy (EELS), Energy Filtered TEM (EFTEM) and energy dispersive X-ray (EDX) analysis are used.

The cubic B2 structure of the matrix has a lattice parameter of a = 0.30121 nm [11], while for the precipitate the hexagonal description will be used with lattice parameters a = b = 1.124nm and c = 0.508 nm [12]. As a result of the decrease in symmetry, eight precipitate variants are possible, with a conventional orientation relationship [1,12]:

$$(1\ 1\ 1)_{B2}\ //\ (0\ 0\ 1)_H\ ;\ [3\ \text{-}2\ \text{-}1]_{B2}\ //\ [1\ 0\ 0]_H$$

In this case the $[111]_{B2}$ direction corresponds to the normal to the central plane of the lens shaped precipitate. In this direction there is a 2.9% contraction in the precipitate with respect to the matrix. TEM images indeed reveal this lens shape and conventional two-beam TEM contrast indicates the presence of a strain field [4]. Figure 1(a)(b) shows schematic top and side views of a precipitate while in figure 1(c) a two-beam bright field (BF) TEM image reveals the stress fields as strong contrast changes in the matrix surrounding the precipitates.

EXPERIMENTAL TECHNIQUES

Two batches of $Ni_{51}Ti_{49}$ samples were prepared; samples A contain precipitates with a diameter between 100 nm and 500 nm, the precipitates in samples B have a diameter smaller than 100 nm. The A and B samples were prepared from the same basic material but received a different heat treatment in vacuum. Both were first annealed at 950°C for 1 hour followed by water quenching and then aged for 4 hours at 500°C (A samples) and 450°C (B samples). TEM specimens were prepared by mechanical grinding followed by twin-jet electropolishing with a solution of 93% acetic acid and 7% perchloric acid at 6°C.

High resolution images are obtained with a top-entry JEOL 4000EX electron microscope equipped with a LaB_6 filament and operating at 400kV. Standard photographic plates were used in order to obtain as large regions as possible on a single image. Lattice deformations or strain are determined by measuring and comparing interplanar spacings by Fast Fourier Transformation (FFT) of different locations in the HR image. In practice, the pixel distance between the central

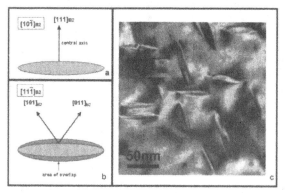

Figure 1. Schematic drawing of the lens shaped Ni_4Ti_3 precipitate in the two zones used for observation: a) the $[10\text{-}1]_{B2}$ and b) the $[11\text{-}1]_{B2}$. c) typical BF image of Ni_4Ti_3 precipitates with surrounding strain contrast.

spot and the spot belonging to the crystallographic plane under consideration is measured with subpixel accuracy by fitting a sinc² function to each spot. The measured interplanar spacing of the corresponding plane in the precipitate is chosen as reference distance. Differences Δd in interplanar spacing are given as percentage in accordance to this reference distance: for example, a Δd of 3% means that the measured interplanar spacing is 3% larger than the corresponding spacing in the precipitate. The accuracy of the method is 0.6% with a spatial resolution of 5 nm, the size of the Fourier window. This error was estimated by applying the technique to the image of an undistorted matrix and taking the standard deviation of the determined interplanar spacings. A second technique used to reveal the presence of strain fields is the geometrical phase image method developed by *Hÿtch et al* [13,14]. This method is based upon calculating the local Fourier components of the digitized high resolution image. After choosing a reference area in the image the local phase (i.e., the complex part of the Fourier components) is calculated for one specific **g** vector in the FFT of the image. The phase field that is received in this way is equivalent to the displacement field [13]. The strain components ε_{xx}, ε_{yy}, ε_{xy}, ε_{yx} as well as the rotation can be deduced from this data by calculating the gradient of the displacement field. The spatial resolution for this method is 2 nm and the error on the strain is 0.5% [15]. Distortions in the image due to the projector lens or introduced by digitizing the image are compensated by using the same reference images as used to determine the error in the first method. It will be shown that both methods give the same result taking into account the respective precisions.

EELS and EFTEM experiments are carried out on a Phillips CM30 field emission TEM equipped with a GIF2000 post column energy filter. When acquiring EELS spectra and EFTEM images, a zone orientation slightly off-axis from $[10\text{-}1]_{B2}$ is chosen in order to reduce diffraction effects and to have a minimal overlap between the lens shaped precipitate and the matrix. EELS spectra are collected in diffraction mode with a camera length of 195 mm and an entrance aperture to the GIF system of 2 mm, which corresponds to a collection semi-angle of 3.35 mrad (much larger than the estimated convergence angle 1.2 mrad). EFTEM images are acquired using the standard three-window method. The post-edge window was positioned right at the threshold of the $L_{3,2}$ edges of Ni and Ti. Energy window widths of 20 eV and 25 eV were chosen for Ti and Ni, respectively, to cover the white lines of each element. Relative drift between successive images was corrected by a cross-correlation technique when computing the elemental intensity map. Energy dispersive X-ray (EDX) analysis is carried out using a Phillips CM20 transmission electron microscope equipped with a Si(Li) Oxford EDX detector.

RESULTS AND DISCUSSION

Part I: Lattice strain measurements

Based on the known lattice parameters and crystallographic relations between the matrix and precipitates, the $[10\text{-}1]_{B2}$ and $[11\text{-}1]_{B2}$ zones are considered to be the most interesting ones since these will reveal the largest deformations. Moreover, in the $[10\text{-}1]_{B2}$ zone the $(111)_{B2}$ central plane is observed edge-on and thus a major part of the interface between the matrix and precipitate can also be considered to be viewed edge-on., however, this plane makes an angle of 19.47° with the incident beam, which results in an area of overlap between matrix and precipitate. On the other hand, the HRTEM images of the $[11\text{-}1]_{B2}$ zone are typically of a better quality due to overall larger lattice spacings.

In samples A precipitates with a diameter between 200 nm and 300 nm are selected for coherency reasons, as explained above. In a $[11\text{-}1]_{B2}$ orientation two of the three edge-on $\{011\}_{B2}$ families of planes $((101)_{B2}$ and $(011)_{B2})$ have a theoretical difference of 2.01% between the interplanar spacings of the precipitate and the unstrained matrix. The difference in interplanar spacing is measured along the $[101]_{B2}$ direction for the $(101)_{B2}$ planes. A typical image is shown in figure 2(a) and the graph in figure 2(b) corresponds with the strain field measured in the direction of the arrow. At a distance of 50 nm away from the centre of the precipitate Δd reaches a maximum of about 4%, which thus implies an expansive strain of about 2% with respect to the unstrained matrix. However, it is unclear from this image whether this value represents a true maximum or whether the expansive strain reaches even higher values when moving further away from the precipitate. Close to the interface the measured Δd is about 2%, i.e. unstrained matrix, and it increases linearly up to the maximum of 4%. On the other hand, at the tip of the precipitate the $(101)_{B2}$ interplanar spacing is found to be smaller than the corresponding value for the matrix thus implying a compression. The interplanar spacing of the third $\{011\}_{B2}$ family of planes, $(1\text{-}10)_{B2}$ which lies perpendicular to the central axis of the precipitate in the present crystallographic zone, has a theoretical mismatch of only 0.38% which is confirmed by the fact that no difference between these interplanar spacings was measured.

For samples B precipitates with a diameter around 50 nm were examined. In this case more precipitates are present and two of the same variant can even be examined in a single micrograph. In the example shown in figure 3 two parallel precipitates are found about 40 nm apart. To obtain a two dimensional image of the strain field with a better spatial resolution of 2 nm instead of 5 nm, the geometrical phase image method is applied. This method allows to determine the strain in any chosen direction. A reference area is taken in the same image at a position where no strain due to the precipitates is expected (dashed rectangle in figure 3(a)). The contour plot in figure 3(a) visualizes ε_{xx}, the strain in the x-direction, which is chosen parallel with the $[101]_{B2}$ direction. The plot in figure 3(b) shows the ε_{xx} profile of the strain field along the $[101]_{B2}$ direction at the arrow in figure 3(a). This profile can directly be compared with the measurements of differences in interplanar spacing for the (101) planes since $\varepsilon_{xx} = \Delta d_{(101)}$. The graph of the Δd measurements is given elsewhere [16], and reveals the same result within the

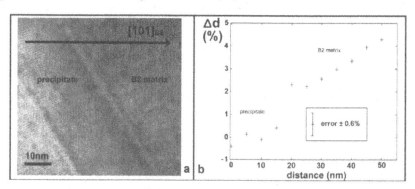

Figure 2. a) High Resolution image of a precipitate and surrounding matrix in $[11\text{-}1]_{B2}$ orientation. The arrow indicates the $[101]_{B2}$ direction along which Δd is measured, as given in (b).

precision of 0.5%. As the strain is now measured with respect to the unstrained matrix (dashed rectangle), the precipitates show a strain of -2%, the eigenstrain for the present direction. The maximum strain in the matrix is 2% and is found at a distance between 5 and 10 nm of the interface between the matrix and the nearby precipitate. When moving further away from the precipitate, Δd decreases sharply. In between both precipitates a region of approximately 20 nm of unstrained matrix is seen indicating that there is no interaction between the strain fields (and related stress fields) arising from these close by precipitates. Black and dark gray areas in the contour plot indicate a negative strain, which means that at the tip of the precipitate the matrix is slightly compressed, as was also the case for the larger precipitates.

As already mentioned above there is an area of matrix-precipitate overlap when looking in the $[11\text{-}1]_{B2}$ zone which implies that the information close to what appears to be the interface might be difficult to interpret. Image simulations (made with Mactempas) indicate that the lattice images observed in the present zone can be expected for samples between 10 and 15 nm in thickness, which corresponds with an area of overlap of 5.3 nm, i.e. certainly closer to the precipitate then the maximum strain location at 50 nm in samples A and only interfering with the first part of the increment towards the maximum in samples B. Moreover, although the image from these areas might be distorted because the two structures give an interfering image, image simulations show that the observed interplanar spacing for lattice images clearly revealing the matrix can never be larger than the real one, as expected. The maximum for the observed Δd, even in the case of the small precipitates, is therefore real and not an artefact. This maximum is reached at a distance d from the precipitate depending on the size of the latter.

Similar results can be obtained when viewing the matrix in the $[10\text{-}1]_{B2}$ zone. In this orientation the interface plane is viewed edge-on and so the region close to the interface can be used in good confidence since there is no region of overlap. Unfortunately, in most cases only line resolution of the (101) planes could be recorded in this zone. Again for samples A it was found that a compression appears at the tips of the precipitates and that the strain is increasing when moving away from the interface.

The experimental results of the strain fields are compared with a theoretical calculation based on the Eshelby approach. The complete solution for the stress field surrounding an elliptical inclusion is found in [17]. An algorithm provided by K. Gall was adapted and used to

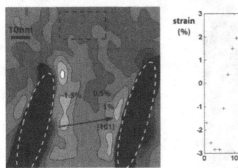
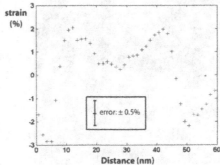

Figure 3. The plot shows the contour lines of the ε_{xx} strain component (x-axis is chosen parallel to $[101]_{B2}$). The graph corresponds with the profile along the arrow.

perform the computation of the strain field. Figure 4 shows a contour plot of $\varepsilon_{[101]}$ for a precipitate with approximately the same size (50 nm) and dimensions of those in Figure 3. This plot confirms the experimental observations on a qualitative level: the matrix is again slightly compressed at the tip of the precipitate and that the maximum of the strain is not localised at the interface but away from it. An important difference, however, is the magnitude of strain. A maximum strain of around $\dot{1}.5\%$ is measured whereas in the computation this is 0.12%, which is more than a factor of ten different. A reason for this might be the assumption of an equal stiffness for precipitate and matrix. Since the elasticity modulus of Ni_4Ti_3 has not yet been measured, this discrepancy between theoretical and experimental values might be an indication that this modulus of the precipitate is not equal to that of the matrix. Interactions between elasticity moduli and shape and orientation of precipitates in a matrix are indeed known [18,19]. Also, the real shape of the precipitates deviates from the ideal ellipsoid although one would only expect some visible differences around the tips. Other features that could affect the calculation of the strain are the actual orientation of the precipitate in the thinned foil in view of the direction of maximum strain and the use of a continuum model for phenomena at the nanoscale. The computation of the strain fields in such a case is more complex and the solution is not considered here.

Part II: Analytical TEM analysis

The formation of Ni_4Ti_3 precipitates not only introduces a coherent strain field in the surrounding matrix, it also affects the composition of the retained matrix since the precipitates are enriched in Ni with respect to the original material with a nominal composition of $Ni_{51}Ti_{49}$. The following EELS and EDX results reveal the effect on the composition in the retained matrix surrounding the Ni_4Ti_3 precipitates in samples A, i.e. those with relatively large precipitates.

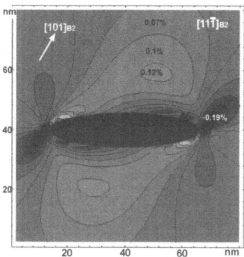

Figure 4. Computed strain field of the $\varepsilon_{[101]}$ component for a precipitate with a diameter of 50 nm.

The local concentration N_A for an element A in the material can be calculated from the EELS spectra by [20],

$$N_A = \frac{I_A(\beta, \Delta)}{I_{low}(\beta, \Delta) \bullet \sigma_A(\beta, \Delta)} \tag{1}$$

where $I_A(\beta, \Delta)$ is the measured ionization edge intensity integrated within an energy range Δ and inside a collection semi-angle β (after background subtraction and plural scattering removal by deconvolution with the Fourier-ratio method), $I_{low}(\beta, \Delta)$ is the intensity of a window of equal β and Δ containing the zero loss, and $\sigma_A(\beta, \Delta)$ is the partial ionization cross-section. The equation (1) can also be used in EFTEM analysis. The effect of plural scattering cannot be removed from the EFTEM intensity map; nevertheless the effect of diffraction contrast and thickness variation can be corrected by dividing the ionization map by the corresponding low loss image and thickness map. In the case of the binary NiTi the Ni/Ti atomic ratio is thus determined as

$$\frac{N_A}{N_B} = \frac{I_A(\beta, \Delta_A) \bullet \sigma_B(\beta, \Delta_B)}{I_B(\beta, \Delta_B) \bullet \sigma_A(\beta, \Delta_A)} = k_{AB}^E \bullet \frac{I_A(\beta, \Delta_A)}{I_B(\beta, \Delta_B)} \tag{2}$$

where $k_{AB}^E = \sigma_B(\beta, \Delta_B)/\sigma_A(\beta, \Delta_A)$ is the so-called k-factor. The stoichiometric proportion for the precipitate in the sample is considered as $N_{Ni}/N_{Ti} = 4/3$, and can be used as a reference standard. The k-factor can thus be determined from the precipitates in the samples and the elemental concentration for each element in the matrix can then be deduced from the measured Ni/Ti atomic ratio. This approach has the advantage that diffraction and thickness variation effects are largely eliminated as all references and measurements are obtained from regions close to one another.

Figure 5 shows a TEM image (a) and the corresponding EELS analysis results (b) on a precipitate and the neighboring matrix. The dark spots on the TEM image are contamination cones because of prolonged nanoprobe electron illumination and can be used as markers for the measured positions. Judging from the TEM image, there is no bending nor significant thickness variation within the investigated region. Further EELS analysis indicates that the thickness is in the range of 0.60-0.75 times of the inelastic scattering mean free path, which is suitable for EELS analysis. The reported data in figure 5(b) are averaged per single distance from the precipitate-matrix interface, thus each time containing three measurements. The calculated Ni/Ti atomic ratio, with a minimum value of 0.95, reveals a region depleted in Ni within a range of about 150 nm from the precipitate-matrix interface and which should be compared with the nominal value of 1.04 for the original $Ni_{51}Ti_{49}$ matrix. The standard deviation for the measurements inside the precipitate is about 1.0%, (although an absolute accuracy for one measurement generally could be considerably worse than 1.0% in absolute quantification by equation (1)). Additionally, the Ni and Ti concentration is calculated and shown in figure 5(b).

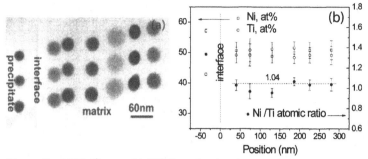

Figure 5. (a) TEM image, (b) EELS results showing the Ni/Ti atomic ratio, Ni and Ti concentrations.

The standard error for the measurements in the adjacent matrix is more than three times larger than in the case of the precipitate, which indicates a certain composition variation among the three averaged positions. Still, the measured depletion of Ni in the matrix can be considered real.

Figure 6 shows a TEM image with four smaller precipitates (1-4) meeting at their tips together with the map and profiles for two rectangular regions. A statistical analysis on the atomic ratio map yields a Ni/Ti ratio of 1.33 ± 0.03 for the precipitates. Profile analysis (c) of region A indicates that there are Ni-depleted regions on both sides of precipitate 3. The right side (with a mean Ni/Ti atomic ratio of 0.96 in a region stretching out over 40 nm) is more depleted in Ni than the left side (with a mean Ni/Ti atomic ratio of 1.00 over a region of 20 nm). Profile analysis (d) on the matrix above precipitate 2 shows a similar result as the left side of precipitate 3. This can be understood by the fact that the measured region on the right side of precipitate 3 could be depleted by the formation of precipitate 3 as well as 2.

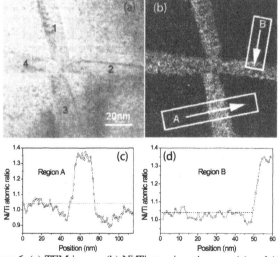

Figure 6. (a) TEM image, (b) Ni/Ti atomic ratio map, (c) and (d) profiles for regions A and B. The dashed lines show the atomic ratio for $Ni_{51}Ti_{49}$.

Precise composition results were obtained using the standard Cliff-Lorimer method for EDX quantification [21]. The relative elemental concentration (C_{Ni} and C_{Ti}) is related to the measured X-ray intensities (I_{Ni} and I_{Ti}) by an equation similar to equation (2). The value of the k-factor for the EDX quantification was again determined from the Ni_4Ti_3 phase in the samples. The measured concentration for Ni and Ti in the matrix was then calculated based on the results from the precipitates. Since the atomic numbers for Ni and Ti are close to one another, a thickness calibration of absorption of X-rays generated by Ni and Ti was not performed, i.e., any effect of thickness variation was neglected in the present EDX quantification. Figure 7 shows the results of an EDX analysis at a region with most of the Ni_4Ti_3 precipitates lying parallel to each other and at distances of around 200 nm. A Ni concentration of less than 51 at% was detected for most measured positions in the matrix. The averaged Ni concentration in the matrix is 50.44 ± 0.61.

From these local analytical measurements it can be concluded that a small but significant Ni depleted region surrounding a Ni_4Ti_3 precipitate can effectively be measured. From the nanoprobe EELS data it is seen that the depletion exists in an area up to 150 nm away from the precipitate-matrix interface. This area perfectly fits with the average value of the depletion when calculating the excess of Ni in the precipitate, i.e. all Ni needed to form the Ni_4Ti_3 structure is obtained from the 150 nm matrix region surrounding it. This immediately implies that the precipitates in the EDX study all lie within one another's range of depletion, explaining the lower mean concentration of Ni as measured in the matrix.

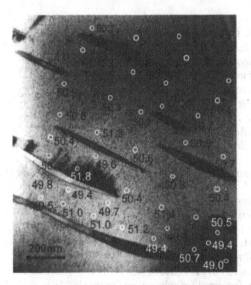

Figure 7. TEM image of a region investigated by EDX. Small circles and numbers show the measured position and the Ni concentration in at%.

CONCLUSIONS

The present work shows that the Ni_4Ti_3 precipitates influence the lattice parameters as well as the concentration in nanoscale regions of the surrounding matrix. For small precipitates the influence region extends to about 20 nm into the matrix for both strain and concentration whereas for large precipitates the concentration gradient can extend over 150 nm while the strain does not yet reach a maximum 50 nm away from the interface. In all cases these regions perfectly compensate for the lattice mismatch and Ni depletion induced by the precipitate which leads to matrix strains of 2% and local concentration changes up to 8%. It should also be noted that no R-phase or martensite was observed surrounding the investigated precipitates.

ACKNOWLEDGMENTS

Z. Yang is supported by the GOA project "Characterisation of nanostructures by means of advanced EELS and EFTEM" of the University of Antwerp. The authors like to thank Martin Hÿtch for providing the necessary software for the geometrical phase image method. Part of this work was supported by the Marie Curie Research Training Network "Multi-scale modelling and characterisation for phase transformations in advanced materials" (MC FP6-505226).

REFERENCES

1. T. Tadaki, Y. Nakata, K. Shimizu, K. Otsuka, *Trans JIM* **27**, 731 (1986)
2. M. Nishida, C.M. Wayman, *Mater. Sci. Eng.* **93**, 191 (1987)
3. M. Nishida, C.M. Wayman and T.Honma, *Metal. Trans. A* **17**, 1505 (1986)
4. L. Bataillard, J.-E. Bidaux, R. Gotthardt, *Phil. Mag. A* **78**, 327 (1998)
5. J. Khalil-Allafi, A. Dlouhy, G. Eggeler, *Acta. Mater.* **50**, 4255 (2002)
6. P. Filip and K. Mazanec, *Scripta Mater.* **45**, 701 (2001)
7. V. Zel'dovich, G. Sobyanina and T.V. Novoselova, *J. Phys. IV France* **7**, 299 (1997)
8. J. Khalil Allafi, X. Ren, G. Eggeler, *Acta. Mater.* **50**, 793 (2002)
9. W. H. Zou, X. D. Han, R. Wang, Z. Zhang, W-Z Zhang, J. K. L. Lai, *Mater. Sci. Eng. A* **219**, 142 (1996)
10. K. Gall, H. Sehitoglu, Y.I. Chumlyakov, I.V. Kireeva, H.J. Maier, *J. Eng. Mat. Tech.* **121**, 19 (1999)
11. D.Y. Li and L. Q. Chen, *Acta. mater.* **45**, 471 (1997)
12. C. Somsen, *"Mikrostrukturelle Untersuchungen an Ni-reichen Ni-Ti Formgedächtnislegierungen"*, Shaker Verlag, (2002)
13. M.J. Hÿtch, *Scanning Microscopy* **11**, 54 (1997)
14. M.J. Hÿtch, E. Snoeck, R. Kilaas, *Ultramicroscopy* **74**, 131 (1998)
15. M.J. Hÿtch, T. Plamann, *Ultramicroscopy* **87, 199** (2001)
16. W. Tirry, D. Schryvers, *Acta. Mater.* (accepted for publication; October 2004)
17. T. Mura, *"Micromechanics of defects in solids"*, Nijhoff : Boston, (1982)
18. Jun-Ho Choy, Jong K. Lee, *Mat. Sci. Eng. A* **285**, 195 (2000)
19. R. Mueller, D. Gross, Comp. *Mat. Sci* **11**, 35 (1998)
20. R. F. Egerton, *"Electron Energy-Loss Spectroscopy in the Electron Microscope"*, New York, (1996).
21. G. Cliff and G. W. Lorimer, *J. Microscopy*, **103**, 203(1975).

Mater. Res. Soc. Symp. Proc. Vol. 842 © 2005 Materials Research Society   S4.1

Sub-nano and Nano-structures of Hydrides of LaNi$_5$ and its related Intermetallics

Etsuo AKIBA, Kouji SAKAKI, Yumiko NAKANURA
Energy Technology Research Institute, National Institute of Advanced Industrial Science and
Technology (AIST), Tsukuba Central 5, 1-1-1 Higashi, Tsukuba, Ibaraki, 305-8565 Japan

ABSTRACT

Sub-nano and nano-structures of intermetallics such as LaNi$_5$ and its related alloys were
studied by in-situ X-ray and neutron diffraction methods.   From the profile shape analysis,
changes in lattice strain and crystalline size during hydrogenation/dehydrogenation were
estimated, while the crystal (sub-nano) structures were refined by the Rietveld method using
diffraction patterns.   The crystallite size of the alloys studied did not change during
hydrogenation and dehydrogenation.   It was found formation of remarkably dense dislocations
and vacancies in the lattice of hydrides of LaNi$_5$ and related intermetallics.

INTRODUCTION

LaNi$_5$ is a representative intermetallic compound which has a reversible hydrogen absorbing
property under ambient conditions.   The alloy phase transforms into hydride phase through
two-phase coexistence region.   The lattice parameters and the cell volume increase by 7 % and
24 %, respectively, in the transformation from the alloy phase to the hydride phase, respectively.
The hydriding reaction with such large lattice expansion in absorption and contraction in
desorption could induce various kinds of lattice defects and strain.

We have found that highly dense dislocations were introduced to the lattice of LaNi$_5$
intermetallic hydride at the first hydrogenation and it did not disappear in the following
desorption reaction using in-situ X-ray diffraction (XRD) technique [1].   Introduction of
dislocation of 10^{12}cm^{-2} was confirmed by Yamamoto et al. using transmission electron
microscopy (TEM) [2].   One of the authors found that both dislocations and vacancies were
introduced into the lattice of intermetallics at the first hydrogenation [3].

This paper will present a brief review of these experiments on LaNi$_5$ and its related
intermetallic compounds such as Al and Sn substituted LaNi$_5$.   In addition, the mechanism of
defects formation will be discussed.

EXPERIMENT

Preparation of alloys

An alloy ingot of LaNi$_5$ was prepared by high-frequency induction melting and alloy ingots
of LaNi$_{4.75}$Al$_{0.25}$ were prepared using the arc melting technique. They were purchased from

Santoku Metal Industry Co., Ltd. The ingot of LaNi$_{4.75}$Al$_{0.25}$ was annealed at 1373 K for 24 h in an Ar atmosphere. Alloy ingots of LaNi$_{4.75}$Sn$_{0.25}$ and LaNi$_{4.78}$Sn$_{0.22}$ were prepared from high purity metals using the arc melting technique by Robert C. Bowman, Jr. of Jet Propulsion Laboratory, USA [4]. They were annealed at 1223K for ~100h in a purified Ar atmosphere. The ingots were crushed into the particle size under 30 μm for X-ray powder diffraction measurements.

XRD measurements and peak profile analysis

In situ X-ray powder diffraction were measured using a horizontal sample stage type diffractometer (Rigaku, RINT-TTR) with a rotating Cu anode, a high-pressure chamber and temperature controller. Hydrogen content in the sample was measured simultaneously by Sieverts' method. The XRD data were analyzed by Rietveld refinement program RIETAN-97beta [5, 6]. A pseudo-Voigt function is used for expressing peak profiles [7]. Strain and crystallite size were calculated from the corresponding parameters using the equations in Ref.7.

RESULTS AND DISCUSSIONS

Change of crystallite size with hydrogenation-dehydrogenation

Fig. 1 shows the change of crystallite size at the first hydrogenation and dehydrogenation of LaNi$_5$ [1]. The crystallite size in the solid solution phase decreased from 200nm to 100nm but that in the hydride phase showed reverse tendency. It means that smaller size crystallite hydrogenates first and the larger crystallite remains un-reacted in the solid solution phase. It should be noted that the size of crystallite kept constant during the first hydrogenation and dehydrogenation, while the crystallite expands and contracts about 24 volume % with hydrogenation and dehydrogenation, respectively.

The crystallite size of Al and Sn substituted LaNi$_5$ did not changed with hydrogenation and the size of crystallite range 100 to 200 nm in these alloys, too [1, 8].

Introduction of dislocation

Fig. 2 shows the lattice strain obtained from profile analysis of X-ray diffraction data of LaNi$_5$ at the first hydrogenation [1]. In the solid solution phase, increase of the strain with hydrogenation was not observed. However, the remarkable strain in the direction of <*hk*0> was

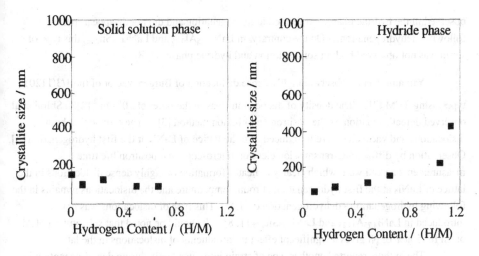

Figure 1. Change of crystallite size with first hydrogenation of LaNi$_5$ at 303K in solid solution and hydride phases [1].

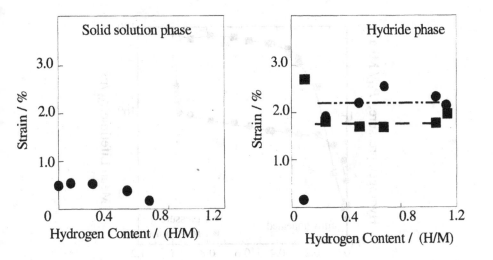

Figure 2. Strains introduced by first hydrogenation of LaNi$_5$ at 303K in solid solution and hydride phases. (Closed circle: isotropic strain, closed square: anisotropic strain) [1].

observed in the hydride phase.   It formed in hydrogenation and did not disappear in the
following dehydrogenation.   On the contrary, in LaNi$_{4.75}$Al$_{0.25}$ and LaNi$_{4.75}$Sn$_{0.25}$, this type of
strain was not observed both in solid solution and hydride phases [1, 8].

Yamamoto et al. observed highly dense dislocation of Burgers vector of the 1/3(1120)
types using TEM [2].   The density of dislocations was in the order of 10^{12}cm^{-2} [2].   Shirai et al.
observed defect formation by the positron annihilation method [3].   They found that both
dislocations and vacancies were introduced into the lattice of LaNi$_5$ at the first hydrogenation [3].
Observation by diffraction, transmission electron microscopy and positron life time
measurements agreed well, which clearly indicates formation of highly dense dislocations in the
lattice of LaNi$_5$ at the first hydrogenation at room temperature and the dislocations remains in the
following hydrogenation - dehydrogenation cycles.   This type of dislocations was not
introduced in LaNi$_{4.75}$Al$_{0.25}$ and LaNi$_{4.75}$Sn$_{0.25}$ [1, 8].   It should be noted that substitution of Al
or Sn for 5 at% of Ni gives a significant effect in introducing of dislocations in the lattice.

The authors reported another type of strain introduced only during dehydrogenation in
the hydride phases of LaNi$_{4.75}$Al$_{0.25}$ and LaNi$_{4.75}$Sn$_{0.25}$ using in-situ XRD [1, 8].   By estimation
of bulk modulus of LaNi$_5$ and its hydride, the hydride phase is more elastic than the alloy phase
[9].   Therefore, the strain is observed only in the hydride phase that is more easily to be
deformed.

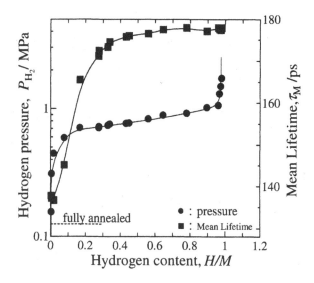

Figure 3.   *In situ* positron lifetime measurement of the first hydrogenation of LaNi$_5$ at room
temperature.

Introduction of vacancies

Fukai et al. reported that a large amount of vacancies were introduced into Ni and Pd systems under 5GPa of hydrogen [10]. Vacancy formation energy is reduced from 1.5eV without hydrogen to -0.50 eV if hydrogen coexists [10].

Positron lifetime measurement is very sensitive for detection of defects in solids [3]. From lifetime of positron, we can distinguish type of defects such as dislocation and vacancy. Figure 3 shows the change of positron lifetime during first hydrogenation of $LaNi_5$. Before hydrogenation, the positron lifetime was about 135 ps that means no defect was introduced in the lattice. However, at the same time to start to hydride formation (the start of the plateau region) the positron lifetime increased dramatically and reached to 180 ps. The lifetime of 180 ps indicates vacancy formation in the metallic lattice [3]. As clearly shown in Figure 3, we found vacancy formation even in the conventional hydrogen absorbing alloys such as $LaNi_5$ at the first hydrogenation process at room temperature. In addition, vacancy formation is more general because Shirai et al. found vacancy formation in other intermetallic hydrogen absorbing alloys such as Laves phase alloys by ex-situ positron lifetime measurements [3].

From the report of Fukai, vacancy-hydrogen bonding, at least stabilization of vacancy by the coexisting hydrogen, is one of the keys to understand introduction of defects into the hydrogen absorbing alloys.

We observed lattice contract after first hydrogenation and dehydrogenation for $LaNi_5$ and its related alloys. Change of lattice parameters after hydrogenation cycles is shown in Table I [11]. The change of lattice parameters of Al substituted $LaNi_5$ was not found, while that of $LaNi_5$ that is far above the standard deviations shown in the table is clearly observed.

Table I. Lattice parameters for $LaNi_5$, $LaNi_{4.5}Al_{0.5}$ and $LaNi_{4.5}Fe_{0.5}$ before and after hydrogenation-dehydrogenation [11].

		before hydrogenation	after hydrogenation-dehydrogenation
$LaNi_5$	a	0.51430(9) nm	0.50103(4) nm
	c	0.397987(7) nm	0.309862(2) nm
$LaNi_{4.5}Al_{0.5}$	a	0.502906(8) nm	0.502921(8) nm
	c	0.400204(7) nm	0.400223(6) nm
$LaNi_{4.5}Fe_{0.5}$	a	0.50372(1) nm	0.50340(3) nm
	c	0.40008(1) nm	0.40039(1) nm

Mechanism of defect formation

Recently, one of authors reported that defects especially vacancies were introduced after the phase transformation in Pd and Pd-Ag system [12].   Pd has a FCC structure and form hydrides of $PdH_{~0.6}$.   The Pd-H system has a critical point at about 573 K and the critical temperature of Pd can be lowered by alloying with Ag [13].   For example, the critical temperature of $Pd_{0.8}Ag_{0.2}$ is about 350K.   They hydrogenated Pd metal and Pd-Ag alloy, and measured positron lifetime after hydrogenation at 373K.   At 373K only Pd goes through two phase region because the critical temperature of Pd is higher than the hydrogenation temperature. In the case of $Pd_{0.8}Ag_{0.2}$ hydrogenation temperature of 373K is higher than the critical point (about 350K).   Therefore, reaction is formation of hydrogen solid solution of $Pd_{0.8}Ag_{0.2}$ [14]. They found vacancy formation in the hydride of Pd that formed through two phase region as shown in Figure 4.   On the contrary, vacancy formation was not observed in the "hydride" of $Pd_{0.8}Ag_{0.2}$ when the hydrogenation occurred above the critical point.   They also hydrogenated Pd-Ag alloys at 296K.   At 296K vacancy formation was observed for both Pd and $Pd_{0.8}Ag_{0.2}$. In this temperature, phase transformation from alloy to hydrides phases occur in both alloys because the critical temperatures of both alloys are higher than that of hydrogenation.   From temperature dependence of vacancy formation, it can be concluded that phase transformation from the alloy phase to the hydride phase closely relates to vacancy formation.

In hydrogen absorbing alloys, crystallite size does not change with hydrogenation but in some cases defects such as dislocation and vacancies were introduced.   The lattice volumes of the alloy and the hydride differ in 20 to 30 % and the hydrogenation is usually completed within a few minutes if the reaction heat is effectively removed.   *Elastic strain* is formed in the boundary between the alloy and the hydride and there should be a mechanism to lower or release

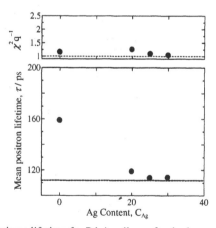

Figure 4.   Positron lifetime for Pd-Ag alloys after hydrogenation at 373K.

this energy. In addition, this two phase boundary, in other words the reaction front, may run by considerable rate in the crystallite at room temperature.

Defects introduced during hydrogenation are due to lower or release the elastic strain energy in or around the boundary of two phases. The defect found in LaNi₅ remains in the crystallite after hydrogenation but another type of defects found in Al and Sn substituted LaNi₅ disappears after the reaction. In addition, it should be noted that hydrogen forms vacancy-hydrogen bonding and it seems to play an important role in stabilizing vacancies, especially in the present alloy-hydrogen systems.

CONCLUSIONS

We have found that the crystallite size of typical intermetallic hydrogen absorbing alloy, LaNi₅, did not change with hydrogenation and dehydrogenation but highly dense dislocations and vacancies are introduced. However, if 5 at % of Ni are replaced by Al or Sn, such type of defects was not observed but another type of defects was found during dehydrogenation in the hydride phase.

Formation of defects in hydrogenation and dehydrogenation is due to elastic strain formed in the boundary of the alloy and hydride phase. Hydrogen seems to play an important role in stabilizing vacancies.

REFERENCES

1. Y. Nakamura and E. Akiba, *J. Alloys Compd.* **308**, 309 (2000).
2. T. Yamamoto, H. Inui and M. Yamaguchi, *Intermetallics*, **9**, 987 (2001).
3. Y. Shirai, H. Araki, T. Mori, W. Nakamura and K. Sakaki, *J. Alloys Compd.* **330-332**, 125 (2000).
4. R.C. Bowman, Jr., C.A. Lindensmith, S. Luo, Ted B. Flanagan and T. Vogt, *J. Alloys Compd.*, **330-332**, 271 (2002).
5. F. Izumi, "Rietveld analysis program RIETAN and PREMOS and special applications" *The Rietveld Method*, ed. R. A. Young (Oxford Univ. Press, 1993) pp.236-253.
6. F. Izumi, http://homepage.mac.com/fujioizumi/
7. A.C. Larson and R.B. Von Dreele, "GSAS-General Structure Analysis System," Report No. LAUR 86-748, Los Alamos National Laboratory (1994).
8. Y. Nakamura, R. C. Bowman, Jr. and E. Akiba, *J. Alloys Compd.* **373**, 183 (2004).
9. L.G. Hector, Jr., J. F. Herbst and T. W. Capehart, *J. Alloys Compd.* **353**, 74 (2003).
10. Y. Fukai and N. Okuma, *Jpn., J. Appl. Phys.*, **32**, L1256 (1993).
11. Y. Nakamura, K. Oguro, I. Uehara and E. Akiba, *J. Alloys Compd.* **298**, 138 (2000).
12. K. Sakaki, M. Mizuno, H. Araki and K. Sakaki, *Mater. Trans.*, **43**, 2652 (2002).
13. M. Nuovo, F. M. Mazzolai and F. A. Lewis, *J. Less-Common Met.*, **49**, 37 (1976).
14. K. Sakaki, R. Date, M. Mizuno, H. Araki and K. Sakaki, to be published.

Mater. Res. Soc. Symp. Proc. Vol. 842 © 2005 Materials Research Society   S3.2

Transformation Behavior of TiNiPt Thin Films Fabricated Using Melt Spinning Technique

Tomonari Inamura, Yohei Takahashi[*], Hideki Hosoda, Kenji Wakashima,
Takeshi Nagase[1], Takayoshi Nakano[1], Yukichi Umakoshi[1] and Shuichi Miyazaki[2].

Precision and Intelligence Laboratory, Tokyo Institute of Technology,
Yokohama, Kanagawa 226-8503, Japan.   (*Graduate student, Tokyo Institute of Technology)
[1]Department of Materials Science & Engineering, Graduate School of Engineering,
Osaka University, Suita, Osaka 565-0871, Japan.
[2]Institute of Materials Science, University of Tsukuba, Tsukuba, Ibaraki 305-8573, Japan.

ABSTRACT

Martensitic transformation behavior of $Ti_{50}Ni_{40}Pt_{10}$ (TiNiPt) melt-spun ribbons were investigated where the heat treatment temperature was systematically changed from 473K to 773K. A hot-forged bulk TiNiPt material with the similar chemical composition was also tested as a comparison. θ-2θ X-ray diffraction analysis and transmission electron microscopy observation revealed that the as-spun ribbons were fully crystallized. The apparent phases of as-spun ribbons at room temperature are both B19 martensite and B2 parent phase instead of B2 single phase for the hot-forged bulk material. No precipitates were found in as-spun and heat-treated ribbons. It was revealed by differential scanning calorimetry that all the specimens exhibit one-step transformation. The martensitic transformation temperatures of the TiNiPt as-spun ribbons are 100K higher than those of the hot-forged bulk material, and the martensitic transformation temperature decreases with increasing heat treatment temperature.

INTRODUCTION

Ti-Ni alloy has a thermoelastic martensitic transformation from B2-structure (parent) to B19'-structure (martensite) and exhibits shape memory effect (SME) and superelasticity (SE) [1]. Actuation temperature of shape memory alloys (SMAs) is determined by the martensitic transformation temperature (M_s) of the alloys. M_s of Ti-Ni binary alloy is at most around 400K [1] and therefore the SME of Ti-Ni alloy cannot be used at high above 400K. In order to expand the applications related with SMAs, development of Ti-Ni based SMAs with high M_s compared to the Ti-Ni binary alloy has been strongly required.

Effects of ternary addition on M_s of Ti-Ni have been extensively investigated and most of additional elements such as Co, Fe, Mn, Cr and V are known to decrease M_s of Ti-Ni. However, it has been known that some ternary additions such as Hf, Nb, Zr, Pd, Au and Pt raise M_s of Ti-Ni [2-4]. Our group has systematically investigated SME of Ti-Ni alloys containing platinum group metals in the pseudobinary systems of TiNi-TiRh, TiNi-TiIr and TiNi-TiPt [4-7].

It is known that microstructure, transformation behavior and mechanical properties of rapidly solidified Ti-Ni based SMAs are different from those of bulk material [8]. However, there is no report concerning phase constitution, transformation behavior and mechanical properties of rapidly solidified Ti-Ni alloy containing Pt. In this paper, phase constitution, crystal structure and transformation temperatures of Ti-Ni-Pt fabricated using rapid solidification technique are presented. Although mechanical properties of the rapidly solidified Ti-Ni-Pt SMA ribbons fabricated in this work were also investigated, details will be presented elsewhere.

EXPERIMENTAL DETAILS

Figure 1 shows M_s of the TiNi-TiPt pseudobinary alloys [7]. It was clear that M_s is raised by the Pt addition with 27K/mol%Pt when the concentration of Pt is more than about 10mol%.

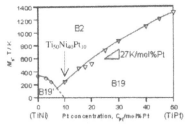 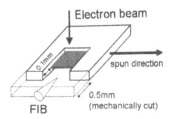

Figure 1. M_s of TiNi-TiPt pseudobinary alloys fabricated by hot-forging [7]

Figure 2. Geometry of FIB machining of thin foils for TEM

The crystal structure of the martensite phase is also affected by Pt addition and is changed from B19'-structure (monoclinic) to B19-structure (orthorhombic) by Pt additions of more than about 10% [6-7]. The composition used in this study was $Ti_{50}Ni_{40}Pt_{10}$ and is additionally indicated in Fig. 1. Master ingots of $Ti_{50}Ni_{40}Pt_{10}$ alloy (termed TiNiPt hereafter) were fabricated by an Ar-arc melting method in an Ar-1%H₂ atmosphere. The ingots were remelted five times for homogeneity. The weight of each ingot was about 5g. No chemical analysis of the ingots was conducted because the change in weight before and after the arc-melting was less than 0.1wt% and was judged to be negligible.

Rapid solidification was conducted using a single roll melt-spinning apparatus. An ingot was cut into several pieces by a spark-cutting machine. Some pieces of the cut ingot were put in a quartz crucible and set in the melt-spinning apparatus. Rapid solidification was carried out by induction melting in an Ar atmosphere followed by ejecting the molten alloy onto a copper wheel rotating at a surface velocity of 42m/s with a gap of 400μm. As-spun ribbons were termed 'As-spun' hereafter. The ribbons were heat-treated at 473, 623 or 773K for 3.6ks in vacuum and then furnace-cooled. The heat-treated ribbons were termed like 'HT(473K)' as shown by the heat treatment temperature, hereafter.

A reference material was prepared by hot-forging at 1173K for 3.6ks followed by furnace cooling, and then it was solution-treated at 1273K for 1.8ks followed by furnace cooling. The reference material is termed 'Bulk-material' hereafter. It should be mentioned that M_s in Fig. 1 was the experimental results for the similar materials fabricated by the hot-forging method [7].

Martensitic transformation temperatures were measured by differential scanning calorimetry (DSC) using SHIMADZU DSC-60 with a heating/cooling rate of 10K/min in a temperature range from 100K to 500K. Phase constitution was examined by CuKα θ-2θ X-ray diffraction (XRD) analysis at RT using Philips X'pert Pro equipped with X'Celerator and a graphite monochromator. Si was used as an external standard material. Transmission electron microscopy (TEM) observation was also carried out using Philips CM200 at an acceleration voltage of 200kV. Thin-foil preparation by a conventional electro-polishing was unsuccessful, probably due to the ternary addition of Pt. Thin-foil specimens for TEM were thus prepared by focused ion beam (FIB) machining using HITACHI FB-2100 with Ga⁺ accelerated at 40kV. Figure 2 depicts the geometry of the prepared thin-foil. FIB was aligned to be almost parallel to the thin foil. The beam current for the final thinning was 0.11nA. Scanning ion microscopy (SIM) observation was also carried out using FIB to observe surface morphology and microstructure of the material.

RESULTS AND DISCUSSION

Crystal structure and microstructure of the ribbon

Melt-spun ribbons of TiNiPt were successfully fabricated under the conditions described above. The appearance of the fabricated ribbons is shown in Fig. 3. The ribbons obtained were 35~40μm in thickness and 1.7~1.8mm in width. Figures 4(a) and 4(b) show the SIM images of the As-spun and HT(773K), respectively. Grain size of As-spun is seen to be inhomogeneous and is in the order of a few tens of μm. Striations are seen in each grain of As-spun and should be due to the presence of martensite plates. No significant change in grain size was recognized even after the heat-treatment at 773K judging from SIM observations. It should be noted that no striations were observed in the grains of HT(773K).

Figure 5 shows XRD patterns taken from the surface of wheel-side of each specimen (except for bulk-material). Only reflections from B2 phase (indicated by filled circles) were observed in the Bulk-material.

As-spun ribbons seemed to be fully crystallized and to contain a large amount of martensite phase (indicated by open triangles). It should be plausible to consider that the possible crystal structure of martensite is B19 (orthorhombic, $a<b<c$ [9]) or B19' (monoclinic, unique axis b, $a<b<c$ [10]). These two structures generate similar XRD profiles. However, a lattice parameter calculation using a linear least-square method in a reciprocal space gives much smaller standard deviation for B19. Therefore, the XRD-profile of As-spun in Fig. 5 is indexed with B19. The lattice parameters of B19 martensite in As-spun was calculated to be

Figure 3. Appearance of the as-spun ribbons

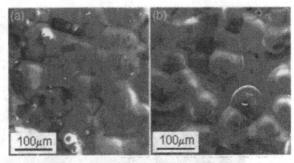

Figure 4. SIM images of wheel side surface of (a) As-spun and (b) HT(773K)

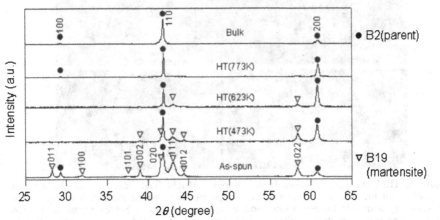

Figure 5. XRD profiles of each specimen taken at RT showing that as-spun is consisted with B2 (parent) and B19(martensite) and the B2-phase is stabilized by the heat-treatments.

$a_{B19}=0.2799\overline{+}0.0002$ nm, $b_{B19}=0.4340\overline{+}0.0001$ nm and $c_{B19}=0.4612\overline{+}0.0001$ nm.

In addition to the XRD analysis, TEM observation was carried out to obtain a decisive conclusion on the crystal structure of the martensite in As-spun. Figure 6(a) shows a bright field image of twinned morphology of martensite in As-spun. The specimen was systematically tilted around some low-indexed poles and selected area diffraction patterns were recorded in a martensite plate (indicated by the single headed arrow in Fig.6(a)). Figures 6(b), 6(c) and 6(d) show selected area diffraction patterns taken with the zone axis of 100_{B19}, 210_{B19} and 410_{B19}, respectively. Subscript 'B19' indicates B19-phase. When the crystal structure of martensite is B19' the angle between broken lines in Fig. 6(c) and 6(d) is a few degrees away from 90degrees depending on the monoclinic angle. Therefore, the crystal structure of the martensite in As-spun was certainly confirmed to be B19 (orthorhombic). No remnant of existence of B19' or R-phase was detected in the TEM observation.

Space group of B19 is *Pmmb* for the notation used in this study [9]. The only extinction rule is that $hk0$ reflections are not observed for odd k in *Pmmb*. Based on the atom positions reported in ref. [9], we can deduce that disordered B19 takes, for the notation in this study, space group of *Bmmb* and then 011 and 100 reflections are not observed. It should be noted that the as-spun ribbons were ordered enough to generate 100_{B2}, 011_{B19} and 100_{B19} reflections clearly in

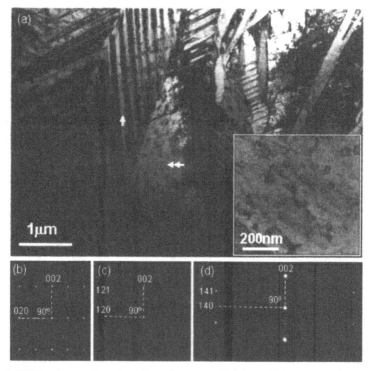

Figure 6. TEM micrographs taken from As-spun. Bright field image (a) shows twinned morphology of the martensite. Dislocation loops are seen around the center of the image (indicated by a double headed arrow) and is enlarged in inset. Selected area diffraction patterns taken from a martensite plate (indicated by a single headed arrow) with zone axis of (b) 100_{B19}, (c) $2\overline{1}0_{B19}$ and (d) $4\overline{1}0_{B19}$ show that the martensite is B19 (orthorhombic).

XRD profiles in Fig. 5.

Inset of Fig. 6(a) shows enlargement of dislocation loops which are seen around the center of the bright field image (indicated by double headed arrow in Fig. 6(a)). This kind of dislocation loops were frequently observed in As-spun. Similar defects were also confirmed even after the heat-treatment at 773K (HT773K). However, the nature of these defects was not clear in the present thin-foils. Edge of the thin-foil was partially amorphous and irradiation damages introduced to the specimen during FIB machining was not negligible in the observed thin-foils. Therefore, details of the lattice defects seen are not further mentioned in this paper. However, it should be noted that no second phase or precipitation such as Ti_3Ni_4 was observed even after the heat treatment.

As the heat-treatment temperature was raised, relative intensity of reflections from B2-phase increased compared to those from martensite in Fig. 5. The apparent phase at RT in HT(773K) was B2 as is observed in Bulk-material. The lattice parameter of B2 phase was calculated for HT(773K) as $a_{B2}=0.30428+0.00003$ nm, where subscript 'B2' indicate B2-phase. The lattice correspondence for B2-B19 martensitic transformation has been proposed in AuCd alloy [11]. Maximum normal strains generated by the lattice deformation of the B2-B19 transformation were evaluated to be +7.2% almost along $<110>_{B2}$ and -8.0% almost along $<001>_{B2}$ in the ribbons.

Transformation temperatures

Figure 7 shows DSC curves obtained during cooling. All the specimens seemed to exhibit one-step transformation from B2-B19 considering with the crystal structure analyzed in the above section. M_s and M_f (martensite transformation finish temperature) are indicated in the DSC curves. M_s of Bulk-material was 233K and that of As-spun was 335K. M_s of As-spun is 102K higher than that of Bulk-material. M_s, M_f, A_s (reverse transformation start temperature) and A_f (reverse transformation finish temperature) are summarized in Table 1. As the heat-treatment temperature is raised, the transformation temperatures become lowered. M_s of HT(773K) is 293K and is 42K lower than that of As-spun. However, M_s is still 60K higher than that of Bulk-material even after the heat-treatment at 773K. These results are in good agreement with the phase constitution determined by XRD analysis.

Changes in transformation temperatures by the heat-treatments should be related with the changes in substructure of the material during the heat-treatments. However, reliable TEM observations of substructures could not be conducted in this study. The origin of the changes in transformation temperatures by the heat-treatment is not clearly understood at present.

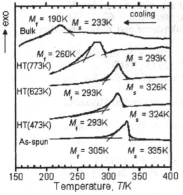

Figure 7. DSC curves during cooling

Table 1. Transformation temperatures determined by DSC

	M_s	M_f	A_s	A_f
Bulk	233K	190K	211K	266K
HT(773K)	293K	260K	250K	313K
HT(623K)	326K	293K	315K	338K
HT(473K)	324K	293K	315K	335K
As-spun	335K	305K	354K	375K

CONCLUSIONS

$Ti_{50}Ni_{40}Pt_{10}$ (TiNiPt) melt-spun ribbons with 30~40μm in thickness, 1.8mm in width was successfully fabricated using single roll rapid solidification technique. As-spun ribbons were fully crystallized. The crystal structure of the martensite was B19-structure (orthorhombic) and that of parent phase was B2-structure. Superlattice reflections which are typical to B19 and B2 were clearly observed by XRD and TEM in as-spun ribbons and indicate that as-spun ribbons were well ordered. The lattice parameters of B19 martensite phase are a_{B19}=0.2799+0.0002 nm, b_{B19}=0.4340+0.0001 nm and c_{B19}=0.4612+0.0001 nm, and that of B2 parent phase is a_{B2}=0.30428+0.00003 nm. Maximum normal strains generated by the lattice deformation of the B2-B19 transformation were evaluated to be +7.2% almost along <110>$_{B2}$ and -8.0% almost along <001>$_{B2}$.

As the heat-treatment temperature was raised, relative intensity of reflections from B2-phase increased compared to those from martensite. The apparent phase at RT in the ribbons heat-treated at 773K was B2 as is observed in the hot-forged material. No precipitates such as Ti_3Ni_4 were found in the as-spun ribbons and heat-treated ribbons.

DSC curves of all the specimens show one-step transformation which should correspond to the martensitic transformation between B2 and B19. The martensitic transformation temperature of the as-spun ribbons is about 100K higher than those of the hot forged bulk-material. Heat-treatments decrease the martensitic transformation temperatures of the ribbons. Lowering of transformation temperatures by the heat-treatment becomes significant by raising the heat-treatment temperature.

ACKNOWLEDGEMENTS

This work was partially supported by Grant-in-Aid for Fundamental Scientific Research (Wakate B: No. 16760566), the 21st COE program from the Ministry of Education, Culture, Sports, Science and Technology, Japan, and Osawa Scientific Studies Grants Foundation.

REFERENCES

1. T. Saburi, "Ti-Ni Shape Memory Alloys", *Shape Memory Materials*, ed. K. Otsuka and C. M. Wayman (Cambridge University Press, 1998) pp. 49-96.
2. J. Van Humbeeck and G. Firstov in *Proc. of The Forth Pacific Rim Int. Conf.*, edited by. S. Hanada et al. (The Japan Ints. Metals, 2001) pp.1871-1874.
3. V. N. Khachin, V. G. Pushin, V. P. Sivokha, V.V. Kondrat'yev, S. A. Muslov, V. P. Voronin, Yu. S. Zolotukhin and L. I. Yurchenko, *Phys. Met. Metall.* **67**, 125 (1989)
4. H. Hosoda, M. Tsuji, M. Mimura, Y. Takahashi, K. Wakashima and Y. Yamabe-Mitarai, *Mat. Res. Soc. Symp. Proc.*, **753**, BB5.51.4 (2003)
5. H. Hosoda, M. Tsuji, Y. Takahashi, T. Inamura, K. Wakashima, Y. Yamabe-Mitarai, S. Miyazaki and K. Inoue, *Mater. Sci. Forum*, **426-432**, 2333 (2003)
6. M. Tsuji, H. Hosoda, K. Wakashima and Y. Yamabe-Mitarai, *Mat. Res. Soc. Symp. Proc.*, **753**, BB5.52.1 (2003)
7. Y. Takahashi, M. Tsuji, J. Sakurai, H. Hosoda, K. Wakashima and S. Miyazaki, *Trans. MRS-J.*, **28**, 627 (2003)
8. E. Cesari, P. Ochin, R. Portier, V. Kolomytsev, Yu. Koval, A. Pasko, V. Soolshenko, Mater. Sci. Eng., **A273-275**, 733 (1999)
9. L. C. Chang and T. A. Read, *Trans. AIME*, **189**, 47 (1951)
10. K. Otsuka, T. Sawamura and K. Shimizu, *Phys. Stat. Sol.*, **5**, 457 (1971)
11. D. S. Lieberman, M. S. Wechsler and T. A. Read, *J. Appl. Phys.*, **26**, 473 (1955)

Factors for Controlling Martensitic Transformation Temperature of TiNi Shape Memory Alloy by Addition of Ternary Elements

Hideki Hosoda, Kenji Wakashima, Shuichi Miyazaki[1] and Kanryu Inoue[2]

Precision and Intelligence Laboratory, Tokyo Institute of Technology, Yokohama, Kanagawa 226-8503, Japan.
[1] Institute of Materials Science, University of Tsukuba, Tsukuba, Ibaraki 305-8573, Japan.
[2] Department of Materials Science and Engineering, University of Washington, Seattle, WA 98195-2120, USA.

ABSTRACT

Correlations between the changes in martensitic transformation start temperature (M_s) by addition of ternary elements X and several factors of the ternary additions were investigated for TiNi shape memory alloy. The change of M_s by addition of 1mol%X is referred to as ΔM_s (in K/mol%), and ΔM_s was systematically evaluated by differential scanning calorimetry experimentally using (Ti, X)$_{50}$Ni$_{50}$ solution-treated at 1273K for 3.6ks where the Ni content was kept constant to be 50mol%. The ternary additions X investigated are the transition metal (TM) elements selected from 4th period group (Zr, Hf) to 10th period group (Pd, Pt). The factors investigated are (1) the number of total outer d- and s-electrons (N_{ele}), and electron hole number (N_V), (2) electronegativity (E_N), (3) atomic volume (V_X) and (4) Mendeleev number (N_M). It was found that the values of ΔM_s are different even in a same period group; ΔM_s of 6th period group are -133K/mol%Cr ($3d$-TM), -152K/mol%Mo ($4d$-TM) and -64K/mol%W ($5d$-TM) for example. The results found in the correlations between ΔM_s and those factors are summarized as follows. (1) ΔM_s depends on N_{ele} and N_V. However, the data are scattered because same N_{ele} and N_V are often given in a same period group. Then, other factors than N_{ele} and N_V are required for clear understanding of ΔM_s. (2) ΔM_s seems to become lowered slightly with increasing E_N. (3) ΔM_s weakly depends on atomic volume V_X. Ternary addition with large V_X increases ΔM_s slightly, and with small V_X decreases ΔM_s largely. Since the stress field must be formed by substitution due to size mismatch, the type of stress field, tension/compression, may be an important role to determine the sign of ΔM_s. (4) ΔM_s shows a good correlation with N_M as -9.4Kmol%$^{-1}$/ΔN_M where ΔN_M is the difference in N_M. This suggests that a ternary alloying element with smaller (larger) N_M stabilizes the B19' martensite (B2 parent) phase. Effect of site occupancy on M_s is also discussed only for Cr.

INTRODUCTION

TiNi alloy is a representative actuator material due to its unique shape memory effect and superelasticity. In order to develop better actuator material based on shape memory alloy (SMA), the precise control of martensitic transformation temperature (M_s) is required. It was already reported that M_s of TiNi is influenced by applied and/or internal stress based on the Clausius-Clapeyron relationship [1]. Besides, M_s depends on chemical compositions; several works have been done for the effects of stoichiometry [2, 3] and ternary additions [2, 4-13]. It was known that most ternary elements such as Fe, Co, Cr reduce M_s [4-7] but that limited elements such as Zr, Hf, Pd, Pt and Au raise M_s [2, 8-12]. Wang [4] and Honma et al. [5] reported that M_s depends on electron atom ratio (e/a) for TiNi containing V, Cr, Mn, Fe and Co

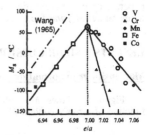

Figure 1. e/a dependence of M_s for TiNi [5].

where e/a is calculated based on the number of total $3d$- and $4s$-electrons, N_{ele} (= $N_{3d}+N_{4s}$, where N_{ele}, N_{3d} and N_{4s} are the numbers of total outer electrons, $3d$-electrons and $4s$-electrons, respectively). The e/a dependence of M_s is shown in Figure 1 [5].

We have reported a good correlation between M_s and electron hole number N_V for $3d$-transition metals [7, 13]. N_V is known to be an important factor to express phase stability of σ-phase formation for Ni-base superalloys in PHACOMP [14]. Besides, we have also found a good correlation between M_s and e/a for TiPd in Figure 2 [15]. It should be noted that N_{ele} used in Fig. 2 is as $N_{ele} = N_{3d} + N_{4s} - 2$ for $3d$-transition elements and $N_{ele} = 10$ for Pd. It is important that the number of d-electrons is different between Ni (=$3d^8$) and Pd (=$4d^{10}$).

Although many works have been done about M_s of ternary TiNi, few systematic works for $4d$- and $5d$-transition elements with a similar heat treatment condition.

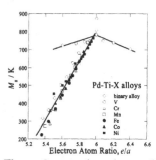

Figure 2. e/a dependence of M_s for TiPd [15], where $N_{ele}=N_{3d}+N_{4s}-2$ for $3d$-elements.

The similar heat treatment condition is needed for quantitative comparison in M_s of ternary TiNi alloys. This is because, the degree of long range order parameter, which depends on heat treatment condition, is suggested to influence M_s of TiNi [6] and the influence was confirmed for NiMnGa SMA [16].

Based on the background, the effects of ternary additions on M_s of TiNi were systematically investigated in this paper. Then, correlations between change in martensitic transformation start temperature (M_s) by addition of 1mol% of ternary elements (X) and the following factors were discussed: the number of electrons (N_{ele}), electron hole number (N_V), electronegativity (E_N), atomic volume (V_X) and Mendeleev number (N_M). N_M is a phenomenological coordinate number proposed by D. G. Pettifor, and the structure mapping, which expresses the appearance of stable crystal structure, can be successfully made using N_M as "Pettifor map" [17, 18].

EXPERIMENTAL PROCEDURE

The chemical compositions selected were $(Ti,X)_{50}Ni_{50}$ where the Ni content was fixed to be 50mol%. X used are V, Cr, Mn, Fe, Co for $3d$-, Zr, Nb, Mo and Pd for $4d$- and Hf, Ta, W and Pt for $5d$-transition metals. The concentrations of X were up to 3mol%. These alloys were fabricated by Ar arc melting method using high purity elemental materials and remelted several times for homogeneity. No chemical analysis was made because of small weight change before and after the melting. The ingots were heat treated at 1273K for 3.6ks in vacuum followed by quenching into iced water. Martensitic transformation temperature was measured by differential scanning calorimetry (DSC) with a heating/cooling rate of 10K/min. In this paper, only martensitic transformation start temperature (M_s) was focused.

RESULTS AND DISCUSSION

Martensitic transformation temperature

Figure 3 shows M_s obtained by DSC as a function of ternary content C_X. M_s changes almost

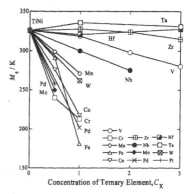

Figure 3. M_s obtained by DSC.

linearly with increasing ternary content C_X in each alloy system. It should be mentioned that two-step DSC peaks appeared in some DSC curves. The two-step DSC peaks stands for the two-step phase transformations from parent to R-phase and from R-phase to martensite phase in cooling. In these cases, the second transformation start temperature from R-phase to martensite phase was employed as M_s. In order to compare effects of ternary addition on M_s, the first derivative dM_s/dC_X which stands for the M_s change by addition of 1mol% of ternary element X was evaluated and referred to as ΔM_s (unit: K/mol%X). ΔM_s obtained are listed in Table 1. It is clear that ΔM_s are small values (4 to -12K/mol%) when X belong to 4th and 5th period groups, and that ΔM_s are negative large values (-55 to -152K/mol%) for the others. The largest increase and decrease in ΔM_s appear for Ta and Mo additions, respectively. It should be emphasized that, even when X belong to a same period group, ΔM_s are different each other: for example, -133K/mol%Cr, -152K/mol%Mo and -64K/mol%W for 6th period group. These results suggest that other factor than e/a is also required to express the dependence of ΔM_s.

Effect of Site Occupancy

It should be mentioned that ΔM_s of -133K/mol%Cr is largely different from the predicted value of -40K/mol%Cr when Cr atoms occupy the Ti-sites [6, 7, 13]. We have proposed that such a deviation from the prediction is caused by antisite defects and actual site occupancy of ternary elements. The ΔM_s experimentally obtained is proposed to be expressed as follows,

$$\Delta M_s = \Delta M_s^{\text{Ni (Ti)}} C^{\text{Ni (Ti)}} + \Delta M_s^{\text{Ti(Ni)}} C^{\text{Ti(Ni)}} + \Delta M_s^{X(\text{Ti})} C^{X(\text{Ti})} + \Delta M_s^{X(\text{Ni})} C^{X(\text{Ni})} \qquad (1),$$

where ΔM_s is the experimental value, and $\Delta M_s^{\text{Ni(Ti)}}$, $\Delta M_s^{\text{Ti(Ni)}}$, $\Delta M_s^{X(\text{Ti})}$ and $\Delta M_s^{X(\text{Ni})}$ are M_s changes by antisite Ti atoms occupying the Ni sites, antisite Ni atoms occupying the Ti sites, ternary X occupying the Ti sites and the Ni sites, respectively. $C^{\text{Ni(Ti)}}$, $C^{\text{Ti(Ni)}}$, $C^{X(\text{Ti})}$ and $C^{X(\text{Ni})}$ are the concentrations of Ni occupying the Ti sites, Ti occupying the Ni sites, X occupying the Ti sites and X occupying the Ni sites, respectively. Hereafter, we consider 49mol%Ti-50mol%Ni-1mol%Cr alloy as an example, and assume that all Ti atoms occupy the Ti sites, $k$mol%Cr actually occupy the Ti sites and that $(1-k)$mol%Cr occupy the Ni sites. k is a distribution coefficient of $0 \leq k \leq 1$ which stands for the site probability. Due to the Cr occupying the Ni sites, excess $(1-k)$mol%Ni occupy the Ti sites. Thus, the actual site occupation is expressed as,

Table 1. M_s change by adding 1mol% of ternary element (ΔM_s) evaluated from Figure 3.

Period	3d	ΔM_s (K/mol%)	4d	ΔM_s (K/mol%)	5d	ΔM_s (K/mol%)
4th	(Ti)	(0)	Zr	-1	Hf	2
5th	V	-12	Nb	-4	Ta	4
6th	Cr	-133	Mo	-152	W	-64
7th	Mn	-55	(Tc)	-	(Re)	-
8th	Fe	-145	(Ru)	-	(Os)	-
9th	Co	-108	(Rh)	-	(Ir)	-
10th	(Ni)	(-150)	Pd	-124	Pt	-136

The parenthetic data were evaluated using M_s of the Ti-Ni binary system in Ref.[3].

Table 2. Values used for the calculation of Eq.(1).

$C^{\text{Ni (Ti)}}$	$C^{\text{Ti(Ni)}}$	$C^{X(\text{Ti})}$	$C^{X(\text{Ni})}$	$\Delta M_s^{\text{Ni (Ti)}}$	$\Delta M_s^{\text{Ti(Ni)}}$	$\Delta M_s^{X(\text{Ti})}$	$\Delta M_s^{X(\text{Ni})}$
mol%	mol%	mol%	mol%	K/mol%Ni	K/mol%Ti	K/mol%Cr	K/mol%Cr
$1-k$	0	k	$1-k$	-150^{*}	0^{*}	-40^{**}	-65^{**}

* listed in Table 1.   ** predicted values [6, 7, 13]

(49mol%Ti, (1-k)mol%Ni, $k$mol%Cr) $_{\text{at Ti sites}}$ ((49+k)mol%Ni, (1-k)mol%Cr) $_{\text{at Ni sites}}$ (2),

where no vacancies and thermal disordering are considered.   Thus, the distribution coefficient k can be calculated if the other values are provided.   By using the ΔM_s of -133K/mol%Cr and the values listed in Table 2, Eq.(1) and k are calculated as,

$$-133 = (-150)(1-k) + (-40)k + (-65)(1-k), \text{ then, } k = 0.47 \tag{3}.$$

According to the calculations, the experimental ΔM_s by Cr addition can be understood if the half of Cr atoms occupy the Ni sites and antisite Ni atoms reduce M_s largely.   This calculated site occupation is in good agreement with the experimental results by ALCHEMI reported by Nakata *et al.* [19].   However, further consideration was not done in this paper, partially because the accurate $\Delta M_s^{X(\text{Ni})}$ and $\Delta M_s^{X(\text{Ti})}$ are not known for 4d- and 5d-transition metal elements.

Correlations between ΔM_s and the factors N_{ele}, N_V, V_X, E_N and N_M

(a) The number of electrons N_{ele} and electron hole number N_V
    Figure 4 (a) shows ΔM_s as a function of the number of electrons N_{ele} where N_{ele} is the sum of the outer d- and s-electrons.   It is seen that ΔM_s shows a tendency to decrease with increasing N_{ele}.   Figure 4 (b) shows ΔM_s as a function of electron hole number N_V.   When considering 3d-transition metal elements only, ΔM_s exhibits a good correlation with N_V as shown previously [7, 13].   However, some data in Figs.4 are scattered, this is because same N_{ele} and N_V are often given for ternary elements belonging to a same period group.   This suggests that other factor(s) than N_{ele} and N_V related with the electron structure also influences on ΔM_s.

(b) Electronegativity E_N
    Electronegativity E_N is a major factor to determine the ionic bonding nature between unlike atoms [20].   It was reported that the larger difference in E_N (ΔE_N) stabilizes the B2 phase in Au(Cd,X) and that M_s decreases with increasing ΔE_N [21].   Eckelmeyer tried to explain the recovery temperature of TiNi by ΔE_N but found no correlation [2].   Figure 5 shows ΔM_s as a function of E_N. ΔM_s slightly seems to become lowered with increasing E_N; ΔM_s is -108K/mol%Co with E_N=1.88 and -152K/mol%Mo with E_N=2.16, for example.   In these cases, however, ΔE_N becomes smaller in comparison with E_N=1.54 of Ti.   The dependence of ΔM_s

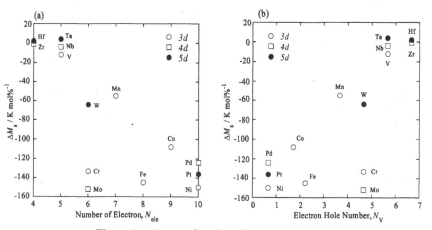

Figure 4. ΔM_s as a function of N_{ele} (a) and N_V.(b).

does not seem to be explained by E_N.

(c) Atomic volume V_X

Atomic size, rather, difference in atomic size, is an important factor for the solid-solution hardening and internal stress is generated by the size mismatch. Then, atomic size should affect M_s at least by means of stress field. Figure 6 shows ΔM_s as a function of V_X. It is seen that ΔM_s has a tendency to become lowered with smaller atomic volume.

Larger elements than Ti, for instance, Zr and Ta, increase ΔM_s slightly. These substitutions must generate compressive stress. On the other hand, smaller atoms decrease ΔM_s largely. In this case, these substitutions must generate tensile stress field. The types of stress field, tension/compression, may be an important role to determine the sign of ΔM_s. The details are not clear at present.

(d) Mendeleev number N_M

Figure 7 shows ΔM_s as a function of N_M. It is seen that a good correlation exist in the figure, and that the increment of ΔM_s is estimated to be -9.4Kmol%$^{-1}$/ΔN_M where ΔN_M is the increment of N_M. It is reliable that the decrease in N_M stabilizes B19' martensite phase, and that the increase in N_M stabilizes B2 parent phase. The correlation should be investigated more precisely from the viewpoint of structure map.

SUMMARY

Correlations between change in martensitic transformation start temperature by addition of 1mol% of ternary elements X (ΔM_s in K/mol%X) and several factors were investigated for $(Ti,X)_{50}Ni_{50}$. The M_s data were obtained experimentally using ternary TiNi alloys solution-treated at 1273K for 3.6ks. It was found that ΔM_s are different even though the numbers of outer electrons are same: -133K/mol%Cr, -152K/mol%Mo and -64K/mol%W for 6th period group, for example. These results suggest that other factor is required to express the dependence of ΔM_s. Effect of site occupancy on ΔM_s was discussed for Cr addition, and the deviation of ΔM_s was quantitatively understood by taking the atomic configuration into account. However, reliable and quantitative data are still limited. The following correlations investigated were summarized.

(1) ΔM_s depends on the number of the electrons N_{ele} and the electron hole number N_V. However, the

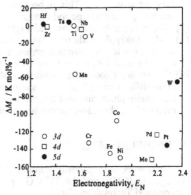

Figure 5. ΔM_s as a function of E_N.

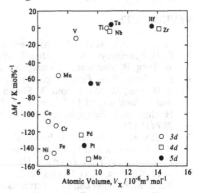

Figure 6. ΔM_s as a function of V_X.

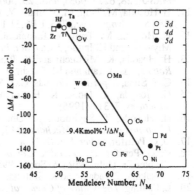

Figure 7. ΔM_s as a function of N_M.

357

data are scattered. This is because same N_{ele} and N_V are often given for ternary elements belonging to a same period group.

(2) ΔM_s seems to become lowered slightly with increasing E_N.

(3) ΔM_s weakly depends on atomic volume V_X. Larger ternary atoms increases ΔM_s slightly, and smaller ternary atoms decrease ΔM_s largely. The tendency may be related with the types of stress field, $i.e.$, tension or compression, generated by size mismatch.

(4) ΔM_s shows a good correlation with Mendeleev number N_M as $-9.4 \mathrm{Kmol}\%^{-1}/\Delta N_M$ where ΔN_M is the difference in N_M. This suggests that a ternary element with smaller N_M stabilizes B19' martensite phase and that a ternary element with larger N_M stabilizes B2 parent phase. The effect of N_M should be further investigated in terms of structure map.

ACKNOWLEDGEMENTS

This work was partially supported by Grant-in-Aid for Fundamental Scientific Research (Houga 16656215) and the 21st COE program from the Ministry of Education, Culture, Sports, Science and Technology, Japan.

REFERENCES

1. G. B. Stachowiak and P. G. McCormick, *Acta Metall.*, **36**, 291 (1988).
2. K. H. Eckelmeyer, *Scripta Metall.*, **10**, 667 (1976).
3. W. Tang, B. Sundman, R. Sandström and C. Qiu, *Acta Metall.*, **47**, 3457-3468 (1999)
4. F. E. Wang, Proc. *First Conf. on Fracture*, BII-103 (1965).
5. T. Honma, M. Matsumoto, Y. Shugo, M. Nishida and I. Yamazaki, *TITANIUM'80 Science and Technology*, ed. O. Izumi, 2, 1455 (1980).
6. H. Hosoda, T. Fukui, K. Inoue, Y. Mishima and T. Suzuki, *Mat. Res. Soc. Symp. Proc.*, **459**, 287 (1997).
7. H. Hosoda, S. Hanada, K. Inoue, Y. Mishima and T. Suzuki, *Intermetallics*, **6**, 291 (1998).
8. J. V. Humbeek and G. Firstov, *The Fourth Pacific Rim Intl. Conf. On Advanced Materials Processing (PRICM-4)*, eds. S. Hanada *et al.*, Jpn. Inst. Metals, 2, 1871 (2001).
9. V. N. Khachin, V. G. Pushin, V. P. Sivokha, V.V. Kondrat'yev, S. A. Muslov, V. P. Voronin, Yu. S. Zolotukhin and L. I. Yurchenko, *Phys. Met. Metall.*, **67**, 125 (1989).
10. Y. Takahashi, M. Tsuji, J. Sakurai, H. Hosoda, K. Wakashima and S. Miyazaki, *Trans. MRS-J.*, **28**, 627 (2003), *ibid.*, **29**, 3005 (2004).
11. H. Hosoda, M. Tsuji, M. Mimura, Y. Takahashi, K. Wakashima and Y. Yamabe-Mitarai, *Mat. Res. Soc. Symp. Proc.*, **753**, BB5.51.1 (2003)
12. H. Hosoda, M. Tsuji, Y. Takahashi, T. Inamura, K. Wakashima, Y. Yamabe-Mitarai, S. Miyazaki and K. Inoue, *Mater. Sci. Forum*, **426-432**, 2333 (2003).
13. H. Hosoda, K. Mizuuchi and K. Inoue, *Intl. Symp. Microsystems, Intelligent Materials & Robots*, eds. J. Tani and M. Esashi, 7th Sendai Intl. Symp., 231 (1995).
14. C. T. Sims, *Superalloys II*, eds. C. T. Sims *et al.*, Chapter 8, 217 (John Wiley & Sons, 1987).
15. H. Hosoda, K. Inoue, K. Enami and A. Kamio, *J. Intelligent Material Systems and Structures*, **7**, 312 (1996).
16. H. Hosoda, T. Sugimoto, K. Ohkubo, S. Miura, T. Mohri and S. Miyazaki, *Intl. J. Electromagnetics and Mechanics*, **12**, 3 (2000).
17. D. G. Pettifor, *Mater. Sci. Technol.*, **4**, 2480 (1988).
18. D. G. Pettifor, *Intermetallic Compounds*, eds. J. H. Westbrook and R. L. Fleischer, **1**, Chapter 18, (John Wiley & Sons, 1995) p.419.
19. Y. Nakata, T. Tadaki and K. Shimizu, *Mat. Trans., JIM*, **32**, 580 (1991).
20. L. Pauling. *The Nature of Chemical Bond*, third edition, (Cornell Univ. Press, 1960).
21. M. E. Brookes and R. W. Smith, *Met. Sci. J.*, **2**, 181 (1968).

Mater. Res. Soc. Symp. Proc. Vol. 842 © 2005 Materials Research Society

Shape Memory Effect through L1$_0$-fcc Order-Disorder Transition

K. Tanaka
Department of Advanced Materials Science,
Kagawa University, Takamatsu 761-0396, Japan

ABSTRACT

Shape memory effect not associated with martensitic transformation but with order-disorder phase transformation is examined in the compounds of AuCu, CoPt and FePd. These exhibit the L1$_0$-fcc transition at 680K, 1060K and 950K, respectively, which is relatively high temperature. Under a uniaxial compressive stress, a reversible shape change (reduction and elongation in edge length) is observed by cooling and heating the specimen temperature. The elongation of the specimen associated with the disordering occurs at their equilibrium transition temperatures, and the reduction along the direction of a compressive stress associated with the ordering occurs at a certain supercooled temperature. A trial actuator constructed with FePd polycrystalline wire and a conventional inconel wire properly works by heating up and cooling down the temperature.

INTRODUCTION

Shape memory alloys are useful functional materials. Ti-Ni alloy is one of such kind and is practically used. Although the general performance of the alloy is excellent, the operative temperature is limited to about 400 K at most. Thus, alloys suitable for use at higher temperatures are sought intensively. In this respect, Ti-Ni-Pd alloys and their variants are very attractive because of good shape memory characteristics over a wide range of temperatures up to 600K [1,2].

Other compounds, for example TaRu and NbRu, have been reported that the compounds exhibit shape recovery with the transition temperatures of about 1400 and 1150 K, respectively [3]. In the case of such materials exhibiting martensitic transformation, however, the low temperature phase is not a stable one. Then some structural change might occur in the low temperature phase due to diffusion, which deteriorates the shape recovery; the shape memory effect of martensitic type materials is insured through diffusionless transformation.

In order to avoid such intrinsic difficulty associated with metastable low temperature phase and to realize stable shape memory effect at higher temperatures, one should seek the shape memory effect originating from the other types of phase transformations. One of the possible phase transformations to be utilized is a diffusional-displacive phase transformation, where the transformation proceeds by a diffusion of the constituent atoms but the lattice correspondence is kept at the transformation. The L1$_0$-fcc order-disorder transformation is the kind of phase transformation accompanying macroscopic shape change. The equiatomic AuCu undergoes such shape change from disordered fcc to the ordered tetragonal structure (L1$_0$) as reported by Hirabayashi [4]. Ohta et al. have reported a shape recovery phenomenon in the compound through order-disorder transition[5]. The author has reported a macroscopic shape change under magnetic or compressive stress fields in ordering of FePd that exhibits the same fcc-L1$_0$ transition [6]. A shape recovery phenomenon in the compound has also been observed [7]. From these results reported, the shape memory effect seems to be a common feature of the L1$_0$-fcc order-disorder transition, but the detail of the phenomenon has not clarified yet.

The aim of this paper is to summarize the shape memory behavior of the compounds of

AuCu, CoPt and FePd those undergo the fcc-L1$_0$ order-disorder transformation, and to demonstrate the performance of a trial actuator as an application of the compounds.

EXPERIMENTAL PROCEDURES

Preparation of single crystal samples

Alloys of equiatomic AuCu, CoPt and FePd were prepared by arc melting from an appropriate mixture of pure metals under an argon atmosphere. Single crystals were grown from the melt by the modified Bridgeman technique. After determination of the crystallographic orientation, rectangular parallelepiped specimens with the each surface parallel to the {100} plane were cut out. The surfaces of the specimens were mechanically polished with diamond pastes. Specimens were annealed at a temperature higher than the transition temperature for 3.6 ks in evacuated quartz tubes. The specimens were slowly cooled so as to obtain the ordered state where the three variants form with an equal volume fraction.

Making a trial actuator

A part of the ingot of FePd was cold rolled with several intermediate anneals to a thin rod. A wire with the diameter of 0.6 mm was drawn from the rod and was shaped to a helix with the diameter of about 8 mm. The helix was annealed at 1073 K (above the transition temperature) for 3.6 ks in evacuated quartz tubes and was quenched in water so as to retain the disordered state. Though the wire may has a texture cased by cold rolling, drawing and annealing, the author has not clarified the texture. A similar helix made by a wire of inconel X-750 with the diameter of 0.5 mm was adopt to apply a bias stress to the FePd helix. The two helices were fixed so as to push a separator from the both sides. The maximum shear stress applied to the disordered FePd wire was set to about 20MPa.

Measurement of the dimension of the specimen under a constant stress

Figure 1 shows a schematic illustration of the apparatus for applying compressive stress. A constant stress was kept applied by a certain amount of weight. The temperature was monitored and controlled by a thermo-couple directly spot-welded to the specimen. The change in the edge length was monitored by a non-contact displacement gauge with the resolution of 1 μm. Thermal expansion of the compression rods (L^l - L^s in figure 1) was determined separately and was subtracted from the obtained change, L^t.

RESULTS AND DISCUSSION

Macroscopic shape change of single crystals under a constant compressive stress

Figures 2-4 show the change in edge length of single crystals, AuCu, CoPt and FePd, normalized to the initial dimension under a constant compressive stress. Magnitude of the compressive stress was 10MPa, 80MPa and 40 MPa for AuCu, CoPt and FePd, respectively, which is applied along the [001] crystallographic direction. The heating and cooling rates were set at 2K / min.

On the first heating (not indicated in the case of AuCu), the edge length gradually reduced. When the temperature reached the transition temperature, the edge length suddenly increased. On

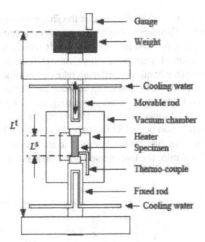

Gauge
Weight
Cooling water
Movable rod
Vacuum chamber
Heater
Specimen
Thermo-couple
Fixed rod
Cooling water

L^t
L^s

Figure 1. Schematic illustration of an apparatus for measuring the change in dimension of the specimen under a constant compressive stress.

cooling, a sharp reduction of the edge length was observed at the temperature lower than the transition temperature. The temperature different to the transition temperature is associated with super cooling for the transformation (ordering). The temperature difference is 70 K, 30 K and 110 K for AuCu, CoPt and FePd, respectively. On the second heating, a sharp elongation was observed at the transition temperature again. Small creep deformation is observed in the case of FePd at the disordered state. There is no evidence of creep deformation in the case of CoPt, even though the applied compressive stress is 80MPa. Note that in the case of AuCu, a small elongation is observed at about 660 K, which is associated with the formation of the L1$_{0-S}$ long period structure.

The magnitude of the change in edge length at the transition temperature is 3.7 %, 1.5 % and 3.0 % for AuCu, CoPt and FePd, respectively. These agree with the reported lattice parameter change (c-axis of L1$_0$ to a of fcc) of the materials [7-9]. This indicates that mono-variant of the L1$_0$ ordered phase forms from a single crystal of the fcc phase under a compressive stress along the [001] crystallographic direction, and the lattice correspondence is perfectly held at the transformation, the deformation matrix ε for single crystals is expressed using the lattice constants a^o and c^o of the ordered L1$_0$ phase and the lattice constants a^d of the disordered fcc phase:

$$\varepsilon = \begin{pmatrix} \dfrac{a^o - a^d}{a^d} & 0 & 0 \\ 0 & \dfrac{a^o - a^d}{a^d} & 0 \\ 0 & 0 & \dfrac{c^o - a^d}{a^d} \end{pmatrix}. \tag{1}$$

This deformation matrix indicates the maximum value of shape change of the materials, the maximum value of polycrystalline materials are much smaller (about one tenth) than the value of single crystals [7].

Table 1 summarizes the characteristic features of the shape memory effect of the compounds obtained. The time required for disordering / ordering process are determined from Figs. 2-4. Note that the time for disordering in FePd is only 10 seconds which is comparable to the required time for austenitic transformation of martensite alloys, though the transition requires diffusion of the constituent atoms. On the other hand, the ordering takes somewhat a long time. This indicates that the shape recovery associated with disordering is more useful like a conventional martensite alloys than that with ordering. From equilibrium phase diagrams, two phase (ordered and disordered) region exists in the $L1_0$-fcc order-disorder system for non-congruent compositions. This implies us that phase separation occurs in the two phase region, however, such phase separation takes much longer time than for disordering. In the present experiment with the temperature change rate of 2K / min, the two phase region does not affect the disordering temperature.

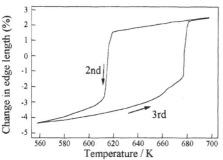

Figure 2. Change in edge length of a single crystal specimen of AuCu as a function of temperature under a constant compressive stress of 10 MPa.

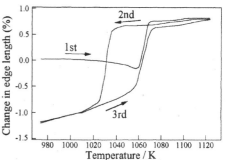

Figure 3. Change in edge length of a single crystal specimen of CoPt as a function of temperature under a constant compressive stress of 80 MPa.

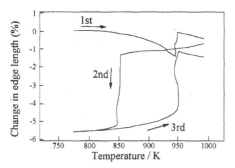

Figure 4. Change in edge length of a single crystal specimen of FePd as a function of temperature under a constant compressive stress of 40 MPa.

Table 1. Characteristic features of the shape memory effect.

Compound	Transition temperature	Transformation strain	Temperature of super cooling	Time required for disordering	Time required for ordering
AuCu	680 K	3.7 %	70 K	120 s	200 s
CoPt	1060 K	1.5 %	30 K	120 s	420 s
FePd	950 K	3.0 %	110 K	10 s	300 s

Evaluation of a trial actuator

Figure 5 shows the movement of the trial actuator by heating up and cooling down the temperature. At the initial state, the helix of FePd is disordered and slightly compressed by the helical spring of inconel. After ordering of FePd, the FePd helix is totally compressed by the inconel spring, and the separator moves to the left. This looks like a normal plastic deformation, however, the deformation is perfectly recovered when the actuator is heated up above the transition temperature (disordering FePd), and the separator moves to the right. These movement can be repeated more than ten times without any damage on the FePd helix. The $L1_0$-fcc order-disorder phase transformation is a material for a high temperature shape memory alloy with both high and low temperature phases being stable.

Shape memory effect through diffusional-displacive phase transformation

It has been clarified that the $L1_0$-fcc order-disorder phase transformation can be utilized for high temperature shape memory materials. In this case, shape memory effect is limited to "one way" shape memory effect that exhibits the effect at heated up the temperature only. If one wants to an advanced "two ways" shape memory effect, other appropriate phase transformations have to be utilized. One of the possible phase transformation if the $D1_a$-$L1_0$ order-order phase transformation, where both phases have a tetragonal symmetry but the axial ratios of these structures are different. In the case of the fcc-$L1_0$ ordering process, there are three variants of the $L1_0$ phase with the c-axis perpendicular to each other. Such symmetrical change violates the

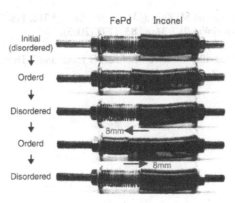

Figure 5. Movement of a trial actuator. A wire of FePd is equipped as a functional material.

shape memory effect at cooled down the temperature. On the other hand, the $D1_a$-$L1_0$ transformation has one to one correspondence of the variants, though the fcc-$D1_a$ ordering has three variants as in the case of the fcc-$L1_0$ ordering. The $D1_a$-$L1_0$ transformation can exist from a view point of thermodynamics, however, no system exhibiting the transformation has been reported. If the system exhibiting the $D1_a$-$L1_0$ or similar order-order phase transformation is discovered, the system will be more useful shape memory alloys for high temperature use.

CONCLUSIONS

The shape memory effect through the $L1_0$-fcc order-disorder phase transformation has been examined in the compounds of AuCu, CoPt and FePd. All the systems exhibit the shape memory effect (shape recover) at the transition temperature, and reversible shape change has been observed under a certain magnitude of compressive stress. Maximum recoverable strains for single crystals are 3.7%, 1.5% and 3.0%, and temperature hysteresis for the reverse motion are 70K, 30K and 110K for AuCu, CoPt and FePd, respectively. An actuator with FePd polycrystalline wire properly works by heating up and cooling down the temperature. The phase transformation of the $L1_0$-fcc order-disorder can be utilized as high temperature shape memory alloys alternative to the conventional martensite alloys.

ACKNOWLEDGEMENTS

This work was partly supported by Grant-in Aid for Scientific Research (B)(No.15360369) from the Ministry of Education, Culture, Sports, Science and Technology of Japan.

REFERENCES
1. K. Otsuka, K. Oda, Y. Ueno, M. Piao, T. Ueki, and H. Horikawa, *Scripta Met. Mater.*, **29**, 1355 (1993).
2. K. Otsuka and X. Ren, *Intermetallics*, **7**, 511 (1999).
3. R. W. Fonda, H. N. Jones and R. A. Vandermeer, *Scripta Mater.*, **39**, 1031 (1998).
4. M. Hirabayashi, *J. Phys. Soc. Jpn.*, **14**, 149 (1959).
5. M. Ohta, T. Shiraishi, R. Ouchida, M. Nakagawa and S. Matsuya, *J. Alloys and Comp.*, **265**, 240 (1998).
6. K. Tanaka, T. Ichitsubo and M. Koiwa, *Mater. Sci. Eng.*, **A312**, 118 (2001).
7. K. Tanaka and K. Morioka, *Phil. Mag.*, **83**, 1797 (2003).
8. B. W. Roberts, Acta Met., 2, 597 (1954).
9. C. Leroux, M. C. Cadeville, V. Pierron-Bohnes, G. Inden and F. Hinz, *J. Phys. F*, **18**, 2033 (1988).

Mater. Res. Soc. Symp. Proc. Vol. 842 © 2005 Materials Research Society   S5.1

Effect of Heat Treatment Conditions on Multistage Martensitic Transformation in Aged Ni-rich Ti-Ni Alloys

Minoru Nishida, Toru Hara[1], Yasuhiro Morizono, Mitsuhiro Matsuda and Kousuke Fujishima
Department of Materials Science and Engineering, Kumamoto University,
2-39-1 Kurokami ,Kumamoto 860-8555, Japan
[1] National Institute for Materials Science, 1-2-1Sengen, Tukuba 305-0047, Japan

ABSTRACT

It has been demonstrated with systematic experiments that the appearance and disappearance of multistage martensitic transformation (MMT) in aged Ni-rich Ti-Ni alloys depend on the heat treatment conditions. No multistage transformation occurs when the evaporation of Ti and Ni and/or the preferential oxidation of Ti in the specimen are prevented and the purification of heat treatment atmosphere in an evacuated quartz tube is achieved. The heterogeneity in precipitation morphology of the Ti_3Ni_4 phase, which is responsible for the multistage transformation can be suppressed with the regulation of heat treatment atmosphere as mentioned above. We have concluded that the multistage martensitic transformation in aged Ni-rich Ti-Ni alloys is extrinsic in nature, and is an artifact during the heat treatment.

INTRODUCTION

Ti-Ni alloys of the near-equiatomic composition are technologically important materials with their superior shape memory and superelastic properties associated with B2 to R-phase and B19' martensitic transformations. The application of the alloys has been spread to not only engineering but also medical and dental fields. It is widely recognized that improvements of shape memory and mechanical properties in the alloys are achieved by thermomechanical and aging treatment. Especially, the aging treatment is an effective process in Ni-rich Ti-Ni alloys due to precipitation strengthening of the parent phase with coherent Ti_3Ni_4 phase. It has been known that the aging treatment induces the R-phase transformation. In addition to the generation of R-phase transformation, it has been also reported that the MMT appears in the aged Ni-rich Ti-Ni alloys. Four mechanisms of multistage transformation have been proposed recently [1-4]. Bataillard et al. have ascribed it to coherent stress fields around Ti_3Ni_4 precipitates [1]. Khalil-Allafi et al. have explained it on the basis of evolving Ni concentration profiles between particles and differences in nucleation barriers between R-phase and B19' martensitic phase [2]. The latest report of Khalil-Allafi et al. has pointed out that the above two mechanisms cannot rationalize the multistage transformation, and thus the heterogeneity in precipitation morphology of Ti_3Ni_4 phase is responsible for the multistage transformation from TEM observations [3]. Fan et al. have ascribed it to a result of competition between preferential grain boundary precipitation of Ti_3Ni_4 particles and a tendency for homogeneous precipitation when supersaturation of Ni is large [4]. We do not intend to argue against those proposed mechanisms, because they have been discussed self-consistently within each experimental condition. However, no reports have been taken into account the heat treatment atmosphere except for the present authors, since we have experienced that the MMT is remarkably influenced by heat-treatment conditions, especially heat-treatment atmosphere [5].

The purpose of the present study is to clarify experimentally whether the MMT associated with the heterogeneity in precipitation morphology of Ti_3Ni_4 phase in the aged

Ni-rich Ti-Ni alloys is an intrinsic nature or an artifact during the heat treatment.

EXPERIMENTAL PROCEDURE

Ti-50.6at%Ni alloy was prepared from 99.7mass% sponge Ti and 99.9mass% electrolytic Ni by a high-frequency vacuum induction furnace using a graphite crucible, followed by casting into an iron mold. The ingot was hot-forged and cold-drawn to rod of 3 mm in diameter. The rod was cut into disks of about 1 mm in thickness for DSC measurements and about 0.2 mm in thickness for TEM observations, respectively. Some of the disks were solely sealed in an evacuated quartz tube of 2.5×10^{-3} Pa as shown in figure. 1, which is referred to as condition A hereinafter. The rest were sandwiched between Ti-Ni sheets of the same chemical composition. Subsequently, they were wrapped with pure Ti foil of 200 μm in thickness and then sealed in an evacuated quartz tube of 2.5×10^{-3} Pa as shown in figure 1, which is referred to as condition B hereinafter. They were solution-treated at 1273K for 3.6 ks, and then quenched into ice water. Subsequently, the specimens were aged at 773 K for various periods, and then quenched into ice water. For instance, the specimen A-A described in the later section indicates that both the solution treatment and the aging are carried out in the condition A. In addition to conditions A and B, conditions C to F shown in figure 1 were appended to establish the further evidence experimentally. The specimen was simply sandwiched between Ti-Ni sheets with the same composition in an evacuated quartz tube, which is referred to as the condition C. The condition D indicates that the specimen was sandwiched between Ti-Ni sheets with the same composition and then wrapped with Ni foil in an evacuated quartz tube, since the evaporation pressure of pure Ni is higher than that of pure Ti. The specimen was evacuated with pure titanium block in the condition E. The condition F indicates that the specimen was sandwiched between Ti-Ni sheets with the same composition and then evacuated with pure Ti block in a quartz tube. The specimens obtained by various heat treatment conditions were lightly mechanically and chemically polished to remove the surface scale. DSC measurements were performed by using a Shimadzu DSC-50 calorimeter with cooling and heating rate of 10 K/min. TEM specimens were electropolished using the twin jet method in an electrolyte of 20% H_2SO_4 and 80%CH_3OH in volume around 270 K. TEM observations were carried out in the JEOL-2000FX microscope operated at 200 kV.

RESULTS AND DISCUSSION

Figures 1 shows six types of heat treatment atmospheres and corresponding DSC cooling curves of the specimens A-A to F-F aged at 773K for 7.2 ks. The first peak denoted as R, which corresponds to the B2 to R-phase transformation, can be seen in all the curves. It is notable that the multistage martensitic transformation peaks denoted as M1 and M2 are clearly separated in the specimens A-A, C-C, D-D and E-E. On the other hand, there are no MMT peaks in the specimens B-B and F-F. From these results one can imagine that the Ti-Ni cover plates prevent the evaporation of Ti and Ni and/or the preferential oxidation of Ti in the specimen, and Ti foil and block act as a getter material to purify the atmosphere in the evacuated quartz tube. It can be concluded that the MMT is suppressible with the prevention of evaporation of Ti and Ni and/or preferential oxidation of Ti in the specimen, and the purification of atmosphere in the evacuated quartz tube. Both factors for the regulation of atmosphere are indispensable for suppressing the MMT.

Subsequently, DSC measurement in specimens A-B and B-A aged at 773K for 7.2ks is

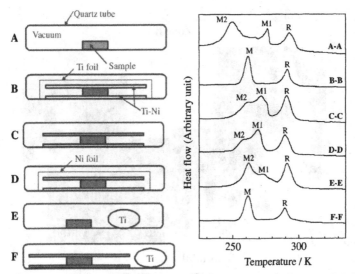

Figure 1. Schematic illustrations of heat treatment conditions and corresponding DSC cooling curves of specimens aged at 773 K for 7.2 ks.

carried out to confirm a controlling step during the heat treatment for the MMT. It is apparent that there is MMT in the specimen A-B in figure 2, although the feature of DSC curve is quite different from that in the specimen A-A aged at 773K for 7.2ks in figure 1. On the other hand, a single stage martensitic transformation is recognized in the specimen B-A as shown in figure 2. The feature of DSC curve and the transformation temperatures are almost the same as those in the specimen B-B in figure 1. These phenomena indicate that the controlling step for the MMT is the solution treatment. It is considered that a kind of compositional

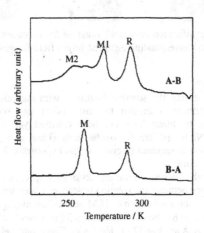

Figure 2. DSC cooling curves of specimens A-B and B-A aged at 773 K for 7.2 ks.

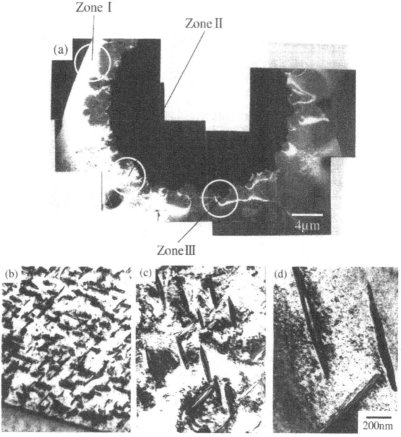

Figure 3. (a) Low magnification dark field image of the specimen A-A aged at 773 K for 3.6 ks. (b), (c) and (d) Corresponding enlarged bright field images of zone I, II and III in (a).

fluctuation is induced during the solution treatment with condition A and is emphasized during the aging treatment. Although the mechanism and process of compositional fluctuation during heat treatment have not been clarified yet, it may be related to the evaporation of Ti and Ni, the preferential oxidation of Ti and so on as described above. The precise analytical method is required to confirm this hypothesis. It is now under study and will be reported in due course.

In order to obtain the microstructural aspects of each specimen, TEM observations were carried out. No difference between the solution treated specimen with the conditions A and B could be recognized on the conventional TEM scale. The average grain size of both the specimens was estimated to be about 40μm. Figure 3 (a) shows low magnification dark field image of the specimen A-A aged at 773 K for 3.6 ks. The bright field image in the zones I to III in (a) is enlarged in figures 3 (b) to (d). The size of Ti_3Ni_4 precipitate at the zones II and

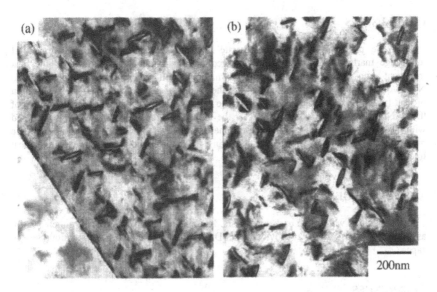

Figure 4. Bright field images at grain boundary and interior of the same grain in the specimen B-B aged at 773 K for 3.6 ks

III is about three and ten times larger than that at the grain boundary, respectively, while the precipitation density at those interior is lower than that at the boundary. The correspondence between M1, M2 and those transformation areas such as grain boundary or interior has not been clarified yet. However, since the martensitic phase has been occasionally observed at grain boundary after jet polishing around 270 K, the transformation area of M1 and M2 may correspond to the grain boundary and interior, respectively. On the other hand, homogeneity in distribution and size of precipitates is observed at grain boundary and interior of the same grain in specimen B-B aged at 773 K for 3.6 ks as shown in figures 4 (a) and (b), respectively. It is recognized that Ti$_3$Ni$_4$ precipitate has the tendency of homogeneous nucleation in TiNi B2 matrix. It is due to similarity of atomic arrangement between TiNi B2 and Ti$_3$Ni$_4$ rhombohedral phases [6,7]. Although we do not reproduce electron micrographs in A-B, B-A, C-C, D-D, E-E and F-F specimens aged at 773 K for 3.6 ks here, the microstructure in the specimen A-B is essentially similar to that in the specimen A-A. The microstructure in B-A and F-F is similar to that in the specimen B-B. In the rest specimens heterogeneous grain boundary precipitation and homogeneous one were randomly observed. Further details will be reported in due course together with the correspondence between M1, M2 and those transformation areas as mentioned above. These results demonstrate that the precipitate morphology is drastically changed with heat treatment atmosphere, especially with solution treatment condition.

Consequently, the appearance of multistage martensitic transformation associated with heterogeneous precipitation of Ti$_3$Ni$_4$ is controlled with heat treatment atmosphere. We can finally conclude that the multistage martensitic transformation in aged Ni-rich Ti-Ni alloys is an extrinsic nature, i.e., a kind of artifact during the heat treatment.

CONCLUSIONS

In the present study we demonstrate that the appearance and disappearance of multistage martensitic transformation in aged Ni-rich Ti-Ni alloys depend on the heat treatment atmosphere. No multistage transformation occurs when the evaporation of Ti and Ni and/or the preferential oxidation of Ti in the specimen are prevented and the purification of heat treatment atmosphere is achieved. The heterogeneity in precipitation morphology of the Ti_3Ni_4 phase, which is responsible for the multistage transformation can be suppressed with the regulation of heat treatment atmosphere as mentioned above. We conclude that the multistage martensitic transformation is extrinsic in nature, and is an artifact during heat treatment in aged Ni-rich Ti-Ni alloys.

ACKNOWLEDGEMENTS

The authors would like to thank Messrs. K. Yamaguchi and K. Tanaka for their helpful assistance. The specimen was provided from Dr. K. Yamauchi of Tokin Corporation, Sendai, Japan.

REFERENCES

1. L. Bataillard, J.-E. Bidaux and R. Gotthardt, Phil. Mag. A **78,** 327 (1998).
2. J. Khalil-Allafi, X. Ren and G. Eggeler, Acta Mater. **50,** 793 (2002).
3. J. Khalil-Allafi, A. Dlouhy and G. Eggeler, Acta Mater. **50,** 4255 (2002) .
4. G. Fan, W. Chen, S. Yang, J. Zhu, X. Ren and K. Otsuka, Acta Mater. **52,** 4351 (2004)
5. M. Nishida, T. Hara, T. Ohba, K. Yamaguchi, K. Tanaka and K. Yamauchi, Mater. Trans. **44,** 2631 (2003).
6. T. Tadaki, Y. Nakata, K. Shimizu and K. Otsuka: Trans. JIM. **27,** 731 (1986).
7. T. Saburi, S. Nenno and T. Fukuda: J. Less-Common Metals **12,** 157 (1986).

Mater. Res. Soc. Symp. Proc. Vol. 842 © 2005 Materials Research Society   S5.2

Application of the CSL Model to Deformation Twin Boundary in B2 Type TiNi Compound

Minoru Nishida, Mituhiro Matsuda, Yasuhiro Morizono, Towako Fujimoto and Hideharu Nakashima[1]
Department of Materials Science and Engineering, Kumamoto University,
2-39-1 Kurokami ,Kumamoto 860-8555, Japan
[1] Interdisciplinary Graduate School of Engineering Sciences, Kyushu University,
Kasuga 816-8580, Japan

ABSTRACT

The deformation structure of B2 type TiNi compound around 573 K has been investigated by transmission electron microscopy (TEM). Serrations are seen in stress-strain curve, which corresponds to the formation of various planar defects with twin relation. The dominant planar defect found in the specimens showing serration is {114} compound twin. The other defects are in mirror symmetry with respect to {113}, {115}, {447} planes and so on. These defects are considered to be <110> symmetric tilt boundaries in bcc structure by ignoring the atomic arrangement of B2 structure and are characterized with Σ value based on coincide site lattice (CSL) model. For instance, the Σ value of {114}, {113}, {115} and {447} boundaries are $\Sigma 9$, $\Sigma 11$, $\Sigma 27$ and $\Sigma 81$, respectively. Numerous (114) defects initially form at grain boundary and grow into grain interior. Some of those deflect to (-1-14) defects. In such case, the {447} $\Sigma 81$ defect is always observed at the interface of (114) $\Sigma 9$ and (-1-14) $\Sigma 9$ defects. This fact indicates that the sigma combination rule of the CSL model can be applied to the triple junction of defects. Similarly, {7710} $\Sigma 99$ boundary forms at the interface of {114} $\Sigma 9$ and {113} $\Sigma 11$ defects. It can be concluded that the ductility of B2 type TiNi compound around 573 K is attributable to the formation of various planar defects with large shear strain and the increment of independent slip system due to the formation of planar defects, and that the arrangement of some planar defects conforms to the sigma combination rule of CSL model.

INTRODUCTION

The near-equiatomic TiNi compound exhibits superior shape memory and superelastic properties. The other peculiar property is anomalous ductility over a wide temperature range in comparison to another martensite memory compound such as NiAl [1-3]. In the present study we have focused on the deformation structure of B2 parent phase around 573 K, since interesting interactions of planar defects have been found on the basis of TEM observations· and electron diffraction experiments. Consequently, the sigma combination rule of CSL model [4-7] can be applied to the triple junction of several defects.

EXPERIMENTAL PROCEDURE

The $Ti_{50}Ni_{48}Fe_2$ alloy used was provided from Furukawa Electric Company, Yokohama 220-0073, Japan. The hot-rolled sheet of 2 mm in thickness was cold-rolled and annealed repeatedly into strip of 0.3 mm in thickness. Tensile specimens with gage portion of 45 X 5 X 0.3 (mm³) were spark-cut from the strip. The specimens were solution-treated in vacuum at 1273 K for 3.6 ks, and then quenched into ice water. They were lightly mechanically and

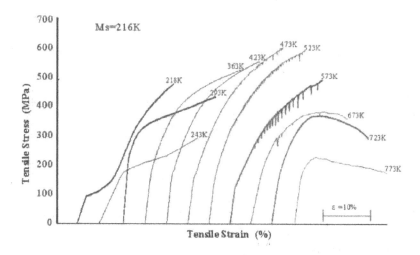

Figure 1. Tensile stress-strain curves in $Ti_{50}Ni_{48}Fe_2$ alloy deformed at various temperatures.

chemically polished to remove the surface scale. Transformation temperatures of the specimen were determined by DSC. M_S', M_S, A_f and A_f' temperatures are 281, 216, 266 and 289 K respectively. Here, M_S' and M_S are the transformation starting temperatures of the B2 to R phases and the R to martensitic phases, respectively. In the same way, A_f and A_f' are the reverse transformation finishing temperatures of the martensitic to R phases and the R to B2 phases, respectively. The advantage of the alloy is to simplify the structure analysis by TEM observations, since no precipitate reaction takes place at elevated temperature during tensile test and the B2 parent phase is stable at room temperature.

The tensile test was carried out in an Instron type machine equipped with cold bath and electric furnace. Testing temperature is about 220 to 773 K. Initial strain rate is about 4.2×10^{-4}/s. Disk specimens of 3 mm in diameter for TEM studies were spark-cut from the gauge portion of fractured tensile specimens. They were mechanically polished and then electropolished with twin jet method. TEM observations were carried out in a JEOL-2000FX operated at 200 kV.

Table 1. Twinning elements of observed deformation twins in B2 type TiNi compounds.

	K_1	η_1	K_2	η_2	S	Σ
{77 10} compound	(77 10)	[55$\bar{7}$]	($\bar{1}\bar{1}$2)	[111]	0.3536	99
{335} compound	(335)	[55$\bar{6}$]	($\bar{1}\bar{1}$2)	[111]	0.5143	43
{447} compound	(447)	[77$\bar{8}$]	($\bar{1}\bar{1}$2)	[111]	0.5657	81
{112} compound	(112)	[11$\bar{1}$]	($\bar{1}\bar{1}$2)	[111]	0.7071	3
{449} compound	(449)	[99$\bar{8}$]	($\bar{1}\bar{1}$2)	[111]	0.8319	113
{113} compound	(113)	[33$\bar{2}$]	(110)	[001]	0.9428	11
{114} compound	(114)	[22$\bar{1}$]	(110)	[001]	0.7071	9
{115} compound	(115)	[55$\bar{2}$]	(110)	[001]	0.5656	27

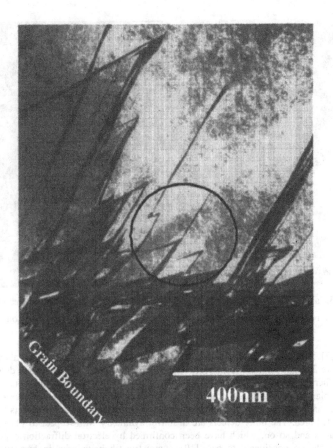

Figure 2. Low magnification dark field image of deformation structure in $Ti_{50}Ni_{48}Fe_2$ alloy deformed at 573 K.

RESULTS AND DISCUSSION

Figure 1 shows stress-strain curves of $Ti_{50}Ni_{48}Fe_2$ alloy at various deformation temperatures. Although the elongation of all specimens at each temperature was more than 30 %, all the curves are illustrated up to 20 %. Anomalous ductility is reconfirmed in the whole temperature range tested in the present study. The microstructure modification in deformed specimens is divided into three types, and is briefly summarized as follows on the basis of TEM observations. Below 400 K, the microstructure is derived from stress and/or strain induced martensitic phase [1,2]. Between 450 and 700 K, serrations are seen in stress-strain curve, which corresponds to the formation of various planar defects with twin relation as described later. Above 750 K, the necking takes place during tensile test and dislocations are only seen in the specimen. Burgers vector of dislocations is identified to

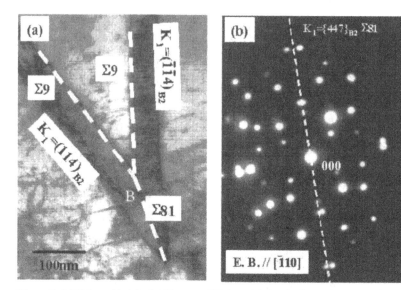

Figure 3. (a) Enlarged bright field image of V-shaped defect in the encircled area in figure 2. (b) Electron diffraction pattern taken from area B in (a) showing {447} Σ81 boundary.

$<100>_{B2}$ [3]. Here, we focus on the deformation structure at 573 K as mentioned above.

Figure 2 shows the low magnification dark field image of $Ti_{50}Ni_{48}Fe_2$ alloy deformed at 573 K. There are various planar defects which develop from the grain boundary. The dominant planar defect is {114} compound twin. Many (114) defects initially form at grain boundaries and grow into grain interiors. Some of those deflect to (-1-14) defects as seen in the encircled area. The other defects are in mirror symmetry with respect to {113}, {115}, {447} planes and so on, which have been confirmed by electron diffraction experiments. These defects are considered to be <110> symmetric tilt boundaries in bcc structure by ignoring the atomic arrangement of B2 structure and are characterized with Σ value based on the CSL model as listed in Table 1. For instance, the Σ value of {114}, {113}, {115} and {447} boundaries are Σ9, Σ11, Σ27 and Σ81, respectively [4-6].

Figure 3 (a) shows the enlarged bright field image of V-shaped defect in the encircled area in figure 2. In this case, a {447} Σ81 defect is always observed at the interface of (1-14) Σ9 and (-114) Σ9 defects as clearly recognized from electron diffraction pattern in figure 3 (b). This fact indicates that sigma combination rule of CSL model can be applied to the triple junction of defects. Similarly, a {7710}Σ99 boundary forms at the interface of {114}Σ9 and {113}Σ11 defects as shown in figure 4. This arrangement is confirmed by not only electron diffraction patterns in figures 4 (b) to (c) but also trace analysis. We also found the combination of {112}Σ3, {114}Σ9 and {255}Σ27 as shown in figure 5, although we do not reproduce the corresponding diffraction patterns. It is likely that there may be the other combinations such as Σ3, Σ27 and Σ81. This is now under study and will be reported elsewhere.

CONCLUSIONS

The deformation structure of B2 type TiNi compound around 573 K has been investigated by TEM. Serrations are seen in stress-strain curve, which corresponds to the formation of various planar defects with twin relation. The dominant planar defect found in the specimens is {114} compound twin. The other defects are in mirror symmetry with respect to {113}, {115}, {447} planes and so on. Many (114) defects initially form at grain boundary and grow into grain interior. Some of those deflect to (-1-14) defect. In this case, a {447} Σ81defect is always observed at the interface. Similarly, {7710} Σ99 boundary forms at the interface of {114} Σ9 and {113} Σ11 defects. These facts indicate that the sigma combination rule of CSL model can be applied to the triple junction of defects. It can be concluded that the ductility of B2 type TiNi compound around 573 K is attribute to the formation of various planar defect with large shear strain and the increment of independent slip system due to the formation of planar defects, and the arrangement of some planar defects conform to the sigma combination rule of CSL model.

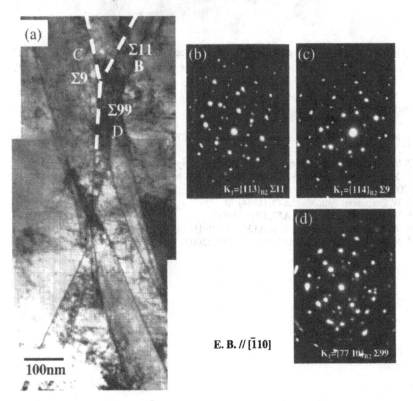

Figure 4. (a) Bright field image, (b), (c) and (d) corresponding electron diffraction patterns taken from areas B, C and D in (a) showing {7710} Σ99 between {113} Σ11 and {114} Σ9 boundaries.

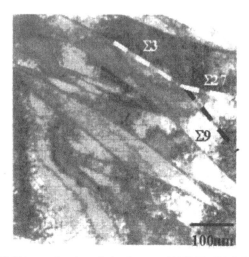

Figure 5. Bright field image showing relation between {112} Σ3, {114} Σ9 and {255} Σ27 boundaries.

REFERENCES

1. M. Nishida, S. Ii, K. Kitamura, T. Furukawa, A. Chiba, T. Hara, K. Hiraga: Scripta Mater. **39,** 1749 (1998).
2. S. Ii, K. Yamauchi, Y. Maruhashi, M. Nishida: Scripta Mater. Vol. **49,** 723 (2003).
3. M. Nishida, K. Tanaka, S. Ii, M. Kohshima, S. Miura, M. Asai,: J. Phys. IV. **112,** 803 (2003).
4. W. Bollmann: Phil. Mag. **A49,** (1984) 73.
5. W. Bollmann: Phil. Mag. **A57,** 637 (1988).
6. W. Bollmann: Mater. Sci. Eng. **A136,** 73 (1991).
7. M. Fray, C. A. Schuh: Acta Mater. **51,** 3731 (2003).

Mater. Res. Soc. Symp. Proc. Vol. 842 © 2005 Materials Research Society

Coarsening Behavior and Coercivity in L1$_0$-Ordered Intermetallic Fe-Pd Ferromagnets with Equiaxed Grain Morphology

Jörg M.K. Wiezorek and Anirudha R. Deshpande
Department of Materials Science and Engineering, University of Pittsburgh, Pittsburgh, PA 15261, USA;

ABSTRACT

Thermo mechanical processing of equiatomic FePd has been used to establish a nearly equiaxed microstructure that consists of mono-variant L1$_0$ ordered grains. The technically important hard magnetic properties, e.g. coercivity, of the uniaxial ferromagnetic L1$_0$-FePd with equiaxed microstructure depend on the grain size. Thus, here, the evolution of grain size during the later stages of annealing has been studied using SEM and computer assisted image analysis with the goal to determine the activation parameters for the relevant coarsening processes. Using a classical grain growth analysis the activation parameter for coarsening of the 5% largest grains in the populations of L1$_0$-ordered grains in FePd has been determined as approximately 30kcal/mol.

INTRODUCTION

L1$_0$-ordered FePd belongs to the class of technologically interesting tetragonal ferromagnetic intermetallics that includes also FePt, CoPt, and MnAl. These intermetallics exhibit uniaxial ferromagnetism with large magnetocrystalline anisotropy and are of significant technological interest for permanent magnet applications [1]. FCC → L1$_0$ ordering in the equiatomic Fe-Pd system is a first-order phase transformation and associated with considerable transformation strain. The system reduces the transformation strain energy through self-accommodating arrays of {101}-conjugated twins. The resulting characteristic lamellar morphology is known as the polytwin microstructural state [2]. The technologically relevant magnetic properties of the tetragonal L1$_0$ ordered γ1-phase in Fe-Pd, e.g. coercivity, are extremely sensitive to microstructure and are quite poor for the polytwinned FePd [3,4]. Previous studies have been able to replace this polytwinned microstructural state by a nearly equiaxed microstructural morphology consisting of mono-variant L1$_0$ ordered FePd grains (Fig. 1), which can exhibit significantly enhanced magnetic properties [5]. A thermomechanical processing route that involves cold deformation of the metastable FCC phase followed by annealing at a temperature lower than the critical ordering temperature can alter scale and morphology of the final ordered microstructure. The microstructural transformation after cold-deformation involves a combined reaction (CR) of FCC→ L1$_0$ phase transformation concomitant with annealing of defects introduced during cold deformation of the metastable FCC phase. During CR-annealing at temperature T < T$_c$ (critical ordering temperature, T$_c$~ 650°C) heterogeneous nucleation and growth of L1$_0$ ordered grains occurs at internal strain heterogeneities in the cold-deformed FCC microstructure favoring the formation of equiaxed mono-variant L1$_0$-ordered FePd grains. The evolution of the coercivity during order-annealing of CR-processed FePd has been studied previously as a function of cold-deformation strain and time and temperature [6]. Consistently, during CR-annealing at a particular temperature (T<T$_c$) the coercivity increases with annealing

time up to a maximum value, coinciding with the completion of the ordering phase transformation [5,6]. Further annealing for times longer than that required to establish the fully ordered state by the CR-mode of transformation leads to decreases in coercivity [5,6]. This evolution of coercivity with annealing time is similar to the age hardening behavior in precipitation hardening alloys for structural applications and has been referred to as the magnetic age hardening curve of CR-processed FePd. The decrease in coercivity in the 'over-aged' state has been associated previously with the evolution of different planar defects in the microstructure, including grain boundaries. Grain boundaries and other planar defects in CR-processed FePd can affect the pinning and nucleation behavior of magnetic domains and influence the magnetization behavior. Thus, a thorough understanding of the grain coarsening process in the fully ordered $L1_0$-FePd with equiaxed morphology is necessary to develop a better quantitative understanding of the effects of grain boundaries on the coercivity of CR-transformed microstructures. In this study microstructures that consist of nearly entirely of equiaxed mono-variant $L1_0$-ordered grains have been generated. Using these fully ordered microstructures as starting condition a classic grain growth analysis has been attempted for different isotherms during the grain growth regime of annealing.

EXPERIMENTAL PROCEDURE

Equiatomic FePd has been prepared by vacuum induction melting of high-purity elemental starting materials. After 50% reduction in thickness, recrystallization at 1173K for 5h and ice-water-quench sections of the consolidated alloy have been cold-deformed by cold rolling to 97% thickness reduction. Transformation of FCC → $L1_0$ by the CR-mode was achieved during isothermal annealing at 400°C, 500°C and 600°C. Microstructures have been characterized by combinations of X-ray diffraction (XRD), scanning and transmission electron microscopy (SEM and TEM) using a Philips X'pert XR diffractometer, a field-emission gun equipped Philips XL30 SEM and a JEOL 2000FX TEM/STEM and JEOL CX200 TEM, respectively. Microstructural metrics such as grain sizes have been measured using computer assisted image analysis using the Scion image analysis software.

RESULTS AND DISCUSSION

Figure -1 SEM backscatter electron (BSE) micrographs a) at the start of grain growth b) at a later stage of annealing involving grain growth and coarsening

Figure-1a is a SEM micrograph of the completely CR transformed microstructure after 3 hours at 600°C after 97% reduction in thickness in the FCC condition, which exhibits the maximum coercivity in the magnetic age hardening curve for this isotherm. Hence, the CR-mode facilitated ordering transformation from FCC to $L1_0$ FePd is completed at this point along the magnetic age hardening curve. This microstructural state represents the starting condition for the grain growth regime of annealing in the fully ordered state at 600°C. Figure-1b depicts a SEM micrograph after 20 hours of annealing at 600°C, which thus is the microstructural state after grain growth or coarsening for 17 hours, (20-3) h. Similar SEM observations of grain growth have also been conducted for isotherms at 400°C and 500°C. Since, they essentially exhibited the same characteristics regarding morphological scale evolution during annealing they have not been shown here. The evolution of the populations of grains, about 800 per annealing condition, have been monitored and some relevant data on the changes of grain size as a function of annealing time and temperature are collated in the table 1.

Isotherm Temp deg C	Time for grain growth hrs	Time (Total annealing hrs)	Dmax (micron)	Dmax-5% (micron)	Davg (micron)	k (using Dmax5%)	ln k	avg ln k
600	0	3	1.99	1.9	0.49	0	nd	
	1.5	4.5	2.53	2.42	0.52	4.16E-12	-26.2	
	5	8	6.03	5.7	1.1	1.60E-11	-24.85	
	9	12	6.42	5.99	1.2	9.95E-12	-25.33	-25.44
	11	14	7.13	6.57	1.36	9.98E-12	-25.32	(-33.07)
	17	20	7.85	7.39	1.79	8.33E-12	-25.51	
	21	24	9.92	8.31	1.93	8.65E-12	-25.47	
500	0	6	1.33	1.13	0.4	0	nd	
	2	8	1.62	1.47	0.42	1.22E-12	-27.42	
	4	10	2.12	1.83	0.6	1.43E-12	-27.26	
	6	12	2.5	1.94	0.66	1.15E-12	-27.49	-27.24
	8	14	2.72	2.41	0.76	1.70E-12	-27.09	(-34.2)
	10	16	3.15	2.86	0.89	1.91E-12	-26.98	
400	0	168	0.90	0.88	0.23	0	nd	
	24	192	1.43	1.24	0.26	8.79E-14	-30.06	
	96	264	1.72	1.53	0.29	4.52E-14	-30.72	-30.5
	144	312	2.11	1.76	0.53	4.47E-14	-30.73	(-38.978)
	216	384	2.47	2.26	0.59	5.56E-14	-30.51	

Table -1 Microstructural metrics (D_{avg}, D_{max}, $D_{max5\%}$) as a function of annealing time. The factor k has been determined using equation 1 for the $D_{max5\%}$ fraction. It has also been determined using D_{avg} of the entire population and has been reported in brackets. ('nd' refers to 'not defined') (An example of calculations – for the 600°C isotherm $Do_{max5\%}$ =1.9 micron at time t = 0hrs (second column) and various $D_{max5\%}$ values for time > 0hrs are 2.42, 5.7, 5.99, 6.57 ...etc. Therefore using Do as 1.9 and D as the $D_{max5\%}$ values for t > 0hrs the different k values have been determined. Similar procedure has been applied to obtain k parameters using D_{avg})

The table reports the average grain size (D_{avg}, obtained using circular cross-sections approximating a grain) for the entire populations. The maximum grain size (D_{max}) and the grain size of the 5% fraction of the largest grains in the microstructure has also been reported. In an attempt to determine activation parameters for the thermally activated process(es) that facilitated the grain growth observed for the equiaxed $L1_0$-FePd grains the empirically established grain growth law has been used:

$$D^n = D^n_0 + k \, t_{expt} \qquad (1)$$

Here D is the average grain size after grain growth for time t_{expt} and D_0 is the grain size at the start of the grain growth experiments. The exponent n has been determined using experimental data to be very close to 2. Furthermore, the factor k has been determined from the experimental data for $D_{max5\%}$ using equation (1) and is reported in table-1. The k factors have also been determined using D_{avg} for all three isotherms using equation (1). The k factors obtained using D_{avg} have been reported in brackets in the last column of table 1. The factor k is related to the activation energy, Q_{GG}, for the elementary process(es) that facilitate(s) grain coarsening by

$$k = k_0 \exp \left(-Q_{GG} / RT \right) \qquad (2)$$

If a single elementary process is dominating the thermally activated microstructural rearrangements during the grain growth regime of annealing of the CR-mode transformed $L1_0$-FePd a good linear fit to the experimental data is expected in an Arrhenius-plot of ln(k) vs. 1/T . The good linear fit would yield a slope of $-Q_{GG}/R$. Such plots have been obtained for the grain growth observed in the 5% fraction of the largest grains in the populations and for D_{avg} of the entire population (Fig. 2). A reasonably good linear fit ($R^2=0.992$ for the least square linear fit) is possible for the data set of the 5% fraction of largest grains in the equiaxed microstructure of $L1_0$-ordered FePd. However, it was not possible to obtain a satisfactory linear fit when the average grain size of the entire populations was used ($R^2=0.930$). The reason for this deviation from linearity is not clear. However, it might be an indication of some type of abnormal behavior in the grain growth process. Examples of the grain size distributions at the onset of grain growth process and after a certain amount of time during the grain growth regime have been depicted in figure-3a and 3b.

With increasing annealing time the average in the grain size does not shift linearly to a larger value. A considerable fraction of grains appear to be present in the smaller size bins even after considerable annealing time. An increase in the population in the largest size bins can also be observed even when the number of the smaller grains in the population does not change significantly. This could imply abnormalities in the grain growth process. Furthermore, the grain size distributions observed at the completion of the CR-mode ordering transformation, establishing the fully-$L1_0$-ordered starting conditions for the attempted grain growth annealing reported here, are only approximately log-normal.

The classic approach to a grain growth analysis employed here can only be satisfactorily used in case of normal grain growth that is for grain size populations that remain self-similar, e.g. log-normally distributed, during annealing. This does not appear to be the case for the equiaxed $L1_0$-ordered FePd studied here. The D_{avg} obtained using the entire population reflects these abnormalities in the grain growth process since a good linear fit cannot be obtained for these data, using the ln (k) vs. 1/T plot. Further work is required to ascertain the origin for the observed abnormalities in the coarsening behavior of the $L1_0$-FePd studied here. The mechanistic details of the process(es) that lead to grain growth in these microstructures are apparently not captured by the metallographic techniques used in this investigation.

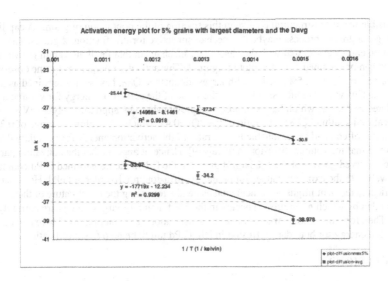

Figure -2 ln k vs 1 / T plot obtained for the 5% fraction of grains with largest diameters and for the Davg of the entire population. A reasonably good fit linear can be obtained for the 5% fraction with the largest diameters. However a considerable deviation from linearity is observed for the plot using Davg.

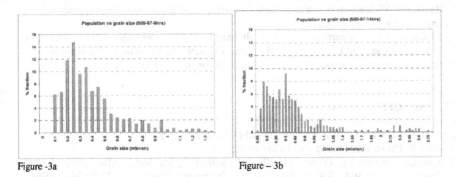

Figure -3a   Figure – 3b

Figure -3a Grain size distribution at the onset of grain growth process after the CR reaction is complete after 6 hrs at 500°C after 97% reduction in thickness in FCC condition.
Figure-3b Grain size distribution after considerable grain growth (after 14 hrs at 500°C after 97% reduction in thickness in FCC condition).

Future experiments using in-situ TEM experiments together with complementary orientation imaging microscopy in the SEM are planned to probe for answers
However, a surprisingly good linear fit can be obtained for the 5% grains with the largest diameters in the microstructure. This observation suggests the presence of a single mechanism

responsible for microstructural change in this fraction of the population of grains. A capillarity driven grain growth process maybe the mechanism active for this fraction of grains. The activation energy associated with the elementary mechanism that facilitates grain growth in this fraction can be obtained from the plot in figure-2. This activation parameter for the observed growth occurring in the 5%fraction with largest diameters, Q_{GG}, has been determined as 29.7 ± 1.4 kcal/mol. This value is much smaller than the value of the activation energy for volume diffusion in ordered FePd, Q_V, which has been proposed previously to be approximately $3.1eV = 71.4kcal/mol$ kcal/mol [7,8]. The ratio of $Q_{GG}/Q_V=0.42$, which lies within the range of 0.33 to 0.50 that is often cited as reasonable for the ratio of the activation energy for boundary diffusion relative to that for volume diffusion in metals [9]. Hence, it appears that the activation parameter for the observed growth of the 5% largest grains in the CR-mode transformed FePd compares quite well with the expected values for boundary diffusion in $L1_0$-ordered FePd. However, this numerical agreement might be coincidental as details regarding the exact nature of the mechanism by which grain growth occurs in the 5% fraction of largest grains are currently not clear. Therefore direct comparison between the activation parameter reported here and the activation energy for boundary diffusion in $L1_0$-FePd might be an oversimplification.

CONCLUSIONS

1.The morphologically equiaxed microstructures of the $L1_0$-ordered FePd do not appear to undergo 'normal' grain growth during annealing after the ordering transformation is complete and the deformation induced strain energy is expended.
2. A simple grain growth model based on capillarity as the dominant driving force cannot be applied to the entire population of grains present in these $L1_0$-ordered FePd microstructures.
3. The largest 5% fraction of grains in these $L1_0$-ordered FePd microstructures appears to undergo a normal grain growth process.
4. An activation energy parameter has been obtained for the grain growth process in the largest 5% fraction of equiaxed $L1_0$-ordered FePd grains, $Q_{GG}=29.7\pm1.4$ kcal/mol. This value is comparable to the value expected for boundary diffusion in $L1_0$ ordered FePd.

REFERENCES
1.Weller D, Moser A. IEEE Trans Magn 1999;35:4423
2.Zhang B, Lelovic M, Soffa WA. Scr. Metall Mater 1991;25:1577
3.Zhang B, PhD thesis, University of Pittsburgh;1991
4.Klemmer T, Hoydick D, Okumura H, Zhang B, Soffa WA. Scr. Metall. Mater 1995;33:1793
5.Klemmer T, Soffa WA. :Solid-Solid phase transformations. Warrendale: TMS; 1994 p969.
6.Deshpande A R, Wiezorek JMK. J Magn Magn Mater (2004), 270:157.
7.Kulovits A, Soffa W A, Puschl, Pfieler W. Materials Research Society Symposium proceedings 2003; vol 753: BB5.37.1.
8.Kulovits A, Soffa W A, Puschl W, Pfeiler W. Proceedings of the Vlll. Seminar, Diffusion and Thermodynamics of Materials 2002; Brno, Czech Republic.
9.Mishin Y, Herzig Chr. Acta Mater 2000;48:589.

ACKNOWLEDGEMENT – Financial support for this research from the National Science Foundation (DMR-0094213) is gratefully acknowledged.

Mater. Res. Soc. Symp. Proc. Vol. 842 © 2005 Materials Research Society   S3.10

Comparison of temperature driven ordering in bulk foil and thin film of L1₀ ordered FePd

Chaisak Issro[1], Wolfgang Püschl[1], Wolfgang Pfeiler[1], Bogdan Sepiol[1], Peter F. Rogl[2], William A. Soffa[3], Manuel Acosta[4] and Véronique Pierron-Bohnes[4]

[1] Institut für Materialphysik, University of Vienna, Strudlhofgasse 4, A-1090 Vienna, Austria
[2] Institut für Physikalische Chemie, University of Vienna, Währingerstraße 42, A-1090 Vienna, Austria
[3] Dept. of Material Science and Engineering, University of Pittsburgh, 842 Benedum Hall, Pittsburgh, PA 15261, U.S.A.
[4] IPCMS-GEMME, CNRS-ULP, 23 rue du Loess, BP 43, F-67034 Strasbourg Cedex 2, France

ABSTRACT

Changes in the degree of long-range order of 10 μm thick FePd foil are presented and compared with results on 50 nm thick FePd films. The films were produced by dc and rf magnetron co-sputtering on Si as well as by molecular beam epitaxy co-deposition on MgO substrates. Long-range order was studied by electrical resistivity measurement, X-ray diffraction and Mößbauer spectroscopy.

INTRODUCTION

It is well known that L1₀ long-range ordered intermetallic compounds show advantageous technological properties such as high corrosion resistance and promising high-temperature mechanical strength. Due to their tetragonal distortion in the ordered state these intermetallics display a very marked mechanical and magnetic anisotropy with the tetragonal c-axis being the easy axis of magnetization. This gives hope for their technical application in high-density magnetic and magneto-optic recording when the c-axis of the ordered domains is oriented perpendicular to the surface, especially if these materials can be stabilised in the form of low-dimensional magnetic structures.

Thermodynamic stability is essential for the technical application and performance of these materials. Therefore, we have already studied in massive FePd the change of long-range order (LRO) during isochronal and isothermal heat treatment starting from the disordered state [1]. Now we use our methods to study the changes in the degree of order in thin films of FePd [2].

In the present paper we present first results on FePd films sputtered on Si and co-deposited on MgO by molecular beam epitaxy (MBE) and compare them with a thin foil of FePd ($\approx$10μm). We studied the as-prepared state and the changes during the subsequent heat treatments.

EXPERIMENTAL

FePd polycrystalline foil containing 49.8±0.5 at% Fe was cold-rolled at room temperature (RT) to a thickness slightly below 10μm.

FePd thin film containing 50±5 at% Fe was non-epitaxially deposited by dc and rf magnetron co-sputtering on Si(100) substrate at room temperature. The base pressure of the sputtering chamber was approximately 9×10^{-9} Torr, and Ar gas used for co-deposition had a pressure of nearly 5×10^{-3} Torr. A Pd target was used with a dc current of 144 mA and a sputtering rate of 1.25 Å/s. The rf power for the Fe target was 203 W at 1.00 Å/s sputtering rate. Epitaxial FePd thin films (composition 50 ± 2 at% Fe) were deposited on MgO(001) substrate by MBE under ultra high vacuum of 10^{-9} Torr at a substrate temperature of 773K. The thickness of the FePd layer was estimated by a quartz monitor and by X-ray reflectometry to be about 50 nm.

During annealing treatments in vacuum below 3×10^{-6} Torr from 473K to 900K, the samples were wrapped up in Nb foil to protect them from oxidation. The state of order was investigated by X-ray diffraction (XRD) using CuK$_\alpha$ radiation. Electrical resistivity was measured with in-plane van-der-Pauw geometry on samples directly immersed in liquid nitrogen (REST). Mößbauer measurements were performed at RT using a 10 mCi ^{57}Co source in a Rh matrix.

RESULTS AND DISCUSSION

FePd thin foil

For a comparison with the bulk limit a thin foil cold-rolled from bulk FePd was studied first. Fig. 1 shows the results of REST during isochronal heating (▲) and cooling (▼, $\Delta T = 20K$, $\Delta t = 20min$). The curve in fig. 1 has a standard interpretation: as soon as atomic mobility is enabled via a vacancy-mediated diffusion process, defect recovery and long-range ordering start simultaneously and lead to a reduction of resistivity. A minimum in the curve is observed when defect recovery is completed and when the state of order of the sample corresponds to the equilibrium value for the current annealing temperature. For higher temperatures the sample remains in thermodynamic equilibrium and the increase of REST with temperature corresponds to a decrease of the degree of order. During subsequent isochronal *reduction* of the annealing temperature (▼), the REST behaviour reflects the change of LRO with temperature in equilibrium as long as atomic mobility is high enough (not frozen in).

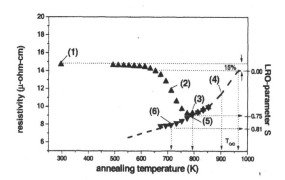

Figure 1. Changes of REST as a function of temperature during isochronal annealing ($\Delta T = 20K$, $\Delta t = 20min$) of a polycrystalline FePd foil (10µm thickness); initial state: as-rolled. Dashed line: 'equilibrium curve' reflecting changes of LRO-parameter in thermodynamic equilibrium (scaling on the right). Arrows correlate with XRD measurements (see text).

These results correspond very well with the results of XRD measurements as shown in fig. 2 describing changes in the degree of long-range order (LRO) of the as-rolled foil (fig.2(1): disordered, S=0) with annealing temperature. After annealing at 713K for 20 min ordering starts with S=0.49 (fig. 2(2)). Subsequent annealings for 20 min at 793K and 900K yield a well developed state of order with S=0.69 and S= 0.73, respectively (figs. 2(3) and 2(4)). Long-time annealing of 40 hours at 793K (minimum of isochronal resistivity curve, see below) changes the LRO-parameter only slightly (S=0.75, fig. 2(5)), whereas a 40-hour anneal at 713K results in a marked increase of order up to a value of S=0.81 (fig. 2(6)).

In fig. 1 we have additionally plotted the X-ray results for the respective annealing temperatures (arrows in fig. 1). Since the relation between REST and the order parameter may be represented by the approximate law $\rho \propto (1-S^2)$ [3], and as we know some reference values we can scale the axis of the LRO-parameter correspondingly. From an earlier investigation [4], we know that the resistivity change due to defect recovery is between 10% and 15%. If we take a value of 15% for our heavily cold-rolled foil and subtract this from the initial REST value, which corresponds to the defected sample in the disordered state, we can take the resulting value for S=0 at $T_{O/D}$. Extrapolating the equilibrium curve (dashed line in fig. 1) to T=0 (S=1), we get the required scale for the order parameter S. The values of S determined by XRD coincide very well with the REST curve, confirming once more the good quantitative correspondence between REST and LRO-parameter [5,6].

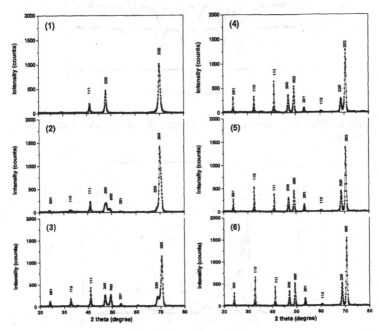

Figure 2. Changes in the degree of LRO of a FePd foil (10μm thickness) during isochronal annealing (ΔT = 20K, Δt = 20min) as measured by XRD. (1) cold-rolled initial state; (2) 713K; (3) 793K; (4) 900K. Additional long-time annealing of 40 h: (5) 793K; (6) 713K.

Mößbauer spectra (fig. 3) were fitted using the RECOIL program with the Voigt-based method for static hyperfine parameter distribution [7]. The maxima of the hyperfine field distribution (fig. 4) can be correlated with the numbers of nearest-neighbour (NN) Fe-Fe atom pairs [8] and compared with REST results (see table I). State (1): as-rolled disordered sample; state (2): L1$_0$-ordered domains with $\langle H_{hf}\rangle \cong$ 26T develop; state (3): maximum of LRO is reached during isochronal annealing, an iron-rich disordered component with $\langle H_{hf}\rangle$>35T appears; state (4): LRO decreases (compare fig. 1) as documented by the decreasing contribution of ordered domains (26T), however, a second ordered component at 27.5T is observed; states (5) and (6): LRO increases, the iron-rich components completely disappear, the ordered components (26T and 27.5T) grow. The splitting of ordered components may be correlated with different ordered structures, twinfree grains and polytwins, respectively [4].

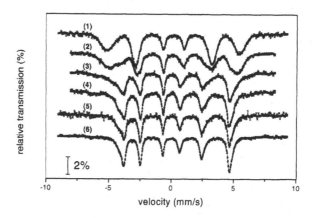

Figure 3. Evolution of Mößbauer spectra of FePd foil after different annealing treatments as indicated (compare table I).

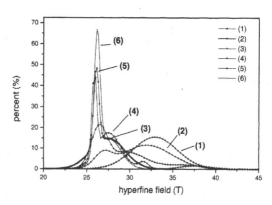

Figure 4. Hyperfine field distribution for annealing treatments 1-6 as described in table I.

Table I. Relation of the hyperfine fields and the long-range order parameter in disordered and ordered FePd thin foil during isochronal heat treatment (see fig.1).

State #	Heat treatment	Hyperfine field $\langle H_{hf} \rangle$ (T) and its contribution (%)	NN Fe-atoms (disordered sites) [8]	Ordered (o), disordered sites (d) [8]	LRO-parameter (X-rays)
(1)	as-rolled	32.9±0.9 (100.)	6.6	d	0.00
(2)	713K	26.9±0.3 (16.0)		o	0.49±0.03
		32.0±0.3 (84.0)	6.0	d	
(3)	793K	26.4±0.1 (42.0)		o	0.69±0.06
		29.5±0.1 (45.0)		o	
		36.2±0.3 (13.0)	8.6	d	
(4)	900K	25.8±0.3 (25.4)		o	0.73±0.05
		27.5±0.1 (62.9)		o	
		31.6±0.8 (4.8)	5.8	d	
		37.6±0.6 (6.9)	9.5	d	
(5)	793K (40 h)	26.1±0.1 (28.0)		o	0.75±0.06
		27.5±0.2 (72.0)		o	
(6)	713K (40 h)	26.2±0.4 (49.0)		o	0.81±0.07
		27.5±0.3 (51.0)		o	

FePd film sputtered on Si substrate

From X-ray diffraction (not shown here) the as-prepared film results as polycrystalline and completely disordered fcc with a very fine initial grain size of about 15nm.

REST measured in-plane during isochronal annealing behaves very similar to the thin foil (fig. 5a). A first stage of resistivity decrease is attributed to grain growth of the fine-grained material, which starts at about 600K; the second, smaller stage at about 750K is interpreted as the onset of L1$_0$ ordering. Annealing at temperatures above 830K increases resistivity drastically. This may be due to the usual onset of disordering with increasing temperature when attaining thermodynamic equilibrium of LRO and an additional influence of interdiffusion between substrate and film [9].

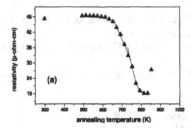 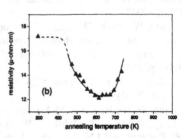

Figure 5. REST measurement on thin films during isochronal annealing (ΔT=20K, Δt=20min). (a) FePd 50nm film sputtered on Si(100); (b) FePd 50 nm film grown by MBE on MgO(001). Note the different scaling of resistivity in (a) and (b).

FePd film epitaxially co-deposited on MgO

X-ray diffraction shows the nearly single crystalline character of the thin film with some mosaic structure and a well developed $L1_0$ LRO with S=0.83 [2].

Due to the comparatively high degree of order of the sample in the as-prepared state, the change during isochronal annealing as obtained from REST (fig. 5b) is much smaller (<½) than that of the sputtered film (fig. 5a). The decrease of REST below 600K reflects a further increase of order towards its thermal equilibrium value, probably brought about by surplus vacancies quenched in during cooling of the substrate after MBE deposition at 773 K. As the sample is practically in thermal equilibrium of LRO, disordering starts already at 650K.

CONCLUSIONS

(i)   A very good correspondence of the degree of LRO of a 10μm FePd foil is obtained as measured by XRD, REST and Mößbauer spectroscopy.
(ii) Sputtering on Si yields disordered nanocrystalline films; above 600K grains grow; above 773K the sample orders (generation of all three variants of ordered domains [2]).
(iii) MBE deposition on MgO produces highly ordered films. From magnetic measurements [2] we know that a c-axis perpendicular to film surface is highly preferred.
(iv) We have shown for the first time that changes in the degree of LRO in thin films can be monitored reliably by REST measurements.

ACKNOWLEDGEMENTS

We acknowledge the financial support by the Materials Dynamics Network (Austrian Ministry for Education, Science and Culture, project GZ 45.529/2-VI/B/7a/2002) and support from the Austrian-French (AMADEUS) programme of scientific-technological co-operation. We thank J. Arabski from IPCMS-GEMME for his kind help during the MBE growth of FePd thin films.

REFERENCES

1. A. Kulovits, W. Püschl, W.A. Soffa, and W. Pfeiler, Mat. Res. Soc. Symp. Proc. **753**, BB5.37 (2003).
2. Ch. Issro, W. Püschl, W. Pfeiler, P.F. Rogl, W.A. Soffa, M. Acosta, G. Schmerber, R. Kozubski and V. Pierron-Bohnes, *Scripta Mat.*, in the press.
3. P.L. Rossiter, *The electrical resistivity of metals and alloys.* Cambridge: University Press (1987), p. 165.
4. A. Kulovits, J.M.K. Wiezorek, W.A. Soffa, W. Püschl and W. Pfeiler, *J. Alloys Compounds*, **378**, 285 (2004).
5. H. Lang, W. Pfeiler and T. Mohri, *Intermetallics* **7**, 1373 (1999).
6. H. Lang, H. Uzawa, T. Mohri and W. Pfeiler, *Intermetallics* **9**, 9 (2001).
7. D.G. Rancourt and J.Y. Ping, *Nucl. Instr. and Meth.* B **58**, 85 (1991).
8. V.A. Tsurin, E.E. Yurchikov, and A.Z. Men'shikov, *Sov. Phys. Solid State* **17**, 1942 (1976).
9. H. Xu, H. Heinrich and J.M.K. Wieczorek, *Intermetallics* **11**, 963 (2003).

Mater. Res. Soc. Symp. Proc. Vol. 842 © 2005 Materials Research Society

Formation of Defect Structures during Annealing of Cold-deformed L1$_0$-ordered equiatomic FePd Intermetallics

Anirudha R. Deshpande and Jörg M.K. Wiezorek

Department of Materials Science and Engineering, University of Pittsburgh, Pittsburgh, PA 15261, USA;

ABSTRACT

Planar defects produced in L1$_0$-ordered FePd during annealing after cold-deformation in the disordered cubic state have been characterized by transmission electron microscopy (TEM). The defects evolving during annealing include arrays of overlapping stacking faults (SF's), {111}-conjugated microtwins (μT's) and thermal antiphase boundaries (APB's). The defect formation mechanisms proposed here are similar to twinning mechanism reported for FCC-metals during annealing. Thus, SF arrays and faulted μT's in the L1$_0$-ordered FePd appear to form during the early stages of annealing by atomic attachment faulting on {111}-facets of the transformation interfaces. During later stages of annealing the reduced amount and the change in nature of the driving forces for the microstructural rearrangement result in changes in the predominant defect formation mechanism. The features of the defect genesis in L1$_0$-FePd are discussed with respect to solid-state transformations during processing of these ferromagnetic intermetallics.

INTRODUCTION

The tetragonal L1$_0$-ordered phase γ_1-FePd exhibits strongly anisotropic, uniaxial ferromagnetic properties, which are very attractive for applications in magnetic memory devices and other ferromagnetic applications [1]. The γ_1-FePd forms below the ordering temperature of about 925K for the equiatomic composition from a concentrated solid-solution with disordered face-centered cubic (FCC) structure via a thermodynamically first-order type phase transformation [2]. Continuous cooling from the fcc-phase field or isothermal annealing of the quenched metastable FCC-phase at temperatures below about 925K results in morphologically lamellar microstructures of the L1$_0$-ordered product that are characteristically twinned on dodecahedral, {101}, planes, [2,3]. The conventional mode of ordering involves the nucleation of coherent precipitates of three possible orientation variants of the tetragonal product on {101} in the FCC-parent phase. Formation of the {101}-conjugated polytwins during strain affected growth of the L1$_0$-precipitates facilitates self-accommodation of the considerable transformation strains [3]. The polytwinned (PT) microstructural state of the L1$_0$-FePd formed by this conventional ordering mode is associated with disappointing technical magnetic properties, low failure strain and brittleness in tensile conditions [3-5]. Room temperature deformation of the FCC-phase prior to order annealing at T<925K can be utilized to inhibit or even prevent the formation of the polytwinned microstructure [6]. The thermomechanically processed FePd intermetallics exhibit improved magnetic properties [7]. Depending on the details of the thermomechanicl treatment both the microstructural scale and the morphology of the L1$_0$-FePd can be altered to comprise varying fractions of A) essentially equiaxed mono-variant L1$_0$-ordered grains and B) emerging polytwin structure with a relatively high dislocation density [7,8]. Examples of FePd microstructures achieved by thermomechanical processing are shown in Fig. 1. Order-annealing of the cold-deformed FCC solid solutions involves concomitant ordering phase transformation and annealing of the defect structure induced by the cold-deformation, i.e.

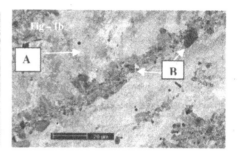

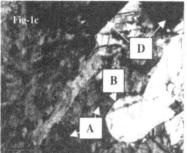

Fig-1 a) Backscatter electron (BSE) SEM micrograph of the completely CR transformed microstructure consists solely of mono-variant $L1_0$ ordered grains.
b) SEM BSE image of the composite like microstructure, consisting of mono-variant $L1_0$ ordered grains (label B) bounded by $L1_0$ ordered regions formed by conventional ordering mode.
c) Bright field TEM micrograph of the composite like microstructure. Emerging PT structure (label A) that has inherited a large dislocation content is also visible. The new mono-variant $L1_0$ ordered grains (labels B & D) also appear to be forming SFs, twins and APBs.

combined solid-state reactions. The presence of the stored strain energy in the deformed microstructure enables the operation of an alternative mode of ordering, which involves heterogeneous nucleation of mono-variant $L1_0$-phase grains at locations with suitable stress-state and their growth by rapid phase boundary migration. This heterogeneous mode of ordering, which is assisted by the stored strain energy of cold-deformation, has been studied in detail previously and was shown to be similar to a massive ordering transformation [6-8]. We refer to this alternative ordering mode in the thermomechanically processed FePd as the combined reaction mode of ordering (CR-mode), as opposed to the conventional ordering mode. The CR-mode ordered product typically contains a high-density population of characteristic defects, such as twin boundaries (between matrix (M) and twin (T) variant γ_1-grains), micro-twins (μT's), stacking faults (SF's), antiphase boundaries (APB's) and dislocations (Fig. 1). Previous investigation of the microstructural and property evolution during the annealing of the cold-deformed FCC solid solution of equiatomic FePd has shown that the peak magnetic properties are obtained at the completion of the $L1_0$-ordering transformation and diminish for longer annealing times [6,7]. The magnetic hysteresis behavior of FePd intermetallics is extremely sensitive to the microstructure and defect structure produced during solid-state processing [7,8]. Thus, TEM experiments are used to study the mechanisms for formation of the defect structure governing the resultant structure-property relationships of these ferromagnetic $L1_0$-intermetallics. The micro-processes at the migrating transformation interfaces during annealing are shown to play a critical role in the generation of the profusion of planar defects characteristic of the FePd ferromagnets. Here, we present preliminary results on the formation mechanisms for twins and stacking faults in $L1_0$-FePd during order-annealing.

EXPERIMENTAL PROCEDURE

Equiatomic FePd has been prepared by vacuum induction melting of high-purity elemental starting materials. After 50% reduction in thickness, recrystallization at 1173K for 5h and ice-water-quenching, sections of the consolidated alloy have been cold-deformed by either a single pass of equal channel angular pressing (ECAP) or cold rolling to 97% thickness reduction. The single-pass ECAP and cold-rolling imparted strains of approximately 0.65 and about 3.0 to the material, respectively. Transformation of FCC $\rightarrow$ $L1_0$ by the CR-mode was achieved during isothermal annealing at 773K and 873K. Microstructures have been characterized by combinations of X-ray diffraction (XRD), scanning and transmission electron microscopy (SEM and TEM) using a Philips X'pert XR diffractometer, a field-emission gun equipped Philips XL30 SEM and a JEOL 2000FX TEM/STEM and JEOL CX200 TEM, respectively.

RESULTS AND DISCUSSION

Depending on the amount of strain imparted during cold-deformation morphologically different microstructures result when the ordering transformation is completed during isothermal annealing of the deformed FCC-(Fe,Pd) solid solution [7,8]. Fig.1 depicts examples of the microstructures established when the FCC $\rightarrow$ $L1_0$ ordering is complete and the peak coercivity is measured for the large strain and smaller strain imparted by cold-rolling to 97% reduction and the single pass ECAP, respectively [7,8]. The cold-rolled sample exhibits an essentially equiaxed morphology of mono-variant $L1_0$-phase grains, the CR-mode product (label B in Fig. 1a). The ECAP sample comprises a minority fraction of the equiaxed grains of the CR-mode product along prior FCC grain boundaries and microbands (label B, Fig. 1b) and a majority fraction of grains with an emerging polytwinned (PT) morphology (label A in Fig. 1b). The multi-beam bright field (MBBF) TEM image, Fig. 1c, shows details of the microstructural differences in the fully ordered FePd consisting of fractions of the emerging PT-grains with a high dislocation density (label A, Fig. 1c) and the mono-variant $L1_0$-grains of the equiaxed CR-product (label B, Fig. 1c). Planar faults characteristic of the defect structure in the mono-variant $L1_0$-grains, such as SF, APB's, μT's and twin boundaries between two $L1_0$-grains (label D in Fig. 1c) are also depicted in Fig 1. The cold rolling of the FCC-phase produced sufficient deformation heterogeneities in the microstructure to enable formation of mono-variant $L1_0$-grains by the alternative CR-mode ordering essentially in the entire microstructure. The single pass ECAP imparts less strain than the cold-rolling and with a more heterogeneous spatial distribution in the microstructure. Hence, the strain energy assisted CR-mode produces mono-variant $L1_0$-grains only at suitable deformation heterogeneities in the microstructure, e.g. prior FCC grain boundaries, along with the lamellar PT-FePd microstructure emerging via the competing conventional ordering mode at locations where the stored strain energy is insufficient to favor a CR-mode (Fig. 1b). Thus, for sufficiently large strain of cold-work an equiaxed 'CR-microstructure' results, while for insufficient strain of cold-work a more complex 'composite-like' microstructure characterizes the thermomechanically processed FePd when the $L1_0$-ordering phase transformation is complete. Further annealing of the fully ordered FePd samples with the different microstructural morphologies involves the classic annealing phenomena of recovery, recrystallization and grain growth or coarsening. The evolution of these different microstructural states of the fully $L1_0$-ordered FePd has been described previously in detail [7-9]. The TEM micrographs of Fig. 2 depict an example of planar defects emerging from the interface between a mono-variant L1$_0$-grain (label B) and an emerging PT-grain (label A) in a FePd sample with the 'composite-like' microstructure. The ordered state of the sample has been confirmed globally by

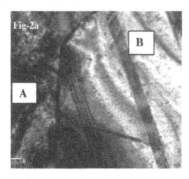

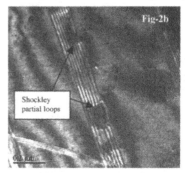

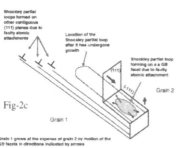

Figure -2 a) BF TEM micrograph depicting the 'composite like' microstructure. Labels A and B mark the deformed emerging PT-FePd and the CR mode mono-variant L1$_0$ grain respectively.
b) DF TEM micrograph using g = [-111] near a [211] zone axis. Individual Shockley partial loops appear in strong contrast along the SF array.
c) Schematic of the modified Dash and Brown mechanism to explain generation of SFs and twins at the mobile 'recrystallization' interface.

XRD (not shown for brevity) and locally by selected area diffraction (SAD) in the TEM [7,8]. The considerable dislocation density present in the conventionally ordered L1$_0$-FePd grain (label A) has been inherited from the cold-deformed FCC phase, indicating that recovery processes would be expected during further annealing in this fraction of the 'composite-like' microstructure. The mono-variant L1$_0$-grain (label B, Fig. 2a) is relatively free of dislocations and exhibits characteristic annealing defects associated with the mobile interface between FePd grains labeled A and B in Fig. 2a. The curvature of the interface between grains A and B suggests mobility towards the center of grain B if capillarity forces would be the dominant driving force, as in grain growth. However, the large difference in dislocation density in grains A and B suggests that recrystallization may be the dominant phenomenon achieving microstructural change. Hence, it would be expected that grain B grows at the expense of grain A via migration of the mobile interface section to the left in Fig. 2a. The growth of grain B is driven by the excess stored strain energy associated with the defect content of grain A and analogous to the growth of new recrystallized grains into a deformed matrix during annealing of disordered solid solutions. The habit planes of the arrays of SF's and apparent μT's produced during annealing in grain B have been determined to be octahedral planes using TEM tilting experiments. The dark field (DF) TEM micrograph of Fig. 2b shows the partial dislocations that bound the overlapping SF's emerging from the apparent segment of a recrystallization front between grains A and B. The contrast features of the planar defects emanating from the interface between grains A and B are consistent with those of cascades of overlapping SF's or SF arrays on octahedral planes of the mono-variant $L1_0$-ordered grain. The most likely Burgers vector of the SF generating Shockley partials would be of the type b=1/6<112], which can generate a proper octahedral twin in the $L1_0$-structure [5]. A suitable mechanism for the formation of the SF arrays at the migrating recrystallization interface between grains A and B of the $L1_0$-FePd can be proposed in analogy to the classic Dash-Brown mechanism for formation of annealing twins in FCC solid solutions [10]. The geometric arrangement of Shockley partial loops bounding stacking faults that emanate from the grain boundary has been previously elucidated by Mahajan et al.[11], which is schematically reproduced in Fig. 2c. The SF cascades are bounded by Shockley partial loops as observed in Fig. 2b, which formed on parallel {111}-facets at the transformation interface and mutually repel due to elastic interactions, dragging out ribbons of SF into the growing grain. This mechanism is also suitable to generate internally faulted µT's on {111} in $L1_0$-ordered FePd if the cascades of parallel SF's emanating from the mobile interface become sufficiently thick and dense. The attachment faulting process for twin formation at mobile interfaces observed here is very similar to that reported previously for SF and twin genesis in massively transformed $L1_0$-ordered τ-MnAl [12].

The formation of twins has also been observed during the later stages of annealing of the heavily cold-rolled material that consists almost entirely of essentially equiaxed mono-variant $L1_0$-grains formed by the CR-mode. In these microstructures the driving force for microstructural changes is diminished relative to that available in the 'composite-like' microstructures, since the strain energy of cold-work associated with the dislocation substructure inherited from the deformed FCC-phase has been dissipated during the CR-mode ordering. Fig 3a shows an example of the twin structures observed in these latter microstructures, which often exhibit twin-related grains and very planar octahedral {111}-twin interfaces that have very low defect density. The SAD pattern (Fig 3b) from regions separated by a very straight boundary in Fig. 3a indicates that octahedrally conjugated twins are present. The DF TEM images shown in Fig. 3c and 3d have been obtained with twin, $g = [1\text{-}1\text{-}1]_T$, and matrix, and $g = [1\text{-}1\text{-}1]_M$, diffraction vectors, respectively, clearly confirming the presence of coarsening twin related grains. The twins observed under these latter annealing conditions are true order twins of the $L1_0$-ordered FePd and formed when the main driving force for the observed microstructural changes was reduced to that associated with reduction of interfacial free energy i.e. that associated generally with the grain growth stage of annealing. It appears reasonable to propose that the larger matrix grain (M) coarsens at the expense of the smaller twin grain (T) predominantly by migration of the non-coherent section of twin boundaries.

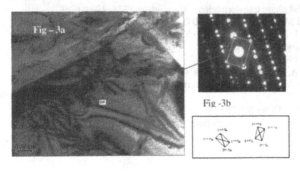

Figure -3a Bright field TEM micrograph depicting twins. The microtwin lamellae observed in this microstructure are much less defected as compared to fig 2b.
Figure -3b Diffraction pattern with [101] as zone axis for both the twin variants. (Relatively mobile non coherent part and immobile coherent part of the twin boundary shown in fig-3a are marked in fig- 3d).

Various mechanisms such as the 'grain growth accident' [13], 'grain boundary dissociation' [14] have been proposed previously, which are suitable to rationalize the twin morphologies observed in the grain growth / coarsening stage of annealing of FePd. Current observations are not conclusive but suggest the dominance of the grain growth accidents mechanism. Additional investigations of this issue are currently underway.

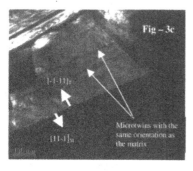 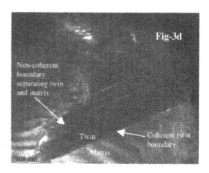

Dark field TEM micrograph using $g = [1\text{-}1\text{-}1]_T$ Dark field TEM micrograph using $g = [1\text{-}1\text{-}1]_M$

Conclusions

1. Twin formation during growth of $L1_0$ ordered grains in the 'composite like' microstructure has been attributed to an attachment faulting mechanism akin to the classic Dash-Brown mechanism for the formation of recrystallization twins in FCC solid solutions.

2. The twins observed during grain growth of the completely CR-mode ordered microstructures are true order twins.

3. Various mechanisms can be applied to rationalize the emergence of these true order twins. Speculatively, the grain growth accidents mechanism has been proposed to explain their generation. Further detailed investigations are required to conclude regarding this issue.

REFERENCES

[1] Weller D, Moser A. IEEE Trans Magn 1999;35:4423
[2] Zhang B, Lelovic M, Soffa WA. Scr. Metall Mater 1991;25:1577
[3] Zhng B, PhD thesis, University of Pittsburgh;1991
[4] Klemmer T, Hoydick D, Okumura H, Zhang B, Soffa WA. Scr. Metall. Mater 1995;33:1793
[5] Rao M, Soffa W A. Scr. Mater.(1997),36:735.
[6] Klemmer T, Soffa WA. :Solid-Solid phase transformations. Warrendale: TMS; 1994 p969.
[7] Deshpande A R, Wiezorek JMK. J Magn Magn Mater (2004), 270:157.
[8] Deshpande A R, Xu H, Wiezorek JMK Acta Mater. (2004) 52:2903
[9] Klemmer T J, PhD thesis, University of Pittsburgh (1995)
[10] Dash S, Brown N, Acta Metall. (1963) 11:1067
[11] Mahajan S, Pande C S, Imam M A, Rath B B. Acta Mater. (1997),45,2633.
[12] Yanar C, Radmilovic V, Soffa W A, Wiezorek J M K. Intermetallics (2001),9,949
[13] Gleiter H, Acta Met. (1969),17:565
[14] Meyers M A, Murr L E, Acta Met. (1978),26:951
Acknowledgement – Financial support for this research from the National Science Foundation (DMR-0094213) is gratefully acknowledged.

Mater. Res. Soc. Symp. Proc. Vol. 842 © 2005 Materials Research Society

Magnetic Properties of E2₁-base Co₃AlC and the Correlation with the Ordering of Carbon Atoms and Vacancies

Yoshisato Kimura[1], Fu-Gao Wei[2], Hideyuki Ohtsuka[2] and Yoshinao Mishima[1]

[1]Materials Science and Engineering, Tokyo Institute of Technology,
4259-G3-23 Nagatsuta, Midori-ku, Yokohama 226-8502, Japan.
[2]National Institute for Materials Science,
1-2-1 Sengen, Tsukuba, Ibaraki 305-0047, Japan.

ABSTRACT

Targeting to develop $E2_1$ Co_3AlC based heat resistant alloys, phase stability of $E2_1$ Co_3AlC and $(Co, Ni)_3AlC$ has been investigated together with the magnetic properties of $E2_1'$ $Co_3AlC_{0.5}$ which is formed by the extra ordering of carbon atoms accompanying anti-phase boundary (APB). The correlation of ferromagnetism with APB in $E2_1'$ $Co_3AlC_{0.5}$ was evaluated using single crystals by high-resolution transmission electron microscopy (HRTEM) and vibrating sample magnetometer. Anti-phase domain (APD) size affects the ferromagnetism: for instance, the saturation magnetization becomes larger as the APD size is smaller. Local atomic configuration at APB was clearly observed by HRTEM image.

INTRODUCTION

The $E2_1$ type intermetallic compound Co_3AlC is a quite hopeful strengthener for a new class of Co-base heat resistant alloys since its ordered crystal structure is almost the same as that of $L1_2$ Ni_3Al [1-6]. The difference between $E2_1$ and $L1_2$ is distinguished by a carbon atom occupying the octahedral interstice at the body center. The carbon content is always less than the stoichiometry, thus it is denoted as Co_3AlC_x where x is around 0.6 [1,5]. Our group reported that the extra ordering of carbon atoms is observed in $E2_1$ Co_3AlC in multi-phase alloys coexisting with α-(Co) [7]. $E2_1'$ $Co_3AlC_{0.5}$ is formed by this ordering most likely due to minimizing the free energy via elastic and magnetic contributions. Unit cells of (a) $E2_1$ and (b, c) $E2_1'$ are represented in Fig. 1, where (b) and (c) are equivalent. The super-cell of $E2_1'$ consists of $E2_1$ and $L1_2$ lattices arranged alternatively in three dimensions, and it can be regarded that carbon atoms and vacancies are ordered in (b). It is described in (c) that carbon atoms are allocated on an fcc sub-lattice without considering vacancies. We also have reported that Co_3AlC alloy exhibits fairly weak ferromagnetism while its curie temperature is unusually high about 1100 K comparing to weak magnetization [8]. We believe this unusual ferromagnetism is correlated with the ordered crystal structure and strongly affected by the extra ordering of carbon atoms. Considering the effects of chemical and magnetic ordering on phase stability as well as

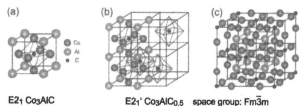

E2₁ Co₃AlC   E2₁' Co₃AlC₀.₅   space group: Fm3̄m

Fig. 1   Unit cells of (a) E2$_1$ type Co$_3$AlC and (b, c) E2$_1$' type Co$_3$AlC$_{0.5}$.

microstructures is quite important to evaluate the potential of strengthener used for high temperature structural application.

In the present work, phase stability of E2$_1$ Co$_3$AlC and E2$_1$' Co$_3$AlC$_{0.5}$ is investigated from the viewpoints of the extra ordering of carbon atoms and magnetic properties. Effect of anti-phase boundary on the ferromagnetism is also investigated and discussed. On the other hand, Co$_3$AlC forms continuous solid solution throughout Co-Ni-Al-C quaternary system with Ni$_3$AlC and even with L1$_2$ Ni$_3$Al [9]. It is quite beneficial to evaluate the phase stability not only for E2$_1$ Co$_3$AlC and E2$_1$' Co$_3$AlC$_{0.5}$, but also for (Co, Ni)$_3$AlC in Co-Ni-Al-C quaternary system for designing attractive heat resistant alloys using (Co, Ni)$_3$AlC as our final goal.

EXPERIMENTAL DETAILS

Single crystals of Co$_3$AlC and (Co, Ni)$_3$AlC (Co:Ni = 0.8:0.2) were successfully grown for the first time using optical floating zone melting (OFZ) [10]. To ensure high carbon content of the alloys, master alloy ingots were prepared by induction heat melting prior to arc melting to prepare seed rods and feed rods used for OFZ. The condition for single crystal growth is; growth rate of 2.0 mm/h, rotational speed of 25 rpm and flowing Ar gas. Several alloys were prepared by arc melting under Ar atmosphere; L1$_2$ (Ni, Co)$_3$Al single-phase alloys and multi-phase alloys consisting of E2$_1$ (Co, Ni)$_3$AlC, α-(Co) and/or graphite. Anti-phase domain (APD) size was controlled by heat treatment; annealing at 1373 K for 24 h or at 1073 K for 24 h, followed by water quenching. Microstructure observation and crystal structure analysis were conducted using conventional and high resolution transmission electron microscopy (TEM, HRTEM). Magnetic properties of alloys, such as curie temperature and magnetization were measured using vibrating sample magnetometer (VSM). Differential scanning calorimetry (DSC) was also used to measure curie temperature, T_C^{mag}, and extra ordering temperature of carbon, T_C^{chem}, in a temperature range from 300 to 1273 K at heating and cooling rate of 10 K/min.

RESULTS AND DISCUSSION
Ordering of carbon and anti-phase boundary in E2$_1$ Co$_3$AlC single crystal

The extra ordering of carbon atoms has been confirmed by TEM to occur in E2$_1$ Co$_3$AlC

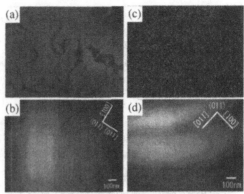

Fig. 2 A lattice image of E2$_1$' Co$_3$AlC$_{0.5}$ single phase observed by HRTEM and corresponding SAD where IB = [110].

Fig. 3 Dark field images of APB in E2$_1$ Co$_3$AlC single crystal: (a,b) as-grown and (c,d) water quenching from 1373 K, and (a,c) [1/2 1/2 1/2] DF and (b,d) [100] DF.

single crystal accompanying the formation of E2$_1$' Co$_3$AlC$_{0.5}$. A lattice image and corresponding selected area diffraction (SAD) pattern taken from the single crystal are shown in Fig.2. The distance 0.36 nm coincides with a half of lattice parameter of Co$_3$AlC$_{0.5}$, i.e., lattice parameter of Co$_3$AlC measured by X-ray diffraction. A half of [111] spots observed in SAD are the extra ordered diffraction peak which indicates the structure of E2$_1$' Co$_3$AlC$_{0.5}$. Dark field images (DF) of anti-phase boundary (APB) formed in the single crystal are shown in Fig. 3. Contrast of APB is only visible in [1/2 1/2 1/2]DF, (a) and (c), but invisible in [100]DF, (b) and (d). It means that this APB is originated from the ordering of carbon atoms being related to E2$_1$' structure. The size of anti-phase domain (APD) is relatively coarse in as-grown sample, (a), while it is very fine in the sample annealed at 1373 K for 24 h followed by water quenching, (c). It suggests that the annealing temperature of 1373 K is just above the chemical ordering temperature, T_C^{chem}, and APB forms with very fine APD size during quenching. In DSC measurements, T_C^{chem} is estimated at about 1325 K.

Magnetic properties of E2$_1$ (Co, Ni)$_3$AlC alloys and E2$_1$' (Co, Ni)$_3$AlC$_{0.5}$ single crystal

Cobalt-nickel concentration dependence of curie temperature, T_C^{mag}, is shown for several E2$_1$-base (Co, Ni)$_3$AlC alloys including E2$_1$' (Co, Ni)$_3$AlC$_{0.5}$ single crystals in Fig. 4. Moreover, T_C^{mag} is also measured for L1$_2$ (Ni, Co)$_3$Al single-phase poly crystal alloys for a comparison. Since some multi-phase alloys contain α-(Co) phase, T_C^{mag} is plotted for α-(Co) as well in the figure. It is noteworthy that T_C^{mag} of E2$_1$' Co$_3$AlC$_{0.5}$ single crystal is evaluated quite high at 1115 K which is just below T_C^{chem}, and Co-rich (Co, Ni)$_3$AlC alloys have high T_C^{mag} around 1100 K. In contrast, T_C^{mag} of L1$_2$ Ni$_3$Al is about 40 K, and it drastically increases with Co content although the solubility limit of Co in Ni$_3$Al is about 21 at% at 1273 K [11]. Addition of C

397

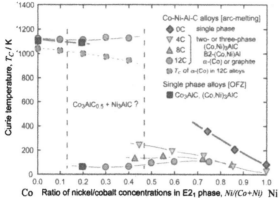

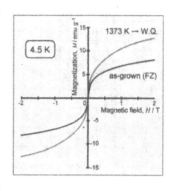

Fig. 4   Co-Ni concentration dependence of curie temperature, T_C^{mag}, in $E2_1$ (Co, Ni)$_3$AlC alloys including $E2_1$' (Co, Ni)$_3$AlC$_{0.5}$ single crystals.

Fig. 5   Effect of anti-phase domain (APD) size on H-M curves observed in $E2_1$' Co$_3$AlC$_{0.5}$ single crystals at 4.5 K.

tends to reduce T_C^{mag} in Ni-rich (Co, Ni)$_3$AlC alloys. The huge gap in T_C^{mag} between Co-rich and Ni-rich alloys would be an indication of two-phase separation of Co$_3$AlC$_{(0.5)}$ and Ni$_3$AlC in a compositional range of Co-Ni ratio from 0.1 to 0.5. Some alloys seem to have two different T_C^{mag}, high temperature at around 1100 K and low temperature at around 40 K.

Magnetization curves, H-M hysteresis, measured at 4.5 K for as-grown and anneal -quenched samples are shown in Fig. 5, and microstructures of those samples are already shown in Figs. 3 (a) and (c). Magnetization of $E2_1$' Co$_3$AlC$_{0.5}$ is strongly affected by APD size. For instance, the saturation magnetization becomes larger as the anti-phase domain (APD) size is smaller, i.e. the total area of APB is larger.

Consideration for ferromagnetism of $E2_1$' Co$_3$AlC$_{0.5}$

The most interesting aspect in Fig. 4 is that T_C^{mag} of $E2_1$' Co$_3$AlC$_{0.5}$ seems to be on the extrapolation from that of L1$_2$ Ni$_3$Al. It suggests that ferromagnetism of $E2_1$' Co$_3$AlC$_{0.5}$ would be raised from its "L1$_2$ Co$_3$Al" like behavior. This can be the key to understand the ferromagnetism of $E2_1$' Co$_3$AlC$_{0.5}$, not only unusually high T_C^{mag} with low magnetization but also the effect of APD size on magnetization.

Atomic arrangement and stacking sequence of (111) close-packed planes are depicted in Fig. 6 for comparing L1$_2$, $E2_1$ and $E2_1$' structures. All these three structures have the same atomic arrangement of Co and Al in common. They are distinguished by C atoms layer; being inserted every (111) planes in $E2_1$, every other (111) planes in $E2_1$' and none in L1$_2$. It is obvious that $E2_1$' is composed of alternative repetition of the $E2_1$ stacking and the L1$_2$ stacking, one after the other. Each C atom in the $E2_1$ stacking is facing to six Co atoms and constructing the Co$_6$C octahedron

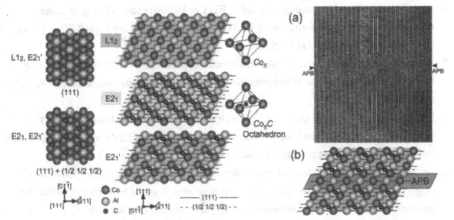

Fig. 6 Schematics showing atomic arrangement and stacking of close packed (111) planes in $L1_2$, $E2_1$ and $E2_1$' structures, and Co_6 and Co_6C octahedron.

Fig. 7 Local atomic configuration at APB in $E2_1$' structure; (a) a schematic drawing and (b) a HRTEM image.

as shown in the figure, the Co_6 octahedron is formed at the $L1_2$ stacking. It can be rationally assumed that the $L1_2$ stacking and the Co_6 octahedron are attributed to the ferromagnetism of $E2_1$' $Co_3AlC_{0.5}$ since the population of Co-Co bonding increases effectively and the overlapping of d-electron orbits changes with exchange integral. Under these assumption, unusually high T_C^{mag} can be explained being on the extrapolation line from $L1_2$ Ni_3Al. Relatively weak magnetization should be essentially much higher if it is correctly normalized by effective volume of the $L1_2$ stacking or the Co_6 octahedron.

Moreover, a local structure of APB, i.e. the shift of (1/2 1/2 1/2) planes at APB, has been successfully observed in HRTEM lattice image as presented in Fig. 7 (a). Bright straight lines correspond to the contrast from C atoms layers being edge-on, and the shift of them at APB can be observed as described in an illustration of Fig. 7 (b). Nevertheless the contrast from C atoms layers is not sufficiently strong, it becomes visible in the case that the foil is properly tilted near a two-beam condition. The formation of APB introduces an additional layer of $L1_2$ stacking so as to result in successive two layers of $L1_2$ stacking, as illustrated in Fig.7 (b). Note that this APB due to the ordering of carbon atoms is essentially equivalent to the stacking fault of C atoms layer, and the atomic arrangement of Co and Al is never affected by this APB.

CONCLUSIONS

Phase stability of $E2_1$ Co_3AlC and $E2_1$' $Co_3AlC_{0.5}$ is investigated and discussed with the relationship between extra ordering of C atoms and magnetic properties. Following conclusions are drawn from the present work.

1. It has been confirmed in the single crystal of $E2_1$ Co_3AlC that extra ordering of C atoms occurs accompanying the formation of $E2_1$' $Co_3AlC_{0.5}$ which has very high T_C^{mag} at 1115 K just below T_C^{chem} evaluated at 1325 K.

2. A huge gap is observed in the Co-Ni concentration dependence of T_C^{mag} between Co-rich and Ni-rich $(Co, Ni)_3AlC$ alloys. It indicates the possibility of two-phase separation of $Co_3AlC_{(0.5)}$ and Ni_3AlC.

3. $E2_1$' $Co_3AlC_{0.5}$ is expected to show $L1_2$ like physical and mechanical behavior as if "Co_3Al" were stabilized since the stacking of close packed (111) planes of $E2_1$' consists of $L1_2$ stacking and that of $E2_1$ which are repeated alternatively.

4. Anti-phase domain (APD) size affects the ferromagnetism of $E2_1$' $Co_3AlC_{0.5}$: the saturation magnetization becomes larger as the APD size is smaller. It would be attributed to that the population of Co-Co bonding increases with $L1_2$ stacking and Co_6 octahedron.

ACKNOWLEDGMENT

The present work was conducted on the base work supported by the PRESTO program, Japan Science and Technology Agency.

REFERENCES

1. L. J. Huetter and H. H. Stadelmaier, *Acta Metal.* **6**, 367 (1958).
2. H. Nowotny, and F. Benesovsky, *Phase Stability in Metals and Alloys*, eds. P. S. Rudman et al., McGraw Hill, New York, 319 (1967).
3. H. H. Stadelmaier, *Developments in the Structural Chemistry of Alloy Phases*, ed. B. C. Giessen, Plenum Press, New York, 141 (1969).
4. H. Hosoda, M. Takahashi, T. Suzuki and Y. Mishima, *High-Temperature Ordered Intermetallic Alloys V*, MRS Symp Proc **288**, 793 (1993).
5. Y. Kimura, M. Takahashi, S. Miura, T. Suzuki and Y. Mishima, *Intermetallics* **3**, 413 (1995).
6. Y. Kimura, C.T. Liu and Y. Mishima, *Intermetallics* **9**, 1069 (2001).
7. F.-G. Wei, K.-Y. Hwang and Y. Mishima, *Intermetallics* **9**, 671 (2001).
8. Y. Kimura, K. Iida and Y. Mishima, *Defect Properties and Related Phenomena in Intermetallic Alloys*, MRS Symp Proc **753**, 433 (2003).
9. H. Hosoda, S. Miyazaki and Y. Mishima, *J. Phase Equilibria* **22**, 394 (2001).
10. Y. Kimura, K. Sakai and Y. Mishima, will be published in J. Phase Equilibria and Diffusion.
11. P. Villars, A. Prince and H. Okamoto, eds., *Handbook of Ternary Alloy Phase Diagrams*, vol. **3**, ASM Intnl., Materials Park, 3052 (1995).

Mater. Res. Soc. Symp. Proc. Vol. 842 © 2005 Materials Research Society   S5.5

Structural Properties and Magnetic Behavior in the Pseudobinary Alloys CoFe-CoAl

Nobutoshi Tadachi, Hiroki Ishibashi and Mineo Kogachi
Department of Materials Science, College of Integrated Arts and Sciences,
Osaka Prefecture University, Sakai, 599-8531, Japan

ABSTRACT

Structural properties and magnetic behavior in pseudobinary alloys $CoFe_{1-x}Al_x$ are studied. The B2-phase exists stably over wide ranges of composition and temperature and the Heusler-phase appears around the stoichiometric composition ($0.4 < x < 0.6$) below about 973K. The mean magnetic moment in the B2-phase region decreases nearly linearly with increase in x in $x \geq 0.2$. Further, it decreases with increase in quenching temperature in $x < 0.4$, while it remains almost constant in $x \geq 0.4$. The mean magnetic moment of Heusler alloy Co_2FeAl ($x = 0.5$) has the value of $1.30\,\mu_B$/atom, nearly the same as in the B2-phase, and shows no remarkable quenching temperature dependence.

INTRODUCTION

B2-type and Heusler ($L2_1$)-type phases with the ordered structure based on a bcc lattice are formed around the stoichiometric compositions, AB and A_2BC, in many binary and ternary alloy systems. Because of the excellent mechanical and physical properties, these phase alloys are thought to be promising candidates for functional as well as structural materials. Particularly, some Heusler-type ferromagnetic alloys such as Co_2MnSi and Co_2MnGe receive considerable attention in the field of magnetoelectronics (spin electronics), since they were recently predicted to be the so-called half-metal ferromagnets which show 100% spin polarization due to a gap at the Fermi level in the minority-spin band [e.g., 1,2].

It is well recognized that various properties are sensitively affected by the point defects (antisite atoms and vacancies) generated intrinsically in the alloy crystals at finite temperature. Since 1992, we have investigated the point defect structure for B2-type binary alloys, NiAl, CoAl, FeAl and AuCd, by means of the density, X-ray and neutron diffraction and positron annihilation measurements [e.g., 3-7]. More recently, we studied a correlation between the point defect behavior and the magnetism in ferromagnetic B2-phase alloys $Co_{1-c}Fe_c$ ($0.4 \leq c \leq 0.6$) [8-10], through the long-range order (LRO) parameter whereby the defect structure can be expressed. As a result, the mean magnetic moment obtained as a function of quenching temperature was found to change linearly with respect to square of the LRO parameter in any compositions, indicating a strong correlation between those.

In the present study, the structural and magnetic properties in pseudobinary alloys, $CoFe_{1-x}Al_x$ ($0 \le x \le 1$), are concerned. Since both terminal alloys, CoFe ($x = 0$) and CoAl ($x = 1$), have the B2-type structure, the B2-phase is expected to appear over considerably wide ranges of composition and temperature [11]. Thus, we can examine in detail how the magnetic behavior in B2 CoFe alloy is changed by substitution of Fe atom for Al atom. Furthermore, at the stoichiometric composition of Co_2FeAl ($x = 0.5$), ferromagnetic Heusler-phase is known to appear [12]. The A_2BC Heusler-type structure can be regarded as the structure consisting of eight bcc unit cells in which eight cube-corner positions (A-site) are occupied by A atom and eight body-center positions are occupied alternately by B atom (B-site) and C atom (C-site). In Co_2FeAl alloy, with increasing temperature, this phase is thought to transform to the B2-phase by atom disordering between the Fe- and Al-sites. It is thus very interesting to study the magnetic behavior in the Heusler-phase region and discuss its correlation with the point defect behavior, including a comparison with that in the B2-phase region.

EXPERIMENTAL PROCEDURE

Eleven ingot samples of $CoFe_{1-x}Al_x$ ($0.1 \le x \le 0.9$) were prepared by arc melting the weighed metals, Co (purity of 99.99%), Fe (99.99%) and Al (99.999%), under an argon atmosphere. Weight losses were less than 0.2%. Each ingot was homogenized at 1273K for 85h under flowing argon gas. From individual ingots, the powdered samples with a particle size less than 60μm for X-ray diffraction and the platelet samples with a size of about $1 \times 1 \times 3$ mm^3 for magnetic measurement using a SQUID magnetometer and those of $1 \times 1 \times 15$ mm^3 for electrical resistance measurement were prepared. They were annealed at 1173K for 2h in silica tubes filled with argon, followed by cooling to room temperature at a rate of 2K/min. Individual samples for the X-ray diffraction and magnetic measurements were again annealed at various temperatures of 773K to 1273K for 2160h to 1h and then water-quenched. The lattice constant at room temperature and the mean magnetic moment at 4.2K were obtained for these samples in a similar way as described in previous work [8-10]. The electrical resistance was measured by a standard dc four-terminal method [10]. Furthermore, for estimation of the vacancy concentration, the density measurements [8] were also performed for three samples with composition of $x = 0.30, 0.50$ and 0.70, which were quenched from 1173K.

RESULTS AND DISCUSSION

Experimental results

The plausible phase relation for the $CoFe_{1-x}Al_x$ system determined in the present study is shown in Fig.1. Three broken curves show the changes in T_0, T_1 and T_C which represent the Heusler to B2 and the B2 to A2 (or B2+A2 phase separation) phase transition temperatures and

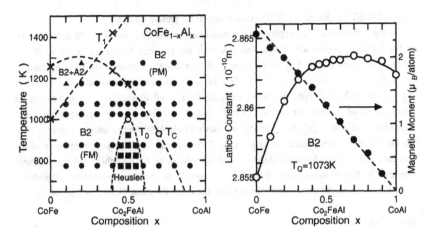

Figure 1. Phase relation in the CoFe$_{1-x}$Al$_x$ system determined based on the present study and the references [10,11]. FM and PM mean the ferromagnetism and the paramagnetism, respectively.

Figure 2. Composition dependences of the lattice constant a and the mean magnetic moment $\langle \mu \rangle$ in the B2-phase CoFe$_{1-x}$Al$_x$ alloys quenched from $T_Q = 1073K$.

the Curie temperature, respectively. These curves were obtained based on the results of electrical resistance (open circles) and X-ray diffraction measurements, including the published data (cross marks) [10,11]. The B2-phase is found to exist stably over wide ranges of composition and temperature, T, as expected. Furthermore, the Heusler-phase appears around the stoichiometric composition ($0.4 < x < 0.6$) below about 973K. Existence of the Heusler-phase is confirmed by detecting the 111 reflection in the X-ray diffraction, but this confirmation was very difficult for the samples merely heat-treated at $T = T_Q$ lower than T_0 (T_Q is the quenching temperature). However, it became possible for $T_Q \leq 873K$, when quenched from 1373K prior to the heat treatment at T_Q. There are thus some ambiguities on judgment on the phase appearing at 923K. The vacancy concentrations determined from the density measurement were very low (less than 0.2 ± 0.1 atomic %), indicating that major defect in the present system is antisite atom, similarly as in B2 Co$_{1-c}$Fe$_c$ alloys [8].

The composition dependences of the lattice constant, a, and the mean magnetic moment, $\langle \mu \rangle$, obtained for the B2-phase alloys quenched from $T_Q = 1073K$ are shown in Fig.2. a and $\langle \mu \rangle$ for $x = 0$ are for the A2-phase which appears at this temperature and their values are quoted from [8,10]. The value of a for $x = 1$ is quoted from [5]. The lattice constant increases rapidly with increase in composition, x, up to $x = 0.4$, beyond which it shows a gradual increase and further, changes to a gradual decrease in the region of $x > 0.7$. The mean magnetic moment decreases rather monotonically with increase in x and its change is approximately linear in the region of $x \geq 0.2$. Definite correspondence between the changes in a and $\langle \mu \rangle$ as observed in B2 $Co_{1-c}Fe_c$ alloys [8-10] is not found in the present system.

In Fig.3, the quenching temperature dependences of a and $\langle\mu\rangle$ for $x=0.20, 0.40$ and 0.60 in the B2-phase region are shown, together with the previous results for $x=0$ [8,10]. A tendency for both a and $\langle\mu\rangle$ to decrease with increasing temperature T_Q is recognized in $x=0.20$, like B2 CoFe alloy ($x=0$). Similar tendency was observed also for alloys of $x=0.10$ and 0.30. However, this is not recognized for $x=0.40$ and 0.60, where they remain almost constant, except for a for $x=0.60$ and $T_Q=773K$. The quenching temperature dependences of a and $\langle\mu\rangle$ for $x=0.50$ are shown in Fig.4, where the lattice constant data in the Heusler-phase are plotted by dividing by 2 for comparison. The lattice constant in the Heusler-phase is found to tend to decrease with decrease in T_Q. On the other hand, the magnetic moment in the Heusler-phase shows no remarkable quenching temperature dependence ($1.30\pm0.01\ \mu_B$ /atom) and is nearly the same as that in the B2-phase ($1.29\pm0.01\ \mu_B$ /atom).

Discussion

Present result showed that the mean magnetic moment in the B2-phase changes nearly linearly over a wide composition range (Fig.2). Recently, Galanakis et al. [2] have studied various Heusler alloys with the half-metallic character, such as Co_2MnSi, Co_2CrAl and Fe_2MnAl, based on the first principle electronic calculation. They found that these alloys show the Slater-Pauling (SP) behavior, i.e., the total magnetic moment, M_t, is given in terms of the total number of valence electrons, Z_t, as $M_t = Z_t - 24$ in μ_B per formula unit. When this rule

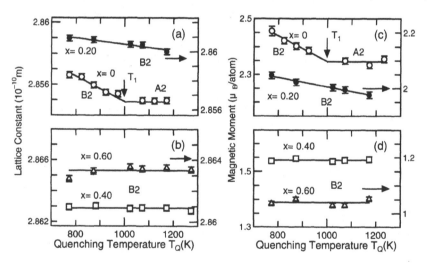

Figure 3. Quenching temperature dependences of the lattice constant a ((a) $x=0$ [8] and 0.20 and (b) $x=0.40$ and 0.60) and the mean magnetic moment $\langle\mu\rangle$ ((c) $x=0$ [10] and 0.20 and (d) $x=0.40$ and 0.60) in the B2-phase $CoFe_{1-x}Al_x$ alloys.

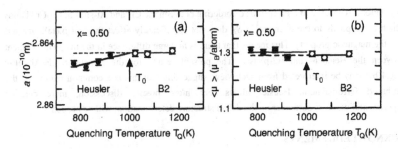

Figure 4. Quenching temperature dependences of the lattice constant a (a), and the mean magnetic moment $\langle\mu\rangle$ (b), in the B2-phase and Heusler-phase Co_2FeAl ($x = 0.50$) alloy.

is applicable to the present alloy Co_2FeAl, the moment M_t becomes $5\mu_B$ ($1.25\mu_B$ per atom), which is close to the present result of $\langle\mu\rangle = 1.30 \mu_B$ / atom (Fig.4(b)). Furthermore, when the SP relation is satisfied for $CoFe_{1-x}Al_x$ system, then the magnetic moment is given by $\langle\mu\rangle = 2.5(1-x)$. This linear relation is shown in Fig.2 by the broken line. Agreement with the observations is good within a deviation of $0.05\mu_B$, except for the region of $x < 0.2$, although these are for the B2-phase (not the Heusler-phase).

The point defect structure in the Heusler-phase is in general described in terms of the four LRO parameters. For simplicity, we here discuss following the defect model presented by Webster [13]. Then, only two LRO parameters are needed, defined as

$$S_1 = x_{Co}^{Co} - x_{Co}^{Fe} = x_{Co}^{Co} - x_{Co}^{Al}, \qquad S_2 = x_{Fe}^{Fe} - x_{Fe}^{Al}$$

where, e.g., x_{Co}^{Fe} represents the concentration of Co atom on the Fe-site. Perfectly ordered state is given by $S_1 = S_2 = 1$, while $S_1 = S_2 = 0$ in a disordered (A2) state. In the B2-phase, always $S_2 = 0$ because this phase attains complete disordering between the Fe- and Al-sites, so that we can discuss by using one parameter S_1, as in the binary alloy case. When the temperature becomes higher, disordering between the Co- and Fe (Al)-sites will be more promoted. Therefore, observed decrease in the magnetic moment with increase in quenching temperature (Fig.3(c)) may be attributed to lowering of the LRO parameter S_1, similarly as in B2 $Co_{1-c}Fe_c$ alloys [9,10]. However, such the decrease is not observed in the region of $x \geq 0.4$ (Fig. 3(d)). When the composition x increases, the transition temperature T_1 becomes much higher. Thus, in the temperature range concerned here, it will be difficult to expect a definite decrease in S_1 due to disordering. The results shown in Fig. 3(d) may come from this situation.

The observed magnetic moment in the Heusler-phase showed no remarkable difference from that in the B2-phase (Fig.4(b)), although the lattice constant decreased with decrease in T_Q (Fig.4(a)). Very recently, Miura et al. [14] studied the atomic disorder effects on the half-metallicity of the Heusler alloy $Co_2(Cr_{1-x}Fe_x)Al$, based on the first principle electronic calculation. According to their results, disordering between the Cr- and Al-sites in Co_2CrAl alloy, which corresponds to the decrease in S_2, does not significantly affect the half-metallicity and also the magnetic property. This means that there is no large difference in the magnetic moment between the Heusler- and B2-phases. The present results for the Heusler Co_2FeAl alloy (Fig.4(b)) may be interpreted from such view, suggesting that this is a common feature on the Co-based ferromagnetic Heusler alloys. For more precise discussion, more detailed experimental work, in particular on the defect structure determination, will be needed.

ACKNOWLEDGEMENTS

The authors wish to thank Professors T. Takasugi of Osaka Prefecture University and T. Tadaki of Osaka Women's University for kind help in sample preparation and Mr. T. Nakanishi and Ms. M. Hirata for devoted help in various measurements. This work was supported by the Grant-in-Aid for Scientific Research from the Japan Society for the Promotion of Science.

REFERENCES

1. S. Fujii, S. Sugimura, S. Ishida and S. Asano, *J. Phys.: Condens. Matter.* **2**, 8583 (1990).
2. I. Galanakis, P.H. Dederichs and N. Papanikolaou, *Phys. Rev.* **B66**, 174429 (2002).
3. M. Kogachi, S. Minamigawa and K. Nakahigashi, *Acta Metall. Mater.* **40**, 1113 (1992).
4. M. Kogachi, T. Tanahashi, Y. Shirai and M. Yamaguchi, *Scripta Mater.* **34**, 243 (1996).
5. M. Kogachi and T. Tanahashi, *Scripta Mater.* **35**, 849 (1996).
6. M. Kogachi and T. Haraguchi, *Mater. Sci. Engineer.* **A230**, 124 (1997).
7. H. Ishibashi, M. Kogachi, T. Ohba, X. Ren and K. Otsuka, *Mater. Sci. Engineer.* **A329-331**, 568 (2002).
8. K. Harada, H. Ishibashi and M. Kogachi, *Mater. Res. Soc. Proc.* **753**, BB5.28.1 (2003).
9. H. Ishibashi, K. Harada, M. Kogachi and S. Noguchi, *J. Mag. Mag. Mater.* **272-276**, 774 (2004).
10. M. Kogachi, N. Tadachi, H. Kohata and H. Ishibashi, *Intermetallics* **13**, (2005) (in press).
11. N. Kamiya, T. Sakai, R. Kainuma, I. Ohnuma and K. Ishida, *Intermetallics* **12**, 417 (2004).
12. K.H.J. Buschow and P.G. Engen, *J. Mag. Mag. Mater.* **25**, 90 (1981).
13. P.J. Webster, *J. Phys. Chem. Solids* **32**, 1221 (1971).
14. Y. Miura, K. Nagao and M. Shirai, *Phys. Rev.* **B69**, 144413 (2004).

Directional Thermoelectric Performance of Ru$_2$Si$_3$

Benjamin A. Simkin, Yoshinori Hayashi, Haruyuki Inui,
Department of Materials Science and Engineering, Kyoto University, Sakyo-ku, Kyoto, 606-8501 Japan

ABSTRACT

The orthorhombic compound Ru$_2$Si$_3$ is currently of interest as a high-temperature thermoelectric material. In order to clarify the effects of crystal orientation on the thermoelectric properties of Ru$_2$Si$_3$, we have examined the microstructure, Seebeck coefficient, electrical resistivity, and thermal conductivity of Ru$_2$Si$_3$ along the three principal axes, using these measured quantities to describe the relative thermoelectric performance as a property of crystal orientation. Ru$_2$Si$_3$ undergoes a high temperature (HT)→low temperature (LT) phase change and polycrystalline Si platelet precipitation during cooling, both of which are expected to effect the thermoelectric properties. The HT tetragonal→LT orthorhombic phase transformation results in a [010]//[010], [100]//[001] two-domain structure, while polycrystalline Si precipitation occurs on the (100)$_{LT}$ and (001)$_{LT}$ planes. The [010] orientation is found to posses superior thermoelectric properties (with the dimensionless figure of merit, ZT$_{[010]}$/ZT$_{[100]}$>4 at 650°C), due principally to the larger Seebeck coefficient along the [010] direction. The effect of the domain structure on the thermoelectric properties is discussed.

INTRODUCTION

Ruthenium sesquisilicide, Ru$_2$Si$_3$, has attracted attention over the past decade [1-10] as a potential thermoelectric material for use at high temperatures. This interest has been stimulated by the prediction of greatly enhanced thermoelectric performance of Ru$_2$Si$_3$ over that currently attainable with conventional Si-Ge alloys for use in thermoelectric generators. The physical properties relevant to a thermoelectric material can be determined by the thermoelectric figure of merit, Z, defined by:

$$Z=\alpha^2/\rho\cdot\kappa, \tag{1}$$

where α is the Seebeck coefficient, ρ is the electrical resistivity, and κ is the thermal conductivity of the material. Because Z has units of 1/K and α, ρ, and κ all tend to vary with temperature, it is often useful to express the thermoelectric performance of a material through the dimensionless figure of merit, ZT, as a function of temperature. All three of the components of Z can be expected to potentially vary depending on crystal orientation, as all three are dependant on either lattice scattering (ρ, κ), or the lattice directional dependence of the band structure (α). Ru$_2$Si$_3$ has two different crystal structures, of which the phase of interest is the orthorhombic low temperature phase. Because of this orthorhombic structure and the dependence of ρ, κ and α on orientation, there is expected to be potential for significant differences in thermoelectric performance depending on Ru$_2$Si$_3$ crystal orientation.

Electrical resistivity behavior for single crystal Ru$_2$Si$_3$ has been reported to depend upon orientation [3]. Thermal conductivity for polycrystalline Ru$_2$Si$_3$ has been extensively reported [2,4,5,8], although with significant variation and as yet directional thermal conductivity has not been reported to our knowledge. Seebeck coefficients for nominally intrinsic polycrystalline

Ru$_2$Si$_3$ have also been widely reported [2,4,5,8,10,11], and are generally reported as either negative at low temperatures and positive at high temperatures, passing through zero at ~300-400°C [2,4,8,10]; or else uniformly positive, with α tending to increase with temperature [5,11]. Reported values for α$_{max}$ vary between ~190μV/K [10] and ~550μV/K [11]. Vining *et al.* [1] make the only report of the directional Seebeck coefficient, where they report α$_{max}$ for both [001] and [010] directions at ~350°C, with α$_{\{001\}}$≈450μV/K and α$_{\{010\}}$≈400μV/K. Both orientations are reported to undergo a sign change of α around 100-200°C.

Modeling of the Ru$_2$Si$_3$ electronic structure [for example 12,13], predicts some variation of the band structure with crystal orientation, with both conduction band minima and valence band minima varying with crystal orientation; from these results, the expected magnitude of directional Seebeck coefficients would be α$_{\{100\}}$ < α$_{\{001\}}$ < α$_{\{010\}}$ for holes, and α$_{\{100\}}$ < α$_{\{010\}}$ < α$_{\{001\}}$ for electrons, based off the positions of the conduction and valence bands relative to the Fermi energy [12-14]. Recent single crystal studies of both un-doped [7] and doped [6,10] Ru$_2$Si$_3$ grown via the floating-zone method have reported thermoelectric or electronic properties without regard to crystal orientation, despite the implied anisotropy of properties due to crystal structure and shown by modeled band structures. In order to clarify the magnitude of what these orientation effects might be, we have chosen to examine the crystallographic orientation effects on the thermoelectric properties of Ru$_2$Si$_3$.

Ru$_2$Si$_3$ has a high-temperature and a low temperature phase, reversibly related through a diffusionless phase transformation [15,16] at ~1000°C. The tetragonal high temperature (HT) Ru$_2$Si$_3$ phase is structurally one of the 'Nowotny chimney-ladder' compounds [17], with a square array of Ru atoms surrounding a spiral arrangement of Si atoms along the [001]$_{HT}$ axis (space group $P\bar{4}c2$, #116). The low-temperature (LT) Ru$_2$Si$_3$ phase (space group *Pbcn*, #60) occurs through the distortion of the Si and Ru arrays from the chimney-ladder configuration (now oriented along the [010]$_{LT}$ axis), resulting in a doubling in the size of the unit cell along the a-axis [15,17]. The LT orthorhombic lattice parameters are **a**=1.1057nm, **b**=0.8934nm, **c**=0.5533nm [15,17,18]. For the low temperature Ru$_2$Si$_3$ phase there are reports of a Ru-rich composition deviation to ~59% Si [7,18]; the interrelationship between the HT↔LT phase transformation and the deviation of Ru$_2$Si$_3$ from 2:3 stoichiometry remains unknown. Both the HT→LT phase transformation and the apparent variation of stoichiometry are expected to have a significant effect on the use of Ru$_2$Si$_3$ as a thermoelectric material.

EXPERIMENTAL PROCEDURE

Polycrystalline ingots of Ru$_2$Si$_3$ were produced by arc-melting of elemental Ru and Si under argon, then re-melted using the optical floating zone method to produce single crystal Ru$_2$Si$_3$ at a growth rate of 6mm/hour. Crystals were oriented using back-reflection Laue, cut via diamond saw, and polished with 2000-grit paper to produce samples for measurement of electrical resistivity, Seebeck coefficient, and thermal diffusivity. Samples for transmission electron microscopy (TEM) were dimple-ground and ion-milled to electron transparency. Electrical resistivity and Seebeck coefficient were measured simultaneously (ZEM-2, Ulvac-Riko), while thermal conductivity was determined by the laser flash method.

RESULTS AND DISCUSSION

As-grown crystals contained a number of defects (Figure 1), the most prominent of which are polycrystalline Si precipitates typically parallel to the (100) and (001) planes. These 'silicon plate-like inclusions' have previously been reported [7]. The polycrystalline nature of these precipitates is not consistent with Si precipitation within either the HT tetragonal phase or the LT orthorhombic phase (both of which would tend to produce single-crystal precipitates), but rather with heterogenous nucleation and growth of individual Si crystals on interior cracks faces. These cracks may result from the thermal strains in the optical floating zone furnace, thermal stresses within the HT phase upon cooling, or strains associated with the HT→LT phase transformation. In addition to the polycrystalline Si platelets, TEM examination also reveals a LT-Ru_2Si_3 domain structure. The a↔c domain relationship is clearly seen in the [010]-oriented foil of Figure 1(b), with a 90° rotational relationship between the [010] diffraction patterns from adjacent domains. The complete orientation relationship between the LT-Ru_2Si_3 domains is illustrated further in Figures 1(c) and 1(d), showing $C_{[100]}//D_{[001]}$, $C_{[010]}//D_{[010]}$ domain relationship via the series of diffraction patterns from the adjacent [001] ('C') and [100] ('D') domains of Figure 1(a) over a series of sample tilts. In addition to domain boundaries with the a↔c axis change, the grown

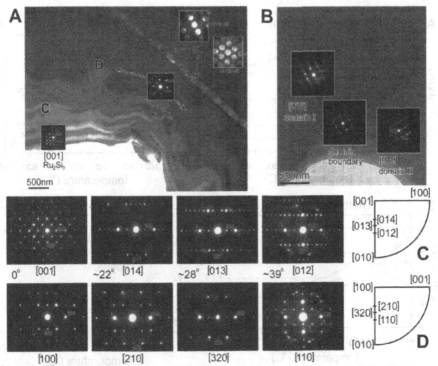

Figure 1. TEM brightfield images and diffraction patterns showing the as-grown Ru_2Si_3 microstructure: A) the [100]/[001] orientation; B) the [010] orientation; C) and D) are diffraction patterns taken from the Ru_2Si_3 domains 'C' and 'D' in (A), and come from tilting parallel to the $(100)_C/(001)_D$ planes. Polycrystalline Si precipitates form parallel to $(100)_{Ru2Si3}$ and $(001)_{Ru2Si3}$, while the Ru_2Si_3 domains form as 90° rotational domains about a common [010] axis.

crystals also contain numerous antiphase boundaries (not shown in Figure 1), which imply a homogenous nucleation of the LT structure within the parent HT crystal. The **a**↔**c** rotational domain orientation relationship is unsurprising, as these two domains represent the two possible variants of the HT→LT phase transformation, with $c_{HT}→b_{LT}$ and $a_{HT}→$ (a_{LT} or c_{LT}) [15]. Silicon precipitation from Ru_2Si_3 implies numerous Si vacancies within the Ru_2Si_3 structure, which is likely to effect both electrical and thermal conductivity.

The measured Seebeck coefficients (α), electrical resistivity (ρ), thermal conductivity (κ), and the overall calculated dimensionless figure of merit, ZT, are plotted as a function of temperature in Figure 2. Due to the domain relationships in Ru_2Si_3 there are only two unique principal axes that can be measured for as-grown Ru_2Si_3, [010]/[010] (henceforth denoted as [010]) and [100]/[001]. As a consequence, the measured properties along the nominal [100] and [001] directions are similar, although some variation can be seen. Both the '[100]' and '[001]' samples would be expected to yield properties characteristic of some combination of the two orientational properties. However, as the [010] samples only contain rotational variants oriented along the same axis, these measurements are expected to more accurately reflect the actual properties along this orientation.

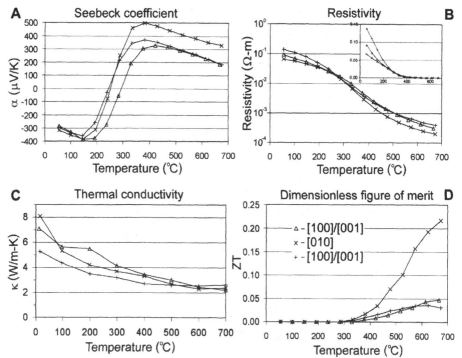

Figure 2. Measured thermoelectric properties of as-grown crystalline Ru_2Si_3: A) Seebeck coefficient; B) electrical resistivity (inset shows a linear scale); C) thermal conductivity; D) calculated dimensionless figure of merit, ZT.

The sign of the measured Seebeck coefficients (Figure 2(a)) show the dominance of hole conduction at high temperatures and electron conduction at low temperatures, with $\alpha > 0$ at high temperatures and $\alpha < 0$ at low temperatures. The magnitudes of α compare as $|\alpha_{[010]HT}| > |\alpha_{[100]HT}|, |\alpha_{[001]HT}|$ and $|\alpha_{[010]LT}| > |\alpha_{[100]LT}|, |\alpha_{[001]LT}|$. While the high temperature effect is expected because the predicted hole $|\alpha_b| > |\alpha_c| > |\alpha_a|$ behavior, the low temperature effect is not, because for electrons the predicted behavior is $|\alpha_c| > |\alpha_b| > |\alpha_a|$. Electrical resistivity measured along the three orientations is shown in Figure 2(b). The lower [010] resistivity may be due to a number of possible effects, from a higher inherent conductivity for the Ru$_2$Si$_3$ lattice along the [010] orientation as compared to the [100]/[001] orientation, to the presence of Si platelets on the (100) and (001) planes serving as high-resistivity interlayers increasing the measured [100]/[001] resistivity. Still another is that the resistivity of the domain boundaries differ markedly, with the $\rho_{[010]-[010]} < \rho_{[100]-[001]}$. The measured thermal conductivity for all three orientations (Figure 2(c)) were similar, decreasing with increasing temperature from a range of ~5.2-8.1W/m·K at room temperature to ~2.2-2.6W/m·K at ~670°C, with no apparent dependence on crystal orientation. Thus, thermal conductivity does not appear to be strongly dependant on sample orientation. However, the trend of decreasing thermal conductivity with increasing temperature contributes to an improved figure of merit at high temperatures. It is worth noting that the range of the measured values of κ are consistent with the values quoted for un-doped polycrystalline Ru$_2$Si$_3$ in [5] from 300-900K, and slightly higher than the values reported for quasi-single crystalline Ru$_2$Si$_3$ in [4] for 600-950K. The combination of higher thermopower and marginally lower resistivity of the [010] orientation at higher temperatures as compared to the [100]/[001] directions leads to a significantly higher dimensionless figure of merit for the [010] orientation, resulting in a ZT$_{[010]}$:ZT$_{[100]/[001]}$ ratio of greater than 4:1 for temperatures over 600°C (Figure 2(d)). This clearly shows that thermoelectric devices oriented along the Ru$_2$Si$_3$ [010] direction can be expected to show superior figures of merit, and requires the consideration of sample orientation during studies of this material.

While the tetragonal→orthorhombic phase transformation of Ru$_2$Si$_3$ seems to constrain the potential for the exploration of the full range of orientation-dependent thermoelectric properties for this structure, it does seem to offer the potential for properties superior to simple random polycrystalline material. The effect of the domain structure on the thermoelectric performance of the [010] orientation appears to be minimal, but the domain structure is expected to moderately degrade thermoelectric performance for the [100]/[001] orientation.

CONCLUSIONS

Floating-zone crystal growth of stoichiometric Ru$_2$Si$_3$ has been shown to produce a domain structure of the Ru$_2$Si$_3$ phase, with polycrystalline Si platelets on the (100) and (001) planes in the two-domain Ru$_2$Si$_3$ matrix, supporting a Ru-rich preferred stoichiometry for LT-Ru$_2$Si$_3$. Single-crystal orthorhombic Ru$_2$Si$_3$ is not achievable for methods of crystal growth from the melt due to the HT tetragonal↔LT orthorhombic phase transformation, however, the LT microstructure resulting from the HT→LT phase transformation forms a [010]//[010], [100]//[001] oriented domain structure, and the measured thermoelectric properties for intrinsic (p-type) Ru$_2$Si$_3$ are superior along the [010] orientation due to larger Seebeck coefficients and marginally lower resistivity. Thermal conductivity appears independent of crystal direction.

ACKNOWLEDGEMENTS

This work has been supported by the Japan Society for the Promotion of Science (JSPS), by a Grant-in-Aid for Scientific Research from the Japanese Ministry of Education, Science and Culture (#16656218), and by the 21st Century COE (Center of Excellence) Program on United Approach for New Materials Science from the Japanese Ministry of Education, Science and Culture.

REFERENCES

1. C.B. Vining, in *American Institute of Physics Conference Proceedings 246* (American Institute of Physics, 1992) p. 338.
2. C.B. Vining, J.A. McCormack, A. Zoltan, L.D. Zoltan, in *Proceedings of the Eighth Symposium on Space Nuclear Power Systems*, edited by M. El-Genk, M. Hoover (American Institute of Physics, 1991) p. 458.
3. U. Gottlieb, O. Laborde, Applied Surface Science. **73** , 243 (1993).
4. T. Ohta, C.B. Vinning, C.E. Allevato, in *Proceedings of the 26th Intersociety Energy Conversion Engineering Conference, 3* (American Nuclear Society, La Grange Park, IL, 1991) pp. 196-201.
5. A. Yamamoto, T. Ohta, Y. Sawade, T. Tanaka, K. Kamisako, in *Proceedings of the 14th International conference on thermoelectrics*, (Iotte-Institute, St. Petersburg, 1995), pp. 264-268.
6. Y. Arita, S. Mitsuda, Y. Nishi, T. Matsui, T. Nagasaki, J. Nuclear Mat. **294**, 202 (2001).
7. D. Souptel, G. Behr, L. Ivanenko, H. Vinzelberg, J. Schumann, J. Crystal Growth, **244**, 296 (2002).
8. Y. Arita, S. Mitsuda, T. Matsui, J. Therm. Anal. and Cal. **69**, 821 (2002).
9. L. Ivanenko, A. Filonov, V. Shaposhnikov, G. Behr, D. Souptel, J. Schumann, H. Vinzelberg, A. Plotnikov, V. Borisenko, Microelect. Eng. **70**, 209 (2003).
10. Y. Arita, T. Miyagawa, T. Matsui, in *Proceedings of the 17th International Conference on Thermoelectrics* (IEEE, Piscataway, NJ, 1998) pp. 394-397.
11. M.A. Hayward, R.J. Cava, J. Physics: Cond. Mat. **14**, 6543 (2002).
12. D.B. Migas, L. Miglio, V.L. Shaposhnikov, V.E. Borisenko, Phys. Stat. Sol. B. **231**, 171 (2002).
13. W. Henrion, M. Rebien, A.G. Birdwell, V.N. Antonov, O. Jepsen, Thin Solid Films **364**, 171 (2000).
14. H. Fritzsche, Solid State Comm. **9**, 1813 (1971).
15. D.J. Poutcharovsky, K. Yvon, E. Parthé, J. Less-Common Met. **40**, 139 (1975).
16. C.P. Susz, J. Muller, K. Yvon, E. Parthé, J. Less-Common Met. **71**, P1 (1980).
17. D.J. Poutcharovsky, E. Parthé, Acta Cryst. B **30**, 2692 (1974).
18. L. Perring, F. Bussy, J.C. Gachon, P. Feschotte, J. Alloys and Compounds **284** , 198 (1999).

Mater. Res. Soc. Symp. Proc. Vol. 842 © 2005 Materials Research Society   S4.7

Mechanical Aspects of Structural Optimization in a Bi-Te Thermoelectric Module for Power Generation

Yujiro Nakatani[1], Reki Takaku[1], Takehisa Hino[1], Takahiko Shindo[1] and Yoshiyasu Itoh[1]
[1]Power & Industrial System R&D Center, Toshiba Corporation,
1-9, Suehiro-cho, Turumi-ku, Yokohama 230-0045, Japan.

ABSTRACT

The thermal stress and strain occurring on a Bi-Te thermoelectric module subjected to variable thermal conditions were estimated based on three-dimensional elastic-plastic finite element method (FEM) analysis. The analysis showed that mechanical integrity of the interface between a Bi-Te thermoelement and electrodes of Al and Mo coatings formed by atmospheric plasma spraying was significantly reduced and that shear strain rose to 0.6~1.1% in the vicinity of the interface. Furthermore, to estimate the sensitivity of configurational parameters of the module to the thermal strain, statistical sensitivity analysis based on the design of experiment (DoE) and response surface method (RSM) was conducted. The statistical analysis revealed that the thickness of electrode coatings of Al and Mo affected the thermal strain and that the thinner Al coating and the thicker Mo coating reduced the thermal strain. In this study, a thermal fatigue test machine was newly developed with a view to verifying these analytical studies.

INTRODUCTION

We are currently developing a high-performance bismuth-tellurium (Bi-Te) thermoelectric module for power generation. This module aims at heat recovery at temperatures of 150°C or less. Although the total amount of waste heat energy below 150°C is considerable, it has not been utilized for heat recovery because the density of energy is low [1]. In addition, the utilization of such low-density energy is not considered economically competitive. Therefore, our objective is to enhance the thermoelectric efficiency and to improve the cost performance of the module. One method of improving the cost performance of the module is to achieve a long-range run without maintenance, based on the enhanced reliability of the module.

Our module consists of the Bi-Te thermoelements and electrodes of Al and Mo formed by atmospheric plasma spraying (APS). In such multilayer coating structures, thermal stress and/or strain is induced owing to the difference in the thermal expansion coefficients of the individual materials, and this can reduce long-period reliability. Although the behavior of the thermal stress and/or strain is significant for designing the module, the influence of configurations and the material characteristics of the thermoelements and the thermally sprayed electrodes on thermal stress or strain have not been ascertained [2,3].

In this study, the effect of the configurational parameters of the thermoelectric module upon the thermal strain is evaluated from numerical analysis, and the optimum coating thickness to improve the integrity of the interface of the thermoelements and electrodes is discussed.

EXPERIMENTAL DETAILS

Finite element method analysis

To evaluate the strain state on a Bi-Te thermoelectric module subjected to variable thermal conditions, thermomechanical elastic-plastic finite element method (FEM) analysis was conducted using the MARC k7.3 code. The developed thermoelectric module is shown in Figure 1-(a). The module consists of Bi-Te elements of the p-type and n-type, diffusion barrier layer of Mo on the elements, electrode layer of Al, and mold resin. A total of 32 couples of the p-type/n-type thermoelements are placed in the module. In this study, a couple of thermoelements was modeled as shown in Figure 1-(b). The finite element mesh used for calculation consisted of 4320 isotropic three-dimensional 8-node hexahedron elements. The elements at the interfaces of each material were 0.05 mm thick. The initial temperature of the whole of the finite element model was 298 K, which simulates room temperature. The upper side then was varied to 398 K, while the lower side was maintained at 298 K. The elastic modulus, Poisson's ratio, the heat expansion coefficient, and plasticity data are listed in Tables I and II.

Thermal strain evaluation method

To evaluate the configurational parameters and material constants affecting the thermal strain, a sensitivity analysis was performed. For the sensitivity analysis, statistical optimization analyses based on the design of experiment (DoE) and the response surface method (RSM) was conducted. The sensitivity analysis process is shown in Figure 2-(a). First of all, design parameters were selected, and their numerical range was determined. In this analysis, the configurational parameters and material constants shown in Figure 2-(b) were selected as $X_1 \sim X_6$. Since the elastic modulus of Al and Mo vary with the condition of the atmospheric plasma spraying, they were selected as variables. The numerical range of these design parameters is listed in Table III. Secondly, the values of the design parameters were divided into 27 patterns in accordance with the DoE. Based

 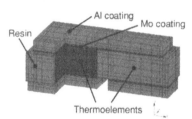

(a) Newly developed thermoelectric module   (b) Cross-sectional illustration of FE model
Figure 1. Developed thermoelectric module and finite element model showing a couple of thermoelements

Table I. Material properties for FEM analysis

	Elastic modulus (GPa)	Poisson's ratio	Thermal expansion coefficient $(10^{-6}/K)$
Bi-Te	73.3	0.32	15.1
Al coating	84.0	0.37	23.9
Mo coating	317	0.29	5.10
Resin	1.93	0.40	60.0

Table II. Plasticity data for FEM analysis

Al		Mo	
Plastic strain (%)	Stress (MPa)	Plastic strain (%)	Stress (MPa)
0	3.06	0	44.4
4.8	15.5	4.8	61.2

414

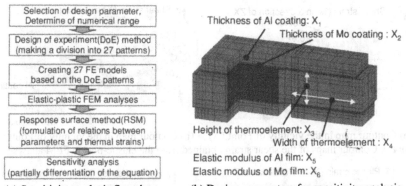

| Selection of design parameter, Determine of numerical range |
| Design of experiment(DoE) method (making a division into 27 patterns) |
| Creating 27 FE models based on the DoE patterns |
| Elastic-plastic FEM analyses |
| Response surface method(RSM) (formulation of relations between parameters and thermal strains) |
| Sensitivity analysis (partially differentiation of the equation) |

Thickness of Al coating: X_1
Thickness of Mo coating : X_2
Height of thermoelement: X_3
Width of thermoelement : X_4
Elastic modulus of Al film: X_5
Elastic modulus of Mo film: X_6

(a) Sensitivity analysis flowchart   (b) Design parameters for sensitivity analysis

Figure 2. Sensitivity analysis process based on the design of experiment and response surface method

Table III. Numerical range of design parameters

	Thickness of Al coating :X_1 (μm)	Thickness of Mo coating :X_2 (μm)	Height of thermoelement :X_3 (mm)	Width of thermoelement :X_4 (mm)	Elastic modulus of Al coating :X_5 (GPa)	Elastic modulus of Mo coating :X_6 (GPa)
Min	250	50	3	5	28.4	130
Max	750	150	6	7	71	325

on the divided patterns, 27 finite element models were created. Elastic-plastic FEM analyses were then performed using these models. Next, the relation between the parameters and the thermal strain was obtained from the RSM using the results of the FEM analyses. RSM is a method of obtaining an interpolation approximation function. In this study, polynomial approximate equations were obtained. By partially differentiating the polynomial equations by the individual parameter variables, the variation in thermal strain accompanied by the variation in parameters was compared.

DISCUSSION

Thermal stress analysis

The strain on the thermoelement and APS coatings may be mainly due to the shear mode because of the temperature difference between the upper and lower surfaces of the thermoelectric module and the difference in the thermal expansion coefficient of the individual materials. Such shear strain would be a driving force of fatigue crack growth. The FEM analysis revealed that the shear strain appeared mainly in the vicinity of Points (A) and (B) shown in Figure 3. Figure 3-(a) and (b) show contour maps of the shear strain obtained from the elastic-plastic FEM analysis. The contour map in Figure 3-(a) shows that the shear strain distributes in the ZX-direction in the vicinity of the upper side of the thermoelements. Figure 3-(b) shows that the shear strain in the YZ-direction distributes in the vicinity of the interface of the Al coating and the Mo coating. The numerical values in the figures are the totals of the plastic strain and elastic strain. The shear strain shown in Figure 3-(a) was not just on the boundary between the thermoelement and the Mo

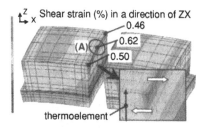

Shear strain (%) in a direction of ZX

(A)
0.46
0.62
0.50

thermoelement

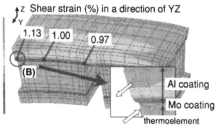

Shear strain (%) in a direction of YZ

(B)
1.13 1.00 0.97

Al coating
Mo coating
thermoelement

(a) Contour map in the vicinity of Point (A)   (b) Contour map in the vicinity of Point (B)

Figure 3. Contour maps of the shear strain obtained from the elastic-plastic FEM analysis

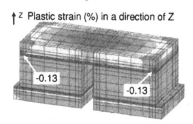

Plastic strain (%) in a direction of Z

-0.13
-0.13

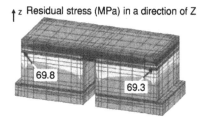

Residual stress (MPa) in a direction of Z

69.8
69.3

(a) Compressive plastic strain in the Z-direction   (b) Residual stress in the Z-direction

Figure 4. Contour maps of the plastic strain and residual stress obtained from cyclic FEM

coating, but also on the thermoelement. The maximum value of the shear strain is 0.62%. Similarly, the shear strain shown in Figure 3-(b) was not just on the boundary between the Al coating and the Mo coating, but also appeared on the Al coating. The maximum value of the shear strain is 1.13%. The FEM analysis reveals that strain arises up to the amount in which fatigue crack growth could occur on the thermoelement and/or the Al coating, although the variation in the temperature of the module is not very great.

Figure 4-(a) shows that there was compressive plastic strain in the Z-direction at the interface of the Al coating and the Mo coating. In general, where a part is deformed plastically while the rest undergoes elastic deformation, the region that is deformed plastically in compression will have tensile residual stress when the external force is removed. Therefore, in this case, tensile residual stress may arise at the interface of the Al coating and the Mo coating when the module cools to room temperature accompanied by stopping the system. Figure 4-(b) shows the residual stress in the Z-direction after the temperature on the upper side of the module was changed from 298 K to 398 K to 298 K. The figure shows that on the part where compressive plastic strain was present in a state of temperature rise, there was tensile residual stress of approximately 70 MPa. Although the interface strength between the Al coating and the Mo coating is not clear here, the tensile residual stress would encourage the likelihood of crack growth or delamination.

Sensitivity analysis

To evaluate the configurational parameters and material constants affecting the thermal strain, sensitivity analysis based on the design of experiment (DoE) and the response surface method (RSM) was carried out. Point (A) and/or (B) in Figure 4 where the possibility of crack growth was shown were evaluated. In accordance with the method shown in Figure 2, the relational equation between the parameters and the thermal shear strain is given by the secondary polynomial equation as shown below:

$$\log \varepsilon_t = A + \Sigma\Sigma(B_{ij}X_i^j) \tag{1}$$

where ε_t is the total shear strain on a certain part, A and Bij are constants, i = 1~6 and j = 1,2. Relational equations were obtained for Points (A) and (B), respectively. Figure 5 shows the change in the thermal shear strain accompanied by the variation in the parameters obtained from these equations. The abscissa in the figure shows the ratio, which is the difference between the minimum value of the parameter and the balance of maximum and minimum. The total strains of $\varepsilon_{t(A)}$ and $\varepsilon_{t(B)}$ indicate shear strain at Point (A) and Point (B), respectively. By partially differentiating these equations, the sensitivity of the parameters to the thermal strains was obtained. Table IV shows the sensitivity of the coating thickness of Al and Mo to the thermal strain. With respect to the coating thickness of the Al, the sensitivity to $_{t(A)}$ and the sensitivity to $_{t(B)}$ are both given by positive values. Therefore, the thicker the Al coating, the higher the thermal strain. Similarly, with respect to the coating thickness of the Mo, the sensitivity to $_{t(A)}$ and $_{t(B)}$ are both given by negative values. Thus, the thicker the Mo coating, the lower the thermal strain. That is to say, for optimum mechanical reliability, the design range for the Al coating should be of minimum thickness and the Mo coating should be of maximum thickness.

To verify the influence of the thickness of the coating on thermal fatigue, a thermal fatigue-testing machine for thermoelectric modules was developed. Figure 6 shows a photograph of the developed testing machine. It has a mechanism to supply a heat cycle onto the upper surface of the module by concurrently cooling the lower surface. In addition, the testing machine can detect changes in the internal resistance of the thermoelectric module accompanied by crack growth. Hereafter, the effect of the optimum coating thickness of the electrodes obtained analytically and

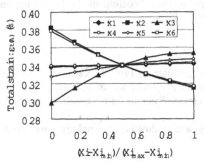 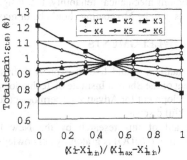

(a) Variation in the thermal shear strain at Point (A)

(b) Variation in the thermal shear strain at Point (B)

Figure 5. Variation in the thermal shear strain accompanied by variation in the parameters

Table IV. Sensitivity of the coating thickness of Al and Mo to the thermal strain

		Thickness of Al coating: X_1	Thickness of Mo coating: X_2
$\varepsilon_{t(A)}/$	x_i	$2.4 \times 10^{-8} \sim 4.3 \times 10^{-8}$	$-2.4 \times 10^{-5} \sim -1.5 \times 10^{-5}$
$\varepsilon_{t(B)}/$	x_i	$7.3 \times 10^{-7} \sim 3.9 \times 10^{-6}$	$-4.5 \times 10^{-6} \sim -1.6 \times 10^{-6}$

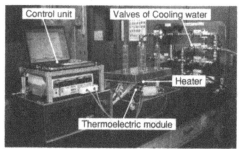

Figure 6. Photograph of the developed thermal fatigue-testing machine for thermoelectric module

the influence of the mechanical properties of the coating will be confirmed by using the developed thermal fatigue-testing machine.

CONCLUSIONS

In this study, the numerical analysis revealed that thermal strain arose at the interface between the thermoelement and the Mo coating of the diffusion barrier layer and at the interface between the electrode Al coating and the Mo coating. The amount of shear strain that appeared was 0.6% and 1.1%, respectively. There was compressive plastic strain at the interface of the Al coating and the Mo coating, and this caused tensile residual stress on the interface due to cyclic heating. The sensitivity analysis also revealed that the sensitivity to strain at the interface between the thermoelement and the Mo coating was a positive value, and revealed that the sensitivity to strain at the interface between the electrode Al coating and the Mo coating was a negative value. Thus, for optimum mechanical reliability, the design range for the Al coating should be of minimum thickness and the Mo coating should be of maximum thickness.

ACKNOWLEDGMENTS

The authors would like to express their sincere gratitude to Prof. T. Kajikawa who is the President of the Shonan Institute of Technology and is the Project Leader of the national project on the Development for Advanced Thermoelectric Conversion Systems for his excellent suggestions and fruitful discussion.

The aid of this research by the New Energy and Industrial Technology Development Organization (NEDO) is gratefully acknowledged.

REFERENCES

1. D. M. Rowe, G. Min, S. G. K. Williams and V. Kuznetsov, Proc. of World Renewable Energy Congress VI, 1499-1504 (2000).
2. L. I. Anatychuk, V. N. Balazyuk, O. J. Luste and V. V. Malyshko, Proc. of 22th Int. Conf. on Thermoelectrics, 619-622 (2003).
3. S. Jorez, S. Dilhaire, L. Lopez, S. Grauby and W. Claeys, Proc. of 20th Int. Conf. on Thermoelectrics, 503-506 (2001).

Mater. Res. Soc. Symp. Proc. Vol. 842 © 2005 Materials Research Society S4.8

Effects of Ga- and In-doping on the Thermoelectric Properties in Ba-Ge Clathrate Compounds

Norihiko L. Okamoto and Haruyuki Inui
Department of Materials Science and Engineering, Kyoto University Sakyo-ku, Kyoto 606-8501, Japan

ABSTRACT

The crystal structures and thermoelectric properties of Ba-Ge type-I clathrate compounds ($Ba_8Ga_xGe_{46-x}$ and $Ba_8Ga_{16-y}In_yGe_{30}$) have been investigated as a function of Ga and In contents. $Ba_8Ga_xGe_{46-x}$ alloys have a crystal structure that contains an ordered arrangement of Ge vacancies, forming a superstructure based on the normal type-I structure until X reaches 3, whereas they have the normal type-I structure when X exceeds 3. The dimensionless thermoelectric figure of merit (ZT) increases with the increase in the Ga content, exhibiting the highest value of 0.49 for $Ba_8Ga_{16}Ge_{30}$. The power factor for $Ba_8Ga_{10}In_6Ge_{30}$ is 1.5 times that for $Ba_8Ga_{16}Ge_{30}$ so that the In containing alloy exhibits a ZT value as high as 1.03 at 700°C.

INTRODUCTION

Clathrate compounds possess polyhedral cages encapsulating guest atoms. The cages consist of group IV and/or III elements while the guest atoms are typically alkali metals or alkali-earth metals. Clathrate compounds have been intensively investigated in hopes of producing more efficient thermoelectric devices because they generally show low thermal conductivity and relatively high electrical conductivity [1-3]. These combined properties of clathrate compounds are explained by Slack with the concept of PGEC (Phonon-Glass, Electron-Crystal) [4]. Among clathrate compounds with a variety of structures, the type-I clathrate compounds have mainly been studied for thermoelectric materials [1] because of their good electrical properties. The type-I clathrate compounds possess in general a composition of M_8X_{46}, where M stands for guest atoms and X for group IV and/or III elements. The corresponding unit cell is cubic with the space group of $Pm\bar{3}n$ [5]. The cage framework consists of 6c, 16i and 24k crystallographic sites, while the guest atoms are located at 2a and 6d sites.

We have shown that the binary type-I clathrate compound Ba_8Ge_{43} possesses a superstructure with the cubic space group of $Ia\bar{3}d$ due to the ordering of Ge vacancies at the 6c sites [6,7]. The superstructure is based on the normal type-I structure and has a lattice constant twice that of the normal type-I clathrate compounds. Ba_8Ge_{43} shows high electrical resistivity resulting in poor thermoelectric properties. On the other hand, $Ba_8Ga_{16}Ge_{30}$ possesses the normal type-I structure and exhibits good thermoelectric properties [8]. In the unit cell of this alloy, the total number of electrons donated by Ba atoms (+2x8) is equal to that accepted by Ga atoms (1x16), implying charge compensation is achieved. In view of this, charge compensation seems to play an important role in determining the thermoelectric properties. In alloys with Ga contents less than 16, charge is not compensated completely and there exist excess electrons in the unit cell because Ge atom has one more electron than Ga atom. We have thus investigated the effect of Ga-doping on the crystal structure and thermoelectric properties of ternary $Ba_8Ga_xGe_{46-x}$ alloys (X=0~16) as a function of Ga content paying special attention to charge compensation.

EXPERIMENTAL PROCEDURES

Alloys with nominal compositions of Ba_8Ge_{43}, $Ba_8Ga_XGe_{46-X}$ (X=0, 1.5, 3, 4.5 6, 12, 16) and $Ba_8Ga_{16-Y}In_YGe_{30}$ (Y=3, 6, 9, 12, 16), were prepared by arc-melting appropriate amounts of elements in an argon atmosphere. The arc-melted samples were annealed at 790°C in vacuum for 12 hours, followed either by oil-quenching or by furnace-cooling. Crystal structure investigations were made by X-ray powder diffraction through structure refinement made by the Rietveld method with a RIETAN-2000 software [9]. Measurements of Seebeck coefficient and electrical resistivity were made with our ULVAC ZEM-2 apparatus while those of thermal diffusivity and specific heat were made by the laser flash method.

RESULTS AND DISCUSSION

Ga-doping effect on the crystal structure of $Ba_8Ga_XGe_{46-X}$

Changes in the crystal structure upon alloying with Ga were investigated by synchrotron X-ray diffraction. Some superlattice diffraction peaks are observed in the patterns obtained for $Ba_8Ga_{1.5}Ge_{43}$ and $Ba_8Ga_3Ge_{43}$ as in the case of Ba_8Ge_{43}. Rietveld analysis inidicates that $Ba_8Ga_{1.5}Ge_{43}$ and $Ba_8Ga_3Ge_{43}$ possess a superstructure similar to that of Ba_8Ge_{43} and that Ga atoms in the ternary alloys preferentially occupy vacancy sites in Ba_8Ge_{43}. On the other hand, no superlattice diffraction peaks are observed in the patterns obtained for samples containing more than 3 Ga, indicating that the 6c sites are randomly occupied by Ga atoms and that the crystal structure changes from the superstructure with the ordered Ge vacancies to that of the normal type-I clathrate compounds.

Thermoelectric properties of $Ba_8Ga_XGe_{46-X}$

Fig. 1(a) shows electrical resistivity of $Ba_8Ga_XGe_{46-X}$ as a function of temperature. The electrical resistivity for Ba_8Ge_{43} is as high as of the order of 10^{-3} Ωm at room temperature and the temperature dependence below 400°C is typical of a semiconductor. The drastic change in electrical resistivity above 400°C is due to the phase decomposition of Ba_8Ge_{43} into $Ba_{24}Ge_{100}$ and Ge phases, which is consistent with the Ba-Ge binary phase diagram recently reported [10]. The electrical resistivity decreases with increasing Ga content until X reaches 4.5 and the temperature dependence changes to of heavily-doped semiconductor. However when X exceeds 4.5, the electrical resistivity increases with the increase in the Ga content until X reaches 16. At the composition of $Ba_8Ga_{16}Ge_{30}$, 16 valence electrons of the eight Ba atoms in the unit formula are considered to be transferred to the cage structure ($Ga_{16}Ge_{30}$) to form $[Ga_{16}Ge_{30}]^{16-}$ in the unit cell. The total number of valence electrons in $[Ga_{16}Ge_{30}]^{16-}$ is then identical to that of Ge_{46} since a Ga atom has three valence electrons while a Ge atom has four. Because Ga and Ge atoms in $[Ga_{16}Ge_{30}]^{16-}$ are tetrahedrally bonded and sp^3-hybrid orbitals may be formed in the cage structure, $[Ga_{16}Ge_{30}]^{16-}$ may be expected to behave similarly to Ge. On the other hand, samples containing Ga contents less than 16 in the unit formula include more electrons than $Ba_8Ga_{16}Ge_{30}$ since a Ge atom has more electrons than a Ga atom. Therefore, the dominant carrier in these ternary alloys is considered to be an electron, which is consistent with the result of our hall coefficient measurement that the ternary alloys exhibit n-type conduction. The number of carriers in the ternary alloys may increase with decreasing Ga content. This is also consistent with the

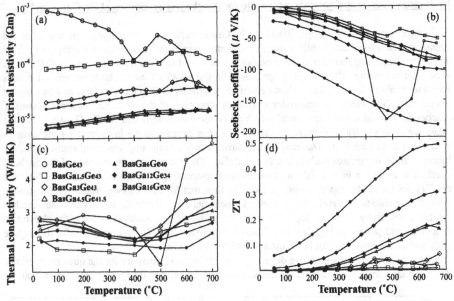

Figure 1. (a) Electrical resistivity, (b) Seebeck coefficient, (c) thermal conductivity and (d) dimensionless figure of merit for $Ba_8Ga_xGa_{46-x}$ (X=0~16) as a function of temperature.

fact that the electrical resistivity increases with the increase in the Ga content.

The Seebeck coefficients for all samples are negative in sign, as shown in Fig. 1(b). Their absolute values increase monotonously with increasing temperature except for Ba_8Ge_{43}. The anomaly in the temperature dependence above 400°C for Ba_8Ge_{43} is again due to the phase decomposition. The absolute values of the Seebeck coefficients decrease with decreasing Ga content. This is also attributed to the increased electrons that lead to the lower electron mobility. The thermal conductivity for Ba_8Ge_{43} shows a decline above 200°C until 500°C, above which it increases rapidly as shown in Fig. 1(c). The increase in thermal conductivity above 500°C also results from the phase decomposition of Ba_8Ge_{43}. The thermal conductivity is as high as 5 W/mK at 700°C for Ba_8Ge_{43} which may be due to the high thermal conductivity of the Ge phase [11]. Upon alloying with Ga, the temperature dependence of the thermal conductivity diminishes with increasing Ga concentration. The thermal conductivity for $Ba_8Ga_{16}Ge_{30}$ is almost temperature-independent with a value at room temperature of 2.1 W/mK.

The figure of merit (Z) is calculated with the following equation,

$$Z = \alpha^2 / \rho \cdot \kappa, \tag{1}$$

where α, ρ and κ stand for the Seebeck coefficient, the electrical resistivity and the thermal conductivity, respectively. The dimensionless figure of merit (ZT) is plotted as a function of temperature in Fig. 1(d). The ZT values for the samples with the vacancy-ordered superstructure are smaller than 0.1, whereas those for the samples with the normal type-I structure increase with the increase in the Ga content and temperature. $Ba_8Ga_{16}Ge_{30}$ exhibits a ZT value as high as 0.49 at 700°C.

In-doping effect on the microstructure and thermoelectric properties of $Ba_8Ga_{16-Y}In_YGe_{30}$

In-doped quaternary alloys ($Ba_8Ga_{16-Y}In_YGe_{30}$) were also investigated in the hope of improving the properties. While a single-phase microstructure is obtained for alloys with Y=3 and 6 after annealing, more than three phases are observed to remain for alloys with Y=9 and 16 even after annealing. The solubility limit of In in the $Ba_8Ga_{16}Ge_{30}$ clathrate compound is thus found to be about Y=6. Fig. 2(a) shows electrical resistivity of $Ba_8Ga_{16-Y}In_YGe_{30}$ as a function of temperature. The electrical resistivity drastically decrease as the In content increases until Y reaches 6. The value for the sample with Y=6 is about one-third that for $Ba_8Ga_{16}Ge_{30}$. On the other hand, the electrical resistivity increases with the increase in the In content for the samples containing more than Y=6. This may be due most probably to the high electrical resistivity of the incorporated other phases such as Ge, In and $BaGe_2$. The temperature dependence of the Seebeck coefficient is shown in Fig. 2(b) as a function of temperature. The absolute values of Seebeck coefficients for the In-doped samples are smaller than that for $Ba_8Ga_{16}Ge_{30}$. However, the sample with Y=6 exhibits relatively large Seebeck coefficient (-150 µV/K) as well as the smallest electrical resistivity of the In-doped samples. Power factor is calculated by using the above two properties and plotted in Fig. 2(c) as a function of temperature. The power factor for the sample with Y=6 is 1.5 times that for $Ba_8Ga_{16}Ge_{30}$ at high temperatures.

Fig. 3(a) shows the temperature dependence of thermal conductivity for the quaternary alloys. The thermal conductivity for the samples with Y=3 and 6 are smaller by 20 % than that for

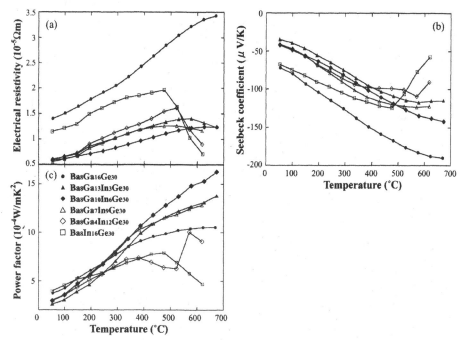

Figure 2. (a) Electrical resistivity, (b) Seebeck coefficient and (c) power factor for $Ba_8Ga_{16-Y}In_YGa_{30}$ (Y=0~16) as a function of temperature.

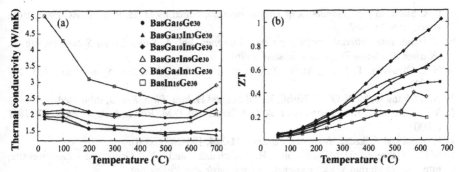

Figure 3. (a) Thermal conductivity and (b) dimensionless figure of merit for $Ba_8Ga_{16-Y}In_YGa_{30}$ (Y=0~16) as a function of temperature.

$Ba_8Ga_{16}Ge_{30}$. The reduction of thermal conductivity may be attributed to the impurity scattering due to alloyed In atoms. ZT values are plotted in Fig. 3(b) as a function of temperature. The ZT values increase with increasing In content until Y reaches 6, whereas values for the samples containing In more than Y=6 decrease with increasing In content. The sample with Y=6 exhibits a ZT value as high as 1.03 at 700°C.

CONCLUSIONS

(1) Upon alloying with Ga, the ternary $Ba_8Ga_XGe_{46-X}$ alloys maintain the superstructure that contains an ordered arrangement of Ge vacancies up to X=3. However, the crystal structure changes from the superstructure to that of the normal type-I clathrate compounds when X exceeds 3. Among the ternary alloys, $Ba_8Ga_{16}Ge_{30}$ exhibits the highest ZT value of 0.49 at 700°C.

(2) The solubility limit of In in the $Ba_8Ga_{16}Ge_{30}$ compound is about 6 per formula units. The power factor for $Ba_8Ga_{10}In_6Ge_{30}$ is 1.5 times that for $Ba_8Ga_{16}Ge_{30}$ at high temperatures so that the In containing alloy exhibits a ZT value as high as 1.03 at 700°C.

ACKNOWLEDGEMENTS

This work was supported by Grant-in-Aid for Scientific Research (a) from the Ministry of Education, Science and Culture (No. 14350369) and in part by the 21st Century COE (Center of Excellence) Program on United Approach for New Materials Science from the Ministry of Education, Science and Culture. One of the authors (NLO) greatly appreciates the supports from Research Fellowships of the Japan Society for the Promotion of Science for Young Scientists.

REFERENCES

1. G.S. Nolas, *Mater. Res. Soc. Symp. Proc.* **545,** 435 (1999).
2. G.S. Nolas, D.G. Vanderveer, A.P. Wilkinson and J.L. Cohn, *J. Appl. Phys.* **91,** 8970 (2002).
3. S.-J. Kim, S. Hu, C. Uher, T. Hogan, B. Huang, J.D. Corbett and M.G. Kanatzidis, *J. Solid State Chem.* **153,** 321 (2000).

4. G.A. Slack, in *CRC Handbook of Thermoelectrics,* edited by D. M. Rowe (CRC Press, Boca Raton, 1995) p. 407.
5. T. Hahn, *International Tables for Crystallography, vol. A, Space-Group Symmetry*, 4th ed. (Kluwer Academic Publishers, Boston, 1996).
6. N.L. Okamoto, T. Nishii, M.W. Oh and H. Inui, *Mater. Res. Soc. Symp. Proc.* **793,** 187 (2004).
7. N.L. Okamoto, M.W. Oh, T. Nishii, K. Tanaka and H. Inui, *Phys. Rev. B*, submitted.
8. V.L. Kuznetsov, L.A. Kuznetsova, A.E. Kaliazin and D.M. Rowe, *J. Appl. Phys.* **87**, 7871 (2000).
9. F. Izumi and T. Ikeda, *Mater. Sci. Forum,* **321-324,** 198-203 (2000).
10. W.C. Cabrera, H. Borrmann, R. Michalak and Y. Grin, on the website; http://pc176.ph.rhul.ac.uk/~grosche/presentations/ba6ge25struc.pdf
11. D. R. Lide, *Handbook of Chemistry and Physics,* 82nd ed. (CRC press, Boca Raton, Florida, 2001).

Crystal Structure and Thermoelectric Properties of Al-containing Re Silicides

Eiji Terada[1], Min-Wook Oh[2], Dang- Moon Wee[2] and Haruyuki Inui[1]
[1]Department of Materials Science and Engineering, Kyoto University,
Sakyo-ku, Kyoto 606-8501, Japan
[2]Department of Materials Science and Engineering, KAIST,
Yuseong-gu, Daejon 305-701, Republic of Korea

ABSTRACT

The microstructure, defect structure and thermoelectric properties of Al-containing $ReSi_{1.75}$ based silicides have been investigated. All the Al-containing alloys investigated contain four differently oriented domains accompanied by the twinned microstructure, as the binary alloy does. However, thin defect layers containing a kind of shear structure are locally and sporadically formed at some of twin boundaries. In the defect layer, shear occurs by the vector of [100] on either ($\bar{1}09$) or (107) planes. Binary $ReSi_{1.75}$ exhibits nice thermoelectric properties as exemplified by the high value of dimensionless figure of merit (ZT) of 0.70 at 800 °C when measured along [001], although the ZT value along [100] is just moderately high. Al-containing Re silicides considerably increase the ZT value along [100] so that the maximum value of 0.95 is achieved at 150 °C for the $ReSi_{1.75}Al_{0.02}$ alloy. The temperature dependence of electrical resistivity changes from of semiconductor for the binary alloy to of metal for the Al-added alloys and the value of electrical resistivity is significantly reduced when compared to the binary counterpart.

INTRODUCTION

Binary rhenium disilicide is of interest owing to its potentials as a promising candidate material for thermoelectric applications [1]. Although the disilicide has been known to be a semiconductor, there is no general agreement on the crystal structure [2]. In our recent research, the crystal structure of the defect disilicide formed with Re ($ReSi_{1.75}$) has been refined by transmission electron microscopy combined with first-principles calculation [3,4]. The crystal structure is monoclinic with the space group Cm (mc44) due to an ordered arrangement of vacancies on Si sites in the underlying (parent) $C11_b$ lattice. The crystal contains four differently oriented domains; two domains related with each other by the 90°-rotation about the c-axis of the underlying $C11_b$ lattice and twin domains for each of the two domains. The twin habit plane is (001) of the underlying $C11_b$ lattice and the thickness of twins is very thin ranging from 100~300 nm. Although several researchers have reported the electrical transport properties of the disilicide, they are not necessarily consistent with each other. Siegrist et al. [5] reported the value of Seebeck coefficients ranging from -90 to -130 µV/K at 310 K. On the other hand, Neshpor et al. [6, 7] reported the value in fair agreement with that reported by Siegrist et al. [5] but the sign is reversed. Our recent measurements on the electrical transport properties as well as thermal transport properties for single crystals of $ReSi_{1.75}$ indicated that the value of electrical resistivity along [001] shows a higher value than that along [100] and that the temperature dependence of electrical resistivity for both orientations shows a semiconducting behavior. The Seebeck coefficients showed highly anisotropic behaviors. The value of Seebeck coefficients along [100] is positive (+230 µV/K at 330 K) while it is negative along [001] (-300 µV/K at 600 K). The values of thermal conductivity at room temperature are 5.02 W/mK and 5.65 W/mK along [100] and [001], respectively and increase gradually as the temperature is increased. The value of dimensionless figure of merit (ZT) along [001] at 1073 K is 0.7 (n-type), which is comparable with the value of Si-Ge alloys currently used as a thermoelectric material for power generation. The ZT value along [001] is further improved by Mo-substitution for Re [3,4]. On the other hand, the value of ZT along [100] is only moderately high, 0.2 (p-type).

In the present study, we investigate the effects of Al substitution for Si on the crystal structure, microstructure and thermoelectric properties of Re silicides in hope that the substitution further improves the properties.

EXPERIMENTAL PROCEDURES

Rod ingots with nominal compositions of $ReSi_{1.75-X}Al_X$ (X=0.02, 0.03) and $ReSi_{1.75-Y}Al_{0.02}$ (Y=0~0.04) and with dimensions of 6 mm in diameter and 50 mm long were prepared by arc melting high purity elements in Ar atmosphere. Single crystals were grown from these rods at a growth rate of 5 mm/h under Ar gas flow, using our ASGAL FZ-SS35W optical floating-zone furnace. The orientations of single crystals were determined by the X-ray back Laue method and were then cut along [100] and [001]. Microstructures were examined by transmission electron microscopy (TEM) with JEM-2000FX and JEM-4000EX electron microscopes operated at 200 kV and 300 kV, respectively. Electrical resistivity and Seebeck coefficient were measured by the four probe method in the temperature range from 330~1073 K. Thermal conductivity was measured by the laser flash method.

RESULTS AND DISCUSSION

Crystal structure and Microstructure

Low-magnification bright-field (BF) images of $ReSi_{1.75}$ and $ReSi_{1.75}Al_{0.02}$ alloys are shown in Fig. 1. The Al-containing alloy contains four differently oriented domains accompanied by the twinned microstructure with the twin thickness of 100~300 nm, as the binary alloy does. But the Al-containing alloy contains thin layers occurring sporadically and locally at some of twin

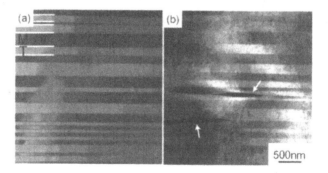

Figure 1. Low-magnification BF images of (a) $ReSi_{1.75}$ and (b) $ReSi_{1.75}Al_{0.02}$.

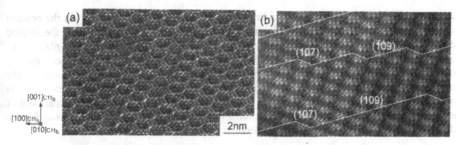

Figure 2. High-resolution TEM images of (a) ReSi$_{1.75}$ and (b) a defect layer in ReSi$_{1.75}$Al$_{0.02}$.

boundaries (dark regions indicated by arrows in Fig. 1(b)). The binary counterpart does not have this feature. These dark areas sometimes exhibit striated contrast parallel to some crystallographic direction and the corresponding selected-area electron diffraction (SAED) patterns often display splitting of some diffraction spots originated from the ordering of Si vacancies in the binary alloy and streaking along the direction of the splitting. In some other dark areas, the contrast is rather uniform without any striation and the positions of diffraction spots originated from the ordering of Si vacancies in the the corresponding SAED patterns are somewhat different from those in the binary counterpart. These indicate the changes in the positions (ordered arrangement) of Si vacancies upon alloying with Al.

High-resolution TEM images of ReSi$_{1.75}$ and the defect layer of ReSi$_{1.75}$Al$_{0.02}$ alloys are shown in Fig. 2. The occurrence of shear operation is evident in the high-resolution TEM image of the defect layer of the ReSi$_{1.75}$Al$_{0.02}$ alloy, as indicated in Fig. 2(b). Shear occurs by the vector of [100] on either ($\bar{1}$09) or (107) planes. Analysis of the atomic arrangement across these shear planes indicates that some atomic planes are inserted or removed in the ratio of Re:Si=1:2 for shears on ($\bar{1}$09) and (107) planes, respectively. These are schematically shown in Figs. 3(a) and (b), respectively. If the shear operation occurs every n unit cell layers on ($\bar{1}$09) and (107), the alloy composition is expected to change to Re(Si,Al)$_{(1.75n-0.25)/(n-0.125)}$ and Re(Si,Al)$_{(1.75n+0.25)/(n+0.125)}$, respectively. In Al-containing alloys, the ($\bar{1}$09) shear is found to be

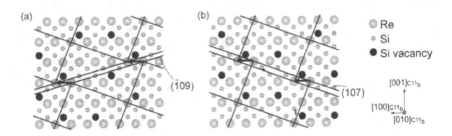

Figure 3. Schematic illustration of shear structures observed in Al-containing Re silicides. Shear occurs by the vector of [100] on (a) ($\bar{1}$09) and (b) (107).

dominant over the (107) shear. This indicates that the density of Si vacancies (in the ordered arrangement) should decrease upon alloying with Al. In the thin dark regions with the striated contrast, the observed number of n ranges from 2 to 6 and varies from place to place in an irregular manner. The crystal structure is thus remained to be monoclinic. On the other hand, the observed number of n in the thin dark regions without any striated contrast is almost unity everywhere. Thus, the crystal structure in this case changes into orthorhombic. When transition metals such as Mo, are substituted for Re, similar shear operations occur. But the occurrence of shear operations in this case is rather uniform throughout the specimen. The reason for the difference in the distribution of shear operations is not clear yet.

Thermoelectric properties

Fig. 4 shows the temperature dependence of electrical resistivity and thermal conductivity of Al-containing alloys along [100]. For Al-containing alloys, electrical conduction changes to of metal and the value of electrical resistivity at room temperature is significantly reduced by an order of magnitude when compared to the binary counterpart. This is because the carrier density of positive charge increases upon substituting Si with Al. The value of thermal conductivity at room temperature is also reduced in spite of the increased carrier density. This indicates that there is some mechanism in which phonons are effectively scattered in spite of the increased carrier density. We believe that the defect layer locally and sporadically formed at some of twin boundaries is responsible for the effective phonon scattering.

The temperature dependence of Seebeck coefficient along [100] is depicted in Fig. 5(a). The values of Seebeck coefficient for all the Al-containing alloys are lower than that of the binary counterpart in the low temperature range below 600 K but they are larger than that for the binary counterpart above 600 K. The lower values of Seebeck coefficient for all the Al containing alloys in the low temperature range is due to the increased carrier density. Performance of

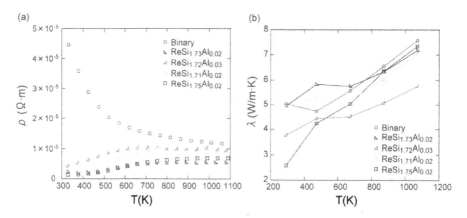

Figure 4. Temperature dependence of (a) electrical resistivity and (b) thermal conductivity along [100] for binary and Al-containing Re silicides.

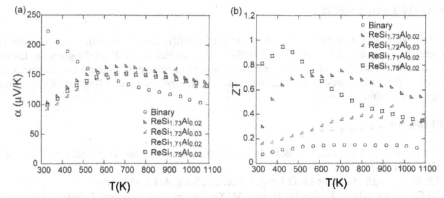

Figure 5. Temperature dependence of (a) Seebeck coefficient and (b) dimensionless figure of merit (ZT) along [100] for binary and Al-containing Re silicides.

thermoelectric materials is usually evaluated with dimensionless figure of merit (ZT),

$$Z = \alpha^2 / \rho \lambda, \tag{1}$$

where α, ρ, λ and T stand respectively for Seebeck coefficient, electrical resistivity, thermal conductivity and temperature (in Kelvin). It is generally believed that the ZT values exceeding unity is needed for practical applications as thermoelectric materials [1]. The temperature dependence of ZT values along [100] is depicted in Fig. 5(b). The ZT values for all the al containing alloys are larger than that for the binary counterpart in the whole temperature range investigated. This is due to the fact that both electrical resistivity and thermal conductivity are improved simultaneously. Further improvement of thermoelectric performance could be achieved by introducing the defect layers at twin boundaries in the optimum distribution through alloying Al in the appropriate amounts.

CONCLUSIONS

(1) All the Al-containing alloys contain four differently oriented domains accompanied by the twinned microstructure,.as the binary alloy does. Thin defect layers containing a kind of shear structure are locally and sporadically formed at some of twin boundaries. In the defect layer, shear occurs by the vector of [100] on either ($\bar{1}$09) or (107) planes.

(2) Al substitution for Si in $ReSi_{1.75}$ leads to the increased positive carrier density, resulting in quite low values of electrical resistivity along [100]. The temperature dependence along this direction changes to of metal. The maximum ZT value along [100] is achieved with the $ReSi_{1.75}Al_{0.02}$ alloy and is 0.95 at 150 °C.

(3) Further improvement of thermoelectric performance could be achieved by introducing the defect layers at twin boundaries in the optimum distribution through alloying Al in the appropriate amounts.

ACKNOWLEDGEMENTS

This work was supported by Grant-in-Aid for Scientific Research (a) from the Ministry of Education, Science and Culture (No. 14350369) and in part by COE 21 Program on United Approach for New Materials Science from the Ministry of Education, Science and Culture. One of authors (M.W. Oh) greatly appreciates the Monbusho scholarship from Ministry of Education, Science and Culture, Japan.

REFERENCES

1. A. Heinrich, H. Griebmann, G. Behr, L. Ivanenko, J. Schumann and H. Vinzelberg, Thin Solid Films **381**, 287 (2001).
2. J.P. Becker. J.E. Mahan and R.G. Long, J. Vac. Sci. Tech. A, **13**, 1133 (1995)
3. J-J Gu, K. Kuwabara, K. Tanaka, H. Inui, M. Yamaguchi, A. Yamamoto, T. Ohta and H. Obara in Defect Properties and Related Phenomena in Intermetallic Alloys, edited by E. P. George, H. Inui, M. J. Mills, and G. Eggeler,(Mater. Res. Soc. Proc. **753**, Boston, Massachusetts, 2002) p. 501-506.
4. Y. Sakamaki, K. Kuwabara, Gu Jiajun, H. Inui, M. Yamaguchi, A. Yamamoto and H. Obara, Materials Science Forum **426-432**, 1733-1738 (2003).
5. T. Siegrist, F. Hulliger and G. Travaglini, J. Less-Common Met. **92**, 119 (1983).
6. V. S. Neshpor and G. V. Samsonov, Izv. Akad. Nauk S.S.S.R., Neorg. Mater. **1**, 655 (1965) [Inorg. Mater. (U.S.S.R.) **1**, 599 (1965)].
7. V. S. Neshpor and G. V. Samsonov, Fiz. Met. Metalloved. **11**, (4) 638 (1961) [Sov. Phys.-Phys. Met. Metallogr. **11**, (4) 146(1961)].

Mater. Res. Soc. Symp. Proc. Vol. 842 © 2005 Materials Research Society

Characterization and Catalytic Properties of Ni₃Al for Hydrogen Production from Methanol

Ya Xu[1], Satoshi Kameoka[2], Kyosuke Kishida[1], Masahiko Demura[1], An-pang Tsai[1,2] and Toshiyuki Hirano[1]

[1]Materials Engineering Laboratory, National Institute for Materials Science, 1-2-1 Segen, Tsukuba, Ibaraki 305-0047, Japan
[2]Institute of Multidisciplinary Research for Advanced Materials, Tohoku University, Sendai 980-8577, Japan

ABSTRACT

The stability of catalytic activity and selectivity of Ni₃Al for methanol decomposition were studied by life test at 633 K on the alkali-leached powder samples. The characterization of the samples was carried out by X-ray diffraction, inductively coupled plasma (ICP) analysis, SEM observation, and surface area measurement. The life test showed that the alkali-leached Ni₃Al exhibits a very stable activity and a high selectivity for methanol decomposition. The surface characterization after reaction suggests that the high selectivity and stable activity may be attributed to the formation of tiny particles and porous structure which increased the surface area significantly during reaction. These results indicate a possibility of Ni₃Al as a catalyst for hydrogen production reaction.

INTRODUCTION

Some intermetallic compounds are known to have a good catalytic selectivity and activity. For example, Ni₃Sn increases the selectivity for hydrogen production [1], and PtGe and CoGe do for hydrogenation[2,3]. In Ni-Al system there are four stable intemetallic compounds, NiAl₃, Ni₂Al₃, NiAl and Ni₃Al. Among them a mixture of NiAl₃ and Ni₂Al₃ (Ni-50 wt% Al) is used as a precursor alloy for Raney nickel catalysts: the Raney nickel catalysts are produced from the precursor alloy by leaching aluminum in alkali hydroxide solution. For NiAl and Ni₃Al, very limited studies have been carried out on the catalytic properties. Probably it has been thought difficult to effectively leach aluminum from them because of their low aluminum concentration [4], and thus high catalytic activity has not been expected, particularly for Ni₃Al. Until now, Ni₃Al was known as promising high-temperature structural materials because of its excellent high temperature strength and corrosion/oxidation resistance and thus many studies have been focused on the mechanical properties and the microstructures [5-7]. Recently we examined the catalytic activity of Ni_3Al powder for hydrogen production from methanol which includes methanol decomposition ($CH_3OH \rightarrow 2H_2+CO$) and steam reforming of methanol ($CH_3OH + H_2O \rightarrow 3H_2+CO_2$) [8-10]. It was found that alkali-leached Ni_3Al shows a high activity and selectivity for methanol decomposition at 633 K. However, the stability of the catalytic activity is not studied yet. In the present study, we carried out a life test of methanol decomposition on the alkali-leached Ni_3Al catalyst in order to examine the stability.

EXPERIMENTAL DETAILS

Ni_3Al (Ni-24 at% Al) alloy ingot was prepared by conventional induction heating and homogenized at 1573 K for 36 ks in high vacuum. Then the ingot was scrapped with a planer machine, crushed to powders by mechanical milling, and sieved less than 150 μm in size. The powder was dipped in a stirring 20 wt% NaOH aqueous solution at 340 K for 18 ks in order to leach aluminum. The amount of leached aluminum was measured by inductively coupled plasma (ICP) analysis. The powder was filtered out and rinsed in deionized water. Finally they were dried at 323 K for 28.8 ks in air. The surface area of the powders was determined by BET (Brunauer-Emmett-Teller) surface area analysis. The crystal structure before and after leaching was analyzed by X-ray diffraction (XRD) analyses using a CuKα source (Rigaku, RINT2500). For comparison, a nickel catalyst was prepared by the same leaching treatment from commercial Ni-50 wt% Al alloy powder (Raney-Nickel, Mitsuwa Chemicals).

The catalytic experiments of hydrogen production from methanol were carried out in a conventional fixed-bed flow reactor as described in the previous reports [8-10]. Prior to reaction, the powder was reduced at 513 K for 3.6 ks in a flowing hydrogen atmosphere. Then, the hydrogen flow was stopped and refilled with nitrogen to flush H_2. Then the temperature was increased to 633 K, and pure methanol was introduced to the reactor at a liquid hourly space velocity of 39 h^{-1} (defined as the volume of methanol passed over the unit volume of catalyst per hour) using a plunger pump together with nitrogen carrier gas. The outlet composition of gas products was analyzed using gas chromatograph (Shimadzu, GC-14B equipped with thermal conductivity detectors (TCD)). The flow rate of gas products was measured by a soap bubble meter. The surface morphology was observed by a scanning electron microscopy with a field emission gun and a energy dispersive X-ray system (EDX) (JEOL, JSM-6500F).

RESULTS AND DISCUSSION

Effects of alkali leaching

The ICP analysis of the NaOH solution after leaching shows that aluminum is selectively leached from the Ni₃Al powder. The amount of leached aluminum is estimated at 13.9 wt% of aluminum concentration in the precursor Ni₃Al (Ni-12.7 wt% Al). No nickel is detected in the solution.

The BET surface area measurement shows that the surface area increases from 2.3 to 5.1 m^2/g by alkali leaching. The size distribution of the surface pore before and after leaching is estimated from the BET measurement. It is found that the pore radius is sharply distributed at approximately 2 nm after leaching, indicating that the observed increase in surface area is due to the formation of these fine pores. The XRD analysis was carried out for the Ni₃Al powder before and after leaching, as shown in Fig. 1. All the diffraction peaks observed are assigned as Ni₃Al. No significant difference is observed between the two samples. These results indicate that leaching of aluminum is limited to the very thin surface layer of the Ni₃Al powder, leaving the bulk as it is.

Life test over alkali-leached Ni₃Al

Life test of methanol decomposition on the alkali-leached Ni₃Al was carried out at 633 K for 65 h. The measurements were carried out at intervals of 20 minutes during the first 3 hours after 3 minutes since introducing methanol to the reactor. The temperature of samples changed in a wide range from 623 to 793 K during the first 1 h, and became stable afterwards. Figure 2 shows the production rates of H_2 and CO as a function of the reaction time. At the beginning, the

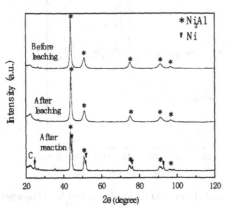

Figure 1. X-ray diffraction patterns of Ni₃Al powder: (1) before alkali leaching, (2) after alkali leaching, (3) after reaction at 633 K for 65 h.

production rates of H_2 and CO increased, and reached a maximum value of 250 ml min^{-1} g-cat^{-1} for H_2, and 125 ml min^{-1} g-cat^{-1} for CO after 1 h. Then, the production rates decreased slightly and became stable during the subsequent reaction. A high production rate of H_2, 190 ml min^{-1} g-cat^{-1}, was obtained after 65 h. The ratio of H_2 and CO production rates was found to be a constant of 2 after the first 1 h (inset in Fig. 2), indicating the methanol decomposition:

$$CH_3OH \rightarrow 2H_2+CO \qquad (1)$$

mainly occurred during the reaction. Figure 3 shows the H_2 and CO selectivity as a function of reaction time. H_2 selectivity (%) is calculated as (molecules H_2 produced/C atoms in gas products)(1/RR)×100, where RR=2, is the H_2/CO decomposition ratio. CO selectivity (%) is calculated as (C atoms in CO/total C atoms in gas products)×100 [1]. Ni_3Al exhibited high selectivities (>97%) for the production of H_2 and CO after the first 1 h, indicating that Ni_3Al is selective to methanol decomposition (1), and permitted the formation of only H_2 and CO. These results show that Ni_3Al exhibits a very stable activity and high selectivity for methanol decomposition at 633 K, indicating a possibility of Ni_3Al as a promising catalyst for hydrogen production reaction.

As shown in Figs. 2 and 3, the production rates and selectivity of H_2 and CO were low at the beginning, mainly during the first 1 h. The reason is considered as follows. Though it was not shown here, some amount of CO_2, CH_4 and H_2O were produced at the beginning of reaction, and decreased quickly with the progress of reaction. Therefore, during the first 1 h, the following

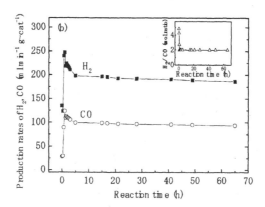

Figure 2. The production rates of H_2 and CO for alkali-leached Ni_3Al powder during the methanol decomposition at 633 K. Inset: the mole ratio of produced H_2 and CO as a function of reaction time.

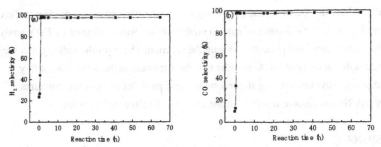

Figure 3. Selectivity to H_2 (a) and CO (b) of alkali-leached Ni_3Al powder for methanol decomposition at 633 K.

reactions also occurred in addition to the methanol decomposition.

Methanation:	$3H_2 + CO \rightarrow CH_4 + H_2O$	(2)
Water gas shift reaction:	$CO + H_2O \rightarrow CO_2 + H_2$	(3)

These reactions resulted in the low selectivity to H_2 and CO at the beginning. The reason why these reactions occurred at the beginning is not clear yet. Further research is in progress.

The XRD analysis of the Ni_3Al powder after reaction was carried out. The diffraction peaks were assigned as Ni_3Al, Ni and graphite, as shown in Fig. 1, indicating that structure change and carbon deposition occurred during the reaction. The BET surface area of reacted Ni_3Al powder was estimated at 20.4 m^2/g, which is much larger than that of the Ni_3Al before reaction (5.1 m^2/g). Figure 4 shows the SEM micrographs of the powder surface before and after reaction at

Figure 4. SEM images of alkali-leached Ni_3Al powder (a) before reaction, (b) after reaction at 633 K for 65 h.

633 K for 65 h. Many tiny particles and porous regions were observed on the powder surface after reaction (Fig. 4b). The existence of carbon on the surface was confirmed by EDX analysis, indicating the carbon deposition occurred during the reaction. These results indicate that the high selectivity and stable activity of Ni_3Al powder may be attributed to the formation of tiny particles and porous structures during the reaction. These particles and porous structures contained Ni_3Al, Ni and carbon, resulting in the increase of active surface area.

CONCLUSIONS

In conclusion, we have performed life test of methanol decomposition over alkali-leached Ni_3Al at 633 K for 65 h. It was found that the alkali-leached Ni_3Al exhibits a very stable activity and a high selectivity for methanol decomposition. A production rate of H_2 above 190 ml min^{-1} g-cat^{-1} was obtained after reaction at 633 K for 65 h. A high selectivity to H_2 and CO (above 97%) was obtained after the first 1 h reaction. The origin of the high selectivity and stable activity may be attributed to the formation of tiny particles and porous structure composed of Ni_3Al, Ni and carbon during reaction.

ACKNOWLEDGMENTS

The authors acknowledge Mr. Kazuhiko Kishi of NIMS for contribution to BET measurement, and Mr. Hiroshi Miyashiro of NIMS for manufacturing reactor tubes.

REFERENCES

1. G. W. Huber, J. W. Shabaker and J. A. Dumesic, *Science* **300**, 2075 (2003).
2. T. Komatsu, S. Hyodo and T. Yashima, *J. Phys. Chem.* **B. 101**, 5565 (1997).
3. T. Komatsu, M. Fukui and T. Yashima, *Studies in Surf. Sci. Catal.* **101**, 1095 (1996).
4. S. Tanaka, N. Hirose, T. Tanaki and Y. H. Ogata, *J. Electrochemical Soci.* **147**, 2242 (2000).
5. D. P. Pope and S. S. Ezz, *Int. Mater. Rev.* **29**, 136 (1984).
6. N. S. Stoloff, *Int. Mater. Rev.* **34**, 153 (1989).
7. M. Yamaguchi and Y. Umakoshi, *Prog. Mater. Sci.* **34**, 1 (1990).
8. Ya Xu, S. Kameoka, K. Kishida, M. Demura, A. P. Tsai and T. Hirano, *Intermetallics*, available online at www.elsevier.com/locate/intermet (in press).
9. Ya Xu, S. Kameoka, K. Kishida, M. Demura, A. P. Tsai and T. Hirano, *Materials Transactions* **45**,(2004) (in press).
10. A. P. Tsai and M. Yoshimura, *Appl. Catal. A: General* **214**, 237 (2001).

Other Intermetallics

Mater. Res. Soc. Symp. Proc. Vol. 842 © 2005 Materials Research Society

Complex Intermetallic Compounds: Defects, Disordering, Details

W. Sprengel[1], F. Baier[1,2], K. Sato[1,3], X.Y. Zhang[1,4], and H.-E. Schaefer[1]

[1] Institute of Theoretical and Applied Physics, Stuttgart University, 70569 Stuttgart, Germany.
[2] Physical Metallurgy, Technical University Darmstadt, 64287 Darmstadt, Germany.
[3] National Institute of Advanced Industrial Science and Technology, Tsukuba 305-8565, Japan.
[4] Key Laboratory of Metastable Materials Science and Technology, Yanshan University, 066004 Qinhuangdao, P. R. China.

ABSTRACT

A short overview will be given on the thermodynamics of the formation of thermal defects in intermetallic aluminides. We focus on thermal vacancies studied by the specific techniques of positron annihilation and time-differential dilatometry and discuss the results together with self-diffusion data. We then demonstrate that these techniques can be employed for studying vacancies in compound semiconductors specifically. Furthermore, structural order-disorder phase transitions can be investigated from an atomistic point of view by making use of positron annihilation as shown in the exemplary case of decagonal Al-Ni-Co quasicrystals.

INTRODUCTION

Intermetallic compounds are promising candidates for high-temperature structural or functional materials [1]. In addition to the great variety of periodic crystalline structures which are observed, for example transition metal aluminides or silicides, even more complex aperiodic quasicrystalline structures can be formed, e. g., in Al-rich compounds with transition metals.

In binary ordered intermetallic compounds as, e. g., in close-packed Ni_3Al with the $L1_2$ structure or in the more open structures such as B2-FeAl or $D0_3$-Fe_3Al the types of thermally formed atomic defects are more complex than in pure metals because two types of vacancies, V_A and V_B, can be formed on the two sublattices A and B as well antisite atoms A_B and B_A may be present. Even more complex defect structures such as the triple defect are postulated in order to maintain the local composition for vacancy formation [2].

The principles of the thermodynamics and the kinetics of high-temperature vacancy formation [3,4] are briefly outlined in the following. The temperature-dependent thermal equilibrium vacancy concentration

$$C_V = \exp\!\left(S_V^F / k_B\right)\exp\!\left(-H_V^F / k_B T\right) \tag{1}$$

is characterized by the enthalpy H_V^F and the entropy S_V^F of vacancy formation. From the temperature-dependent isothermal equilibration of the mean value $\overline{C}_V(t)$ in dependence of time t after a fast temperature change at high temperatures the time constant t_E for equilibration can be derived from the equation

$$\frac{\overline{C}_V(t) - C_i}{C_V - C_i} \cong 1 - \exp\left(-\alpha^2 D_V t\right) = 1 - \exp\left(-t / t_E\right), \tag{2}$$

where C_i is the initial and C_V the final vacancy concentration and where

$$D_V = \frac{1}{6} d^2 z \, v_D \exp\left(S_V^M / k_B\right) \exp\left(-H_V^M / k_B T\right) \tag{3}$$

is the diffusion coefficient of the vacancy with H_V^M the vacancy migration enthalpy assuming regularly distributed vacancy sources or sinks characterized by the paramter α. Here, d is the jump distance, z the coordination number, v_D the Debye frequency and S_V^M the vacancy migration entropy.

By measuring the variation of $\overline{C}_V(p)$ with the pressure p the vacancy formation volume

$$\Delta V_V^F = -k_B T \left(\frac{d \ln C_V}{dp}\right)_T \tag{4a}$$

can be derived whereas the pressure dependence of the characteristic relaxation time $t_E(p)$ yields the vacancy migration volume

$$\Delta V_V^M = -k_B T \left(\frac{d \ln(1 / t_E(p))}{dp}\right)_T . \tag{4b}$$

characterizing the volume change in the saddle-point configuration during the vacancy jump.

EXPERIMENTAL TECHNIQUES

The experimental requirements for a specific detection of defects with high sensitivity at high temperatures within short measuring times can be fulfilled only by a few measuring techniques as, e. g., positron annihilation [5] or time-differential dilatometry [6].

In the case of positron lifetime measurements the formation of thermal vacancies can be specifically detected by the reversible sigmoidal change of the positron lifetime with temperature upon positron trapping at thermally formed vacancies with a reduced valence electron density compared to the value probed by delocalized positrons in the defect-free lattice (see Figure 1a). From this change in the positron lifetime the vacancy formation enthalpy H_V^F is deduced [7].

By positron annihilation in addition to the valence electron density the electron momentum distribution up to the high momenta of core electrons can be sampled by means of coincident measurements of the two electron-positron annihilation photons in order to study the electron-momentum induced Doppler broadening of the photon line (see Figure 1b and Ref. 8).

This technique allows for the detection of the chemical nature of the atoms neighboring a vacancy or the sublattice on which thermal vacancies in an ordered intermetallic are formed as exemplified in the case of MoSi$_2$ below [8].

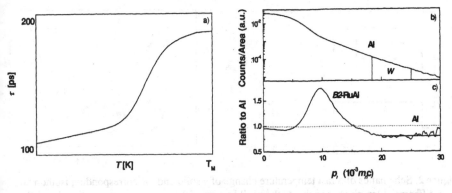

Figure 1. (a) Schematics of the sigmoidal change of the positron lifetime τ due to trapping of positrons with the specific trapping rate at thermally formed vacancies. (b) Example of the Doppler broadening spectrum of the positron-electron annihilation photon line in Al with strong background suppression resulting from the coincident measurement of the energies E_1 and E_2 of the two annihilation quanta with two high-resolution Ge-detectors and by making use of the energy conservation $E_1+E_2=1022$ keV. This spectrum represents the electron momentum distribution where the data at high momenta characterized by the W parameter (hatched area) are due to positron annihilation with core electrons and can be used for a chemical analysis of the local atomic surrounding of the annihilated positron. The Doppler broadening spectrum of Al is fully understood theoretically [9]. (c) Ratio curve of a Doppler spectrum of ordered B2-RuAl quenched from high temperatures devided by the spectrum of pure, annealed Al. The curve in the high momentum region (solid line) is parallel to that of pure Al (dotted line) which indicates an Al dominated vacancy environment, i.e., vacancy formation on the Ru sublattice [9a].

Another specific technique for observing thermal vacancy formation and migration has been established recently with time-differential dilatometry [6]. With high resolution optical interferometry the isothermal time-dependent length change due to the equilibration of the vacancy concentration was observed in B2-intermetallics after fast temperature changes [11]. From the temperature variation of this effect the enthalpies H_V^F and H_V^M are determined (see Figure 2).

VACANCY PROPERTIES

The thermodynamic and kinetic data of vacancies in intermetallic compounds are summarized in Table I [4]. From there it is evident that the vacancy formation enthalpy H_V^F is high (low C_V equilibrium vacancy concentration) in close-packed structures as, e. g., L1$_2$-Ni$_3$Al whereas H_V^F is low (high C_V) in the open structured B2 compounds. This can be understood according to ab-initio calculations (see Table I). The vacancy migration enthalpy H_V^M in B2 compounds such as, e. g., FeAl appears to be high so that the equilibration of C_V to low values is slow as demonstrated recently for NiAl [10] in long-term low-temperature annealing experiments and confirmed theoretically (see Table I).

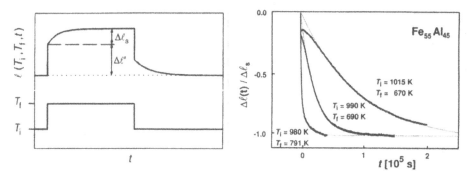

Figure 2. Schematics of a fast temperature change of a solid and the corresponding isothermal time-differential length change due to the equilibration of the vacancy concentration. From the temperature variations of the amplitude and of the time constant t_E values for H_V^F and H_V^M can be derived, respectively [6].

Table 1 Effective vacancy formation enthalpies H_V^F derived from the temperature dependent positron trapping rate $\sigma C_V (T)$ with a temperature-independent specific trapping rate σ, vacancy equilibrium concentrations C_V at half the melting temperature $0.5 T_M$, and activation enthalpies Q^{SD} for self-diffusion or foreign diffusion and effective vacancy-migration enthalpies H_V^M. The values for S_V^F are taken from differential thermal expansion data on FeAl and Fe₃Si [11] or from positron lifetime data for the other compounds assuming $\sigma = 4\times10^{14}\,\mathrm{s}^{-1}$. The σ-values are derived from a comparison of the positron trapping rates and of differential thermal expansion data [11]. Values for the vacancy formation volume, V_V^F, and the vacancy migration volume, V_V^M, are given for Fe₆₁Al₃₉. The detailed references for the data are given in Ref. 4.

Compound	Structure	T_M (K)	H_V^F (eV) exp.	H_V^F (eV) theor.	C_V (10⁻⁷)	S_V^F (kB)	Q^{SD} (eV)	H_V^M (eV) exp. theor.	V_V^F (Ω)	V_V^M (Ω)	σ (10¹⁵ s⁻¹)
Ni₇₅Al₂₅	L1₂	1663	1.82	V$_{Ni}$:1.87 V$_{Al}$:2.65	0.014	4.9	Ni: 3.15				0.9
Ti₄₈.₅Al₅₁.₅	L1₀	1750	1.4		0.083	0.7	Ti: 3.03				
Ti₆₆.₄Al₃₃.₆	DO₁₉	1930	1.55		0.21		Ti: 3.10				
Fe₇₆.₃Al₂₃.₇	DO₃ B2,A2	1770	1.18	V$_{Fe}$:1.25	240	5	Fe: 2.44				3.6
Fe₇₅Si₂₅	DO₃	1500	0.80		100	3.7	Fe: 1.65 Ge: 3.25				
Fe₆₁Al₃₉	B2	1660	0.98 1.0ᵃ⁾	V$_{Fe}$:1.06 V$_{Al}$:3.71	1200	3.7	Fe : 2.76	1.7 2.14 1.5	1.7	4.6	1.3
NiAl	B2	1911	1.5ᵃ⁾	V$_{Ni}$:0.93	7000		Ni : 3.01	1.8 2.1			

It turns out that the wide range of self-diffusivities in intermetallics (see Ref. 3) can be understood by the vacancy properties given in Table I. For the activation enthalpy of self-diffusion the relationship of $Q^{SD} = H_V^F + H_V^M$ holds for the data of the ^{59}Fe tracer diffusivity measured in FeAl (Table I) indicating that this diffusion process is exclusively governed by the temperature dependences of vacancy formation and migration. The extremely high Fe diffusivity in Fe$_3$Si can thus be understood by the low values H_V^F and H_V^M of vacancy formation and migration.

Further specific information on vacancies is available from measurements in dependence of pressure yielding for B2-FeAl the activation volumes $V_V^F = 1.7\,\Omega$ for vacancy formation [4, 12] or the large value $V_V^M = (4.6 \pm 1.5)\,\Omega$ for vacancy migration [4] where Ω is the value for the mean atomic volume. The large value of V_V^F has been tentatively discussed in terms of divacancy formation. The vacancy migration volume, which has been measured for the first time at high temperatures, is high and demonstrates that the vacancy jump requires a high volume change and is, therefore, difficult to be performed as also demonstrated by the high value of the potential barrier H_V^M to be overcome by the jumping vacancy. It should be mentioned here that the experimentally found vacancy activation volumes are not fully consistent with the low activation volumes V^{SD} of 1.4 Ω [13, 14] or 0.8 Ω [15] for Fe tracer diffusion in B2-FeAl when the relation $V^{SD} = V_V^F + V_V^M$ is considered. However, a high value of $V^{SD} = 2.8\,\Omega$ has been reported for the diffusion of Fe in pure Al [16].

One question about thermal vacancies has been left open until recently: On which sublattice are vacancies formed in a chemically ordered multinary crystal? According to theoretical predictions thermal vacancies are, e. g., in FeAl exclusively formed on the Fe transition metal sublattice because the formation enthalpy H_V^F on the Al sublattice is much higher (see Table I).

An experimental demonstration on which sublattice vacancies are formed has been given recently in the case of MoSi$_2$ with a C11$_b$ structure [8]. In this structure vacancies on the Si sublattice are surrounded by Si *and* Mo atoms (see Figure 3) whereas vacancies on the Mo sublattice are exclusively surrounded by Si atoms.

The coincident Doppler broadening data now exhibit the characteristic signature of the high-electron-momentum slope of Mo (see Figure 3) characteristic for vacancy formation on the Si sublattice. This has been confirmed theoretically [8]. Additional evidence originates from self-diffusion studies in MoSi$_2$ showing that the Si diffusivity [17] is much higher than the Mo diffusivity [18] which is fully consistent with a high concentration and mobility of thermal vacancies on the Si sublattice.

It should be emphasized that the identification of the sublattice on which vacancies are located by Doppler broadening measurements of the positron-electron annihilation radiation is not restricted to ordered alloys. A recent study of the wide band-gap semiconductor SiC [19] showed that vacancies can be selectively generated on either the C or the Si sublattices by irradiation with electrons of appropriate energy and identified by Doppler broadening measurements of the positron-electron annihilation radiation. This may be of interest for the localization of the two types of vacancies in the band-gap.

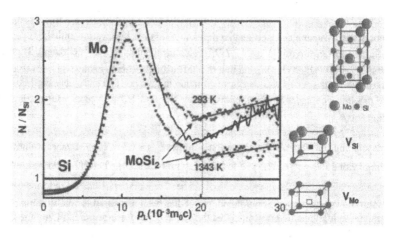

Figure 3. Ratio curves of coincidently measured Doppler broadening spectra of $MoSi_2$ with positron trapping at thermally formed vacancies ($T = 1343$ K), for pure Mo at room temperature (293 K) vs. pure Si (293 K). In the case of the $MoSi_2$ curve the positive high-momentum slope similar to that of pure Mo shows that the vacancy-trapped positrons partially sample the high-momentum core electrons of Mo neighboring the vacancy as expected for vacancies, V_{Si}, on the Si sublattice (see structures at the right side).

ATOMISTIC VIEW OF PHASE TRANSITIONS BY POSITRON ANNIHILATION STUDIES

Structural phase transitions in solids are conventionally studied by diffraction or scattering techniques where coherence lengths over many atomic spacings are required. Recently it has been demonstrated in an exemplary study on the decagonal quasicrystal d-$Al_{71.5}Ni_{14}Co_{14.5}$ that structural phase transitions can be studied from an atomistic point of view by making use of positron annihilation as a nuclear probe technique [20].

In d-Al-Ni-Co quasicrystals with quasicrystalline aperiodic layers of atoms stacked periodically a wealth of structural modifications are present as depicted in the phase diagram [21, 22]. Chemical ordering between Al and the transition metal atoms Co and Ni [22 – 26] may play a pivotal role for the understanding why quasicrystals with their aperiodic structure form. The ordered quasi-unit-cell decagons (see Figure 4) exhibit a central cluster rich in transition metals which eventually may give rise to a symmetry breaking of the decagon structure [23, 26, 27] which may be a prerequisite for perfect quasi-periodic tiling [23, 28].

In the positron annihilation data shown in Figure 5 we first consider the prominent change of the Doppler broadening W parameter centered at 1140K (Figure 5), a temperature range where structural changes in decagonal Al-Ni-Co quasicrystals were detected by dilatometry, x-ray diffraction, and by neutron scattering (see Ref. 20). From the width of the diffuse neutron scattering peak in decagonal $Al_{72}Ni_{12}Co_{16}$ [29] ordered domains with a diameter of 3.7 nm can be derived within the decagonal planes at 293 K.

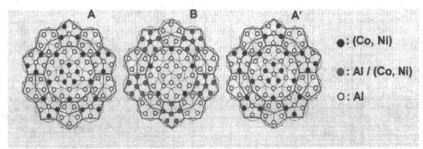

Figure 4. Atomic decoration of the quasi-unit-cell decagons of a decagonal Al-Ni-Co quasicrystal as derived from Z contrast high-resolution electron microscopy. The aperiodic planes are periodically stacked (ABA'B) [22].

In a stage at 1140 K reversible chemical disordering is observed by a shrinkage of the domain size d and the disappearance of the neutron scattering peak [29].

In a simple model, comprising both the atomic-scale positron data as well as the temperature-dependent domain size from neutron scattering, the two data sets are compared. In this model the mean volume-averaged Doppler broadening parameter

$$W = f_o W_o + f_d W_d \tag{5}$$

is characterized by the Doppler broadening parameter in the ordered (W_o) and the disordered (W_d) structure with the corresponding volume fractions

$$f_o = d^2 / d_{max}^2 \tag{6}$$

and f_d=1-f_o where for the fully ordered state the relations d=d_{max} and f_o=1 hold. When we consider the temperature dependence of the scattering data within this model (Figure 5, solid line) then it is evident that they coincide with the present positron annihilation data. This means that by the positron annihilation Doppler broadening data we can specifically detect structural phase transitions as discussed in the following.

The positron lifetime of 198 ps measured at 293 K, which is substantially higher than the positron lifetime in a defect-free case as expected from the mean valence electron density, demonstrates that positrons are trapped at structural vacancies of an atomic concentration of 5×10^{-4}. At room temperature the positrons sample an atomic environment dominated by transition metal atoms as deduced from the high-momentum Doppler broadening ratio curves of d-$Al_{71.5}Ni_{14}Co_{14.5}$ (Figure 6) located between those of pure Ni and pure Co. This structure may be identified with the central cluster of transition metals in the 2.0 nm decagon observed by Z-contrast electron microscopy [22, 23] and x-ray diffraction [25] (see Figure 4).

At high temperatures at, e. g., 1173 K the Doppler broadening ratio curve is found to be very similar to that of pure Al (see Figure 6). This is ascribed to disordering yielding a central cluster with – according to the mean composition of $Al_{71.5}Ni_{14}Co_{14.5}$ – an Al concentration higher than in the ordered TM rich case.

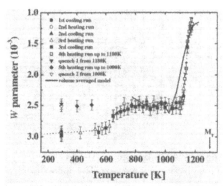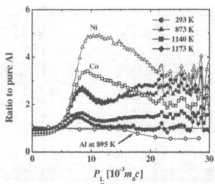

Figure 5. Reversible temperature dependence of the Doppler broadening W parameter (see Fig. 1) in a decagonal $Al_{71.5}Ni_{14}Co_{14.5}$ quasicrystal for heating and cooling runs. In addition, W parameters at 293 K after quenching from 1180 K and 1000 K are given together with isochronal annealing data (full diamonds). The bold line indicates the model for a volume averaged W parameter in the ordered domains and in the disordered structure (see Eqs. 5,6) making use of the temperature dependent domain size from neutron scattering [29]. The dotted line is drawn for guiding the eye.

Figure 6. Ratio curves of the Doppler broadening spectra measured in coincidence for a decagonal $Al_{71.5}Ni_{14}Co_{14.5}$ quasicrystal at various temperatures together with those for pure Al, pure Ni, and pure Co. All spectra are normalized to that of pure Al, the ratio spectrum of pure Al at 895 K where positron saturation trapping at thermal vacancies occurs is additionally plotted.

This reversible temperature-dependent change of the high-momentum W parameter thus demonstrates that structural phase transitions can be detected by the local chemical changes on an atomistic level by making use of positron annihilation techniques.

It may be mentioned that in the present studies in addition to the 1140 K phase transition in d-Al-Ni-Co a reversible transition is observed by the W parameter measurements (see Figure 5). In this stage disordering between Ni and Co atoms may occur since the Ni-Co ordering energy is predicted to be much smaller than the Al-TM ordering energy [23] of the 1140 K stage. Short-range atomic transport required for disordering is well available as concluded from atomic diffusivity experiments (see Ref. 20). We may add here that the stages of the W parameter at 650 K and 1140 K cannot be attributed to thermal vacancy formation because they are by far too narrow for a thermal vacancy formation process [20].

In conclusion we summarize that atomic processes in binary and complex solids, which are correlated to thermal vacancies, can be studied by specific techniques such as positron annihilation or time-differential dilatometry. By making use of coincident Doppler broadening measurements of the positron-electron annihilation photons structural phase transitions can be studied on an atomistic level which is an extension of and complementary to scattering techniques.

ACKNOWLEDGEMENTS

H.-E. Schaefer is indebted to Prof. Y. Shirai for his outstanding hospitality during the teaching and research period at Osaka University, Japan. The financial support of the 21st Century COE Program "Center of Excellence for Advanced Structural and Functional Materials Design", Osaka University, Japan is acknowledged.

REFERENCES

1. J.H. Westbrook and R.L. Fleischer (eds.), *Intermetallic Compounds: 4 Vols.*, J. Wiley and Sons, Chichester, UK, (2000).
2. R.J. Wasilewski, S.R. Butler, and J.E. Hanlon, *J. Appl. Phys.* **39**, 4234 (1968)
3. H.-E. Schaefer, K. Frenner, and R. Würschum, *Intermetallics* **7**, 277 (1999)
4. W. Sprengel, M.A. Müller, and H.-E. Schaefer, Diffusion and Defect Structures, in: *Intermetallic Compounds, Vol. 3, Principles and Practice*, (eds.) J.H. Westbrook and R.L. Fleischer, J. Wiley and Sons, Chichester, UK, (2002), 275-293
5. W. Sprengel, F. Baier, K. Sato, X.Y. Zhang, K. Reimann, R. Würschum, R. Sterzel, W. Assmus, F. Frey, and H.-E. Schaefer, in: *Quasicrystals: Structure and Properties*, Wiley-VCH, Weinheim, Germany (2003), p. 414-429
6. H.-E. Schaefer, K. Frenner, and R. Würschum, *Phys. Rev. Lett.* **82**, 948 (1999)
7. H.-E. Schaefer, *Phys. Stat. Sol.* (a) **102**, 47 (1987)
8. X.Y.Zhang, W. Sprengel, T.E.M. Staab, H.Inui, and H.-E. Schaefer, *Phys. Rev. Lett.* **92**, 155502 (2004)
9. P.E. Mijnarends, A.C Kruseman, A van Veen, H Schut and A Bansil, *J. Phys. – Cond. Matter* **10** , 10383 (1998)
9a. E. Partyka, W. Sprengel, H. Weigand, H.-E. Schaefer, F. Krogh, and G. Kostrorz, to be published
10. X. Y. Zhang, W. Sprengel, K.J. Reichle, K. Blaurock, R. Henes, and H.-E. Schaefer, *Phys Rev. B* **68**, 224102 (2003)
11. R, Kerl, J. Wolff, and Th. Hehenkamp, *Intermetallics* **7**, 301 (1999)
12. J. Wolff, A. Broska, M. Franz, B. Köhler, and Th. Hehenkamp, *Mater. Sci. Forum* **255-257**, 593 (1999)
13. M. Eggersmann, Dr. rer. nat. thesis, Münster University, Germany (1998)
14. M. Eggersmann and H. Mehrer, *Metallofizika I Noveshie Tekhnologii* **21**, 70 (1999)
15. R. Nakamura and Y. Iijima, *Philos. Mag.* **84**, 1906 (2004)
16. G. Rummel, T. Zumkley, M. Eggersmann, K. Freitag, and H. Mehrer, *Z. Metallkd.* **86**, 131 (1996)
17. M. Salamon, A Strohm, T. Voss, P. Laitinen, I. Riihimäki, S. Divinski, W. Frank, J. Räisänen, and H. Mehrer, *Philos. Mag.* **84**, 737 (2004)
18. M. Salamon and H. Mehrer, *Defect Diffus. Forum* **216-217**, 161 (2003)
19. A. A. Rempel, W. Sprengel, K. Blaurock, K.J. Reichle, J. Major, and H.-E. Schaefer, *Phys. Rev. Lett.* **89**, 185501 (2002)
20. K. Sato, F. Baier, W. Sprengel, R. Würschum, and H.-E. Schaefer, *Phys. Rev. Lett.* **92**, 127403 (2004)

21. S. Ritsch, C. Beeli, H.-U. Nissen, T. Gödecke, M. Scheffer, and R. Lück, *Philos. Mag. Lett.* **78**, 67 (1998)

22. K. Hiraga, T. Ohsuna, W. Sun, and K. Sugiyama, *Mater. Trans., JIM* **42**, 2354 (2001)

23. Y. Yan and S. Pennycook, *Phys. Rev. Lett.* **86**, 1542 (2001)

24. H. Takakura, A. Yamamoto, and A. P. Tsai, *Acta Crystallogr. Sect. A* **57**, 576 (2001)

25. A. Cervellino, T. Haibach, and W. Steurer, *Acta Crystallogr. Sect. B* **58**, 8 (2002)

26. C. L. Henley, M. Mihalkovič, and M. Widom, *J. Alloys Compd.* **342**, 221 (2002)

27. M. Mihalkovič, I. Al-Lehyani, E. Cockayne, C. H. Henley, N. Moghadam, J. A. Moriarty, Y. Wang, and M. Widom, *Phys. Rev. B* **65**, 104205 (2002)

28. Y. Yan and S. Pennycook, *Nature* **403**, 266 (1999)

29. F. Frey, E. Weidner, K. Hradil, M. de Boissieu, A. Letonblon, G. McIntyre, R. Currat, and A.P. Tsai, *J. Alloys Compd* **342**, 57 (2002)

Mater. Res. Soc. Symp. Proc. Vol. 842 © 2005 Materials Research Society   S1.5

Effect of Heat Treatments on Microstructure of Rapidly Solidified TiCo Ribbons

Kyosuke Yoshimi, Akira Yamauchi, Ryusuke Nakamura, Sadahiro Tsurekawa[1] and Shuji Hanada
Institute for Materials Research, Tohoku University,
Sendai, Miyagi 980-8577, JAPAN
[1]Department of Nanomechanics, Tohoku University,
Sendai, Miyagi 980-8579, JAPAN

ABSTRACT

The effect of heat treatments (aging or annealing) on microstructure was investigated for rapidly solidified ribbons of near-stoichiometric TiCo. In as-spun ribbons, it was observed by TEM that an equiaxed grain structure was developed and its crystal structure had been already B2-ordered, while a small amount of a second phase, Ti_2Co, finely disperses in grains and along grain boundaries. Some grains were dislocation-free but others contained curved or helical dislocations and prismatic loops having a Burgers vector parallel to <100> directions. By annealing the as-spun ribbons at 700°C for 24h, the dislocation density was obviously increased compared with that of the as-spun ribbons, while grain growth appears to occur slightly. The increase of the dislocation density in the annealed ribbons is believed to result from the condensation and/or absorption of supersaturated vacancies. Therefore, the TEM observation results indicate that a large amount of supersaturated thermal vacancies were retained in the TiCo ribbons by the rapid solidification.

INTRODUCTION

In our recent study on rapidly solidified Fe-45mol%Al [1], it was found that supersaturated vacancies of about 1.4 % are introduced into B2-ordered FeAl ribbons by the melt-spinning process and a large amount of mesopores are formed near surfaces as well as inside the ribbons by the condensation of supersaturated vacancies. FeAl is one of B2-type, Berthollide compounds and well known for containing an extremely high concentration of thermal vacancies at high temperature. Würschum et al. reported thermodynamic data of thermal vacancies in B2-type FeAl measured by positron lifetime analysis [2], that is, a low vacancy formation enthalpy (0.89eV), high vacancy formation entropy ($4.2k_B$, where k_B is the Boltzmann constant) and a high vacancy migration enthalpy (1.7eV) with the consideration of a vacancy binding enthalpy of 0.5eV estimated by Fu et al. [3]. The low vacancy formation enthalpy and high vacancy formation entropy give a much higher concentration of thermal vacancies at high temperature. The trend that the vacancy migration enthalpy is higher than the vacancy formation enthalpy is oppositeto that usually seen in pure metals and alloys. Therefore, such peculiar thermal vacancy behavior leads to a high concentration of supersaturated vacancies and the mesostructure changes in the rapidly solidified FeAl ribbons. On the other hand, TiCo is also of the B2-type, Berthollide compounds. Takasugi and Izumi [4] studied defect structures in Co-rich TiCo phase and proposed that structural defects with the deviation from stoichiometry are the anti-structure type in this compositional range. Unfortunately, thermally activated defect behavior is still unknown for TiCo at present. If it has similarities to that of FeAl, mesostructure changes would be also possible in TiCo using supersaturated vacancies. Therefore, the purpose of this study is to investigate the effect of heat treatments (aging and annealing) on microstructure of rapidly solidified TiCo ribbons in order to clarify the existence of a high concentration of supersaturated vacancies.

EXPERIMENTAL DETAILS

A TiCo ingot having the nominally stoichiometric composition was made by Ar arc-melting from commercially available high purity Ti and Co. The ingot was re-melted by arc-melting and drop-cast into a rod-shape Cu mold of about 6mm in diameter for rapid solidification. Rapidly solidified ribbons were produced by a twin-roll melt-spinning method without a nozzle to avoid contaminating the molten TiCo with a nozzle. The gap between two rolls was 40μm, the rotation speed was 3600rpm, and the atmosphere was a vacuum of about 10^{-4}Pa. Chemically analyzed compositions of the arc-melted ingot and the rapidly solidified ribbons are tabulated in Table I. Some of the obtained ribbons were aged at 200°C for 100h or annealed at 700 or 900°C for 24h

Table I. Chemical compositions of the arc-melted ingot and as-spun ribbons. (mol%)

	Ti	Co	Fe	C	N	O	H
Ingot	50.06	49.80	0.01	0.01	0.006	0.09	0.02
Ribbons	49.79	49.93	0.02	*	0.005	0.1	0.15

*not analyzed yet

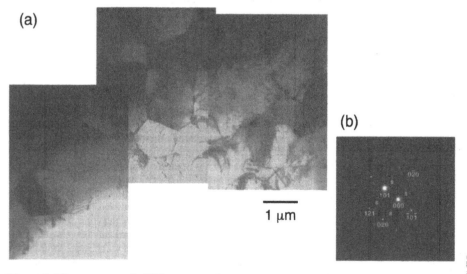

Figure 1. Microstructure of a TiCo as-spun ribbon. (a) A bright field image. (b) Electron diffraction pattern along the $[\bar{1}01]$ zone axis.

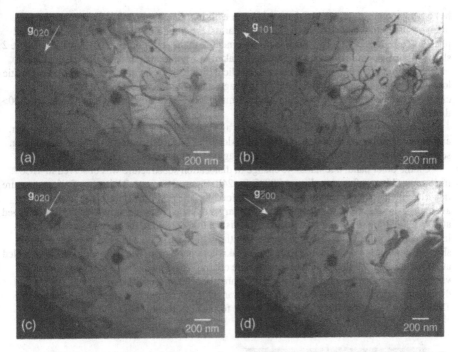

Figure 2. Bright field images of dislocation substructure in TiCo as-spun ribbons. (a) and (b) $B \approx$ [$\bar{1}$01], (c) and (d) $B \approx$ [001].

in a vacuum of better than 5×10^{-4}Pa. Thin foils for transmission electron microscopy (TEM) were prepared by twin-jet electropolishing using a Struers Tenupol-5 in a solution of 20vol%HClO$_4$ + 80vol%CH$_3$COOH at 15kV at room temperature. TEM was carried out using a JEM-2000EX II operating at 200kV for microstructure observation and a JEM-3000F operating at 300kV for energy dispersive X-ray spectroscopy (EDX). Differential scanning calorimetry (DSC) was performed for the rapidly solidified ribbons using a Netzsch DSC404 at heating and cooling rates of 30K·min^{-1} in an atmosphere of the Ar gas flow of 75ml·min^{-1}.

RESULTS AND DISCUSSION

Figure 1(a) shows a bright field (BF) image of microstructure in a TiCo as-spun ribbon. It is found that an equiaxed grain structure is developed in the ribbons immediately after melt-spinning. Electron diffraction patterns taken along the [001] or [011] zone axis (figure 1(b)) indicate that all grains have the B2-ordered structure, although it is very difficult to identify the B2 ordering in TiCo by X-ray diffractometry (XRD) because of the small difference in the atomic scattering factors between Ti and Co. There is no evidence of the existence of anti-phase domains (APDs) or meso-scaled voids in the as-spun ribbons. As shown in figure 1, some grains are dislocation-free, but others contain an amount of dislocations. In addition, fine dispersions of a second phase are often observed in grains and along grain boundaries. From the TEM-EDX analyses, the second phase is considered as Ti_2Co, which is in agreement with the result obtained by Vitta [5]. Figure 2 shows BF images of dislocation substructures taken in a grain of an as-spun ribbon at a higher magnification. In this grain, some curved or helical dislocations are observed as well as prismatic loops and spherical dispersions. From these images taken with four different g vectors, it is clarified that the observed dislocations and prismatic loops have Burgers vectors parallel to <100> directions.

Microstructure in a TiCo melt-spun ribbon aged at 200°C for 100h was observed. However, any marked grain growth or development of dislocation substructure and dispersion distribution seems not to occur even through the long term aging treatment.

On the other hand, a remarkable change in dislocation substructure was observed in TiCo melt-spun ribbons after annealing at 700°C for 24h. Figure 3 shows a BF image of microstructure in a TiCo melt-spun ribbon annealed at 700°C for 24h. The grain structure is equiaxed and its average grain size seems to become somewhat larger than that of as-spun ribbons and ribbons aged at 200°C for 100h. It should be noted that the dislocation density in the ribbon is considerably increased by the heat treatment compared with those of the as-spun and aged ribbons. Figure 4 shows BF images of dislocation substructures taken in a grain of a TiCo melt-spun ribbon annealed at 700°C for 24h at a higher magnification. The contrast experiments with several different g vectors indicate that the observed dislocations are also <100>-type. The significant increase in dislocation density after annealing would be strange, if it occurred in deformation substructures. Thus, it is believed that those dislocations were formed through the

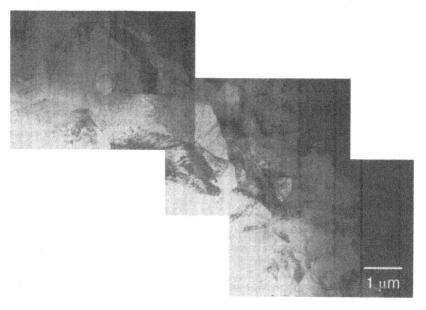

Figure 3. Bright field image of TiCo melt-spun ribbons annealed at 700°C for 24.

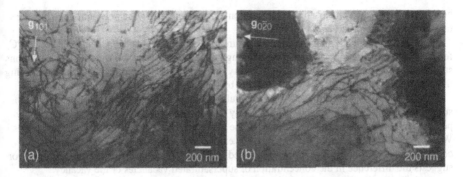

Figure 4. Bright field images of dislocation substructure in TiCo melt-spun ribbons annealed at 700°C for 24h. $\mathbf{B} \approx [\bar{1}01]$.

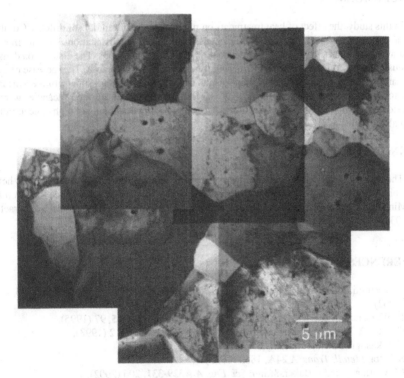

Figure 5. Bright field image of a TiCo melt-spun ribbon annealed at 900°C for 24h.

heat treatment by the condensation or absorption of supersaturated vacancies introduced by the rapid solidification process.

Figure 5 shows a BF image of microstructure in a TiCo melt-spun ribbon annealed at 900°C for 24. Grain growth obviously proceeded through this heat treatment, whereas Ti_2Co dispersions still remain. Furthermore, the dislocation density in the ribbons is definitely decreased, indicating that the dislocation substructure was recovered by the heat treatment.

Therefore, the results obtained above undoubtedly demonstrate that a large amount of supersaturated vacancies are retained in TiCo by the rapid solidification process. However, any peaks were not detectable in DSC profiles of the TiCo melt-spun ribbons in the same manner as Wittmann and Baker reported before [6], while exothermic peaks due to excess vacancy condensation were observed in rapidly quenched FeAl [1, 6, 7]. This difference in DSC behavior suggests the difference in the concentration of supersaturated vacancies or the vacancy condensation process between FeAl and TiCo. Further studies are required to understand thermal and supersaturated vacancy behavior in B2-type TiCo.

CONCLUSIONS

In this study, the effect of heat treatments on microstructure of rapidly solidified TiCo ribbons was investigated by TEM. In the as-spun ribbons, some grains are dislocation-free but others contain an amount of curved or helical dislocations and prismatic loops. The dislocation density is obviously increased in the ribbons by the heat treatment at 700°C for 24h. The increase of the dislocation density in the annealed ribbons results from the absorption of supersaturated vacancies. Therefore, the obtained results indicate that a large amount of supersaturated vacancies were retained in TiCo by the rapid solidification process, suggesting that there is a high concentration of thermal vacancies in TiCo at high temperature.

ACKNOWLEDGEMENTS

The authors thank T. Matsuzaki of Dept. Nanomechanics, Tohoku University, for his help in the rapid solidification. This work was supported by the Grant-in-Aid for Science Research from the Ministry of Education, Culture, Sports, Science and Technology of Japan under Contract No. 16360339.

REFERENCES

1. K. Yoshimi, S. Hanada, T. Haraguchi, H. Kato, T. Itoi and A. Inoue, *Mater. Trans.* **43**, 2897 (2002).
2. R. Würschum, C. Grupp and H.-E. Schaefer, *Phys. Rev. Lett.* **75**, 97 (1995).
3. C.L. Fu, Y.Y. Ye, M.H. Yoo and K.M. Ho, *Phys. Rev. B* **48**, 6712 (1993).
4. T. Takasugi and O. Izumi, *Phys. Stat. Sol. (a)* **102**, 697 (1987).
5. S. Vitta, *Metall. Trans. A* **24A**, 1869 (1993).
6. M. Wittmann and I. Baker, *Mater. Sci. Eng. A* **A329-331**, 206 (2002).
7. S. Zaroual, O. Sassi, J. Aride, J. Bernardini and G. Moya, *Mater. Sci. Eng. A* **A279**, 282 (2000).

Mater. Res. Soc. Symp. Proc. Vol. 842 © 2005 Materials Research Society   S4.4

In-situ Observation of Surface Relief Formation and Disappearance during Order-Disorder Transition of Equi-atomic CuAu alloy using Laser Scanning Confocal Microscopy

Seiji Miura[†], Hiroyuki Okuno*, Kenji Ohkubo and Tetsuo Mohri
Division of Materials Science and Engineering, Graduate School of Engineering,
Hokkaido University, Kita-13, Nishi-8, Kita-ku, Sapporo 060-8628, JAPAN
*) Graduate Student, Division of Materials Science and Engineering,
Graduate School of Engineering, Hokkaido University.
[†]) corresponding author. Tel. & Fax: +81-11-706-6347
E-mail address: miura@eng.hokudai.ac.jp (S. Miura)

ABSTRACT

In-situ observation of the formation and disappearance of the surface relief associated with the twinning during the order-disorder transitions among CuAu-I (L1$_0$), CuAu-II (PAP) and disordered fcc phases was conducted using Confocal Scanning Laser Microscopy equipped with a gold image furnace. The Retro effect was confirmed in poly-crystal samples, however no evidence was found in single-crystal samples. Also observed in poly-crystal samples are that the disordering temperature detected by the disappearing of relieves is different from grain to grain, and that grain boundary cracking alleviates the Retro effect. The observed phenomena were explained based on the crystallographic orientation relationship among grains investigated by FESEM/EBSD in terms of the elastic strain effect around grain boundaries induced by transition. It was confirmed that in each grain the surface relieves correspond to a set of two {011} planes having a <100> axis perpendicular to both planes in common. It was also found that the larger the average strain of two neighboring grains is, the lower the transition temperature. This observation was explained by the stress effect on the stability of a phase.

INTRODUCTION

Many studies have been dedicated to understand the nature of order-disorder transition in various alloy systems. In the first-order phase transition it has been accepted that the phase transition temperature goes down with increasing a cooling rate, which is explained in terms of the delay of nucleation. This is also expected for the case of heating, while for the order-disorder transition of equi-atomic CuAu it was found, by both DSC analysis and electrical conductivity measurements, that the transition temperature decreases with increasing the heating rate [1-5]. This phenomenon has been called "Retro effect". It has been known that there are three phases for equi-atomic CuAu alloy; high temperature disorder fcc, low temperature CuAu-I with L1$_0$ structure and CuAu-II with a periodic antiphase (PAP) structure in the middle [6]. By extending the Johnson-Mehl-Avrami model to competitive nucleation and growth of three phases, it was shown that the Retro effect can be explained qualitatively [7]. However, this study includes several unknown factors, which obscures the applicability to the general case.

Traditional macroscopic experimental technique such as the DSC (or DTA), electrical conductivity or thermal expansion measurements for the investigation of order-disorder transition have no spatial resolution and detected signals provide information only on homogeneous transition. However, some of the transitions accompany a mesoscopic phenomenon such as a twinning and the progress of such transition is essentially heterogeneous. It has been reported that the twinning during the transition in CuAu accompanies surface relief formation and its disappearance. The surface roughness formed by transition is reported to be of the order of several μm [8]. Therefore, in the present study a laser scanning confocal microscopy (LSCM) equipped with a gold image furnace is adopted. A monoclinic laser which is used for the observation by LSCM provides a high resolution image even at the temperature as high as 1500 °C. With this apparatus the surface change associated with transition was observed in each grain at various heating rates.

Both single- and poly-crystal CuAu samples were subjected to the in-situ observations using LSCM to detect the change of surface morphology during the order-disorder transition. FESEM/EBSD is also utilized to determine the crystallographic orientation of each grain to understand the effect of local strain generated by the transition on the difference of temperature of the disappearance of surface relief.

EXPERIMENTAL PROCEDURES

An alloy with an equi-atomic composition of 50 at.% Cu – 50 at.% Au was arc-melted under an argon atmosphere. After the homogenization at 1123 K for 72 h, poly-crystal samples were quenched to retain the high temperature disorder phase. Single crystal was grown in the Bridgeman-type furnace with high purity graphite crucible sealed in an evacuated silica tube, followed by the homogenization at 1123 K for 120 h, then quenched into water. Therefore, surface relief is formed by the ordering of the quenched disordered alloy during heating prior to the order-disorder transition, and the start of disordering can be detected as the disappearing of surface relief.

The determination of the crystallographic orientation of each grain in poly-crystal sample was performed by FESEM-EBSD analysis with TexSem Laboratories-OIM software. For the in-situ observation using LSCM, both the samples were set in the Al_2O_3 pan in the gold image furnace. The observed surface change was recorded by a VCR system. The heating rate ranges from 2 to 20 K/min, and cooling rate is 10 K/min. Maximun and minimum temperatures for the heating and cooling patterns are 450°C and 300°C, respectively.

DTA and TMA analysis were also performed on the poly-crystal samples with the dimension of about 2 x 2 x 2 mm and 3 x 3 x 15 mm, respectively. All the samples were subjected to the same heating and cooling patterns.

RESULTS AND DISCUSSION

Figure 1 shows the results of the in-situ observation of the relief on a single-crystal, and the disappearance of the relief in a poly-crystal during the heating. It is noteworthy that in the single-crystal sample the directions of the traces of relief are almost the same in a large area. In the poly-crystal sample traces of relief observed are different from grain to grain. Crack formation which was reported previously [9] is also confirmed in some of poly-crystal samples.

Their cyclic opening and closing observed during heating and cooling strongly suggests the introduction of internal stress across the transition temperature.

Shown in Figure 2 are the temperatures at which the surface relief disappeared, as observed by LSCM, and the transition temperatures determined by TMA and DTA as a function of the heating rate for both poly- and single-crystal samples. One sees that the Retro effect was only found in the poly-crystal sample, and single-crystal samples show ordinary superheating behavior. Although the absolute values of measured temperatures are different among the measurement apparatus and the results in different samples provide different heating rate dependence, some of the results of DTA and TMA have similar dependences on the heating rate. The scattering of the transition temperatures in the poly-crystal curve indicates the difference of the disappearing temperatures of relieves at each grain. It was also found that the area near a crack tends to have a relatively high disappearing temperature, i.e., the Retro effect is weak in such areas. This suggests the effect of internal stress.

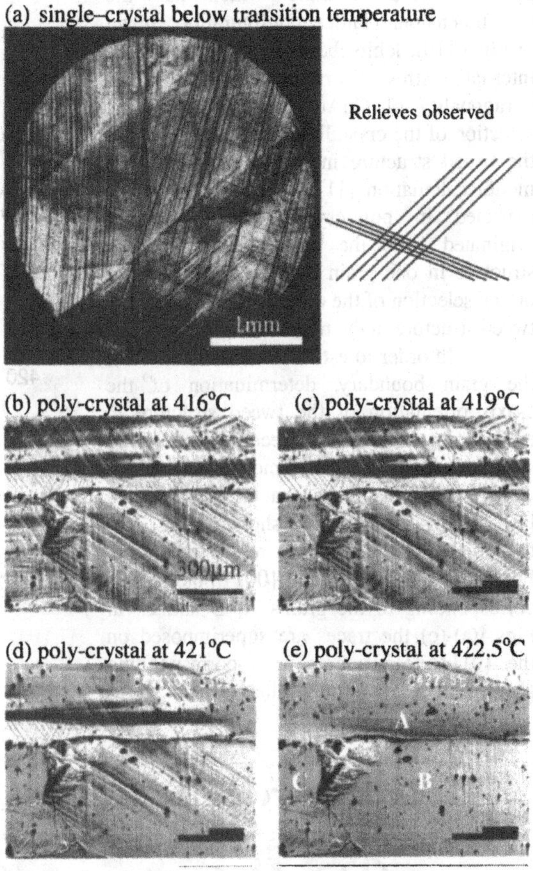

Shown in Figure 3 is a twin structure proposed by Hirabayashi and Weissmann which alleviates the tetragonal strains by fcc-$L1_0$ transition [10]. They suggest that the tetragonal strain in

(a) single–crystal below transition temperature

Relieves observed

(b) poly-crystal at 416°C

(c) poly-crystal at 419°C

(d) poly-crystal at 421°C

(e) poly-crystal at 422.5°C

Figure 1   Results of the LSCM in-situ observation of the relieves of (a)single- and (b)-(e)poly-crystal samples. A, B and C are the grains of concern. The disappearing of relief during the heating is clearly observed in the poly-crystal sample.

the tweed structure is smaller than that of a single $L1_0$-CuAu-I due to a regular alignment of c-axis in CuAu-I in which twin interfaces are assumed to be parallel to {011}. However, still the two kinds of strains are not relaxed in a poly-crystal: one originates from the difference of the unit cell volume between ordered and disordered structures, and the other from the tetragonality of the tweed structure which has the two-dimensional lattice configuration shown in Fig.3 [10]. Ichitsubo et al. showed that the internal stress originated from the tetragonality of CuAu-I is crucial for the selection of the crystallographic orientation of the tweed structure in a single crystal by a model calculation [11]. Therefore, it can be expected in poly-crystal that the strain originated from the existence of a tweed structure in one grain may have some effect on the selection of the c-axis orientation of the tweed structure in the neighboring grains.

In order to estimate the strain around the grain boundary, determination of the c-axis orientation of the tweed structure in each grain is required because the tweed structure has a two-dimensional lattice configuration resulting in a tetragonal strain [10]. In most of the grains shown in Fig. 1(b) two kinds of the trace of relief are found. Fig.4 shows the {011} and {001} pole figures corresponding to the grains A, B and C. In Figs.4(a)-(c) the traces are superimposed on the {011} pole figures of a corresponding grain, and it was found that all the traces

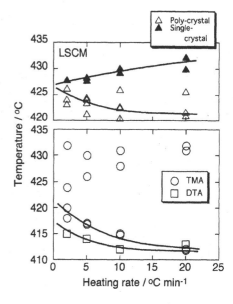

Figure 2 The temperatures at which disappearing of relieves finished or the transition temperature by TMA and DTA are plotted as a function of the heating rate for poly- and single-crystal samples.

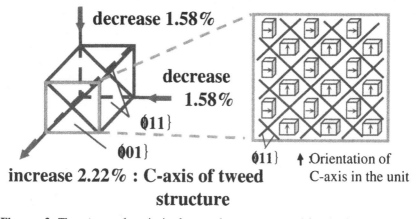

Figure 3 The tetragonal strain in the tweed structure caused by regular alternation of c-axis direction in CuAu-I. Twin interfaces are parallel to {011}.

coincide with two of six different {011} planes. These two {011} planes in each grain have plane normals perpendicular to a common <001> as shown in Figs.4 (d)-(f). The identified <001> is nothing but the c-axis direction of the tweed structure shown in Fig. 3. It is obvious that the crystallographic orientation of tweed structure in grain A is close to that of grain C.

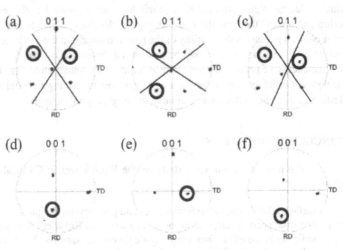

Figure 4 The {011} pole figures ((a)-(c)) and {001} pole figures ((d)-(f)) corresponding to the grains A, B and C, respectively. The traces are superimposed as lines on the {011} pole figures.

Table 1 Strains and the disappearing temperatures of relieves during heating at observed grain boundaries in Figs.1(b)-(e).

Grain boundary	Average strain ε_{ij} / x 10^{-2}	Difference of strains $\varepsilon_{ij} - \bar{\varepsilon}_{ij}$ / x 10^{-2}	Disappearing temperature
A-B	$\begin{pmatrix} -0.69 & -0.90 & -0.26 \\ -0.90 & -0.50 & -0.31 \\ -0.26 & -0.31 & 0.23 \end{pmatrix}$	$\begin{pmatrix} -0.37 & -0.90 & -0.26 \\ -0.90 & -0.18 & -0.31 \\ -0.26 & -0.31 & 0.55 \end{pmatrix}$	422.5 °C
A-C	$\begin{pmatrix} -1.45 & 0.12 & -0.59 \\ 0.12 & -1.49 & -0.64 \\ -0.59 & -0.64 & 1.98 \end{pmatrix}$	$\begin{pmatrix} -1.13 & 0.12 & -0.59 \\ 0.12 & -1.17 & -0.64 \\ -0.59 & -0.64 & 2.30 \end{pmatrix}$	420.0 °C

By taking the crystallographic orientation of tweed structures in each grain into account, the strains around the grain boundaries are estimated and listed in Table 1. The average strain of neighboring grains originates from both the difference of volumes between the unit cells of ordered and disordered phases and the difference of the c-axis of tweed structures. By subtracting dilatation terms caused by the volume change during the order-disorder transition ($\bar{\varepsilon}_{ij} = -0.32 \times 10^{-2} (i = j)$, $\bar{\varepsilon}_{ij} = 0 (i \neq j)$), the latter contribution, i.e., the elastic strain of the area around the grain embedded in an isotropically strained medium of CuAu, can be estimated

459

as the third column of the Table 1. The Retro-effect is manifested at the grain boundary A-C which holds the higher strain. Although this large strain might be relieved by slip deformation within and around the area, these results suggest that the constraints caused by the presence of grain boundary changes the stability of the phase, resulting in the lowering of the order-disorder transition temperature, which may be understood through Clausius-Clapeylon contribution on the free energy of phases. On the other hand, as was pointed out in Fig. 1(a), the directions of the traces of relief are almost the same in a large area in the single-crystal sample. Therefore, no elastic strain remains after the transition in the single crystal sample.

CONCLUDING REMARKS

The present experimental study on the Retro effect in CuAu alloy is summarized as follows.

(1) Surface relief is formed in both single- and poly-crystal samples.
(2) The disappearing temperature of relief shows the Retro effect in poly-crystal samples, while no Retro effect was found in single crystal samples. Weak Retro effect near the grain boundary cracks in poly-crystal samples was also found.
(3) Strains introduced by the presence of the grain boundary associated with the transition are estimated by analyzing the crystallographic orientation of each grain. A tendency was found that the higher constraints result in the larger Retro effect. This may be understood through Clausius-Clapeylon contribution.

ACKNOWLEDGMENTS

This research was supported by Grant-in-Aid from the Ministry of Education, Science, Sports and Culture, Japan, No. 13305043, and Grant for Experimental Apparatus from Hokkaido University.

REFERENCES

1. B.Sprušl, V.Šíma and B.Chalupa, Z. Metallk., 84, 118-123, (1993).
2. B. Sprušl, and W. Pfeiler, Intermetallics, 5, 501-505, (1997).
3. B.Sprušl and W. Pfeiler, Intermetallics, 8, 81-83, (2000).
4. B. Sprušl and B. Chalupa, Intermetallics, 8, 831-833, (2000).
5. B. Sprušl and B. Chalupa, Intermetallics, 10, 58-61, (2002).
6. J. Boneaux and M. Guymont, Intermetallics, 7, 797-805, (1999).
7. H. Uzawa: Master Thesis, Hokkaido University, (2000).
8. M. Ohta, T. Shiraishi, R. Ouchida, M. Nakagawa and S. Matsuya, J. Alloys and Comp., 265, 240-248, (1998).
9. P. Masek, F. Chmelik, V. Šíma, A. Brinck and H. Neuhauser, Acta mater., 47, 427-434, (1999).
10. M. Hirabayashi and S. Weissmann, Acta Met., 10, 25-36, (1962).
11. T. Ichitsubo, K. Tanaka, M. Koiwa and Y. Yamazaki, Physical Review B, 62, 5435-5441, (2000).

Mater. Res. Soc. Symp. Proc. Vol. 842 © 2005 Materials Research Society   S5.37

Phase Equilibria and Lattice Parameters of Fe₂Nb Laves Phase in Fe-Ni-Nb Ternary System at Elevated Temperatures

Masao Takeyama, Nobuyuki Gomi[1], Sumio Morita[1], Takashi Matsuo
Department of Metallurgy and Ceramics Science, Tokyo Institute of Technology
2-12-1 Ookayama, Meguro-ku, Tokyo 152-8552, JAPAN; [1]Graduate Student

ABSTRACT

Phase equilibria in Fe-Ni-Nb ternary system at elevated temperatures have been examined, in order to identify the two-phase region of γ-Fe (austenite) and ε-Fe₂Nb (C14). The ε single phase region exists in the range of 27.5 to 35.5 at.% Nb in the Fe-Nb binary system, and it extends toward the equi-niobium concentration direction up to 44 at.% Ni in the ternary system at 1473 K, indicating that more than half of the Fe atoms in Fe₂Nb can be replaced with Ni. Thus, the $\gamma+\varepsilon$ two-phase region exists extensively, and the solubility of Nb in γ phase increases from 1.5 to 6.0 at.% with increase in Ni content. The lattice parameters of a and c in the C14 Laves phase decrease with increasing Ni content. The change in a axis is in good agreement with calculation based on Vegard's law, whereas that of c axis is much larger than the calculated value. The result suggests that atomic size effect is responsible for a-axis change and the binding energy is dominant factor for the c-axis change. To extend these findings to development of new class of austenitic steels strengthened by Laves phase, an attempt has been made to control the c/a ratio by alloying. The addition of Cr is effective to make the c/a ratio close to the cubic symmetry value (1.633).

INTRODUCTION

In electric power generation systems, increase in steam pressure and temperature is necessary for high energy efficiency. Currently ferritic heat resistant steels with 9~12 wt.% Cr have been used for ultra super critical (USC) power plants at around 873 K [1-3]. However, beyond USC, it is necessary to develop austenitic heat-resistant steels capable to use above 973 K for steam turbine component. Takeyama [4] suggests that A₂B TCP (Topologically close-packed) compounds are potential strengthener superior to carbides because of their high thermal phase stability at elevated temperatures. Fe₂Nb (ε) Laves phase with hexagonal C14 structure is the most attractive because of the high congruent melting temperature (1914 K) and wide composition homogeneity region. In addition, among the transition metal Laves phase Fe₂M, this Laves phase is in equilibrium with fcc γ-Fe phase above 1228 K, so that there is a possibility to extend the $\gamma+\varepsilon$ two-phase field by the addition of γ-former elements. In general, Laves phase has been considered as harmful phase for ferritic heat-resistant steels because it precipitates massively at grain boundaries and deteriorates strength and toughness. However, no systematic study on the effect of Laves phase on precipitation kinetics as well as morphology precipitated in austenitic steels.

In this study, thus, phase equilibria in Fe-Ni-Nb ternary system at elevated temperatures have been examined. Since we found the extended $\gamma+\varepsilon$ two-phase region, we have studied the change in lattice parameters of the Laves phase with Ni content as well as its precipitation morphology, and briefly discussed the morphology control of the Laves phase.

EXPERIMENTAL PROCEDURES

The Fe-Nb binary and ternary alloys in composition ranges of Fe-(15-38) Nb-(10-40%) Ni (at.%) were used for phase diagram study. In addition, the three quaternary alloys of Fe-2Nb-(25, 30, 35)Ni-20Cr were also prepared. These alloys were prepared from 2N9 iron, 3N niobium, 3N7 nickel and 4N chromium as 35 g button ingot by arc-melting in argon atmosphere with a non-consumable tungsten electrode. The ingot was cut to pieces with a size of 10 x 8 x 6 mm^3, and they were sealed off in silica capsules under argon back-filled after evacuating to 2 x 10^{-3} Pa, followed by equilibration treatment at 1473 K for 72-240 h. The quaternary alloys were homogenized at 1473 K/240 h, followed by aging at 1073 K/1200 h. Microstructure was examined by scanning electron microscope (SEM). Compositions of the phases present in the equilibrated samples were analyzed by an electron probe microanalyzer (EPMA) equipped with a wavelength dispersive spectrometer under an operating condition of 20 kV and 2 x 10^{-8} A. The phase identification and lattice parameter measurement were done by powder X-ray diffraction (XRD). The lattice parameters of precipitated Laves phase in the quaternary alloys were measured by transmission electron microscope (TEM). TEM discs with 0.15 mm in thickness and 3 mm in diameter were machined, and then mechanically polished, followed by twin-jet electropolishing in an electrolyte of ethanol with 20 vol.% perchloric acid at 253 K.

RESULTS AND DISCUSSION

Phase equilibria in Fe-Ni-Nb ternary system (1473 K)

Figure 1 shows back scattered electron images (BEI) of some of the alloys equilibrated at 1473 K. The phases present and their analyzed compositions of all alloys are summarized in Table I. The binary Fe-33.3Nb exhibits ε-Fe$_2$Nb single phase microstructure (Figure 1 (a)). Fe-33.3Nb-20Ni also shows ε single phase microstructure, although a very few amount of μ-Fe$_7$Nb$_6$ phase are observed at the grain boundaries (Figure 1 (b)). In contrast, the binary Fe-15Nb consists of ε (bright) and γ-Fe (dark) phases (Figure 1 (c)). Note that observed dark phase at room temperature is α-Fe with very fine lath morphology due to γ→α martensitic transformation. The cracks observed in ε phase are also caused by the transformation. Fe-15Nb-40Ni also consists of the two phases of ε and γ with almost equal volume fraction. However, unlike the Fe-15Nb, the γ phase is stable at room temperature, and no cracks are observed within the ε phase (Figure 1 (d)).

Figure 2 shows an isothermal section of Fe-Ni-Nb ternary system at 1473 K. In the binary system the ε single phase region exists in the range of 27.5 to 35.5% Nb, extending about 6% in Fe-rich side and only 2% in Nb-rich side from the stoichiometry [5, 6]. This region extensively penetrates toward the equi-niobium concentration direction up to 44% Ni, and widely in equilibrium with γ-(Fe, Ni) phase. With increasing Ni content, the Nb concentration of the ε phase in equilibrium with γ phase slightly decreases to 26% and the solubility of Nb in γ phase increases from 1.5 to 6.0%. The extension of ε single phase region in the ternary system suggests that the more than half of the Fe atoms in Fe$_2$Nb can be replaced by Ni atoms.

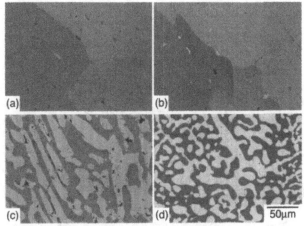

Figure. 1 BEIs of Fe-Nb binary and Fe-Nb-Ni ternary alloys equilibrated at 1473K/240h: (a) 33.3Nb, (b) 33.3Nb-20Ni, (c) 15Nb, (d) 15Nb-40Ni.

Table I EPMA analysis of the phases present in the alloys equilibrated at 1473K/240h.

Alloy	Phases present	Phase composition (at.%)			
		Nb	Ni	Cr	Fe
Fe-15Nb	ε	27.5	-	-	72.5
	γ*	1.5	-	-	98.5
Fe-29Nb	ε	28.8	-	-	71.2
Fe-33.3Nb	ε	32.5	-	-	67.5
	μ	50.0	-	-	50.0
Fe-38Nb	ε	35.5	-	-	64.5
	μ	48.5	-	-	51.5
Fe-15Nb-20Ni	ε	26.7	15.9	-	57.3
	γ	2.6	23.8	-	73.6
Fe-15Nb-40Ni	ε	25.7	33.1	-	41.2
	γ	6.0	44.2	-	49.8
Fe-25Nb-60Ni	ε	26.1	44.5	-	29.4
	δ-Ni₃Nb	24.3	67.2	-	8.5
Fe-2Nb-25Ni-20Cr	ε	*	*	*	*
	γ	0.47	24.9	20.9	53.7
Fe-2Nb-35Ni-20Cr	ε	*	*	*	*
	γ	1.30	34.6	20.7	43.4

*The phase is γ-Fe at 1473 K, but α-Fe at room temperature. *: not analyzed.

Note: the quaternary alloys were equilibrated at 1073 K/240 h.

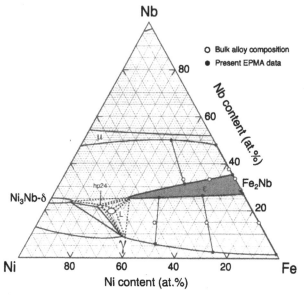

Figure 2. Isothermal section of Fe-Ni-Nb ternary system at 1473 K.

Change in lattice parameters of ε-Fe₂Nb

The effects of Ni content on lattice parameters a and c for stoichiometric ε-Fe₂Nb and Fe-rich ε phase in equilibrium with γ phase are shown in Figure 3, together with their calculated values (thin lines) based on Vegard's law. In the binary stoichiometric Fe₂Nb, a and c are 4.838 and 7.885 Å, respectively. These values become small to 4.810 and 7.852 Å in the Fe-rich ε phase. This is because the excess Fe atoms with radius (r_{Fe}=1.274 Å) smaller than that of Nb atom (r_{Nb}=1.468 Å), would occupy Nb sublattice sites. With increasing Ni content, the a decreases gently and monotonously regardless of Fe content. On the other hand, the c in stoichiometric ε phase remains almost unchanged up to 10% Ni but it decreases sharply beyond that. In Fe-rich ε phase, however, the c monotonously but sharply decreases with increasing Ni content. The change in lattice parameter of a with Ni content is in good agreement with the calculated value based on hard

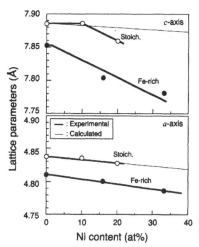

Figure 3. Change in lattice parameters a and c of ε-Fe₂Nb Laves phase with Ni content (thin line: calculation).

464

sphere model [7, 8, 9], assuming that all Ni atoms with radius (r_{Ni}=1.246 Å) smaller than that of Fe atom occupy Fe sublattice site in the stoichiometric Fe$_2$Nb. This indicates that atomic size effect is responsible for a-axis change of the C14 Laves phase. However, the change in c becomes much larger than that expected from the size effect when Ni content exceeds more than 10%. In should be noted that the basal planes in C14 structure consist of the same kind of atoms with their arrangements of one *kagome*-net and three 3^6-nets, and the different kind of atom pairs exists only along their stacking direction. In addition, the binding energy between Ni-Nb (-135 kJ/mol) is larger than that of Fe-Nb (-70 kJ/mol) [10]. All these facts suggest that the c-axis change is somehow related to the binding energy between the atoms occupying the different sublattice sites.

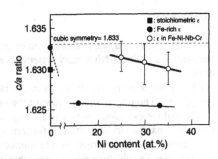

Figure 4. Change in lattice parameter ratio of Laves phase in equilibrium γ phase with Ni and Cr content, together with stoichiometric Laves phase.

Figure 4 shows change in c/a ratio of the ε phase with Ni content. The ratio of the stoichiometric Fe$_2$Nb is 1.630, and it increases close to the cubic symmetry value (1.633) in Fe-rich ε phase. The value drops sharply to 1.625 by 16 at.% Ni in solution, although it remains almost unchanged with further Ni addition.

In considering the Laves phase as a strengthener for austenitic heat-resistant steels, the c/a ratio close to the cubic symmetry should be better for homogeneous precipitation since the elastic strain field caused by the nucleation in the matrix becomes more isotropic. Based on the results on lattice parameter change, addition of Cr would be effective in increasing the c/a ratio of the ε-(Fe, Ni)$_2$Nb, since Cr atom tends to occupy Fe sublattice site [11], and its binding energy with Nb (-32 kJ/mol) is smaller than those of Fe-Nb [10], although its atomic size (1.282 Å) is slightly larger than that of Fe. Then, based on the present phase diagram together with our previous studies [5,6], an attempt has been made to prepare the Fe-Nb-Ni-Cr quaternary alloys in order to examine the morphology as well as c/a ratio of the ε phase precipitated within the γ

matrix. As expected, the c/a ratio of the ε phase measured by selected area diffraction patterns obviously increases by Cr addition and it becomes more closer to cubic symmetry value with reducing Ni content, as shown in Fig. 4. A backscattered electron image of Fe-20Cr-2Nb-25Ni aged at 1073 K/240 h shown in Figure 5 clearly demonstrates very fine ε particles homogeneously precipitated in the matrix. Although the ε phase is too fine to analyze the composition by EPMA (Table I), we recognized that the morphology becomes more fine and uniform in the alloy with lower Ni content. The detailed study on the morphology control of the Laves phase is now under progress.

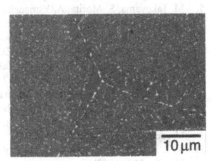

Figure 5. BEI of Fe-2Nb-25Ni-20Cr aged at 1073 K/240 h.

SUMMARY

Phase equilibria and lattice parameters of Fe_2Nb (ε) with C14 structure at elevated temperatures in Fe-Ni-Nb ternary system have been examined, and the following conclusions are drawn:

(1) The ε single phase region exists in the range of 27.5 to 35.5 at.% Nb in the binary Fe-Nb system. The region extends toward the equi-niobium concentration direction up to 44 at.% Ni and the extended $\gamma + \varepsilon$ two-phase region exists in Fe-Nb-Ni ternary system.

(2) Lattice parameters of a and c in the ternary ε phase decrease with increasing Ni content. The change in a is governed by the atomic size factor since Ni with radius smaller than Fe occupies Fe sublattice site, whereas that in c is due to the binding energy of Ni-Nb atom pair stronger than that of Fe-Nb one.

(3) The c/a ratio of the ε phase can be controlled through alloying, and Cr addition is effective in increasing the value close to cubic symmetry (1.633), making it possible to refine the morphology of ε phase precipitated in γ matrix. Thus, it is reasonable to conclude that the Laves phase is promising strengthener for to develop a new class of austenitic heat-resistant steels.

ACKNOWLEDGEMENT

This research was supported by the grant-in-aid (14205102) for Scientific Research from Ministry of Education, Culture, Sports, Science and Technology.

REFERENCES

1. F. Masuyama, Report of JSPS 123rd Committee on Heat-Resisting Materials and Alloys, **38**, 211 (1997).
2. M. Igarashi, A. Iseda and T. Kan, ib (id)., **44**, 205 (2003).
3. Y. Kadoya, ib (id)., **44**, 215 (2003).
4. M. Takeyama, ib (id)., **45**, 51 (2004).
5. M. Takeyama, S. Morita, A. Yamauchi, M. Yamanaka and T. Matsuo, Superalloys 718, 625, 706 and Various Derivatives, Eds. by E. A. Loria, TMS, 333 (2001).
6. M. Takeyama, H. Yokota, M. M. Ghanem and T. Matsuo, Journal of Materials Processing Technology (Proceedings of THERMEC'2000) **117**, 3 (2001). [CD-ROM], Session D7.
7. N. Gomi, S. Morita, T. Matsuo, M. Takeyama, Report of JSPS 123rd Committee on Heat-Resisting Materials and Alloys, **45**, 157 (2004).
8. D.J. Thoma and J. H. Perepezko, Journal of Alloys and Compounds, **224**, 330 (1995).
9. K. C. Chen, E. J. Peterson and D. J. Thoma, Intermetallics, **9**, 771 (2001).
10. A. K. Niessen, F. R. de Boer, R. Boom, P. F. de Chatel, W. C. M. Mattens and A. R. Miedema, CALPHAD, **7**, 51 (1983).
11. N. I. Kaloev et al., Russian Metallurgy, Translated from Izvestiya Academii Nauk SSSR, Metally, 202 (1988).

Mater. Res. Soc. Symp. Proc. Vol. 842 © 2005 Materials Research Society   S5.38

Sputtered Coatings Based on the Al₂Au Phase

Christian Mitterer, Helmut Lenhart, Paul H. Mayrhofer, Martin Kathrein[1]
Department of Physical Metallurgy and Materials Testing, University of Leoben, Franz-Josef-Strasse 18, A-8700 Leoben, Austria
[1]CERATIZIT Austria GmbH, A-6600 Reutte, Austria

ABSTRACT

Transition metal nitride-based, wear-resistant hard coatings on cutting tools and other substrates often lack distinct colorations allowing product differentiation and self-lubricating properties. In the present work, the possibility of achieving these objectives for sputtered coatings based on the purple-red Al_2Au phase within the Al-Au system was investigated. Coatings were characterized with respect to morphology, chemical and phase composition, hardness, optical, oxidation and tribological properties. Al_2Au-containing coatings were deposited with dense, fine-grained structures yielding a hardness of 4 GPa and pink coloration. The coatings were stable up to about 850°C, where the onset of oxidation occurs. Low friction coefficients against alumina were achieved between 500 and 700°C. The concept of applying Al_2Au-containing coatings as a colored self-lubricating layer on top of a hard coated cemented carbide tool warrants further investigations.

INTRODUCTION

Hard coatings based on transition metal nitrides are widely applied to improve the lifetime and the performance of cutting and forming tools. Besides the functionality of the coating itself, provided by hardness, toughness, oxidation and wear resistance, there is also an increasing demand for aesthetic factors like attractive colorations of a cutting tool [1]. Fortunately, the widely applied titanium nitride TiN is characterized by a bright golden-yellow color making determination of used cutting edges of a cutting insert in poor lighting conditions of a machine shop easier. However, for other hard compounds attractive bright colors allowing product differentiation and to distinguish used and unused cutting edges are lacking.

Consequently, this study is focused on the development of a new type of colored coatings based on the intermetallic Al_2Au phase. Representing a Zintl phase with the cubic CaF_2 structure [2], Al_2Au shows a distinct reflectivity minimum at 545 nm [3,4], resulting in a purple-red color. Being relatively hard and brittle at room temperature, it shows plastic deformation and possibly self-lubrication at high temperatures [5]. With a melting temperature of 1060°C, Al_2Au is the thermally most stable intermetallic phase within the Al-Au phase diagram, whereas the other phases $AlAu_2$, Al_2Au_5 and $AlAu_4$ show melting temperatures between 525 and 625°C. No less-Au containing phases besides Al_2Au have been reported [6].

EXPERIMENTAL DETAILS

The coatings investigated in this work were deposited using an unbalanced d.c. magnetron sputtering system described in detail in [7]. Different Al/Au atomic ratios were adjusted by placing different numbers of Au pieces ($20\times20\times1$ mm³) on the erosion track of an Al target

(Ø150×6 mm). Coatings were deposited onto cemented carbide cutting inserts and Inconel 718 substrates (Ø5×1 mm) in Ar atmosphere at 0.2 Pa. The substrate temperature, bias voltage and sputter power was 300°C, -50 V and 380 W, respectively. Prior to deposition, all substrates used were metallographically grounded, polished and ultrasonically cleaned with ethylene and acetone. After target pre-cleaning and ion etching of the substrates within the deposition chamber, coatings in the thickness range between 0.1 and 7 μm were deposited.

Coating morphology was investigated using scanning electron microscopy (SEM) on cross-sections of coated cemented carbide inserts prepared by brittle fracturing in liquid nitrogen. The chemical composition of the coatings was determined by energy-dispersive X-ray analysis (EDX) using Au and Al elemental standards. X-ray diffraction (XRD) patterns were recorded using a Bragg-Brentano diffractometer and Cu-Kα radiation. High-temperature XRD (HT-XRD) investigations using a heating rate of 0.6 K·min^{-1} were performed in He atmosphere to obtain information on the thermal stability of the coatings.

Coating hardness was characterized using a Vickers microhardness tester at a load of 25 mN. Optical properties were determined using a spectroscopic rotating analyzer ellipsometer (angle of incidence, 75°; wavelength spectrum, 370...770 nm). Friction coefficients and their evolution at temperatures from 25°C up to 700°C in ambient air were studied using a ball-on-disc tribometer and alumina balls with Ø 6 mm. A load of 1 N and a sliding speed of 0.1 m·s^{-1} were used. Thermo-gravimetric analysis (TGA) was conducted for coatings deposited onto Inconel 718 substrates in order to characterize the oxidation behavior of the coatings deposited. Dynamical TGA runs were performed for thermal ramping from 400 to 1000°C at 5 K·min^{-1} in Ar/O$_2$ atmosphere (Ar/O$_2$ flow ratio of 2.5).

RESULTS AND DISCUSSION

Varying the number of Au pieces on the Al target, the Al/Au atomic ratio could be adjusted between 0.5 and 3.2. Figure 1 shows representative SEM micrographs of cross-sections of Au-Al coatings with different Al/Au atomic ratios. Low Al contents in the coating result in the formation of rough, open morphology (Figure 2(c)), whereas increasing Al contents yields a significant densification (Figures 2(a) and (b)). This could be a result of the different melting temperatures of Al and Au causing different thermal activation of the growth process for the individual condensing atoms [8], different diffusion rates of Al and Au leading to the formation of Kirkendall voids [9], the different phases formed in the coating (see below), or the difference between the atomic masses of the atoms involved where various collision behavior in the transport phase from the target to the substrate and during ion-assisted growth could be assumed [10].

Figure 1: SEM micrographs of the fracture cross-sections of Al-Au coated cemented carbide. (a) Al/Au = 3.2, (b) Al/Au = 1.6, (c) Al/Au = 0.6.

XRD results show that the Al$_2$Au phase is present in all coatings deposited within the Al/Au atomic ratio range between 0.5 and 3.2. For high Al or high Au contents, respectively, the presence of metallic Al (see Figure 2, Al/Au = 3.2) or Au (see Figure 2, Al/Au $\leq$ 1.0) is also detected. For Al/Au $\leq$ 1.0, the monoclinic AlAu phase is additionally formed. No evidence for the other phases present within the Al-Au system, i.e. AlAu$_2$, Al$_2$Au$_5$ and AlAu$_4$ can be found. It should be mentioned that the formation of a single-phase coating consisting of the Al$_2$Au phase results in the densest coating morphology as shown in Figure 1.

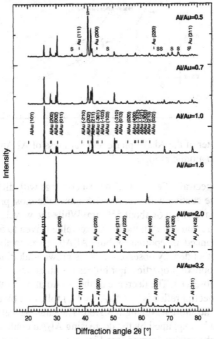

Figure 2: XRD patterns of Al-Au coated cemented carbide substrates for different Al/Au atomic ratios (s...substrate).

The lattice parameter of the Al$_2$Au phase is plotted in Figure 3 against the Al/Au atomic ratio. For the stoichiometric composition (i.e. Al/Au = 2), the measured lattice parameter is only slightly below the range of the literature values given in [2]. Increasing the Au content, i.e. decreasing Al/Au, results in an increasing lattice parameter with a maximum for Al/Au = 0.7 and decreasing tendency for the lower Al/Au ratio studied. The increasing lattice parameter could be interpreted by the incorporation of additional Au atoms in the Al$_2$Au phase where these Au atoms cause an expansion of the lattice. For Al/Au $\leq$ 0.7, metallic Au is present in the coating (see Figure 2), enabling stress relaxation and probably reducing the excess Au content within the Al$_2$Au phase. Analogously, for increasing Al/Au atomic ratios a decrease of the lattice parameter is found, which could be attributed to an Au deficiency within the Al$_2$Au lattice.

The measured coating hardness shown in Figure 3 is in good agreement with the content of the Al$_2$Au phase within the coatings. For Al/Au = 1...2, the hardness maximum with a value of about 4 GPa is obtained. Increasing the Al or the Au content beyond this range results in decreasing hardness, caused by the formation of metallic Al or Au phases, respectively, and by the open coating structure at low Al/Au atomic ratios. Most probably due to the formation of the intermetallic AlAu phase (see Figure 2), the range of coatings with maximum hardness is somewhat expanded to Al/Au atomic ratios below the value of stoichiometry of Al_2Au.

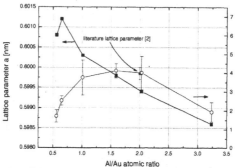

Figure 3: Lattice parameter of the Al_2Au phase and hardness of Al-Au coated cemented carbide as a function of the Al/Au atomic ratio.

Figure 4 shows the spectral reflectivity of Al-Au coatings with different Al/Au contents. The spectral reflectivity for the coating with Al/Au = 1.6 shows the most pronounced minimum, which is obtained for a wavelength of about 550 nm. While the wavelength position of this minimum is in excellent agreement with the literature value (given to 545 nm in refs. [3,4]); the minimum is not as pronounced as reported for bulk Al_2Au (given to about 10 % at 545 nm [3]). This is due to the high defect density observed in films grown under intense ion irradiation, which often deteriorates optical properties. For higher Al/Au atomic ratios, a flattening of the reflectivity minimum is observed. The decrease of the Al/Au atomic ratio results first in a general decrease of the spectral reflectivity (Al/Au = 1.0) followed by a significant increase of the reflectivity values in the short-wavelength region. The latter as well as the flattening of the reflectivity minimum can be explained by the decreasing Al_2Au content within the coating (see Figure 2) and by the rough surfaces formed (see Figure 1).

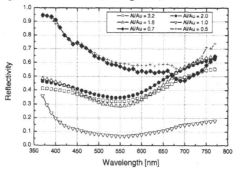

Figure 4: Spectral reflectivity of Al-Au coated cemented carbide in dependence of Al/Au atomic ratio.

The mass gain observed during dynamical TGA is considerably lower compared to both, transition metal nitride-based hard coatings measured using the same equipment [12] and the Inconel substrate. With increasing Al_2Au content, i.e. Al/Au atomic ratio approaching 2, the mass gain decreases (see Figure 5). This indicates that the metallic Al present in the coating deteriorates the oxidation resistance. In addition, the open structure of the coating (see Figure 1) is assumed to contribute to the lower oxidation resistance of high-Al containing coatings. XRD investigations after TGA up to 700°C only show the presence of the Al_2Au phase for all coatings; there is no evidence for crystalline oxide phases formed, which is in agreement with the negligible mass gain. After TGA up to 1000°C, no evidence for the Al_2Au phase is found by XRD; only peaks of α-Al_2O_3 and elemental Au are detected.

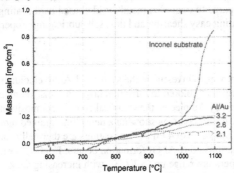

Figure 5: TGA curves for thermal ramping of Al-Au coated Inconel samples compared to the uncoated Inconel substrate.

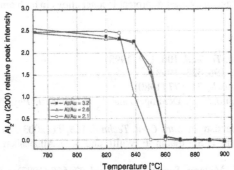

Figure 6: Relative intensities of the (200) XRD peak of the Al_2Au phase as a function of temperature for different Al/Au atomic ratios.

Figure 6 shows the dependence of the Al_2Au (200) relative peak intensity obtained during HT-XRD investigations for coatings deposited onto cemented carbide as a function of temperature. For all coatings investigated, the Al_2Au peaks vanish above 830...860°C, with a slightly higher thermal stability of the phase for coatings with higher Al/Au atomic ratios. It should be noted here that this temperature range is substantially lower than the melting temperature of Al_2Au ($T_m = 1060°C$ [2,3]), which points towards oxidation of the Al_2Au phase. Although using He atmosphere, α-Al_2O_3 is detected after HT-XRD investigations up to 1100°C, since the residual oxygen in the heating chamber is sufficient for oxidation. In addition to the WC substrate peaks, Co and Au peaks are found. Thus, we assume that above 830...860°C α-Al_2O_3 is slowly formed [13] at the top of the Al-Au coating as indicated by the continuous slight mass gain during TGA. This is confirmed by the increasing mass gain for the higher Al content indicating pronounced Al segregation and α-Al_2O_3 formation (see Figure 5). The remaining Au might be incorporated in Au-rich and Co-rich solid solutions formed in the Al-Co phase diagram.

The average friction coefficient obtained at room temperature against alumina balls is 0.41 ± 0.14. At 500 and 700°C, the friction coefficients are determined to 0.54 ± 0.26 and 0.46 ± 0.32, respectively. Although for both temperatures large scattering of the friction coefficients is observed indicating stick-slip effects, the average friction coefficients are considerably lower compared to transition metal nitride-based hard coatings tested at the same temperatures [14] indicating easy shearing and thus self-lubricating properties of the coating.

CONCLUSIONS

The aim of this work was to investigate the system Al-Au to elucidate the potential of Al_2Au-containing coatings as a colored low-friction layer on tools. Using co-sputtering from an Al target partially covered by Au pieces, the chemical coating composition could be adjusted to obtain single-phase pink Al_2Au coatings showing a dense fine-grained morphology. The coating hardness yields a maximum value of 4 GPa. The coatings are thermally stable up to 850°C, where the onset for oxidation and α-Al_2O_3 formation occurs. Friction coefficients of 0.4 to 0.5 are obtained for the temperature range up to 700°C, with increasing stick-slip effects at higher temperatures. It is concluded that Al_2Au coatings could be potential candidates for thin colored and self-lubricating layers deposited onto coated cemented carbide tools. In addition, these coatings may be advantageously used to protect Inconel alloys in high temperature applications.

REFERENCES

1   B. North, *Surf. Coat. Technol.* **106**, 129 (1998).
2   R. W. Cahn, *Nature* **396**, 523 (1998).
3   H. C. Shih, *Z. Metallkde.* **71**, 577 (1980).
4   L.-S. Hsu, G.-Y. Guo, J. D. Denlinger, and J. W. Allen, *J. Phys. Chem. Solids* **62**, 1047 (2001).
5   M. Eskner and R. Sandström, *Surf. Coat. Technol.* **165**, 71 (2003).
6   C. Xu, C. D. Breach, T. Shritharan, F. Wulff, and S. G. Mhaisalkar, *Thin Solid Films* **462-463**, 357 (2004).
7   P. Losbichler and C. Mitterer, *Surf. Coat. Technol.* **97**, 568 (1997).
8   J. A. Thornton, *J. Vac. Sci. Technol.* **11**, 666 (1974).
9   T. Raymond, *Semicond. Int.* **12**, 152 (1989).
10 C. Mitterer, *J. Solid State Chem.* **133**, 279 (1997).
11 H. S. de Waal, R. Pretorius, V. M. Prozesky, and C. L. Churms, *Nucl. Instrum. Meth.* **B130**, 722 (1997).
12 P. H. Mayrhofer, H. Willmann, and C. Mitterer, *Surf. Coat. Technol.* **146-147**, 222 (2001).
13 H. Svensson, J. Angenete, and K. Stiller, *Surf. Coat. Technol.* **177-178**, 152 (2004).
14 K. Kutschej, P. H. Mayrhofer, M. Kathrein, P. Polcik, R. Tessadri, and C. Mitterer, *Surf. Coat. Technol.*, in press.

Mater. Res. Soc. Symp. Proc. Vol. 842 © 2005 Materials Research Society

Solidification Processing and Fracture Behavior of RuAl-Based Alloys

Todd Reynolds and David Johnson
Materials Engineering, Purdue University
West Lafayette, IN 47907-1289

ABSTRACT

Alloys of RuAl-Ru were processed using various solidification methods, and the fracture behavior was examined. The fracture toughness values for RuAl-hcp(Ru,Mo) and RuAl-hcp(Ru,Cr) alloys ranged from 23 to 38 MPa√m, while the volume fraction of RuAl ranged from 22 to 56 percent. Increasing the volume fraction of RuAl resulted in a decrease in fracture toughness. The hcp solid solution was shown to be the more ductile phase with a fracture toughness approaching 68 MPa√m, while the B2 solid solution (RuAl) was found to have a fracture toughness less than 13 MPa√m. An alloy of Ru-7Al-38Cr (at.%) that consisted of a hcp matrix with RuAl precipitates had the highest room temperature toughness and the greatest hardness.

INTRODUCTION

Ruthenium aluminide with a CsCl (B2) crystal structure has been identified as a possible high temperature aluminide (T_{mp}>2000 °C) for structural applications [1]. Processing RuAl alloys from the melt is difficult due to the high vapor pressure of Al. The resulting Al loss results in the occurrence of an intergranular eutectic film (RuAl-Ru). Wolff et al. reported that such an intergranular eutectic film may enhance the fracture toughness by acting as a ductile compliant layer between the more brittle intermetallic phase [2-4]. Thus, eutectic alloys consisting of RuAl may potentially have good fracture toughness. However, due to the cost of Ru, producing ingots of the RuAl-Ru eutectic with a composition of Ru-24Al (at.%) [5] is prohibitively expensive.

To lower the cost of RuAl alloys, ruthenium must be substituted with a less expensive element. Additions of Mo to the RuAl-Ru eutectic have been investigated and good room temperature fracture toughness values were measured for these alloys [6]. However, the oxidation resistance of these Ru-Al-Mo alloys is poor [6,7]. Additions of chromium instead of molybdenum may improve oxidation resistance while retaining the good fracture toughness. In this paper, the fracture toughness of RuAl-Ru(Mo) and RuAl-Ru(Cr) alloys will be reported.

EXPERIMENTAL DETAILS

Alloys containing a combination of Ru, Al, Mo, and Cr with compositions as listed in Table I were arc-melted with a non-consumable tungsten electrode into approximately 5 g buttons. Buttons of same composition were then welded together to produce an ingot for directional solidification. These ingots were then processed using one of five different processing techniques. Ingot-1 was directionally arc-melted with a growth speed of 45 mm/h, while ingot-2 remained in an as arc-melted condition. Ingot-3 was directionally solidified at 500 mm/h by using a cold crucible Czochralski tri-arc melter, and ingot-4 was zone melted in an optical floating zone furnace at 20 mm/h. Lastly, ingots 5-7 were directionally solidified at 20 mm/h in a modified Bridgman furnace using MgO crucibles. A flowing Ar/5% H_2 atmosphere was used in all the solidification processes except the modified Bridgman furnace where a flowing 100% Ar atmosphere was used.

Ingot-7 was cut into two halves with a water cooled diamond wheel. One half (ingot-7(a)) was kept in the as-processed condition, and the other half (ingot-7(b)) was heat treated in air at 1100°C for 200 hours (Table I). A protective oxide scale formed on the heat treated half, and an oxide free microstructure was found within the heat treated ingot.

The microhardness of all the ingots was measured with a Vickers indentor using a load of 5 kg, and at least five readings were taken from each ingot. The average microhardness values are listed in Table I. Bend bars were machined with dimensions of 3 x 4 x 40 mm^3 from the ingots listed in Table I. Two bars were machined from the ingots 1, 7(a)&(b). The bars were notched with a liquid cooled diamond saw to a depth of approximately 0.8 mm. A fatigue notch was not initiated at the notch root; instead a micro-notching technique was used in which a razor blade with 1 μm diamond paste was oscillated with constant pressure to create a sharp notch. After micro-notching, the total notch length was approximately 1.5 mm and the crack tip radius was no greater than 20 μm. Four point bend tests with a 15 mm inner span and a 30 mm outer span were performed on a screw driven test frame using a displacement rate of 0.1 mm/min. Fracture toughness values were calculated using the K calibration for pure bending [8].

RESULTS
Ru-Al-Mo Alloys

Figure 1 shows the RuAl-Ru(Mo) eutectic microstructure for the three bars of Ru-20Al-25Mo (ingots 2-4). As listed in Table I, the three bars were each produced differently resulting in different microstructures. All three bars had primary RuAl dendrites surrounded by RuAl-Ru(Mo) eutectic, but the volume fraction of RuAl was different for each (Table I). The difference in the amount of RuAl was probably due to the difference in aluminum loss during solidification processing. For example, ingots that were produced using the cold crucible Czochralski tri-arc technique had the most aluminum loss due to the large amount of superheating and the relatively slow growth rate when compared to simple arc-melting.

Table I. Composition, hardness, fracture toughness, and volume fractions of total RuAl and primary RuAl dendrites are listed for Ingots 1-8. Ingot-7 was as-processed while ingot-8 was heat treated at 1100°C for 200 hours.

Ingot	Composition	Hardness	K_Q (MPa√m)	Volume Fractions RuAl	Volume Fractions Primary RuAl
1[a]	Ru-35Al-15Mo	512 ± 38	13, 14	56	48
2[b]	Ru-20Al-25Mo	465 ± 65	23	51	41
3[c]	Ru-20Al-25Mo	555 ± 36	28	45	30
4[d]	Ru-20Al-25Mo	507 ± 27	37	34	19
5[e]	Ru-10Al-35Cr	412 ± 31	38	22	16
6[e]	Ru-7Al-38Cr (single phase)	425 ± 67	68	0	0
7(a)[e]	Ru-7A-138Cr	659 ± 33	61, 63	-	-
7(b)[e]	Ru-7Al-38Cr (heat treated 1100 °C)	636 ± 21	64, 65	-	-

[a]Zone arc-melted (45 mm/hr), [b]Float zone (20 mm/hr), [c]As arc-melted, [d]Tri-arc (500 mm/hr), [e]Modified Bridgman furnace (20 mm/hr)

Figure 1. Backscattered SEM images taken from ingots process by (a) the cold crucible Czochralski tri-arc technique (ingot 2), (b) arc-melting (ingot-3), and (c) the float-zone technique (ingot-4).

The fracture toughness values for the three Ru-Al-Mo bend bars (machined from ingots 2-4) are listed in Table I. The fracture toughness for these specimens varied from 23 to 37 MPa√m, and this variation is probably due to differences in the RuAl volume fraction. From SEM images of the fracture surface, Fig. 2, regions corresponding to the primary RuAl dendrites are clearly visible. Furthermore, the zone arc-melted ingot of Ru-35Al-15Mo alloy (ingot-1, Table I) had the highest volume fraction of RuAl and the lowest fracture toughness. Therefore, increasing the volume fraction of RuAl decreases the fracture toughness.

RuAl-Ru(Cr) Alloys

The resulting microstructure of the directionally solidified Ru-10Al-35Cr alloy (ingot-5) is shown in Fig. 3(a). The Ru-10Al-35Cr ingot resulted in a two phase microstructure consisting of a Ru(Cr) matrix with RuAl precipitates. The measured fracture toughness of 38 MPa√m was higher than any of the RuAl-Ru(Mo) ingots, but the volume fraction of the RuAl phase was also less (Table I). The fracture surface of ingot-5 is shown in Fig. 3(b). Similar to the RuAl-Ru(Mo) alloys, the RuAl phase appears to have cleaved with the surrounding Ru(Cr) matrix showing evidence of plastic deformation.

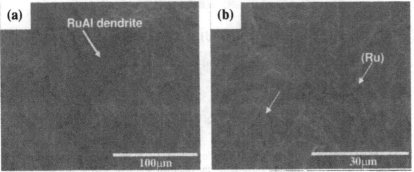

Figure 2. Secondary SEM images of fracture surfaces taken from an arc-melted Ru-20Al-25Mo (at.%) ingot. The arrows in Fig. 2(b) mark the ductile fracture of the hcp-metal phase.

With less aluminum (ingot-6, Ru-7Al-38Cr), a single phase (Ru-hcp solid solution) ingot was obtained upon directional solidification. The bend bar machined from this ingot had the highest fracture toughness of 68 MPa√m. However due to the small specimen size, plane strain conditions were not satisfied. Therefore the plane strain fracture toughness of this alloy must be lower than 68 MPa√m. Regardless the fracture toughness of this alloy is still greater than alloys 1-5 listed in Table I. Additional tests performed on specimens machined from ingot 7(a) had similar fracture toughness values.

Heat treating ingots (6 or 7) of composition Ru-7Al-38Cr resulted in microstructure consisting of fine RuAl precipitates embedded in Ru(Cr) matrix with a representative microstructure shown in Fig. 3(c). The RuAl phase was verified from XRD diffractometer scans. The fracture toughness of heat treated Ru-7Al-38Cr alloy was found to be near 64 MPa√m. The corresponding fracture surface is shown in Fig. 3(d) which shows evidence of ductile fracture.

DISCUSSION

A survey of alloys containing molybdenum by Manzone and Briggs [9] and Geach, Knapton, and Woolf [10] reported that molybdenum additions significantly increase the ductility and formability of ruthenium with a ductility peak near 40 to 50 at.% Mo. This compositional range is similar to the composition of the Ru-rich solid solution found in Ingots 1-4 listed in Tables I [6,7]. These observations are consistent with the high fracture toughness values (K_Q>25 MPa√m) measured from the RuAl-Ru(Mo) alloy having a composition of Ru-20Al-25Mo at.%.

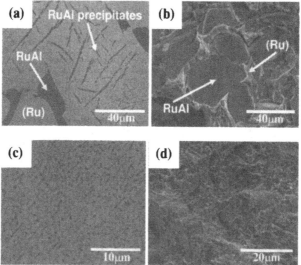

Figure 3. SEM images taken from the Ru-Al-Cr alloys showing (a) the microstructure from ingot-5, (b) the corresponding fracture surface, (c) the microstructure of ingot-7(b), and (d) the corresponding fracture surface.

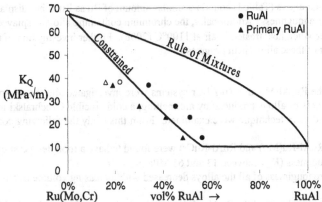

Figure 4. Fracture toughness as a function of RuAl volume fraction (Table I and the models by Chan and Davidson [11]). Closed symbols are from RuAl-Ru(Mo) alloys and open symbols are from RuAl-Ru(Cr) alloys.

The RuAl-Ru(Cr) alloys had the highest fracture toughness and the highest hardness. The Ru-rich solid solution contained about the same amount of Cr as the other alloys contained Mo, and the fracture toughness values for both alloys were in the same range. Chromium has been reported to embrittle RuAl [11]. However, high fracture toughness values can be obtained for fine RuAl precipitates in the Ru matrix.

To estimate the fracture toughness of the RuAl and Ru solid solutions, the fracture toughness data was compared to two models derived by Chan and Davidson for in-situ composites [11]. The first model from the rule of mixtures assumes no interaction between the two phases (equation-1). However, the second model accounts for the plastic constraint imposed by the more brittle phase (equation-2). For both these models, K_d and K_b are the fracture toughness of the ductile and brittle phases, while V_f is the volume fraction of the brittle phase.

$$K_c = K_b \left[V_f + \left(1 - V_f\right)\left(\frac{K_d}{K_b}\right)^2 \right]^{1/2} \qquad (1)$$

$$K_c = K_b \left(1 + \sqrt{1 - V_f} \left[\left(\frac{K_d}{K_b}\right)^2 \exp\left(-\frac{8q}{3}\left(\frac{V_f}{1-V_f}\right)\right) - 1 \right] \right)^{1/2} \qquad (2)$$

Due to the large RuAl dendrites in ingots 1-5, a size factor of q = 1 corresponding to spherical shaped particle was used for the constrained model (equation-2). Figure 4 shows the trend predicted by these two models and the corresponding experimental values. A fracture toughness of 10 MPa√m was estimated for the RuAl solid solution and fracture toughness from ingot-6 (68 MPa√m) was used for Ru solid solution. Using these values, the experimental data closely follows the trend predicted by equation-2. The single phase solid solution RuAl was estimated at 10 MPa√m, since it must be less than 13 MPa√m (ingot-1, Table I). Additional tests with a higher volume fraction RuAl are needed to better estimate the fracture toughness of the RuAl solid solution.

The Ru-Al-Cr alloys (Table I) contain a similar amount of Ru as RuAl but display a much higher fracture toughness. Additionally, the chromium containing alloys displayed good oxidation resistance when heat treated in air at 1100°C (200h). Further investigation of the oxidation behavior of these alloys is in progress.

CONCLUSIONS

Alloys in the Ru-Al-Mo and Ru-Al-Cr systems were investigated and the microstructure and fracture toughness of alloys produced by arc melting, a cold crucible Czochralski technique, and a modified Bridgman technique were examined. From this study the following conclusions are made.

(1) Alloys of RuAl-Ru(Mo) and RuAl-RuCr) were found to have a room temperature fracture toughness (K_Q) between 13 and 65 MPa√m.
(2) The fracture toughness of all the alloys decreased with increasing volume fraction of RuAl.
(3) The fracture toughness of the RuAl solid solution was found to be less then 13 MPa√m.
(4) The Ru-Al-Cr alloys consisting of an Ru(Cr) matrix with RuAl precipitates had the highest room temperature fracture toughness and the greatest hardness.

ACKNOWLEDGEMENTS

We wish to acknowledge Patricia Metcalf for help with the Czochralski tri-arc furnace and modified Bridgeman furnace, and the National Science Foundation for financial support (DMR-0076219). We would also like to acknowledge NSF/JSPS Summer Program for supporting a visit to Kyoto University.

REFERENCES

1. Fleischer RL, Zabala RJ, *Met Trans A*, **21A**, 2709-2715 (1990).
2. Wolff IM, Sauthoff G, *Acta Mater*, **45**, 2949-2969 (1997).
3. Wolff IM, Sauthoff G, Cornish LA, DeV. Steyn H, Coetzee H, Structural Intermetallics 1997, edited by MV Nathal, et al., (TMS, Warrendale, 1996), pp. 815-823.
4. Fleischer RL, Field RD, Briant CL, *Met Trans A*, **22A**, 403-414 (1991).
5. Ilic N, Rein R, Göken M, Kempf M, Soldera F, Müchlich F, *Mater Sci and Eng A*, **A329-331**, 38-44 (2002).
6. Todd Reynolds and David Johnson, *Intermetallics*, **12**, 157-164 (2004).
7. Todd Reynolds and David Johnson, *Mat Res Soc Symp Proc*, **753**, BB5.34 (2003).
8. Brown WF, Srwaley JE, *ASTM Special Publication No. 410*, (ASTM, Philadelphia, PA, 1966), pp. 13.
9. Manzone MG, Briggs JZ, Less-common alloys of Molybdenum, Climax Molybdenum Co., 1962, pp. 124.
10. Geach GA, Knapton AG, Woolf AA, *Powder Metallurgy in the Nuclear Age*, pp. 750-758 (1962).
11. Chan KS, Davidson DL, *Met and Mater Trans A*, **32A**, 2717-2727 (2001).

Mater. Res. Soc. Symp. Proc. Vol. 842 © 2005 Materials Research Society S6.3

Formation and Morphology of Kurnakov Type D0$_{22}$ Compound in Disordered fcc γ-(Ni, Fe) Matrix Alloys

Akane Suzuki[1] and Masao Takeyama[2]
[1]Materials Science and Engineering, University of Michigan, Ann Arbor, MI 48109, USA
[2]Metallurgy and Ceramics Science, Tokyo Institute of Technology, Tokyo 152-8552, Japan

ABSTRACT

Formation and morphology of D0$_{22}$ compound Ni$_3$V in γ-fcc alloys were investigated as a model case in order to understand fundamentals for microstructure control of new class of austenitic steels strengthened by Kurnakov-type GCP intermetallic compound. The formation process of the D0$_{22}$ phase in γ matrix varies, depending on the composition and heat treatment temperature, and precipitation of D0$_{22}$ takes place at temperature above T$_0$ in the γ+D0$_{22}$ two phase region. The morphology of coherent precipitates of the D0$_{22}$ compound is sensitive to misfit strains against matrix and can be controlled by alloying addition. The habit planes between the two phases become irrational, parallel to both directions of the invariant line and an a-axis of D0$_{22}$, since the lattice misfit becomes negative along a-axis (δ_a) and positive along c-axis (δ_c) of the D0$_{22}$. The calculation based on the lattice invariant theory as well as experimental results clearly demonstrate that the misfit strain ratio δ_c/δ_a is a dominant factor to determine the habit plane. In addition, the shape of D0$_{22}$ phase, either prism or plate, depends strongly on the magnitude of $|\delta_a|$. These findings will extend to Ni-Fe-Nb-V quaternary system, by partial replacement of Ni with Fe and that of V with Nb.

INTRODUCTION

There is a growing demand for a new class of heat-resistant austenitic steels for the steam turbine components of USC (ultra-super critical) power plants due to the turbine inlet temperature which is going to exceed 700°C. From the viewpoints of Ni-base superalloys strengthened by GCP intermetallic compounds and austenitic heat-resistant stainless steels containing carbides, a new concept of strengthening austenitic steels by intermetallic compounds has been recently proposed by Takeyama [1]. In this study, we focus on Kurnakov-type compounds, which have order-disorder temperature T$_C$, as potential strengtheners, because they will bring the following advantages; (1) the easiness of fabrication through hot working processes in the high-temperature disordered region above T$_C$, and (2) the feasibility of microstructure control including volume fraction of the compound (0~100%) through phase transformations.

Among the Fe or Ni based alloy systems having Kurnakov-type compounds, Ni-V system which forms Ni$_3$V was selected as a model case, because of the relatively high T$_C$ of Ni$_3$V (1318 K [2]) and its fcc-based tetragonal D0$_{22}$ structure. In this system, γ-fcc→D0$_{22}$ allotropic transformation takes place, accompanied by formation of "multi-variant structure (MVS)" consisting of periodically aligned three variants of D0$_{22}$ phase, by quench from γ-fcc region to below T$_0$ temperature (allotropic phase boundary between γ and D0$_{22}$ in the two-phase region) [3-5]. By subsequent aging in γ-fcc+D0$_{22}$ two-phase region, the MVS decomposes into two phases via precipitation of γ-fcc or continuous phase separation, depending on the alloy composition and the aging temperature [4,5]. However, there is not enough knowledge about precipitation of D0$_{22}$

in γ-fcc matrix.

In this study, the region where precipitation of D0$_{22}$ phase takes place has been firstly identified in Ni-V binary system, and then, the morphology of coherent D0$_{22}$ precipitates and its controlling factor have been examined with paying attention to the misfit strains between the tetragonal precipitates and cubic matrix phase. To extend the obtained knowledge to austenitic steel, morphology control of D0$_{22}$ phase has been attempted in γ-(Ni, Fe) matrix alloys.

EXPERIMENTAL PROCEDURES

The alloys investigated in this study have compositions of Ni-(17~23)V (all compositions are given in atomic percent). Ternary and quaternary alloys containing Nb and/or Fe were also examined. They were prepared by arc melting in Ar atmosphere with a non-consumable tungsten electrode as 35 g button ingots, using 3N7 nickel pellets, 3N vanadium tablets, 3N niobium grains and electrolytic iron. Each ingot was melted five times by turning over each time to avoid segregation. These ingots were cut to pieces with a size of 6×6×10 mm by an electro-discharged machine. All samples were homogenized at 1523~1573 K (γ single-phase region) for 3.6~10.8 ks, followed by water quench. Subsequent aging treatments were employed at 1073 or 1273 K (two-phase region) for 0.9~8640 ks by encapsulating the samples in silica tube with back-filled Ar.

Microstructures were examined by transmission electron microscope (TEM). TEM discs with 0.15 mm in thickness and 3 mm in diameter were machined and then mechanically polished, followed by twin-jet electro-polishing in a solution of ethanol with 12 vol.% perchloric acid at 253 K. The misfit strains between D0$_{22}$ and matrix phases were measured using microbeam diffraction patterns obtained from each phase region [4-6].

RESULTS AND DISCUSSION

Region where precipitatuin of D0$_{22}$ occurs

Figure 1 shows a part of Ni-V binary diagram with γ/γ+D0$_{22}$ and γ+D0$_{22}$/D0$_{22}$ coherent boundaries [2,4-8]. In alloys with V content more than 20%, MVS is formed by γ-fcc→D0$_{22}$ allotropic transformation during quench from γ-fcc region (Ni-23V) or during short time aging at 1073 K after quench (Ni-20V) [4,5]. On the other hand, in Ni-18V and 17V aged at 1073 K, coherent precipitates of D0$_{22}$ phase were observed as shown in Figure 2, and the volume fraction of D0$_{22}$ phase is approximately 30% and 3%, respectively, after 864 ks aging. Thus, precipitation of D0$_{22}$ occurs in the composition range less than 19%V at 1073 K, and the region can be understood as above T$_0$ temperature as indicated by hatching in Figure 1. In this region, γ-fcc phase is energetically stable against D0$_{22}$ phase, and γ matrix remains untransformed by quench [9].

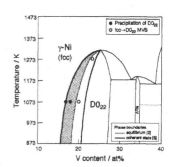

Figure 1. Ni-V binary diagram showing the region where D0$_{22}$ precipitation occurs by hatching.

480

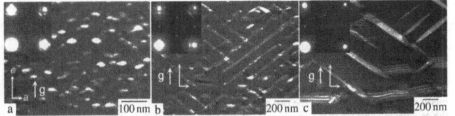

Figure 2. TEM DFIs showing change in the morphology of DO_{22} precipitates with aging at 1073 K ($B=100$, $g=002_{DO22}$): (a) 3.6 ks, (b) 86.4 ks and (c) 864 ks.

Morphology of DO_{22} precipitates

The morphology of the coherent DO_{22} precipitates in Ni-18V is octahedral shape with parallelogram-shaped (100) cross section at the initial stage of aging [Figure 2(a)], and it changes into plate shape with aging [Figures 2(b) and (c)]. The plate-shaped DO_{22} phase has an irrational habit plane close to $\{304\}_\gamma//\{308\}_{DO22}$. Thus, the DO_{22} precipitates preferentially grow and coarsen along only one habit plane of the initial octahedron. The plate morphology of DO_{22} in Ni-18V is quite similar to that in Ni-20V, in which two-phase microstructure is formed via phase separation of DO_{22} [4,6], except for volume fraction of DO_{22}.

In order to understand the dominant factor determining the habit plane and the morphology, misfit strains between the two phases were measured using microbeam diffraction patterns, and they are -0.0038 along a-axis (δ_a) and +0.0086 along c-axis (δ_c) of DO_{22} against the matrix for the sample aged for 864 ks. The negative and positive values indicate the existence of invariant line between the two phases [10]. Figure 3 is a schematic two-dimensional projection of the matrix and DO_{22} lattices depicted as a circle and an ellipse, respectively, onto [100] and [001] axes. Against the circle with a radius of 1, the radii of the ellipse are $1+\delta_a$ and $1+\delta_c$. The two-dimensional description was used to simplify, because every habit plane is parallel to one of the a-axes of DO_{22}. Defining the original point and the cross point of the circle and the ellipse as O and T, respectively, the direction OT is an undistorted direction before and after the formation of DO_{22} in the matrix. Once a part of matrix transforms to DO_{22}, any radius vector will change its length and direction due to the lattice parameter change. Therefore, there exists one specific direction OC in the matrix lattice that becomes OT by lattice rotation of α after the transformation, and OC becomes an invariant line.

To examine the relationship between the habit plane and the invariant line, the angle θ ($=\tan^{-1}(y/x)$) between the invariant line and $[100]_{DO22}$ was calculated geometrically as the cross point of the circle and the ellipse, and is given as the following equation as a function of misfit strains;

$$\theta = \tan^{-1} \sqrt{\frac{-\left(1+2\delta_a^{-1}\right)\left(\delta_c/\delta_a - \delta_a^{-1}\right)^2}{\left(\delta_c/\delta_a\right)\left(\delta_c/\delta_a + 2\delta_a^{-1}\right)\left(1+\delta_a^{-1}\right)^2}} \qquad (1)$$

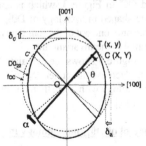

Figure 3. Projection of matrix and DO_{22} lattices depicted as an ellipse and a circle, showing an invariant line between the two phases.

It should be noted that θ depends on both the misfit strain δ_a and the misfit ratio δ_c/δ_a. Then, the coherent habit planes were measured as angles θ between the habit planes and $(001)_{D0_{22}}$ plane, and plotted against δ_c/δ_a as shown in Figure 4, together with the data for $D0_{22}$ phase formed by phase separation of MVS [6]. The observed angle apparently approaches to $45°$ as the misfit ratio approaches to -1. The two curves represent the calculated value using the equation (1) at two fixed values of δ_a: the thick line is the case of -0.003 close to the actual value, and the thin line is an extreme case of -0.030. All experimental data fall into the calculated lines, regardless of the formation mechanism of $D0_{22}$. The difference of θ between the two calculated lines is as much as $2°$, even though the δ_a is changed by an order of magnitude. Thus, it is reasonable to conclude that the plane parallel to the invariant line and a-axis becomes habit and that the misfit ratio is the key factor determining habit plane.

On the other hand, the transformation from γ-fcc to $D0_{22}$ can be described using the lattice transformation and rotation matrices for the invariant line vector $(X, Y)=OC$ in Figure 3;

$$\begin{pmatrix} 1+\delta_a & 0 \\ 0 & 1+\delta_c \end{pmatrix} \begin{pmatrix} \cos\alpha & \sin\alpha \\ -\sin\alpha & \cos\alpha \end{pmatrix} \begin{pmatrix} X \\ Y \end{pmatrix} = \begin{pmatrix} X \\ Y \end{pmatrix} \quad (2)$$

The rotation angle α for the transformation is given as;

$$\alpha = \cos^{-1}\left\{ \frac{(1+\delta_a)(1+\delta_c)+1}{(1+\delta_a)+(1+\delta_c)} \right\} \quad (3)$$

The value α in Ni-18V was calculated as $0.3°$ using the

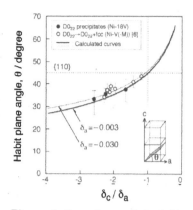

Figure 4. Dependence of the habit plane on the misfit strain ratio δ_c/δ_a, together with calculated invariant line angles at fixed δ_a.

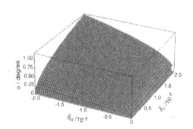

Figure 5. The calculated lattice rotation α as a function of misfit strains.

misfit strains. The lattice rotation α to coincide OT with OC increases the angle gap between OT' and OC' in Figure 3, which is another possible direction to become an invariant line. Thus, the plate-shaped morphology of $D0_{22}$ after prolonged aging can be attributed to the lattice rotation α, since only one out of the four equivalent planes of $D0_{22}$ in the initial octahedron becomes habit plane parallel to the invariant line.

Figure 5 shows the calculated α plotted as a function of δ_a and δ_c. The α becomes close to 0 with decrease in either $|\delta_a|$ or $|\delta_c|$. This result suggests the possibility to result in another morphology with more than one habit planes, other than plate-shaped one by reducing misfit strain, $|\delta_a|$ or $|\delta_c|$.

Control of D0$_{22}$ morphology in γ-(Ni, Fe) matrix alloys

In this section, Ni-V binary system was extended to Fe containing alloy system by partially replacing Ni with Fe. To understand the effect of Fe addition, the γ+D0$_{22}$ two-phase microstructure of Ni-V-Fe ternary alloys was examined. In Ni-19V-4Fe, the morphology of D0$_{22}$ phase is plate-shape [Figure 6], and the misfit strains are $\delta_a = -0.0060$ and $\delta_c = +0.0077$. Regardless of Fe content, DO_{22} phase has plate-shaped morphology because of larger $|\delta_a|$ than Ni-V binary alloys [5,6]. Thus, the plate is only one possible morphology in Fe containing ternary alloys.

Table I shows the effect of alloying elements, Fe, Nb and Co, on the lattice parameters of γ-matrix and DO_{22} phases in Ni-25V quenched from 1473 K and 1273 K [11]. Fe and Co replace Ni, and Nb replaces V in Ni_3V. The value of $|\delta_a|$ slightly increases with Fe addition, but decreases with addition of Nb or Co. These data would give some insight into the morphology control of DO_{22} precipitates in two-phase alloys. When Ni is partially replaced with Fe, a combined addition of Nb or Co is favored for reduction of $|\delta_a|$, so as to change the morphology of DO_{22} other than plate.

Based on the above knowledge, an alloy with a composition of Ni-15V-8Fe-2Nb, where 8% of Ni and 2% of V in the binary Ni-17V was replaced with Fe and Nb, respectively, was examined. As shown in Figure 7, the alloy exhibits so-called "*chessboard* structure", where two variants of diamond-shaped DO_{22} phase (dark contrast) and rectangle-shaped γ-fcc phase (blight contrast) are alternately aligned. The morphology of DO_{22} phase is prism elongated along a-axis (perpendicular to the figure) with diamond-shaped cross section, and two irrational habit planes remain even after 86.4 ks aging. The measured misfit strains in this sample are $\delta_a = -0.0011$ and $\delta_c = +0.0085$. The prism-shaped morphology of DO_{22} phase can be attributed to the reduction of lattice rotation α due to the smaller $|\delta_a|$ than that of Ni-18V. Compared with plate-shaped morphology in Ni-18V, the coarsening of prism-shaped DO_{22} phase along the irrational habit planes is significantly suppressed, and thus, the prism shape is more favorable than the plate shape as precipitation morphology of strengthener in heat-resistant alloys.

Figure 8 shows morphology map of DO_{22} phase plotted against misfit strains, together with

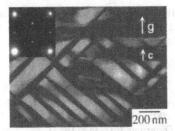

Figure 6. TEM DFI of Ni-19V-4Fe aged at 1073 K / 864 ks.

Figure 7. TEM BFI of Ni-15V-8Fe-2Nb aged at 1073 K / 86.4 ks.

Table I. Effect of alloying elements on the lattice parameters of γ and DO_{22} phases in Ni-25V [11].

Alloy composition / at%	Lattice Parameter /Å			Misfit strain	
	γ-fcc[*]	DO_{22}[**]			
	a	a	c /2	δ_a	δ_c
Ni-25V	3.562	3.543	3.608	-0.0054	0.0130
Ni-25V-3Fe	3.564	3.544	3.611	-0.0058	0.0129
Ni-23V-2Nb	3.567	3.554	3.618	-0.0038	0.0141
Ni-25V-7Co	3.564	3.549	3.604	-0.0042	0.0112

[*] Water quenched from 1473 K, [**] Water quenched from 1273 K

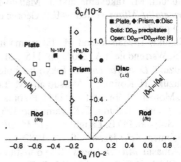

Figure 8. Morphology map of DO_{22} phase plotted onto the axes of misfit strains.

data for DO_{22} phase formed by phase separation of MVS [6]. The morphology becomes plate or prism in the region of $\delta_a<0<\delta_c$ and $|\delta_a|<|\delta_c|$, and the morphology changes from plate to prism by decreasing $|\delta_a|$ to less than 0.002, regardless of the formation mechanism of microstructure. Outside of this region, the morphology becomes disc or rod, depending on the relative values of $|\delta_a|$ and $|\delta_c|$ [5,12,13]. These results indicate that the morphology of DO_{22} phase is strongly depends on the misfit strains, and therefore, in alloys where Ni is partially replaced with Fe, the morphology is changeable into either plate or prism by combined addition with Nb.

SUMMARY

1. Precipitation of DO_{22} phase takes place in the region above T_0 temperature in the $\gamma+DO_{22}$ two-phase region.
2. The habit plane of DO_{22} phase is parallel to one of the a-axes of DO_{22} and the invariant line direction which strongly depends on the misfit ratio δ_c/δ_a. The small lattice rotation α to form the invariant line causes the preferential growth and coarsening of DO_{22} phase along one habit plane, resulting in plate-shaped morphology.
3. The morphology of DO_{22} phase strongly depends on the misfit strains, regardless of alloying elements and formation mechanism of DO_{22} phase, and it changes from plate to prism shape by reducing $|\delta_a|$ less than 0.002. When Ni is partially replaced by Fe, the combined addition with Nb makes it possible to control the morphology into either plate or prism.

ACKNOWLEDGEMENT

This research was supported by the grant-in-aid (13011570) for Japan Society for Promotion of Science (JSPS) Fellows and the grant-in-aid (14205102) for Scientific Research from Ministry of Education, Culture, Sports, Science and Technology.

REFERENCES

1. M. Takeyama in *Integrates and Interdisciplinary Aspects of Intermetallics*, edited by M.J. Mills *et al.*, (Mater. Res. Soc. Proc., Warrendale, PA, 2005) to be published.
2. J.F. Smith, O.N. Carlson and P.G. Nash, *Alloy Phase Diagrams* 3(3), 342 (1982).
3. A. Suzuki, M. Takeyama and T. Matsuo in *Defect Properties and Related Phenomena in Intermetallic Alloys*, edited by E.P. George *et al.*, (Mater. Res. Soc. Proc. **753**, Warrendale, PA, 2003) pp. 363-368.
4. A. Suzuki, T. Matsuo and M. Takeyama in *Superalloys 2004*, edited by K.A. Green *et al.*, (TMS, Warrendale, PA, 2004) pp. 115-124.
5. A. Suzuki, *Doctoral Thesis* (Tokyo Institute of Technology, Tokyo, Japan, 2003).
6. A. Suzuki, H. Kojima, T. Matsuo and M. Takeyama, *Intermetallics* **12**, 969 (2004).
7. H.A. Moreen, R. Taggart and D.H. Polonis, *Metall. Trans.* **5**, 79 (1974).
8. S. Kobayashi, T. Sumi, T. Koyama and T. Miyazaki, *J Japan Inst Metals* **60**, 22 (1996).
9. W.A. Soffa and D.E. Laughlin, *Acta Metall.* **11**, 3019 (1989).
10. U. Dahmen, *Acta Metall.* **30**, 63 (1982).
11. M. Okihashi, *Master Thesis* (Tokyo Institute of Technology, Tokyo, Japan, 2002).
12. J.M. Oblak, D.F. Paulonis and D.S. Duvall, *Metall. Trans.* **5**, 143 (1974).
13. J.-C. Zhao and M.R. Notis, *Acta Mater.* **46**, 4203 (1998).

Mater. Res. Soc. Symp. Proc. Vol. 842 © 2005 Materials Research Society   S5.25

Parameters of Dislocation Structure and Work Hardening of Ni₃Ge

N.A. Koneva, Yu.V. Solov'eva, V.A. Starenchenko, E.V. Kozlov
Tomsk State University of Architecture and Building,
Department of Physics,
Solyanaya sq. 2, Tomsk 634003, Russia

ABSTRACT

Orientation dependence of the yield stress temperature anomaly in Ni₃Ge single crystals with the L1₂ structure was investigated during compression tests. The measurements were carried out in the 4.2 K-1000 K temperature interval for two orientations of single crystals, [001] and [$\bar{2}$34]. The dislocation structure was studied by TEM. Quantitative measurements of different parameters of dislocation structure were carried out. The values of the scalar dislocation density, ρ, were determined for different temperatures in the deformation interval from the yield stress up to fracture. Temperature dependence of the friction stress $\tau_F(T)$ and the interdislocation interaction parameter $\alpha(T)$ were also obtained. The change in the fraction of straight dislocations as a function of temperature was analyzed.

INTRODUCTION

Positive temperature dependence of the flow stress appears to be a characteristic property of the intermetallics with the L1₂ structure. The yield stress and the flow stress are increased considerably with increasing temperature in some temperature intervals. Generally, such behavior is associated with self-locking mechanisms of individual superdislocations, namely with the formation of Kear-Wilsdorf locks. In this connection, a lot of attention is directed towards the experimental studies of the superdislocation fine structure using TEM [1]. Most investigations are carried out on a Ni₃Al alloy and its ternary alloys [2]. To understand the mechanisms of deformation and thermal strengthening, it is important to study the evolution of the dislocation structure in different intermetallics. In connection with this, the present work is devoted to the experimental study of the evolution of the parameters of the dislocation structure in a Ni₃Ge single crystals with the L1₂ structure, where a considerable temperature anomaly of mechanical properties was observed [3, 4].

EXPERIMENTAL DETAILS

Single crystals of the Ni₃Ge were grown using a Chokhralsky technique. Samples with the size of 3.0× 3.0×6.0 mm³ were cut from single crystal ingots using the spark erosion method. The facet orientations were determined by the XRD Laue method. The samples were homogenized at T=953 °C for 48 hrs. The uniaxial compression tests were carried out for two orientations of single crystals [001] and [$\bar{2}$34] with the constant rate of 3.3×10^{-4}s⁻¹ using an Instron type test machine in the 4.2-1000 K temperature interval. The shear stresses in the octahedral slip planes were calculated using the τ=m×σ relationship, where m is the Schmid factor of the primary octahedral slip system, σ is the flow stress. The critical resolved shear stress, τ_0, was considered equal to the flow stress value at ε =0.2 %. Different foil sections were examined, particularly in [111], [100], [110] and other orientations near these zone axes. The foils were cut using spark erosion and mechanically thinned down to 0.1 mm. Final thinning was done in by electropolishing. The dislocation structure was studied in a transmission electron microscope, operating at 125 kV, with a goniometer capable of up to ±30° tilt angle. The scalar dislocation density and straight dislocation density were measured by the secant method. The active slip systems were identified using slip trace analyses.

RESULTS AND DISCUSSION

Fig 1. shows the $\tau - \varepsilon$ dependencies of the Ni$_3$Ge single crystals for two orientations of the deformation axis: [0 0 1] and [$\bar{2}$ 3 4]. At low temperatures (Fig. 1 a, b), the work hardening curves have a form of an arc with curvature. The initial temperature increase straightens the curves and the further increase in temperature changes their curvature to the opposite one. In some cases, the additional temperature increase leads to the formation of the deformation sections with a negative work hardening coefficient ($\theta = d\tau/d\varepsilon < 0$) on the deformation curves.

Studies of the yield stress temperature dependence showed that for the [0 0 1] orientation, the yield stress τ_0 is increased with the temperature increase practically in all investigated temperature intervals (Fig. 2a). The value τ_0 reaches its maximum at the temperature of T$_p$=873 K [5]. The study of the slip trace pattern showed that the single crystals of this orientation are deformed by multiple octahedral slip. The phenomenon of super-localization of plastic deformation was observed at temperatures equal to and above T$_p$ [6]. In the case of the [$\bar{2}$ 3 4] orientation (Fig. 2 b) the maximum of τ_0 appears to be greatly displaced towards the low temperature region (T$_p$=265 K). At temperature T$_p$, the yield stress value for this orientation is half of the τ_0 value at T$_p$ for the [0 0 1] orientation. In the [$\bar{2}$ 3 4] orientation, the anomalous increase of the yield stress in the 4.2 K-265 K temperature interval corresponds to the slip in two octahedral systems: (111) [10$\bar{1}$] and (111) [1$\bar{1}$0] (Fig 2b). Around room temperature, mixed slip takes place: the active slip systems are the octahedral slip system (111)[110] and the cubic

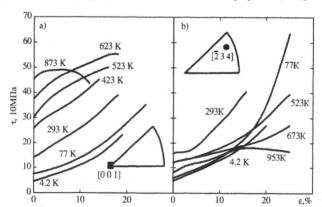

Figure 1. Stress-strain curves of the Ni$_3$Ge single crystals with (a) [001], (b) [$\bar{2}$ 3 4] compression axes.

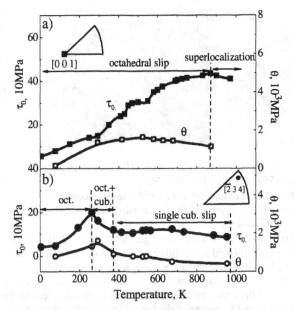

Figure 2. Temperature dependencies of the yield stress τ_0 and work hardening coefficient θ for 5% strain of the Ni₃Ge single crystals with different orientations of the compression axis (a) [0 0 1]; (b) [$\bar{2}$ 3 4].

slip system (001)[10$\bar{1}$] with the maximum Schmid factor. In the [$\bar{2}$ 3 4] orientation, after the first anomalous peak, weak increase of τ_0 was also observed in the 473-673 K temperature interval. This increase occurs at the conditions of the single cubic slip in the (001)[10$\bar{1}$] system with the maximum Schmid factor (Fig.2 b). The investigation of mechanical properties of the Ni₃Ge single crystals carried out in this paper confirmed that unlike the Ni₃Al alloys [7, 8], the position of the maximum on the temperature dependence of the yield stress for the Ni₃Ge alloy depends greatly on the orientation of the deformation axis [4, 5, 9]. The work hardening coefficient, θ, also exhibits the anomalous temperature dependence for both orientations (Fig. 2). The maximum on the temperature dependence of the Ni₃Ge work hardening coefficient is displaced towards the lower temperature region compared to the τ_0 maximum for the [0 0 1] orientation (Fig 2a), which is similar to the Ni₃Al alloy. In the case of the [$\bar{2}$ 3 4] orientation, temperature dependence of the deformation work hardening qualitatively repeats the yield stress behavior with temperature.

The study of the dislocation structure in the Ni₃Ge single crystals with the octahedral slip in the [0 0 1] orientation showed that starting from the temperature of 4.2 K two systems of straight screw dislocations with <110> Burgers vector are present in the structure. Fig. 3a shows the dislocation structure pattern observed in the foil cut parallel to the (1 1 1) slip plane from a sample deformed at room temperature. Numerous straight dislocations in the dipole configuration that represent the Kear-Wilsdorf locks [10] are observed. The study of the dislocation structure in the region of the temperature anomaly peak (Fig. 3 b) showed that the number of the straight dislocations decreases noticeably. The dislocations are heavily curved and there are a large number the dislocation loops. Furthermore, at the temperature of T$_p$=873 K, there are almost no straight dislocations in the dislocation structure. It should be mentioned that the yield stress level remains as high as before. These experimental data show that at high temperature thermal strengthening is possible with no Kear-Wilsdorf locks present. Numerous curved dislocations may be formed as a result of interactions between the dislocations with edge and mixed orientations and point defects. Therefore, in order to understand the thermal work

Figure 3. Dislocation structure of the Ni₃Ge single crystals with [0 0 1] compression axis. [1 1 1] zone axis. (a) T=293 K, $\varepsilon = 5\%$; (b) T=873 K, $\varepsilon = 15\%$.

hardening phenomenon in the high temperature interval one should take into account the mechanisms related to diffusion precipitation of point defects on the dislocations of the edge orientation resulting in the formation of locks.

The analysis of the dislocation structure in the case of a single cubic slip at T>20 °C in the single crystals with the [$\bar{2}$ 3 4] orientation showed that in this case most of the dislocations are close to the edge orientation (Fig. 4). The straight screw dislocations are not observed in the structure. The preferential accumulation of the dislocations with the edge orientation observed in this case may indicate the fact that, under the cubic slip condition, diffusion processes lead to the reduction in the edge dislocation mobility and formation of the locks from the edge dislocations. The latter are locked due to separate climb of the superpartial dislocations [11].

Measurements of the dislocation density (ρ) as a function of temperature and strain were carried out for the two investigated orientations of the Ni₃Ge single crystals. Fig. 5 shows the $\rho(T)$ dependencies for different values of strain. These dependencies show that the change in the dislocation density with temperature correlates with the change in the yield stress and the flow stress (see Fig. 2). The same results were obtained earlier for Ni₃(Al, Hf) [12]. It then follows that the temperature dependence of the dislocation density also exhibits the anomalous behavior. This means that one of the contributions to the increase in the deformation resistance with temperature is caused by overcoming the forest dislocation resistance and the elastic fields of the dislocation structure.

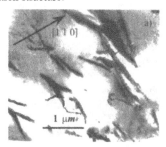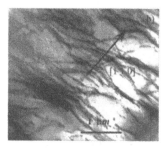

Figure 4. Dislocation structure of the Ni₃Ge single crystals with [$\bar{2}$ 3 4] compression axis. [1 0 0] zone axis. T=523 K (a) 5%, (b) 10%.

488

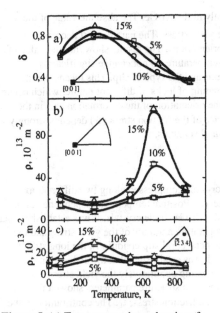

The decrease in the fraction of the straight dislocations in the single crystal with [0 0 1] orientation in the high temperature region was quantitatively confirmed. The fraction of the straight dislocations was determined as $\delta = \rho_{sd} / \rho$, where ρ_{sd} is density of the straight dislocations. It turned out that the fraction of the straight dislocations increases at first with the temperature increase from 77 K to 293 K (fig. 5 a). The following temperature increase leads to a decrease in its value. Simultaneously, increase in the shear stress is observed.

On the basis of the dislocation density measurements, it was determined that the connection between the shear stress, τ, and the dislocation density, ρ, may be described by the following relationship:

$$\tau = \tau_F + \alpha Gb\rho^{1/2},$$

Figure 5. (a) Temperature dependencies of the fraction of the straight dislocations (δ) in the Ni₃Ge single crystals with [0 0 1] orientation of the compression axis at different strains. Temperature dependences of the scalar dislocation densities of the Ni₃Ge single crystals with different orientations of the compression axis (b) [0 0 1]; (c) [$\bar{2}$ 3 4] at different strains.

where τ_F is the friction stress, α is the interdislocation interaction parameter, G is the shear modulus, b is Burgers vector. Using the experimental dependencies of $\tau = f(\rho^{1/2})$, the values of the parameters α and τ_F were obtained as a function of temperature. These data are presented in Fig. 6. The temperature dependence of the τ_F correlates well with the yield stress temperature dependence. This means that the superdislocation self-locking is

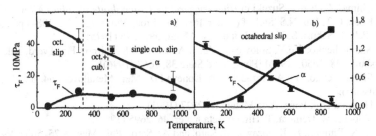

Figure 6. Temperature dependence of the parameters α and τ_F for the Ni₃Ge single crystals with different orientations of the compression axis (a) [$\bar{2}$ 3 4], (b) [0 0 1].

the main mechanism of the τ_0 temperature anomaly in the Ni_3Ge. The absolute value of the friction stress is equal approximately to 0.8 of the yield stress. The parameter α decreases linearly with temperature for both single crystal orientations. The results show that the value of the parameter α and the rate of its decrease with temperature do not change significantly depending on the orientation of the deformation axis and the type of slip. This means that the mechanisms related to the thermoactivated overcoming of locks of different nature, which occur in pure metals as well, are weakly dependent on the orientation of the deformation axis in the Ni_3Ge. On the contrary, the temperature dependence of the friction stress (τ_F) depends strongly on the orientation of the deformation axis in the single crystals.

CONCLUSIONS

The $\tau - \varepsilon$ dependence of the Ni_3Ge single crystals compressed along two different axes was investigated in the 4.2-923 K temperature interval. Positive temperature dependence of the yield stress appears at cryogenic temperatures in the Ni_3Ge single crystals. It was established that there is strong orientational dependence of the temperature anomaly of the yield stress and the work hardening coefficient in this alloy. Under the octahedral slip conditions, the long straight dislocations (Kear-Wilsdorf locks) were observed. It was established that the dislocation density exhibits the anomalous temperature dependence repeating qualitatively the yield stress temperature dependence. The estimates of the dislocation structure parameters showed that the strongest orientational dependence is observed for the friction stress, τ_F. The contribution to the flow stress, related to the value of the interdislocation interaction parameter α, is weakly dependent on the orientation of the single crystal deformation axis and the type of slip. It was determined that the superdislocation self-locking is the main contribution to the positive temperature dependence of the yield stress in the Ni_3Ge phase with the $L1_2$ structure.

REFERENCES

1. P. Veyssière, and G. Saada, 1996, in *Dislocations in Solids*, edited by F. R. N. Nabarro and M. S. Duesbery (Amsterdam: North-Holland), p. 255.
2. F. R. N. Nabarro, and H. L. de Villiers, 1995, *The Physics of Creep. Creep and Creep-resistant Alloys* (Taylor & Francis Ltd, London), p. 413.
3. Y. Mishima, Y. Oya, T. Suzuki in *High-temperature Ordered Intermetallics Alloys,* edited by C.C. Koch, C.T. Liu, N.S. Stoloff (Mater. Res. Soc. Proc., Pittsburgh, PA, 1984) pp. 263-279.
4. H.-R. Pak, T. Saburi, S. Nenno, J. Jap. Inst. Metals **39**, 1215 (1975).
5. V.A. Starenchenko, Yu.V. Solov'eva, Yu.A. Abzaev, B. I. Smirnov Fiz. Tverd. Tela (S.-Petersburg) **38**, 3050, (1996) [Phys. Solid State **38**, 1668 (1996)].
6. V.A. Starenchenko, Yu.A. Abzaev, N.A. Koneva Fiz. Met. Metalloved. **64**, 1148, (1987) (in Russian).
7. F. E. Heredia, D.P. Pope, Acta Mater. **39**, 2027 (1991).
8. D. Goldberg, M. Demura, T. Hirano Acta mater. **46**, 2695 (1998).
9. S. Ueta, K. Jumonji, K. Kanazawa, M. Kato, and A. Sato, Phil. Mag. A **75**, 563 (1997).
10. J. Fang, E.M. Schulson, and I. Baker Phil. Mag. A **70**, 1013, (1994).
11. P.A. Flinn Trans. Met. Soc. AIME **218**, 145 (1960).
12. T. Kruml, V. Paidar, S.L. Martin Intermetallics **8**, 729, (2000).

Identification of the chirality of intermetallic compounds by electron diffraction

S. Fujio, H. Sakamoto, K. Tanaka[1] and H. Inui
Department of Materials Science and Engineering, Kyoto University,
Sakyo-ku, Kyoto 606-8501, Japan
[1]Department of Advanced Materials Science, Kagawa University,
Takamatsu 761-0396, Japan

ABSTRACT

A new CBED method is proposed for the identification of the chirality of enantiomorphic crystals, in which asymmetry in the intensity of the reflections of Bijvoet pairs in an experimental symmetrical zone-axis CBED pattern is compared with that of a computer-simulated CBED pattern. The intensity difference for reflections of these Bijvoet pairs results from multiple scattering among relevant Bijvoet pairs of reflections, each pair of which has identical amplitude and different phase angles. With the method, a single CBED pattern is sufficient and chiral identification can be made for all possible enantiomorphic crystals that are allowed to exist in crystallography. The method is successfully applied to some chiral intermetallic compounds.

INTRODUCTION

Enantiomorphism is usually referred to and used to describe objects that are lacking of improper rotation (rotoinversions and rotoreflections). Because of the absence of a center of symmetry ($\bar{1}$), a mirror plane ($m = \bar{2}$) and a $\bar{4}$ axis, such enantiomorphic (chiral) crystals or molecules can occur in two different forms that are related as a right hand and a left hand and these crystals are mirror related and are not superimposable with each other. These enantiomorphically-related crystals belong to either of the 11 crystal classes (point groups), as summarized in Table 1 [1,2]. Enantiomorph identification is usually made by X-ray diffraction methods utilizing anomalous scattering phenomenon [3,4]. However, a relatively large-sized single crystal of high quality and a sufficiently strong X-ray beam are needed in many cases [5,6]. If the distinction can be made by electron diffraction in the transmission electron microscope (TEM), such difficulties can be avoided owing to the capability of the TEM of using a nanometer-sized electron probe [7,8]. Although a few methods have so far been proposed to determine the chirality of enantiomorphic crystals such as quarts [9,10] and MnSi [11] by electron diffraction, none of them can easily be extended to all possible enantiomorphic crystals.

We have recently proposed a new CBED (convergent-beam electron diffraction) method for identification of chirality of enantiomorphic crystals, in which asymmetry in the intensity of the reflections of Bijvoet pairs in an experimental symmetrical zone-axis CBED pattern is compared with that of a computer-simulated CBED pattern. The intensity difference for reflections of these Bijvoet pairs results from multiple scattering (dynamical nature of electron diffraction) among relevant Bijvoet pairs of reflections, each pair of which has identical amplitude and different phase angles. With the present method, chiral identification can be made for all the possible enantiomorphic crystals that are allowed to exist in crystallography. In the present study, we apply the CBED method to an intermetallic silicide, $TaSi_2$ that belongs to the enantiomorphic space group pair of $P6_222$ and $P6_422$.

PRINCIPLES

In the proposed method, an appropriate zone-axis orientation has to be first chosen so that Bijvoet pairs of reflections, which are Bragg reflections of space group symmetry equivalents to the two members of a Friedel pair, are observed symmetrically in a single CBED pattern. For this purpose, any reflections belonging to ZOLZ (zero-th order Laue zone) and FOLZ (first order Laue zone) can be used. Since which of *hkl* reflections correspond to Bijvoet pairs depend on point group (Table 1), the appropriate symmetrical zone-axis orientations depend on point group of crystal. Appropriate zone-axis orientations can thus be found by plotting the distribution of Bijvoet pairs for a particular type of *hkl* reflections (for example, twelve $12\bar{3}4$-type reflections for crystals with the hexagonal symmetry) located in the standard triangles in the stereographic projection. Since Bijvoet pairs for crystals with the point group of 622, to which the space groups of $P6_222$ and $P6_422$ belong, are

$$hkil = \bar{h}\,\bar{k}\,i\,l = hik\bar{l} = \bar{h}\,i\,\bar{k}\,\bar{l} \quad \text{and} \quad \bar{h}\,\bar{k}\,i\,\bar{l} = hki\bar{l} = \bar{h}\,i\,\bar{k}\,l = hikl, \qquad (1)$$

Bijvoet pairs (expressed I_A-I_B) for any of particular types of reflections are distributed in the stereographic projection as shown in Fig. 1. Then, the appropriate zone-axis orientations at which Bijvoet pairs of FOLZ (and/or ZOLZ) reflections disks appear symmetrically in a single CBED pattern are readily known as those on the $<11\bar{2}0>$, $<1\bar{1}00>$ and [0001] zone-circles for crystals with the point group of 622 (see, Figs. 2(a) and (b)). Therefore, the appropriate zone-axis orientations for the symmetrical observation of Bijvoet pairs are of the $<h\,\bar{h}\,0l>$-, $<hh\,\overline{2h}\,l>$- and $<hki0>$-types. Similarly, the appropriate zone-axis orientations are determined in Table I for all the enantiomorphic point groups.

Table I: Crystal system, crystal class (point group), equivalent indices of Bijvoet pairs of reflections in the asymmetric unit and appropriate zone-axis orientations for chiral identifications for crystals belonging to all possible enantiomorphic point groups. The nomenclature "ZOLZ" in the last column means that Bijvoet pairs of reflections that appear in ZOLZ as ±g are utilized for chiral identification. * The point group 32 in the rhombohedral coordinate system is divided into two space groups, 321 and 312 in the hexagonal coordinate system.

Crystal system	Point group	Bijvoet pairs Equivalent indices	Equivalent indices	Appropriate zone-axis
Triclinic	1	hkl	$\bar{h}\bar{k}\bar{l}$	ZOLZ
Monoclinic	2	hkl = $\bar{h}k\bar{l}$	$\bar{h}\bar{k}\bar{l}$ = h$\bar{k}$l	$<h0l>$
Orthorhombic	222	hkl = $\bar{h}\bar{k}l$ = $\bar{h}k\bar{l}$ = h$\bar{k}\bar{l}$	$\bar{h}\bar{k}\bar{l}$ = h$k\bar{l}$ = h$\bar{k}$l = $\bar{h}k$l	$<hk0>$, $<0kl>$, $<h0l>$
Tetragonal	4	hkl = $\bar{k}$hl = $\bar{h}\bar{k}$l = k$\bar{h}$l	$\bar{h}\bar{k}\bar{l}$ = k$\bar{h}\bar{l}$ = hk$\bar{l}$ = $\bar{k}$h$\bar{l}$	$<hk0>$
	422	hkl = khl = $\bar{h}$kl = $\bar{k}$hl = h$\bar{k}\bar{l}$ = $\bar{k}\bar{h}\bar{l}$ = $\bar{h}$kl = khl	$\bar{h}\bar{k}\bar{l}$ = k$\bar{h}\bar{l}$ = hk$\bar{l}$ = $\bar{k}$h$\bar{l}$ = h$\bar{k}$l = $\bar{h}\bar{k}$l = $\bar{h}$kl = $\bar{k}$hl	$<hk0>$, $<h0l>$, $<hhl>$
Trigonal	3	hkil	$\bar{h}\bar{k}\bar{i}\bar{l}$	ZOLZ
	32*			
	321	hkil = hik$\bar{l}$	$\bar{h}\bar{k}\bar{i}$ = $\bar{h}ik$l	$<h\bar{h}0l>$
	312	hkil = $\bar{h}ik$l	$\bar{h}\bar{k}\bar{i}\bar{l}$ = hik$l$	$<hh\overline{2h}l>$
Hexagonal	6	hkil = $\bar{h}\bar{k}\bar{i}$l	$\bar{h}\bar{k}\bar{i}\bar{l}$ = hki$l$	$<hki0>$
	622	hkil = $\bar{h}\bar{k}\bar{i}$l = hik$\bar{l}$ = $\bar{h}ik\bar{l}$	$\bar{h}\bar{k}\bar{i}\bar{l}$ = hki$\bar{l}$ = $\bar{h}\bar{k}\bar{i}$l = hik$l$	$<h\bar{h}0l>$, $<hh\overline{2h}l>$, $<hki0>$
Cubic	23	hkl = h$\bar{k}\bar{l}$ = $\bar{h}$k$\bar{l}$ = $\bar{h}\bar{k}$l	$\bar{h}\bar{k}\bar{l}$ = $\bar{h}$kl = h$\bar{k}$l = hk$\bar{l}$	$<hk0>$
	432	hkl = h$\bar{k}\bar{l}$ = $\bar{h}$k$\bar{l}$ = $\bar{h}\bar{k}$l = $\bar{h}$lk = h$\bar{l}$k = hl$\bar{k}$ = $\bar{h}\bar{l}\bar{k}$	$\bar{h}\bar{k}\bar{l}$ = $\bar{h}$kl = h$\bar{k}$l = hk$\bar{l}$ = h$\bar{l}\bar{k}$ = $\bar{h}$l$\bar{k}$ = $\bar{h}\bar{l}$k = hlk	$<hk0>$, $<hhl>$

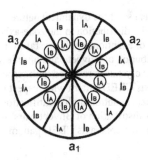

Fig. 1: Distribution of Bijvoet pairs of reflections located in the standard triangles for the point groups of 622. Bijvoet pairs of reflections are expressed as I_A-I_B and the symbols expressed with a circle indicate reflections that appear in the reverse side of the triangles.

In the kinematical approximation, structure factors for any hkl reflections are given as follows [7,8].

$$F(hkl) = \sum_i f_i(\theta) \exp\{-2\pi i\,(hx_i + ky_i + lz_i)\}$$

$$= |F|\,\exp(i\varphi), \tag{2}$$

where $f_i(\theta)$ is atomic scattering factor of i-th atom, θ is scattering angle and (x_i, y_i, z_i) are atomic coordinate of i-th atom. Since members of a Bijvoet pair have equal amplitude ($|F|$) and opposite phase (φ) as those of a Friedel pair (hkl-$\overline{hkl}$) do, the phase distribution of these Bijvoet pairs of reflections that appear symmetrically in a CBED pattern taken along any of the appropriate zone-axis orientations is not symmetrical with respect to the symmetry line m-m' while their positions are symmetrical. Since the right-handed crystal can be converted to the left-handed one by changing the coordinates of atom positions from (x,y,z) to $(-x,-y,-z)$, the following relationships analogous to the Bijvoet relations are obtained,

$$F_R(hkl) = |F_R(hkl)|\exp(i\varphi), \quad F_R(\overline{hkl}) = |F_R(\overline{hkl})|\,\exp(-i\varphi) \quad \text{and}$$

$$F_L(hkl) = |F_L(hkl)|\exp(-i\varphi), \quad F_L(\overline{hkl}) = |F_L(\overline{hkl})|\,\exp(i\varphi), \tag{3}$$

where the subscripts R and L stand for right- and left-handed crystals, respectively. This indicates that the asymmetrical phase distribution with respect m-m' in a zone-axis CBED pattern for each of the two member of enantiomorphic crystals is reversed when the handedness is reversed, as shown in Figs. 2(a) and (b). In other words, the phase distributions of these Bijvoet pairs of reflections for the two enantiomorphic crystals are asymmetric with respect to m-m' and are related to each other by a mirror reflection through the symmetry line m-m'. In such a zone-axis CBED pattern with the asymmetrical phase distribution for Bijvoet pairs of reflections, the breakdown of the Friedel's law occurs as a result of multiple scattering among reflections in the CBED pattern because of the dynamical nature of electron diffraction [12-14]. This leads to the asymmetric intensity distribution in the zone-axis CBED pattern with respect to m-m' for each of the two members of enantiomorphic crystals. Of importance to note here is that since the asymmetrical phase distribution of Bijvoet pairs of reflections in a zone-axis CBED pattern for a pair of enantiomorphic crystals is reversed with respect m-m', the asymmetric intensity distribution in the zone-axis CBED pattern with respect to m-m' is also reversed for the pair of enantiomorphic crystals. Then, enantiomorph identification can be made easily by inspecting the asymmetric intensity distribution of Bijvoet pairs of reflections that are arranged symmetrically with respect to m-m' in a zone-axis CBED pattern.

RESULTS

We show a typical example of chiral identification made for intermetallics based on the above mentioned method. $TaSi_2$ is known to have enantiomorphic crystals related to each other with respect to the reverse screw axis parallel to the hexagonal c axis [15]. The space groups of $TaSi_2$ are $P6_222$ and $P6_422$. Since the point group is 622, the appropriate zone-axis orientations are of the $<hh\,0l>$-, $<hh\,2h\,l>$- and $<hki0>$-types (Table 1). Here we choose $[3\,\overline{3}\,01]$ as an appropriate zone-axis orientation to observe Bijvoet pairs of reflections symmetrically. The phase distribution of Bijvoet pairs of reflections that appear symmetrically in a $[3\,\overline{3}\,01]$ CBED pattern is shown in Figs. 2(a) and (b) for $TaSi_2$ with the space groups of $P6_222$ and $P6_422$, respectively. In this case, the phase distribution of HOLZ reflections is asymmetric with respect to m-m' for each of the two patterns in Figs. 2(a) and (b), and is reversed with respect to m-m' when the space group is changed from one to the other. Strictly speaking, the relative values (instead of the absolute values) of the phase angles of the Bijvoet pairs of reflections are reversed when the space group is changed. On the other hand, the phase distribution of the ZOLZ reflections is symmetric with respect to m-m' and identical for both the two patterns. This is because of the fact that the l index for all ZOLZ reflections in the $[3\,\overline{3}\,01]$ CBED patterns is $l = 3n$ (n: integer), which does not produce different phase angles for Bijvoet pairs of reflections. We thus utilize the intensity asymmetry of FOLZ reflections in the $[3\,\overline{3}\,01]$ CBED patterns for enantiomorph identification. Amplitude-phase diagrams for $(1\,\overline{2}\,1\,8)$ and $(2\,\overline{1}\,\overline{1}\,8)$ FOLZ

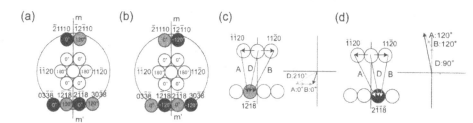

Figs. 2: Phase distributions for ZOLZ and FOLZ reflection disks that appear in a $[3\,\overline{3}\,01]$ zone-axis CBED pattern of $TaSi_2$ with the space groups of (a) $P6_222$ and (b) $P6_422$. Amplitude-phase diagrams for $(1\,\overline{2}\,1\,\overline{8})$ and $(2\,\overline{1}\,\overline{1}\,\overline{8})$ FOLZ reflection disks for the space group of $P6_222$ are shown in (c) and (d), respectively.

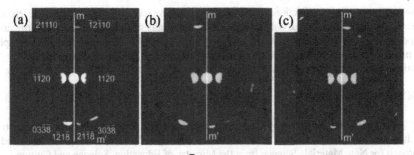

Figs. 3: (a) Experimental and calculated [3$\bar{3}$01] zone-axis CBED patterns for TaSi$_2$ with the space groups (b) P6$_2$22 and (c) P6$_4$22.

reflection disks in the [3$\bar{3}$01] zone-axis CBED pattern of TaSi$_2$ with the space group of P6$_2$22 are shown in Figs. 2(c) and (d), respectively. In addition to the direct scattering route (D), multiple scattering routes (A) and (B), which are the origin of the breakdown of the Friedel's law, are taken into account. The 90° phase difference of the diffracted beam with respect to the transmitted beam is taken into account in constructing these amplitude-phase diagrams [12]. As seen in Fig. 2, the intensity (amplitude) of the (2$\bar{1}$1$\bar{8}$) FOLZ reflection is expected to be stronger than that of the (1$\bar{2}$1$\bar{8}$) FOLZ reflection for the space group of P6$_2$22. Amplitude-phase diagrams constructed for the space group of P6$_4$22 indicate that the opposite is true. The intensity distribution of the FOLZ reflection disks in this case is expected to be reversed with respect to m-m' when the space group changes from one to the other. For other Bijvoet pairs of reflections, the intensity asymmetry with respect to m-m' is expected to occur, as the stronger and weaker reflections are indicated as darker and less darker disks, respectively, in Figs. 2(a) and (b). An experimental [3$\bar{3}$01] zone-axis CBED pattern of TaSi$_2$ is shown in Fig. 3(a). Two CBED patterns of Figs. 3(b) and (c) with the same incidence are those calculated based on the space groups of P6$_2$22 and P6$_4$22, respectively [16]. The asymmetric intensity distribution of Bijvoet pairs of reflections for these calculated patterns is consistent with the result of analysis of the amplitude-phase diagrams shown in Figs. 2(a) and (b) respectively for the space groups of P6$_2$22 and P6$_4$22. When the intensity for Bijvoet pairs of FOLZ disks, (2$\bar{1}$1$\bar{8}$)-(1$\bar{2}$1$\bar{8}$) and (03$\bar{3}$$\bar{8}$)-(30$\bar{3}$$\bar{8}$) in the experimental pattern of Fig. 3(a) is compared with each other, the intensity of the latter disks is stronger than that of the former disks. This indicates that TaSi$_2$ in this case belongs to the space group of P6$_4$22. There are twelve equivalent <3$\bar{3}$01>-type directions for crystals belonging to the point group of 622. All these twelve <3$\bar{3}$01> incidences produce the identical CBED patterns for each of the two space groups, indicating that only a single CBED pattern is sufficient to identify the chirality (either P6$_2$22 or P6$_4$22) of TaSi$_2$.

CONCLUSIONS

A new CBED method is proposed for chiral identification of enantiomorphic crystals, in which asymmetry in the intensity of reflections of Bijvoet pairs in an experimental symmetrical zone-axis CBED pattern is compared with that of an computer simulated CBED pattern. The intensity difference for reflections of these Bijvoet pairs results from multiple scattering among relevant Bijvoet pairs of reflections, each pair of which have identical amplitude and different phase angles. With the method, only a single CBED pattern is sufficient and chiral identification can be made for all possible enantiomorphic crystals that are allowed to exist in crystallography. The method is successfully applied to an intermetallic silicide, $TaSi_2$ that belongs to the enantiomorphic space group pair of $P6_222$ and $P6_422$.

ACKNOWLEDGEMENTS

This work was supported by Grant-in-Aid for Scientific Research (a) from the Ministry of Education, Science and Culture (No. 14350369) and in part by COE 21 Program on United Approach for New Materials Science from the Ministry of Education, Science and Culture.

REFERENCES

1. T. Hahn, (ed), "International Tables for Crystallography, Volume A: Space-Group Symmetry 4th revised edition", The International Union of Crystallography by Kluwer Academic Press, Dordrecht (1996) pp. 786-792.
2. G. Burns and A.M. Glazer, " Space Groups for Solid State Scientists, 2nd edition", Academic, Boston, MA (1990).
3. G.H. Stout and L.H. Jensen, "X-Ray Structure Determination A Practical Giide, 2nd edition", John Wiley & Sons, New York (1989).
4. J.P. Glusker, M. Lewis and M. Rossi, "Crystal Structure Analysis for Chemists and Biologists", VCH, New York (1994).
5. A. McPherson, "Preparation and Analysis of Protein Crystals", John Wiley & Sons, New York (1982).
6. J. Drenth, "Principles of Protein X-Ray Crystallography", Springer-Verlag, Berlin (1994).
7. L. Reimer, "Transmission Electron Microscopy", Springer-Verlag, Berlin (1984).
8. J.C.H. Spence and J.M. Zuo, "Electron Microdiffraction", Plenum, New York (1992).
9. Goodman. P and Secomb. T. W, *Acta Cryst*, A33, 126-133 (1977).
10. Goodman. P and Johnson. T.W, *Actacryst*, A33, 997-1001 (1977).
11. Tanaka. M, Takayoshi. H, Ishida. M and Endoh. Y, *J. Phys.Soc.Jpn*, 54, 2970-2974 (1985).
12. J.M. Cowley, "Diffraction Physics", North-Holland, Amsterdam (1986).
13. J.M. Cowley and A.F. Moodie, *Acta Cryst*, **12**, 360-367 (1959).
14. P. Goodman and G. Lehmpfuhl, *Acta Cryst*, **A24**, 339-347 (1968).
15. P. Villars and L.D. Calvert, " Pearson's Handbook of Crystallographic Data for Intermetallic Phases", American Society for metals, Metals park, OH (1985).
16. K. Ishizuka, "Proc. Int. Symp. on Hybrid Analyses for Functional Nanostructure", Ed. by M. Shiojiri and N. Nishio, Japanese Society of Electron Microscopy, Tokyo (1998) pp. 69-74.

Mater. Res. Soc. Symp. Proc. Vol. 842 © 2005 Materials Research Society   S5.52

Mechanical Behavior of a Pt-Cr Jewelry Alloy Hardened by Nano-Sized Ordered Particles

Kamili M. Jackson, Miyelani P. Nzula, Silethelwe Nxumalo, and Candace I. Lang
University of Cape Town, Mechanical Engineering Department, Centre for Materials
Engineering

Abstract
The materials engineering of platinum jewelry is interesting because only 5wt% can be used for
alloying in order to maintain hallmarking. However, pure platinum is very soft and must be
alloyed in order to be used effectively as jewelry. In several binary systems an increase in
hardness has been found after cold working and annealing at low temperatures. The hardening
in these alloys has shown to be a result of nano-sized ordered particles. In particular, the
existence of the ordered particles has previously been confirmed for a Pt-Cr alloy by TEM.
Extensive work has been done on the Pt-Cr alloy to understand the crystal structure and
mechanisms of the ordered phase. Hardness tests were performed to measure mechanical
properties after various heat treatments. In addition, tensile tests were conducted using a small-
scale tensile testing machine. An 8mm long specimen is used, which significantly reduces the
cost of the specimens while providing necessary properties. Tensile tests on the Pt-Cr alloys at
various post deformation heat treatments show an increase in tensile strength with no effect on
ductility. They confirm results of the hardness tests while providing additional properties data.
In addition, the results show a fairly good relationship between strength and hardness.

Introduction
Pure platinum is much too soft for the mechanical demands of jewelry, having a Vickers
hardness of only 40-41 and UTS of 117-159 MPA [1,2]. While this softness (and the resulting
ductility of 32% strain to failure) is ideal for mechanical forming of jewelry [2] the challenge is
to harden jewelry alloys while maintaining the 95% purity hallmark required by most countries.
There are various ways of hardening metals but an ideal one for jewelry would provide
additional strengthening after forming by cold work. Thus, a useful hardening method for
jewelry alloys might be ordering by post-deformation heat treatment, particularly where cold
deformation enhances the kinetics of ordering [3]. This type of hardening is being investigated
in platinum-chromium alloys at UCT. The Pt-Cr phase diagram shows evidence of ordering
from 20-65 at. % chromium [4-6]; however, at the platinum rich end (greater than 95 wt. % Pt)
of the phase diagram no evidence of ordering has previously been reported. Here we present the
effect of ordering on hardness and tensile properties in these alloys.

Experimental Methods
A range of techniques was used to explore the ordering and hardening behavior of two platinum
rich alloys, Pt 10 at.%Cr and Pt 11 at.% Cr (Pt 3.0 wt.% Cr and Pt 3.2 wt.% Cr), after various
heat treatments. These techniques include Vickers hardness measurements, optical microscopy,
transmission electron microscopy (TEM), and small sample tensile testing. Alloys were cold
rolled to 90% deformation and then heat treated for three hours at temperatures between 200°C
and 800°C. Hardness tests were conducted using a 100g load producing indents of 20-25 micron
diagonals. Electron diffraction patterns were obtained using JEOL 200CX and Philips CM20

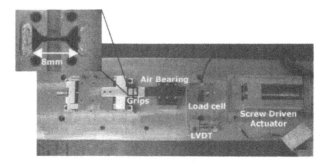

Figure 1. Small sample tensile specimen and apparatus for testing

TEMs in the selected area diffraction mode (SAD) with the beam fully spread for parallel illumination. Tensile testing was conducted with small samples of 8 mm in total length and a nominal gage width of 0.5mm. They were tested in a custom-built apparatus shown in Figure 1. Hardness and tensile specimens were polished to a mirror finish for consistency.

Results and Discussion
Hardness tests showed an increase in hardness after post-deformation heat treatment at 300°C and 400°C as shown in Figure 2. This increase in hardness cannot be accounted for by any of the usual hardening mechanisms. Micrographs show no change in microstructure after heat treatment up to 700°C, as shown in Figure 3.

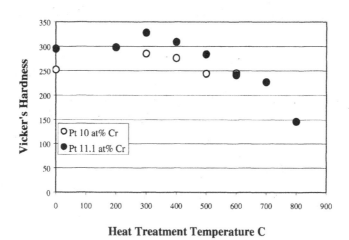

Figure 2. A graph of hardness vs. heat treatment temperature for the Pt 10 at.% Cr and Pt 11.1 at.% Cr heat-treated for 3 hours.

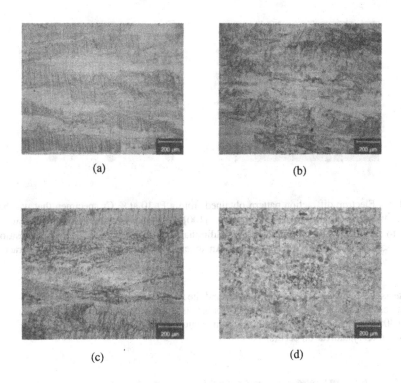

(a)   (b)

(c)   (d)

Figure 3. Micrographs of Pt 11 at.% Cr after (a) 90% cold work; and subsequent heat treatment of (b) three hours at 300°C (c) three hours at 700°C (d) three hours at 800°C

There is no evidence of the precipitation of a second phase, nor is a second phase predicted in the phase diagram. Solid solution hardening and strain hardening have already taken place prior to heat treatment. Further investigation by electron diffraction of cold rolled specimens heat treated at 400°C for 6 hours show additional superlattice reflections besides the expected reflections for FCC, as shown in Figure 4. These additional reflections were not seen in the cold rolled specimens that were not heat treated, nor in specimens heat treated above 400°C. The additional supelattice reflections were attributed to ordered regions with the structure Pt_8Cr, of which Pt_8Ti is the prototype [7]. The Pt_8Ti ordered structure was first identified using XRD by Pietrokowsky [8] and confirmed using electron diffraction by Schryvers and Amelinckx [9]. The dark field image in Figure 5 shows that ordered regions are nano-sized and are evenly dispersed throughout the cold worked and heat-treated material.

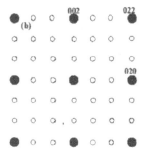

Figure 4. (a) Electron diffraction pattern obtained from a Pt 10 at.% Cr specimen that was heat treated at 400°C for 6 hours, viewed along the [100]$_{fcc}$ zone axis of the matrix. It shows, in addition to the main fcc lattice reflections, additional reflections at every ⅓ and ⅔ positions along the <002> and <022> directions. (b) An indexed schematic diagram of the diffraction pattern shown in (a).

Tensile tests were carried out on Pt 11 at.% Cr and the results in Figure 6 show that there is a steady increase in the ultimate tensile strength as heat treatment temperature increases from 200°C to 400°C. The heat treatments appear to have no significant effect on ductility until 800°C, when recrystallization occurs.

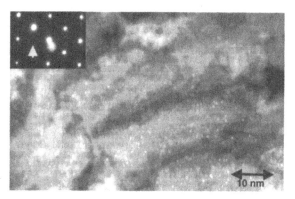

Figure 5. Dark field image of a Pt 10 at.% Cr specimen after heat treatment at 400°C for 6 hours, imaged using the reflection indicated by the arrow (inset [110]fcc diffraction pattern).

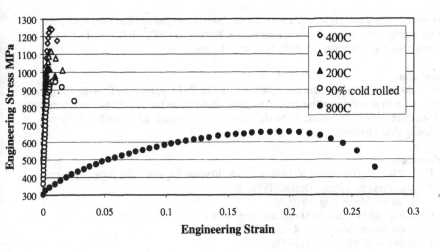

Figure 6. Typical stress strain curves from Pt 11 at.% Cr that was cold worked to 90% and then heat treated for three hours

It is interesting that the peak in UTS apparently occurred at 400°C and not at 300°C; however, the difference is within the standard deviation, as shown in Table 1.

Between 8 and 10 specimens were tested for each case but it can be seen from Table 1 that the standard deviation for the ultimate tensile strength gives a coefficient of variation between 5-9%. While 5% is acceptable for coefficient of variation in ultimate tensile strength, the changes we are seeing are subtle and the difference between the 300°C and 400°C is not readily distinguishable [10]. The undesirably large amount of scatter in the results could have been due to the unintentional heat treating during specimen preparation and inconsistencies in the

Table 1. Results of Vicker's hardness and tensile tests of Pt 11 at.% Cr after cold rolling 90% and additional heat treatments at temperatures from 200 °C -800°C

Heat Treatment Temperature °C	Vicker's Hardness	Ultimate Tensile Strength MPa	Fracture Strength MPa	%Strain to Failure
0	210 ± 7	1031 ± 64	973 ± 122	2.6 ± 0.8
200	303 ± 5	1113 ± 55	1027 ± 68	2.0 ± 1.1
300	324 ± 8	1153 ± 54	1042 ± 83	1.9 ± 0.6
400	300 ± 7	1167 ± 86	1102 ± 82	1.8 ± 1.0
500	278 ± 18	1084 ± 56	1032 ± 70	2.9 ± 1.4
800	170 ± 2	669 ± 58	474 ± 68	34 ± 3.0

specimen dimensions. However, it is clear from both the tensile and hardness results that strengthening is occurring after heat treatment at 300°C and 400°C. It is most like due to the ordering that was characterized in the Pt 10 at.% Cr alloy.

Conclusion

The results of the both hardness and tensile tests of the Pt 11 at.% Cr confirm the change in strength due to post deformation heat treatments. Pt_8Cr is the most likely ordered composition and structure since it was found in the slightly off-stoichiometric alloy that has the same hardening characteristics.

References:

1. E. Savitsky, V. Polyakova, N. Gorina, and N. Roshan, Physical Metallurgy of Platinum Metals, (Pergamon Press, Oxford, 1978), p. 76.
2. K. H. Miska, Mater. Eng. 1976, 65.
3. C.I. Lang and M.P. Shaw, Mater. Sci. Eng. A **164,** 180 (1993).
4. R.M. Waterstrat, Met Trans. **4,** 1585 (1973).
5. L. Muller, Ann. Phys. **15,** 9 (1930).
6. J. Baglin, F. d´Heurle and S. Zirinsky, J. Eletrochem. Soc. **125,** 1854 (1978).
7. M. Nzula, Title, Ph.D. thesis, University of Cape Town, Cape Town, South Africa, 2004.
8. P. Pietrokowsky, Nature **206,** 291 (1965).
9. D. Schryvers and S. Amelinckx, Acta Metall. **34,** 7 (1986).
10. N. Dowling, Mechanical Behavior of Materials, (Prentice Hall New Jersey, 1999), p. 803.

Acknowledgements

The authors would like to acknowledge funding from the Innovation Fund (South Africa) and the Claude Harris Leon Foundation.

Modeling of Intermetallics

FIRST-PRINCIPLES STUDY OF STRUCTURAL AND DEFECT PROPERTIES IN FeCo INTERMETALLICS

M. Krcmar, C.L. Fu and J.R. Morris
Metals and Ceramics Division, ORNL, Oak Ridge, TN 37831-6114

ABSTRACT

Employing *ab-initio* calculations and statistical thermodynamic modeling, we investigated the structural stability, defect energies, and ordering of B2 FeCo intermetallics. We find that FeCo in the B2 structure is a marginally stable and weakly ordered system, with a high density of antisite defects on both sublattices and low APB energies for the <111> slip on both {110} and {112} planes. The structural stability of B2 FeCo is very sensitive to the change in local atomic environment, as the system transforms to a lower-symmetry $L1_0$ phase under the effects of reduced dimensionality or applied shear stress. We suggest that internal stresses near dislocation cores might be closely connected with the intrinsic brittleness of ordered FeCo, as it is likely to induce a local structural transformation from the B2 structure to the $L1_0$ structure.

INTRODUCTION

The FeCo alloys possess exceptional magnetic properties (e.g., high Curie's temperature) for a variety of soft magnetic materials applications. The FeCo alloys are in the disordered bcc structure from 1000 K to 1200 K, but transform into the ordered B2 structure below 1000 K. The order-disorder transformation profoundly affects the mechanical properties of FeCo alloys, as they change from ductile (disordered) to brittle (ordered). The poor ductility makes ordered FeCo alloys difficult to process. Even though the brittleness can be suppressed by alloying with other elements, their presence usually deteriorates the FeCo magnetic properties. Therefore, key issues regarding the applicability of Fe-Co alloys would need to address the nature of structural stability [1,2], especially since the origin of B2 FeCo brittleness is not yet understood [3,4]. Here, we report results of first-principles calculations on the defect energies of B2 FeCo and the prediction of the B2 to $L1_0$ phase transformation under reduced dimensionality (i.e., bulk to thin films) and applied shear stresses.

RESULTS AND DISCUSSIONS

B2 FeCo structural stability

The stabilization of FeCo in the ordered B2 structure originates from the existence of ferromagnetic ordering. Magnetism stabilizes the B2 structure by a drastic decrease in the electronic energy: in the majority spin channel, the d-band is nearly fully occupied, thus chemically inactive; in the minority spin channel, the Fermi level (E_F) lies near the pseudo-gap separating the bonding and non-bonding states. The calculated magnetic moments are large, 2.83 μ_B for Fe and 1.82 μ_B for Co, suggesting that the Fe-Co exchange interaction exceeds that of either Fe-Fe or Co-Co in bulk Fe and bulk Co. The calculated heat of formation is small (-0.065 eV/atom), suggesting marginal stability of the ordered B2 FeCo. This marginal stability, as we will discuss below, provides grounds for a structural transformation when bonding is disrupted.

Point defect structure and ordering

Experimentally, B2 FeCo is known as a weakly ordered system. To predict the order-disorder phase transition temperature from first-principles, we start by calculating the point defect energies in the dilute defect limit. Then, we obtain point defect structure based upon statistical thermodynamic modeling with the inclusion of configurational entropy [5-7]. We find that B2 FeCo has abundance of antisite defects on both sublattices at stoichiometry; formation energies of vacancies exceed those of antisite defects by order of magnitude, thus vacancy concentrations are negligible. The site-exchange energy between Fe and Co in the dilute defect limit is found to be very low (0.34 eV). The site-exchange energy, however, should depend on defect concentration, since the interaction between defects becomes increasingly significant as the defect concentration increases. To include effects of partial disordering on the defect energy in the B2 structure, we assume that the site-exchange energy decreases linearly with antisite defect concentration. This assumption is justified since the interatomic interaction in the B2 structure is dominated by the nearest-neighbor interactions. The model based on this assumption enables us to predict precisely the defect concentration as a function of temperature, as well as the order-disorder transition temperature T_c as a function of alloy composition.

The calculated results are shown in Fig. 1. The Arrhenius plot for defect concentration vs. 1/T does not exhibit standard linear behavior due to the presence of defect interactions that affect the site-exchange energy. The order-disorder transition temperature is determined by the temperature at the critical point which corresponds to a defect concentration of 0.25. The calculated T_c at stoichiometry is 980 K, in excellent agreement with the experimental value of 1000 K. Furthermore, the order-disorder transition temperature decreases parabolically with the amount of alloy composition deviating from stoichiometry, as observed experimentally.

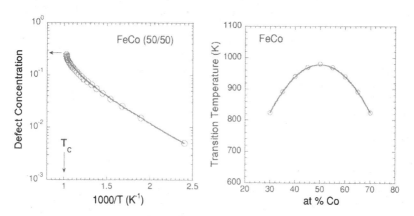

Figure 1. Left panel: antisite defect concentration at stoichiometry as a function of temperature in B2 FeCo. Right panel: order-disorder transition temperature as a function of composition of B2 FeCo.

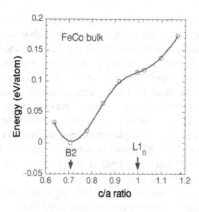

Figure 2. Energy versus c/a-ratio of bulk FeCo in the face-centered-tetragonal structure. Here the $L1_0$ structure denotes the structure with ideal c/a=1 ratio. See text for details.

APB energies

Since B2 FeCo is weakly ordered with low site-exchange energy, we would expect low anti-phase boundary (APB) energies in these alloys. Indeed, our calculations find that the APB energies are low with magnitudes of 129 mJ/m^2 and 169 mJ/m^2 for <111> slip on the {110} and {112} plane, respectively. The fact that FeCo has low APB energies, however, is insufficient to warrant room temperature B2 ductility. The origin for the brittleness of B2 FeCo remains poorly understood.

Structural instability of B2 FeCo

It has been observed recently that an ordered Fe-Co superlattice of $L1_0$-based structure can be stabilized on the fcc Cu (100) substrate [2]. The layer-by-layer growth in $L1_0$ structure, however, exists only up to 6 monolayers (ml), and significant surface roughening occurs as Fe-Co thickness increases beyond 6 ml. To understand the exact cause of this unexpected quick collapse of the FeCo $L1_0$ phase, we have studied the structural stability of both bulk FeCo and (001)-oriented thin Fe/Co slabs using first-principles calculations. Figure 2 shows the total energy of bulk FeCo in the ordered face-centered-tetragonal (fct) structure as a function of the c/a ratio. Our attention is focused on two particular c/a ratios of the fct structure, i.e., c/a=1 (fcc-based ideal $L1_0$ structure) and $1/\sqrt{2}$ (bcc-based B2 structure). For bulk FeCo, although the total energy is minimized at c/a = $1/\sqrt{2}$, there is a 'shoulder' in the energy-versus-c/a curve near c/a=1.0. Such a 'shoulder' indicates that a structural transformation is possible when atoms are in a different environment than in the bulk.

1. Effect of reduced dimensionality (bulk to thin-film transition)

We have calculated the total energies of seven, five and three atomic layer slabs with both Fe- and Co-terminated surfaces. The averaged energy per atom in the slabs as a function of c/a ratio is also shown in Fig. 3, in which the energy of the B2 phase is set as a reference in each case. A truly remarkable effect is found regardless of the type of surface termination. For the seven-layer slabs, a local energy minimum develops near c/a =1.0; however, the slab maintains the absolute energy minimum at the c/a ratio close to that of the bulk B2 structure. For the five-layer slabs, the system stabilizes near the c/a = 1.0 region, whereas the minimum corresponding to the B2 structure begins to disappear. For the three-layer slabs, the slab is stabilized near c/a =1.05-1.10, whereas the structure near c/a = 1/√2 becomes unstable. These calculated results provide a clear picture for the experimental observations. At low thickness, the system prefers, energetically, to be in $L1_0$ structure, which has a lattice constant (3.58 Å for c/a=1.0) close to that of Cu (3.61 Å), allowing good layer-by-layer growth. At higher thickness, the system can minimize its energy by transforming into the more stable B2 phase. However, because the B2 structure has a large lattice mismatch with Cu (~12%), layer-by-layer growth breaks down and a significant surface roughening is thus observed experimentally.

As the ordered bulk B2 FeCo structure is stabilized by the effect of magnetism, the B2 to $L1_0$ structural phase transformations in reduced dimensionality can be understood from electronic structure. As an example, we have analyzed the layer-projected density of states (DOS) of Fe atoms in Fe-terminated B2 FeCo slab. For the bulk Fe atom, the E_F is located near the valley of the pseudo-gap separating the bonding and non-bonding states; for the surface Fe atom, the effect of band narrowing due to reduced coordination number smears out the pseudo-gap, and the surface DOS at E_F is significantly increased from its bulk value. Therefore, the structural stability of Fe-Co slabs depends on the surface-to-bulk volume ratio. It is far more costly to create a surface on the B2 than on the $L1_0$ lattice. In other words, in reduced dimensionality, ordered FeCo systems can be more stable in the $L1_0$-like structure than in B2-structure, with the transition occurring at a slab thickness of 6 ml.

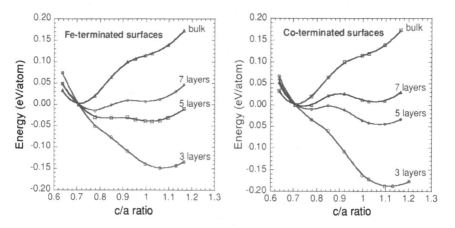

Figure 3. The total energy of Fe-terminated (left) and Co-terminated (right) slabs in the fct structure as a function of c/a ratio. The energy of the B2 structure (at c/a=0.707) is set as the reference in each case (i.e., as zero energy). For thin slabs, the structure transforms to a $L1_0$-like phase with c/a ~ 1.

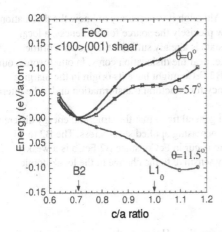

Figure 4. Energy vs. c/a ratio of bulk fct FeCo under <100>(001) sheer stress. For each shear angle (θ), the energy of the sheared B2 structure is taken as the reference (i.e., set as energy zero) .

2. Effect of shear stress

Next, we explore if the shoulder in the energy vs. c/a-ratio curve of bulk fct FeCo (Fig. 2) can develop into a true energy minimum even if the system retains its dimensionality. We find that B2 FeCo shows a tendency to transform into the $L1_0$ structure under applied shear stress. In Fig. 4 we present the energy vs. c/a-ratio curves for FeCo under the C_{44}-type shear stress in the (001) plane (i.e., <100>(001) shear) for several shear angles (θ). As the shear angle increases, the shoulder region (c/a~1) becomes more pronounced; at θ =5.7°, there already exists a meta-stable region in the vicinity of the ideal $L1_0$ structure; and to emphasize the effect under large (although unrealistic) shear angles, at θ =11.5°, the deformed $L1_0$-like structure has a much lower energy than the B2 structure. Note that since $L1_0$ phase has a order-disorder transition temperature above 300° C, the $L1_0$ structure is a lower symmetry (and brittle) phase.

We note that the C_{44} stress reduces the nearest-neighbor distance more significantly in the B2 structure than in the $L1_0$ structure. Since the bonding in the B2 structure is dominated by the nearest-neighbor interactions, the structural stability of the B2 structure is clearly disrupted by the changes in the nearest-neighbor bonding. This is evident in the increase in the DOS at E_F in the minority spin channel of B2 FeCo under C_{44}-type stress. On the other hand, in the close-packed $L1_0$-like structure, there is no clear separation between the bonding and non-bonding states in the DOS profile, thus the change in the electronic energy under shear stress is less than that of the B2 structure.

From Figure 4, it is clear that the combination of tetragonal distortion (i.e. C_{11}-C_{12} type shear stress) and C_{44} type shear stress can locally bring FeCo system into a low symmetry meta-stable $L1_0$ structure with c/a ~1. Note that, at small shear angles ($\theta \leq 5.7°$), the energy difference between B2 and $L1_0$ (with c/a ~1) structures is small: for an energy difference of 0.05 eV/atom, the corresponding B2 to $L1_0$ transformation stress is only ~ 10 GPa (with strain given by the amount of contraction in the (100) direction). Although it is necessary to know the dislocation core structure and dislocation mobility to know precisely the source for brittleness, a local transition into the low symmetry, meta-stable state is already sufficient to produce brittle fractures in the vicinity of the stress sources, i.e. near the dislocation cores. In other words, our results suggest that the intrinsic brittleness of B2 FeCo might have its origin in the marginal stability of the ordered B2 phase and its tendency for structural transformation under the internal stress.

For the B2 structure, we find that it is a general trend that the structural energy difference between B2 and $L1_0$ structures decreases with increasing applied shear stress. The B2 to $L1_0$ structural phase transformation is more likely to occur in FeCo, since B2 FeCo is a weakly ordered system and its electronic energy is very sensitive to the change in the local atomic bonding environment.

SUMMARY

B2 FeCo is stabilized by the effect of magnetism. However, the system is weakly ordered, marginally stable and has a tendency for structural transformation to a lower symmetry $L1_0$ phase under reduced dimensionality or applied shear stresses. Tendency for structural transformation can have a significant influence on the mechanical properties of FeCo, and might be closely connected with the intrinsic brittleness of alloys.

ACKNOWLEDGEMENTS

This work was sponsored by the Division of Materials Sciences and Engineering, the U.S. Department of Energy, under contract DE-AC05-00OR-22725 with UT-Battelle, LLC.

REFERENCES:

1. I.A. Abrikosov et. al., *Phys. Rev. B* **54**, 3380 (1996).
2. G.A. Farnan et. al., *Phys. Rev. Lett.* **91**, 226106-1 (2004).
3. L. Zhao, I Baker and E.P. George, *Mat. Res. Soc. Symp. Proc.* **288**, 501(1993).
4. E.P. George et. al., *Mat. Sci. Eng.* **A329-331**, 325 (2002).
5. Y. Mishin and Chr. Herzig, *Acta Mater.* **48**, 589 (2000).
6. C.L. Fu, *Phys. Rev. B* **52**, 3151 (1995).
7. M. Neumayer and M. Fähle, *Phys. Rev. B* **64**, 132102-1 (2001).

Mater. Res. Soc. Symp. Proc. Vol. 842 © 2005 Materials Research Society   S2.7

Dislocation Structure, Phase Stability and Yield Stress Behavior of Ultra-High Temperature L1₂ Intermetallics: Combined First Principles-Peierls-Nabarro Approach

Oleg Y. Kontsevoi, Yuri N. Gornostyrev and Arthur J. Freeman
Department of Physics and Astronomy, Northwestern University, Evanston, IL 60208-3112

ABSTRACT

We present results of comparative studies of the dislocation properties and the mechanical behavior for a class of intermetallic alloys based on platinum group metals (PGM) which are being developed for ultra-high temperature applications: Ir_3X and Rh_3X (where X = Ti, Zr, Hf, V, Nb, Ta). For the analysis of dislocation structure and mobility, we employ a combined approach based on accurate first-principles calculations of the shear energetics and the modified semi-discrete 2D Peierls-Nabarro model with an ab-initio parametrization of the restoring forces. Based on our analysis of dislocation structure and mobility, we provide predictions of temperature yield stress behavior of PGM-based intermetallics, show that their dislocation properties are closely connected with features of the electronic structure and the $L1_2 \rightarrow D0_{19}$ structural stability, and demonstrate the dramatic difference in dislocation structure and the mechanical behavior between PGM alloys with IVA and VA group elements.

INTRODUCTION

Modern two-phase γ/γ' nickel-based superalloys, designed to be used under extreme conditions (high temperatures, aggressive environment and high operating stress), have found a wide range of high temperature applications such as jet turbines and power generators [1]. At present, however, the potential for a further increase of the temperature capability of these materials has been exhausted; the current γ/γ' alloys operate at temperatures up 85% of their melting temperature.

A new approach for the development of high–temperature superalloys is based on the use of platinum group metals (PGM) with their higher melting temperatures and superior environmental properties. For example, Ir-Nb alloys can sustain high strength properties up to 1800 °C [2], which is the highest temperature resistant alloy of all the platinum group alloys (PGA) studied so far. Rh-based alloys, while having lower melting points compared to Ir alloys, possess a large advantage in density and have superior oxidation resistance [3]. Although there are few doubts as to the potential of the PGA for ultra-high temperature applications, further progress in their development is hindered due to an insufficient understanding of the fundamental factors that control their mechanical properties. In particular, there is no reliable information about dislocation structure and mobility in these alloys.

One important characteristic affecting the potential of alloys for high temperature applications is the temperature dependence of the yield stress, $\sigma_y(T)$. It is now well established that the yield stress temperature anomaly (YSA), or the increase of $\sigma_y(T)$ with temperature, is an intrinsic property of many $L1_2$ alloys resulting from features of their superdislocations which can lose mobility due to thermally activated transformation into a non-coplanar configuration (Kear-Wilsdorf locks) [4]. There are only a few $L1_2$ systems (e.g., Fe_3Ge alloys), where the YSA has not been observed; for PGA, existing experiments give contradictory results. A comprehensive study of this problem should necessarily include an analysis of the dislocation properties.

In this study, we employ a theoretical approach that allows us to analyze the structure, energetics and mobility of dislocations on the solid foundation of highly accurate first-principles calculations. It combines *ab initio* calculations of the generalized stacking fault (GSF) energetics and the modified semi-discrete two-dimensional (2D) Peierls–Nabarro (PN) model – that was proven highly successful in our previous studies of the dislocation structure of metals and alloys in different structures [5].

METHODOLOGY

Our combined *ab initio*–PN approach, which we briefly outline below, offers the advantage of describing dislocation structure starting from accurate electronic structure calculations of GSF energetics. In the framework of the PN model, the total energy functional E_{tot} of the crystal with an infinite straight dislocation can be written as [6]

$$E_{tot}(\mathbf{u}(x)) = E_{el}(\mathbf{u}(x)) + E_{mis}(\mathbf{u}(x)) \tag{1}$$

with a linear elastic (E_{el}) and a non-linear (sinusoidal) atomistic misfit energy term (E_{mis}), which depend on relative displacements $\mathbf{u}(x)$ of atomic rows below and above the cut plane (here x is the distance from the dislocation axis in the slip plane). The misfit energy

$$E_{mis} = h \sum_{n} \Phi(\mathbf{u}(nh-l)), \tag{2}$$

is defined by a periodic energy profile $\Phi(\mathbf{u}(x))$ which we assume to be equal to the GSF energy [7] (here, n numbers atomic rows, h is the distance between them, and l the position of the dislocation center in the lattice). This so-called local approximation for $\Phi(\mathbf{u}(x))$ is sufficiently accurate for the purpose of determining the dislocation structure [8]. In the PN model framework, the distribution of $\mathbf{u}(x)$ is determined from the condition of a balance between elastic stresses on the cut plane intersecting the elastic medium into two half-spaces, and the atomistic forces originating from the interaction potential between these half-spaces (see [9] for details).

To determine the superdislocation core structure, we employ the modified 2D-PN model which was described earlier [5]. We perform a minimization of the total energy functional, Eq.(1), with a discrete representation of the misfit energy, Eq.(2), using trial functions, $\mathbf{u}(x)$, defined from the Laurent expansion [10] of their derivatives

$$\rho_{\beta}(x) = \frac{du_{\beta}(x)}{dx} = Re \sum_{k=1}^{N} \sum_{n=1}^{p_k} \frac{A_{nk}^{\beta}}{(x - z_k^{\beta})^n}, \tag{3}$$

where N is the maximal number, p_k is the maximal order of the poles z_k^{β}, and the A_{nk}^{β} are expansion coefficients. The poles $z_k^{\beta} = x_k^{\beta} + i\omega_k^{\beta}$ have a clear meaning: x_k^{β} gives the position, and ω_k^{β} gives the width of the partials for the screw ($\beta = 1$) and edge ($\beta = 2$) components of the displacement in the partial cores.

Using these trial functions, the energy functional, Eq. (1), can be transformed into a function of several essential dislocation structure parameters, $E(d_k, \omega_k, l, \mathbf{b_k})$ where d_k are the distances between partials, ω_k is the partials width , and $\mathbf{b_k}$ is the Burgers vector. Then, these sets of parameters are determined from minimization of the energy functional Eq. (1). The dislocation splitting scheme is not introduced *ad hoc* but appears naturally within this theoretical approach as the so-called "splitting path" [11] – or dependence of the screw (u_1) on the edge (u_2) component of displacements (e.g., function $u_1 = f(u_2)$).

The GSF energy $\Phi(\mathbf{u})$, which is defined as the energy associated with a rigid shift of one-half of an ideal infinite crystal with respect to another half on a fault vector $\mathbf{u}$ in a certain slip plane [7], was calculated using the first-principles all-electron full-potential linearized augmented plane wave (FLAPW) method [12] without any shape approximation for the potential and charge density. The generalized gradient approximation was used for the exchange-correlation potential [13]. To simulate the faults, we used supercell geometry with six atomic layers and homogeneous periodic boundary conditions [14]. The calculations were performed for theoretical equilibrium lattice constants, which were found to agree with experimental values to within 1% for all alloys.

GSF energies on the {111} slip plane were calculated for several displacement vectors **u** in this plane including antiphase boundary (APB), complex stacking fault (CSF), and superlattice intrinsic stacking fault (SISF) vectors. The full GSF for arbitrary fault vectors, also known as a γ-surface, was then restored from these data using Fourier expansions over vectors in the 2D lattice which is reciprocal to the 2D lattice of the slip plane.

RESULTS

In Fig. 1, the section of the calculated {111} γ-surface along the APB–SISF direction for Ir(Rh)₃X alloys is shown. One can see that there is a strong correlation between the SISF energy in Ir₃X and the position of the alloying element X in the Periodic Table: (i) SISF energy gradually *increases* with the increase of the period number of the element within the same group, e.g. Ti→Zr→Hf (group IVA) and V→Nb→Ta (group VA); (ii) SISF energy dramatically *decreases* with the increase of the group number for the elements of the same period, Ti→V, Zr→Nb, and Hf→Ta and suggests that all Ir(Rh)-based alloys considered can be divided into two groups. The first group consist of Ir(Rh) alloys with IVA elements (Ti, Zr, Hf); these alloys have a "normal" SISF of 0.3-0.4 J/m² and a "normal" ratio of the SISF to the APB$_{\{111\}}$ energies (0.3-0.5) typical for most of L1₂ alloys like Ni₃Al with the YSA. To the second group belong Ir(Rh) alloys with VA elements (V, Nb, Ta), with an abnormally small SISF energy of 0.1 J/m² or less, and SISF/APB$_{\{111\}}$ << 1. It should be noted, that for Rh-based alloys the difference between these two groups is not as dramatic as for Ir-based ones, and Rh₃Ta can be considered as a boundary case.

The current dislocation theory (see, e.g., [15]) suggests that such a low ratio should result in a change of the dislocation structure and a lack of anomalous $\sigma_y(T)$ behavior. Indeed, a similar small SISF/APB$_{\{111\}}$ ratio was previously found for Fe₃Ge – one of only a few L1₂ alloys without the YSA. Thus, based on the results of the stacking fault calculations one would expect that dislocation structure and mechanical behavior of each alloy group should be different.

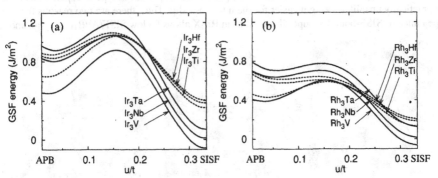

Figure 1 Section of the {111} GSF surface in (a) Ir₃X and (b) Rh₃X alloys along ⟨112⟩ direction corresponding to the path from ideal APB to SISF positions.

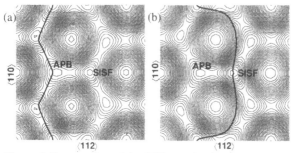

The dislocation structures were calculated by minimization of the energy functional Eq. (1) using the full γ-surfaces as an input to Eq. (2). Several initial configurations with arbitrary parameters were chosen as starting points for mimimization. The dislocation structures with lower energies, are presented in Figs. 2 and 3 in the form of "splitting paths" together with the corresponding γ-surfaces (shown as a contour plot).

Figure 2 Contour plots of the GSF energies for (a) Ir₃Zr and (b) Ir₃Nb; the structures of the superdislocations are shown as "splitting paths" (bold curves).

For Ir₃X alloys, two representatives of their group are shown: Ir₃Zr from the Ir-IVA (Fig. 2(a)) alloys and Ir₃Nb from the Ir-VA group (Fig. 2(b)). We found that superdislocations with type II core structure (Kear splitting scheme $\langle 110 \rangle \rightarrow 1/3\langle 2\bar{1}\bar{1} \rangle + SISF + 1/3\langle 121 \rangle$ with a SISF ribbon between partials) are strongly preferable energetically for Ir₃Nb, and type I superdislocations (Shockley splitting scheme $\langle 110 \rangle \rightarrow 1/2\langle 110 \rangle + APB + 1/2\langle 110 \rangle$ with an APB ribbon between partials) are predicted for Ir₃Zr. We also found that type I superdislocations are possible in Ir₃Nb as a metastable solution with higher energy in addition to the type II superdislocations; at the same time, type II superdislocations are impossible in Ir₃Zr.

For Rh₃X we found similar tendencies in dislocation structure as for Ir₃X intermetallics: the dislocation structure is qualitatively different between Rh-IVA and Rh-VA alloys. However, the Rh₃X intermetallics are distinguished by the small value of the energy barrier between APB and SISF, which practically disappear for Rh₃Hf and Rh₃Zr (Fig. 1(b)). Due to shallow APB minima, type I dislocations become unstable, and only SISF-bounded superdislocations are realized in these two alloys (Fig. 3) despite relatively high value of the SISF energy, analogously to what was previously found for Fe₃Ge [5]. These SISF-bounded superdislocations are of an uncommon kind, called type II' [5]: they have a different sequence of the Shockley partials (c.f., Fig. 3(b) and 3(c)) compared with type II. Note also that screw components of the partials in type II' superdislocations are practically unsplit, making it easier for them to cross-slip. Thus, there are two reasons for appearance of SISF-bounded superdislocations in Rh₃X alloys (i) low SISF/APB$_{\{111\}}$ ratio, and

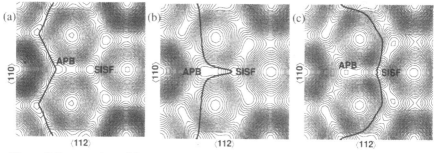

Figure 3 Contour plots of the generalized stacking fault energies for (a) Rh₃Ti, (b) Rh₃Zr and (c) Rh₃Nb; the structures of the superdislocations are shown as "splitting paths" (bold curves).

(ii) unstable APB energy, which result in type II and type II' superdislocations, correspondingly.

The unusually low SISF energy found in Ir(Rh)-VA alloys could manifest a possible structural instability of the L1$_2$ phase with respect to the hexagonal D0$_{19}$ structure. Indeed, the twin-like shear associated with the {111} SISF results in the formation of a local *hcp*-like atomic coordination equivalent to that in D0$_{19}$. We calculated the total energies of Ir(Rh)$_3$X alloys in both L1$_2$ and D0$_{19}$ structures, and plotted the SISF energy in Fig. 4 as a function of the $E(D0_{19}) - E(L1_2)$ energy difference. One can see that there is an almost ideal linear correlation between the SISF energy and the structural stability of the L1$_2$ phase, and the Ir(Rh)-VA alloys show a low stability.

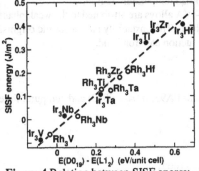

Figure 4 Relation between SISF energy and energy difference $E(D0_{19}) - E(L1_2)$

It is well established that the structural stability of L1$_2$ intermetallics is governed by the electronic structure and is connected with electron band filling (see, e.g., [16]) and pair interaction strengths [17]. An incompletely filled valence band favors the L1$_2$ structure, further filling of the works in favor of the D0$_{19}$ structure. Our calculations show that for all Ir(Rh)-VA the Fermi level (E_F) lies at the deep pseudogap in the L1$_2$ electron density of states (DOS) dividing bonding from antibonding states thus indicating that the L1$_2$ structure is unstable. The exact E_F position, however, is a little bit different: slightly below the DOS minima in the bonding region for Ir(Rh)$_3$Ta, which is an evidence that the L1$_2$ structure is still more stable; at the minimum for Ir(Rh)$_3$Nb, suggesting that this alloy is at the point of the loss of the L1$_2$ ordering preference; and slightly above the DOS minimum in the antibonding region for Ir(Rh)$_3$V, which signifies that the L1$_2$ structure is unstable. These band filling tendencies correlate exactly with the calculated energy difference $\Delta E = E(D0_{19}) - E(L1_2)$

DISCUSSION AND CONCLUSIONS

Since it is considered that only type I superdislocations can lead to the YSA [4,15], our results suggest that normal $\sigma_y(T)$ behavior should be expected in Ir(Rh)-VA alloys where type II superdislocations are found, as well as in Rh$_3$Zr, and Rh$_3$Hf, where we predict type II' superdislocations. In Ir-IVA group alloys and in Rh$_3$Ti we predict that the type I superdislocations will operate. However, the APB energy difference between {111} and {001} planes was found to be too small and the condition APB$_{\{111\}}$/APB$_{\{001\}} > \sqrt{3}$ which is usually considered as necessary for the YSA (see [4,15]) is not satisfied. Thus, based on the results of our calculations, we expect the normal (negative) yield stress temperature behavior in both Ir(Rh)$_3$X alloy groups.

The structure of dislocations in Ir(Rh)$_3$X alloys has not been studied experimentally. There is only one TEM observation [18] of a few superdislocations in Ir$_3$Nb; type II superdislocations were observed and no evidence of type I was found. These observations are in excellent agreement with our results, thus proving that type II superdislocations are the major slip mode in Ir$_3$Nb.

The results of experimental investigations of the $\sigma_y(T)$ dependence in Ir$_3$X alloys are contradictory. The decrease of strength with temperature was found in Ir$_3$V, Ir$_3$Ti [19], and in Ir$_3$Zr and Ir$_3$Nb [20]. In Ref. [21], however, the YSA was observed in Ir$_3$Zr and Ir$_3$Nb above room temperature and normal yield stress behavior was found in Ir$_3$Ti and Ir$_3$Hf. Thus, there is a strong disagreement between the results obtained by different experimental groups. Nevertheless, all of them agree that at low temperatures all Ir_3X intermetallics show a decrease of σ_y with temperature. Rh_3Ti, Rh_3Nb and Rh_3Ta also have normal yield stress behavior up to 1000 K [22]. Since the results of our dislocation analyses are based on *ab initio* GSF calculations corresponding to the ground state at $T = 0$ K, they are most applicable to the properties of Ir_3X alloys in the low temperature region, where our conclusions about the normal $\sigma_y(T)$ dependence are in agreement with experiment. A temperature increase can change the relation between the $APB_{\{111\}}$ and $APB_{\{001\}}$ energies (for example, due to temperature-induced local disorder), and result in the appearance of the YSA in Ir-IVA alloys and Rh_3Ti. At the same time, we expect normal behavior in the Ir(Rh)-VA alloys and Rh_3Zr and Rh_3Hf over a broad temperature region.

The unusual dislocation properties found in Ir(Rh)-IVA alloys are attributed to the weak phase stability of the $L1_2$ phase in with respect to the $D0_{19}$ phase. This instability has electronic origins, resulting from the degree of filling of the bonding conduction electron band.

ACKNOWLEDGMENTS

This work was supported by the AFOSR (under grant FA9550-04-1-0013) and computer time grants at NAVO, ARSC, and ARL.

REFERENCES

1. D.G. Backman and J.C. Williams, *Science* **255**, 1082 (1992).
2. Y. Yamabe-Mitarai, Y. Ro, T. Maruko, and H. Harada, *Metall. Mater. Trans. A* **29**, 537 (1998).
3. Y. Yamabe-Mitarai, Y. Koizumi, H. Murakami, Y. Ro, T. Maruko, and H. Harada, *Scripta Mater.* **36**, 393 (1997).
4. P. Veyssiere and G. Saada, in *Dislocations in Solids*, ed. F.R.N. Nabarro, and M.S. Duesbery, (Elsevier, Amsterdam, 1996), vol. 10, pp. 254-441.
5. O.N. Mryasov, Yu.N. Gornostyrev, M. van Schilfgaarde, and A.J. Freeman, *Acta Mater.* **50**, 4545 (2002), and references therein.
6. G. Schoeck, *Phil. Mag. A* **69**, 1085 (1994).
7. V. Vitek, *Crystal Lattice Defects* **5**, 1 (1974).
8. R. Miller and R. Phillips, *Phil. Mag. A* **73**, 803 (1996).
9. J.P. Hirth and J.Lote, *Theory of Dislocations* (McGraw-Hill, New York, 1968).
10. L. Lejcek, *Czech. J. Phys.* **26**, 294 (1976).
11. G. Schoeck, *Phil. Mag. A* **81**, 1161 (2001).
12. E. Wimmer, H. Krakauer, M. Weinert, and A.J. Freeman, *Phys. Rev. B* **24**, 864 (1981).
13. J.P. Perdew, K. Burke, and M. Ernzerhof, *Phys. Rev. Lett.* **77**, 3865 (1996).
14. A.T. Paxton and Y.Q. Sun, *Phil. Mag. A* **78**, 85 (1998).
15. V. Paidar, D.P. Pope, and V. Vitek, *Acta Metall.* **32**, 435 (1984).
16. J.-H. Xu, W. Lin, and A.J. Freeman, *Phys. Rev. B* **48**, 4276 (1993), and references therein.
17. A. Bieber and F. Gautier, *Solid State Commun.* **38**, 1219 (1981).
18. Y. Yamabe-Mitarai, M.-H. Hong, Y. Ro, and H. Harada, *Phil. Mag. Lett.* **79** 673 (1999).
19. D.M. Wee and T. Suzuki, *Trans. JIM* **20**, 634(1979).
20. A.M. Gyurko and J.M. Sanchez: *Mater. Sci. Eng. A* **170**, 169 (1993).
21. Y. Yamabe-Mitarai, Y. Ro, and S. Nakazawa: *Intermetallics* **9**, 423 (2001).
22. S. Miura, K. Honma, Y. Terada, J.M. Sanchez, and T. Moria, Intermetallics **8**, 785 (2000).

Mater. Res. Soc. Symp. Proc. Vol. 842 © 2005 Materials Research Society   S4.10

Atomic, Electronic, and Magnetic Structure of Iron-Based Sigma-Phases

Pavel A. Korzhavyi,[1] Bo Sundman,[1] Malin Selleby,[1] and Börje Johansson[1,2]
[1]Department of Materials Science and Engineering, Royal Institute of Technology (KTH), SE-100 44 Stockholm, SWEDEN;
[2]Condensed Matter Theory Group, Department of Physics, Uppsala University, SE-751 21 Uppsala, SWEDEN.

ABSTRACT

A combination of *ab initio* total energy calculations with Calphad approach is applied to model the site occupancy and thermodynamic properties of the Fe-Cr, Co-Cr, Fe-V, and Fe-Mo binary sigma-phases as a function of composition and temperature. For each binary sigma-phase the parameters of the model are the *ab initio* calculated total energies of so-called end-member compounds, which represent all the $2^5=32$ variants of complete occupancy of each of the five crystallographically inequivalent sites by one or the other alloy component. The paramagnetic state of the sigma-phases has been taken into account within the disordered local moment approach. The Fe and Co atoms are found to retain high spin moments when they occupy high-coordination-number sites in the structure. Using our model we were able to reproduce the experimentally observed site occupancy in the FeCr sigma-phase. The calculated site occupancies in the Co-Cr, Fe-V, and Fe-Mo sigma-phases are also presented and discussed.

INTRODUCTION

Sigma-phases are complex tetragonal structures [1] (structure type $D8_b$) that form in transition metal alloys having the average number of $s+d$ electrons in the interval 6.2 – 7.4. These phases, which may form during solidification or heat treatment, cause embrittlement of transition metal alloys such as stainless steels or Ni-based superalloys. A well known example is the FeCr σ-phase, whose formation has a deleterious effect on the mechanical properties of stainless steels. Like other σ-phases, the FeCr phase possesses a relatively high degree of chemical disorder among its five crystallographically inequivalent sites. Among these lattice sites (see table I) one finds two kinds of icosahedrally coordinated sites (IC sites 1 and 4) and three types of sites with high coordination numbers (high-CN sites 2, 3, and 5). It has been found experimentally that the group VIIB and VIII elements tend to occupy the IC sites (10 out of the 30 atomic positions in the structure), while the remaining 20 atomic positions of high-CN sites are preferentially occupied by the group VB and VIB elements [2]. Accurate site-occupancy data have been obtained [3] for the Fe-Cr binary σ-phase using X-ray diffraction experiments on single crystals, whereas for the majority of σ-phases only qualitative ordering schemes are available. The aims of the present theoretical study were to investigate the electronic structure and to derive the site occupancy in the Fe-Cr, Co-Cr, Fe-V, and Fe-Mo sigma-phases, as a function of composition and temperature, from *ab initio* calculated total energies for the complete set of $2^5=32$ end member compounds (all the variants of integral site occupancy compatible with space group $P4_2/mnm$) for each binary system. The compound energy formalism [4] was employed to perform thermodynamic modeling, where the calculated total energies were used as the model parameters.

Table I. Structural parameters of the five inequivalent crystallographic sites in the σ-phase structure, Pearson symbol tP30, described using centrosymmetric space group $P4_2/mnm$.

Site	1	2	3	4	5
Multiplicity (Wyckoff index)	2(a)	4(f)	8(i)	8(i)	8(j)
Coordination number (CN)	12	15	14	12	14

METHODOLOGY

Combination of *ab initio* calculations with thermodynamic modeling provides an access to the temperature- and composition-dependent properties of solid phases. This combination has been successfully applied to study the site occupancy in the W-Re σ-phase on the basis of *ab initio* calculated total energies, using the cluster variation method (CVM) [6] and the compound energy formalism (CEF) [5]. The two modeling techniques have yielded almost identical results, which also agree with experimental data [6]. However, applications of these techniques to the Fe-Cr case have not been very successful [7, 8]. Here we show that a good description of Fe- and Co-based σ-phases can be achieved if their paramagnetism is properly taken into account.

In the present work we use the compound energy formalism to perform finite-temperature modeling. The configurational space of a binary A-B σ-phase (where A={V,Cr,Mo} and B={Fe,Co}) is equivalent to that of five independent binary alloys on the five inequivalent sites (sublattices), as given by formula $(A,B)_2(A,B)_4(A,B)_8(A,B)_8(A,B)_8$. In the simplest version of the CEF, alloy on each sublattice is modeled as a regular solution with a zero energy (and an ideal entropy) of mixing. The total energy is represented in the configurational space by the surface of reference:

$$E = \sum_{ijklm} y_i^{(1)} y_j^{(2)} y_k^{(3)} y_l^{(4)} y_m^{(5)} E_{ijklm} , \qquad (1)$$

where E_{ijklm} are the calculated total energies of the end-member compounds *ijklm* ($i...m$ ={A,B}) and $y_i^{(s)}$ is the mole fraction of component *i* on site *s*. The configurational entropy for the set of random solid solutions on the five sites is given by

$$S_{conf} = -k_B \sum_s a^{(s)} \sum_{i=A,B} y_i^{(s)} \ln\left(y_i^{(s)}\right) , \qquad (2)$$

where k_B is the Boltzmann constant and $a^{(s)}$ is the multiplicity of site *s*, as given in table I. It will be shown below that in the paramagnetic state of σ-phases the B type atoms (Fe and Co) possess non-zero local magnetic moments, $M_B^{(s)}$, thus giving rise to a magnetic entropy, which can be approximated as:

$$S_{magn} = k_B \sum_s a^{(s)} y_B^{(s)} \ln\left(M_B^{(s)} + 1\right) . \qquad (3)$$

The electronic structure and total energy calculations were based on density functional theory (DFT) and employed the Korringa-Kohn-Rostoker (KKR) method, in conjunction with the multipole-corrected atomic sphere approximation (ASA+M) [9]. The local density approximation (LDA) [10] was used in the DFT self-consistency, whereas the total energies were calculated from the LDA charge densities using the local Airy gas (LAG) exchange-correlation potential [11]. A basis set with the angular momentum cutoff $l_{max}=2$ was used, non-spherical components of the electron density up to $l=4$ were taken into account. In order to minimize the overlap of atomic spheres, their radii were chosen to be slightly different for the five lattice sites, with relative ratios 1:1.034:1.013:1:0.956, respectively. The experimentally reported [3] internal structural parameters of the FeCr σ-phase, as well as the tetragonal ratio c/a=0.518, were used in all calculations, except for the Fe-Mo binary system where, due to the large size mismatch between the Fe and Mo atoms, we had to optimize the c/a ratio for every end member using the total energy. The volume of each considered structure was relaxed to its equilibrium value.

To model the paramagnetic state we employed the disordered local moment (DLM) [12] approach based on the coherent potential approximation (CPA). The Fe and Co atoms were in these calculations allowed to have collinear, but random, spin up or spin down orientations of their local spin moments. For example, the paramagnetic state of the $ijklm$=BAABA end member compound was modeled as $(Fe{\uparrow}_{0.5},Fe{\downarrow}_{0.5})_2(Cr)_4(Cr)_8(Fe{\uparrow}_{0.5},Fe{\downarrow}_{0.5})_8(Cr)_8$. Here the up and down arrows denote the direction of the total spin moment of an atom.

In order to test our method, we have calculated the energies of pure Fe and Cr in the σ-phase structure relative to the ground state structures of these elements, ferromagnetic α-Fe and antiferromagnetic α-Cr. Within the generalized gradient approximation (GGA) [13], and for non-spin-polarized treatment of σ-phases, the calculated energy differences are 38.2 and 13.3 kJ/mol for Fe and Cr, respectively, which compare well with the corresponding GGA values of 40.6 and 20.1 kJ/mol obtained using a full-potential method [8]. Within the spin-polarized DLM treatment of the paramagnetic state of σ-Fe, magnetic moments of 2.1 and 1.1 Bohr magnetons (μ_B) are found for the Fe atoms at high-CN and IC sites, respectively, The Fe σ-α structural energy difference reduces to 21.0 kJ/mol (GGA) or 18.0 kJ/mol (LAG). This result clearly demonstrates the importance of spin-polarization for thermodynamic description of iron-based sigma phases.

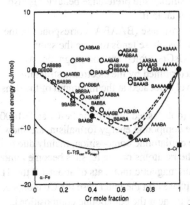

Figure 1. Energies of end-member compounds (circles) for the Fe-Cr binary σ-phase. Ground-state structures are indicated by filled symbols; circles (σ-phase) and squares (α-phase). The free energy and total energy, calculated at T=1000 K, are shown by solid and dashed lines, respectively.

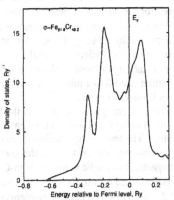

Figure 2. Total density of electron states in σ-$Fe_{51.8}Cr_{48.2}$ with experimental occupancies of the five inequivalent sites [3]. Electronic structure calculations were performed using the KKR-CPA method [9].

Table II. Lowest-energy configurations in the Fe-Cr, Co-Cr, Fe-V, and Fe-Mo binary σ-phases. The trivial cases of standard state configurations AAAAA and BBBBB, whose energies are taken to be zero, are not included. Index in parentheses indicates whether the configuration is in the ground state (Y), close to it (C), or much higher in energy (N).

At.% A	Configuration	Energy relative to σ-B and σ-A, kJ/mol			
		B=Fe, A=Cr	B=Co, A=Cr	B=Fe, A=V	B=Fe, A=Mo
13.3	BABBB	−2.18 (Y)	−4.85 (Y)	−8.56 (C)	−4.38 (Y)
26.7	BBABB	−5.27 (Y)	−6.67 (C)	−17.39 (Y)	+0.46 (N)
40.0	BAABB	−8.07 (Y)	−10.75 (Y)	−23.44 (Y)	−10.46 (Y)
53.3	BBABA	−7.91 (C)	−8.54 (N)	−27.44 (C)	+3.17 (N)
66.7	BAABA	−11.70 (Y)	−12.68 (Y)	−32.16 (Y)	−9.02 (Y)
73.3	AAABA	−9.43 (C)	−9.71 (C)	−27.54 (Y)	−4.53 (C)
93.3	BAAAA	−2.96 (Y)	−3.20 (Y)	−7.72 (Y)	−3.22 (Y)

RESULTS AND DISCUSSION

The calculated energies of all 32 end member compounds for the binary Fe-Cr σ-phase are plotted in figure 1, where the σ-Fe and σ-Cr are chosen as standard states. One can see a strong tendency towards atomic ordering in the σ-FeCr system. At zero temperature, the constrained ground-state of the system (indicated by a broken dashed line) is represented by the end-member configurations (solid circles) that tend to have as many as possible IC sites occupied by Fe and as many as possible high-CN sites occupied by Cr. Very similar site preference behavior is also obtained for the other considered σ-phases, as shown in table II.

The minimum energy configuration for the Fe-Cr σ-phase (BAABA) corresponds to the overall composition $FeCr_2$. Thus, at low temperatures this phase tends to have the same composition as its isoelectronic analogues $RuCr_2$ and $OsCr_2$, which are non-magnetic. An increased stability of these σ-phase compositions is determined by the electronic structure contribution that takes a minimum when the Fermi level is located at the minimum of the density of states shown in figure 2.

The total energy (solid line) and the free energy (dot-dashed line) of the Fe-Cr at T=1000 K, obtained using free energy minimization within the compound energy formalism, are shown in figure 1. The free energy minimum occurs near the equiatomic composition, mainly due to the magnetic entropy contribution. At this composition, the Fe atoms on the IC sites become almost non-magnetic, but the Fe atoms on high-CN sites retain magnetic moments of about 1.8 μ_B. The same is also valid for the Fe-V, Fe-Mo, and Co-Cr σ-phases (in the latter case the Co atoms on high-CN sites possess magnetic moments of about 0.8 μ_B near the equiatomic composition). Once again, we see that magnetism plays an important role in the thermodynamics of σ-phases containing $3d$ transition elements like Fe and Co. However, as figure 1 also shows, the sum of configurational and magnetic entropy contributions is not enough to make the σ-FeCr phase thermodynamically stable at 1000 K, which points to a significance of the vibrational contribution to the free energy of sigma phases [14, 15].

The Fe-Mo case is slightly different from the other cases, in that the total energy minimum does not occur at composition $FeMo_2$. Our calculations show that occupation of site 5 by Mo is energetically very costly, due to a large atomic radius of Mo and a very short interatomic distance along the c-axis direction for this site. This also accounts for the anomalously high c/a ratio in Fe-Mo, which increases in order to accommodate the large Mo atoms at site 5.

The site occupancies in all four considered σ-phases, obtained by free energy minimization at $T=1000$ K, are plotted in figure 3 as a function of composition. Experimental data for σ-FeCr [3] are also shown in figure 3a as symbols; for the other binaries our theoretical results are compared with existing qualitative models [2]. The Fe-Cr, Co-Cr, and Fe-V σ-phases exhibit similar ordering behavior: First, the high-CN sites are (almost simultaneously) occupied by element A, but occupation of the IC sites by this element takes place only at very A-rich compositions. This is due to the fact that the three high-CN sites and the two IC sites, respectively, are very similar in their electronic structure and, as a result, in energy. This allows one to reduce the presently used five-sublattice model to a model with smaller number of sublattices, by considering the similarly occupied sites as completely equivalent. We have used this opportunity in the Fe-V case, as shown in figure 3c, by merging site 1 with site 4, as well as site 2 with site 3.

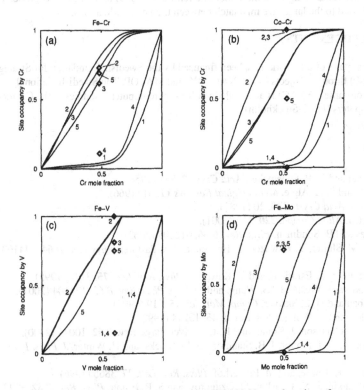

Figure 3. Calculated equilibrium site occupancies at 1000 K as a function of composition in the (a) Fe-Cr, (b) Co-Cr, (c) Fe-V, and (d) Fe-Mo σ-phases (lines) as compared with experimental data and qualitative models (symbols).

In contrast, the occupation of different sites by Mo in the Fe-Mo σ-phase is calculated to occur gradually, one by one. This effect is caused by the large size mismatch between the Fe and Mo atoms: The sites that have similar coordination numbers, but different nearest neighbor distances, split in energy considerably when they are occupied by Mo. As a result, Mo exhibits a decreasing tendency to occupy sites 2, 3, and 5. Therefore, in the ternary Fe-Cr-Mo σ-phase one can expect a separation of Cr and Mo among these three sites.

SUMMARY

Using a combination of *ab initio* calculations with Calphad modeling we have investigated the electronic, magnetic, and atomic structure of four binary σ-phases. Our calculations have demonstrated that magnetism of Fe(Co) atoms increases the stability of σ-phases containing these elements, especially at Fe(Co)-rich compositions. In general, all the considered systems exhibit similar ordering trends: The A-type elements prefer high-coordination-number sites and the B-type elements prefer icosahedrally coordinated sites. We find, however, that Mo shows different affinity towards different high-coordination-number sites in the Fe-Mo σ-phase. This effect is attributed to the large size mismatch between the Fe and Mo atoms.

ACKNOWLEDGEMENTS

Different parts of this work have been financed by the Swedish Foundation for Strategic Research (SSF) through projects CCT, INALLOY, and MATOP. The Swedish National Allocation Committee (SNAC) is acknowledged for supercomputer resources at the Center of Parallel Computers (PDC, Stockholm).

REFERENCES

1. G. Bergman and D. P. Shoemaker, *Acta Cryst.* **7**, 857 (1954).
2. E. O. Hall and S. H. Algie, *Metallurgical Reviews* **11**, 61 (1966).
3. H. L. Yakel, *Acta Cryst.* **B39**, 20 (1983).
4. M. Hillert, *J. Alloy Compd.* **320**, 161 (2001).
5. S. G. Fries and B. Sundman, *Phys. Rev. B* **66**, 012203 (2002).
6. C. Berne, M. Sluiter, Y. Kawazoe, T. Hansen, and A. Pasturel, *Phys. Rev. B* **64**, 144103 (2001).
7. M. H. F. Sluiter, K. Esfarjani, and Y. Kawazoe, *Phys. Rev. Lett.* **75**, 3142 (1995).
8. J. Havránková, J. Vřeštál, L. G. Wang, and M. Šob, *Phys. Rev. B* **63**, 174104 (2001).
9. A. V. Ruban and H. L. Skriver, *Comp. Mat. Sci.* **15**, 119 (1999).
10. J. P. Perdew and Y. Wang, *Phys. Rev. B* **45**, 13244 (1992).
11. L. Vitos, B. Johansson, J. Kollàr, and H. L. Skriver, *Phys. Rev. B* **62**, 10046 (2000).
12. B. L. Gyorffy, A. J. Pindor, J. B. Staunton, G. M. Stocks, and H. Winter, *J. Phys. F* **15**, 1337 (1985).
13. J. P. Perdew, K. Burke, and M. Ernzerhof, *Phys. Rev. Lett.* **77**, 3865 (1996).
14. S. I. Simdyankin, S. N. Taraskin, M. Dzugutov, and S. R. Elliott, *Phys. Rev. B* **62**, 3223 (2000).
15. D. B. Downie and J. F. Martin, *J. Chem. Thermodynamics* **16**, 743 (1984).

B2 Phases and their Defect Structures: Part I. Ab Initio Enthalpy of Formation and Enthalpy of Mixing in the Al-Ni-Pt-Ru System

Sara Prins[1,2], Raymundo Arroyave[1], Chao Jiang[1,3] and Zi-Kui Liu[1]

[1]Department of Materials Science and Engineering, Pennsylvania State University, University Park PA-16802, USA
[2]CSIR-NML, PO Box 395, Pretoria, 0001, South Africa
[3(now at)] Department of Materials Science and Engineering, Iowa State University, Ames, IA-50011, USA

ABSTRACT

The enthalpies of formation of the bcc phases in the Al-Ni-Pt-Ru system, particularly in the Al-Ru binary and Pt-Al-Ru ternary subsystems, were calculated by first principle methods. The enthalpies of formation for stoichiometric bcc-B2 phases have been calculated using both the GGA and LDA approximations, while the enthalpies of formation for B2 phases with large amounts of constitutional defects (both vacancies and anti-site atoms) were calculated using the Special Quasirandom Structures (SQS) approach. The enthalpies of mixing for the disordered bcc-A2 phases have also been calculated with SQS by mimicking the random bcc alloy with the local pair and multisite correlation functions. The calculated B2 lattice parameters for the different defect structures were compared with experimental results. These results are used as input values for the CALPHAD modified sublattice model to describe the A2/B2 phases with one Gibbs energy function.

INTRODUCTION

In the Al-Ni-Pt-Ru system, investigated for its importance in the jet turbine engine industry, the B2 phase stabilities (NiAl and RuAl are the two stable binary B2 structures in this system) are important from several perspectives. NiAl is used as protective bond coat to γ/γ' Ni-based superalloy (NBSA) matrix as it is structurally and chemically more stable at high temperatures than the γ/γ' matrix. Using a Pt-modified aluminide, (Ni,Pt)Al, the stability of the bond coat can be improved significantly [1]. RuAl exhibits the same good high-temperature properties as NiAl while showing exceptional room-temperature toughness and ductility [2]. Because of this, B2-RuAl has also been proposed as an alternative to B2-NiAl as a bond coat material [3], as it has a higher melting temperature, and Ru is an important matrix strengthening alloying element in the 4th generation NBSAs .

The B2 phases in the Al-Ni-Pt-Ru system can exist over a wide composition range due to the presence of anti-site (ASD), vacancy (Va) defects and even a combination of both. The defect structure in these phases affects their mechanical properties and thermodynamic stability [4-7]. Although the stable B2 phases in this quaternary system have been investigated for quite a long time, there is still uncertainty regarding the defect structure resulting from adding other constituents, such as Pt, to the structure. A modification of the defect structure can lead to dramatic changes in both the thermodynamic (thermal stability) and kinetic properties (interdiffusion, creep) of these phases. Although there have been several investigations focused on this topic, there are still several issued that need to be addressed, such as the stability ranges of the ternary and quaternary B2 phases, existence of high-order miscibility gaps, nature of point defects, ternary solubility ranges, site occupancy of ternary dopants, and so forth [3, 5, 8-16].

In this work a systematic study of the Al-Pt-Ru ternary B2 phases is carried out using first principles methods. The resulting enthalpies of formation for the B2 phases with ASD and Va defects, as well as Pt substitutions on the Ru sublattice in RuAl and the disordered A2 (random bcc) energies were estimated by using Special Quasi-random Structures. These ab initio results will later be used as input values for CALPHAD (CALculation on PHAse Diagrams) modelling. The calculated lattice constants as a function of defect structure and composition were also evaluated as they yield important information regarding the stability of the structures.

COMPUTATIONAL DETAILS / THEORY

CALPHAD modelling has become a widely used tool in predicting multi-component phase diagrams and is also widely used in assisting new materials development, as the minimisation of the Gibbs free energy is the fundamental principle for phase stability. Advanced models have been developed to better relate the phase descriptions to their actual crystal structures.

The Modified Sublattice Model (MSL) [17] has been used in the past to describe the B2 phases in the binary subsystems of the Al-Ni-Pt-Ru system. This model considers three sublattices, $(A,B,Va)_{0.5}(A,B,Va)_{0.5}(Va)_3$, and requires the Gibbs energies for all the possible combinations of atoms in all the sublattices (the so-called end members) as well as interaction parameters between species within each sublattice. A shortcoming of this model is that, due to the empirical nature of the CALPHAD approach, some model parameters describing the interactions between vacancies and substitutional atoms have been arbitrarily fixed [17, 18]. Since the binary systems have been modeled independently, serious inconsistencies (Al-Ni [19], Al-Ru [20] and Al-Pt [21]) have resulted when these sub-systems were combined to make ternary extrapolations. Thanks to recent progress in ab initio methods, it is envisaged that first principles energies can be used to give a more physical basis for the model parameters in the MSL model, making it less empirical while facilitating extrapolations to high-order systems from independent assessments.

The enthalpies of formation for the stoichiometric B2 phases (no defects) were calculated using the Vienna ab initio simulation package (VASP) [22-25] with Plane Augmented Wave potentials and the generalised gradient approximation (GGA). The energy cut-off was set to 350 eV and a mesh of at least 1000 k points per atom in the unit cell was used. The structures were relaxed three times with the cell shape and internal atomic coordinates were allowed to vary in order to minimize the energies of the structures. A final self-consistent static calculation was performed after the structure optimisation.

The concept of Special Quasirandom Structures (SQS's) was proposed by Zunger et al.[26] to describe random solid solutions. This approach was able to overcome the limitations of mean-field theories, but without the prohibitive computational cost associated with directly constructing large supercells with random occupancies of substitutional atoms. Jiang et al. [27] applied the SQS approach to bcc-A2 phases, and developed the supercell structures to mimic the B2 alloys with defect structures. These SQS's are specially designed small-unit-cell periodic structures with only a few (2~32) atoms per unit cell, which closely mimic the most relevant, nearest-neighbour pair and multisite correlation functions of the random substitutional alloys.

For this work it was assumed that only one type of point defect is allowed on any one sublattice. Admittedly, the SQS calculations only considered the temperature-independent energetics of constitutional defects, but in part II of this work, vibrational entropy effects for the perfectly stoichiometric and stable B2 structures in the Al-Pt-Ru ternary system are considered.

In addition to the constitutional defects, thermally activated defects are also present at higher temperatures. However, their concentrations are considered, for the moment, to be negligible.

RESULTS

The enthalpies of formation for stoichiometric B2-RuAl, PtAl and PtRu were calculated with VASP. These phases were also calculated with a 32-atom supercell using the SQS approach, and, as can be seen in Table 1, the results were in agreement with VASP 2-atom calculations as well as experimental and other calculated values.

For RuAl (this is the only stable B2 phase at stoichiometry compositions in this ternary system, the PtAl B2 structure being stable only at higher Pt compositions and temperatures), it can be seen in Table 1 and Figure 1 that there is not a significant difference in the enthalpies of formation for an Al ASD or a Va defects on the Ru lattice. This agrees with the recent suggestion that RuAl does not favour either defect structure, but instead probably contains a mix of Al-ASD and Va on the Ru sublattice [28]. The steep slopes of enthalpy of formation in the Ru-rich side suggests that the presence of point defects in the Al lattice is unlikely, which agrees with observations that RuAl decomposes to the eutectic compositions of RuAl and (Ru) at compositions above ~50.5 at. % Ru [29].

The enthalpies of formation for randomly mixed Al and Ru in the bcc-A2 structure at compositions of 0.25, 0.5 and 0.75 at. % Ru were calculated with an 8-atom SQS supercell. As mentioned above, the MSL model requires the energies of formation for the stable and metastable end-members of the model. In order to reduce the empirical nature of the MSL formalism, the metastable end-member structures for A2-Al, A2-Ru, B2-AlAl, B2-RuRu, B2-AlVa and B2-RuVa were also calculated. Two approaches were used for the calculation of the formation enthalpies of the B2-AlVa and B2-RuVa end-members. The supercell can be allowed to relax completely with regard to atomic positions and volume (denoted AlVa and RuVa in Figure 1), or these can be fixed and a static calculation (denoted AlVa* and RuAl* in Figure 1) can be performed. The physical meaning of these two calculations is the following: the static calculations corresponds to the formation energy of Al-Va and Ru-Va 'B2' structures at the lattice parameters of the original B2 RuAl phase, while the fully relaxed calculations correspond to the lattice stability of bcc Al and Ru. As long as the models are consistent, any definition can be used as reference state for the CALPHAD assessments. Further corrections can be done by adding interaction parameters between species in each sublattice.

Fig 1 also shows that, at Ru concentrations > 0.52 at. %, the disordered bcc-A2 phase is more stable than the B2 structures with defects. However, the bcc-A2 structure is not stable in the overall Al-Ru system, which implies then the instability of RuAl on the Ru-rich side, as is observed experimentally.

In Figure 2, the calculated lattice parameter for the RuAl phase, with various possible defect structures, are compared to experimental results from Fleisher [4] and Groban *et al.* [28]. While the calculated values are slightly higher than the experimental values, the same trend is observed. Near-stoichiometry on the Al-rich side of B2, ASD and Va have sililar effects on the lattice constant, it is only at much higher Al-contents, where the B2 phase is unstable, that the effect of the different defects become significant.

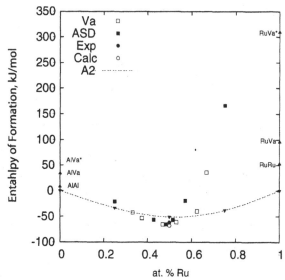

Figure 1. Enthalpies of formation for B2-RuAl with Va and ASD, and disordered A2 enthalpies.

Table 1. Enthalpies of formation and lattice parameters for stable, metastable and unstable B2 phases, Al-, Ru- and Pt-ASD in RuAl and Va defects in RuAl.

B2 phase (A)(B)	Enthalpy of Formation [kJ/mol]			Lattice parameters [nm]		
	This work	Experimental	Other calc	This work	Experimental	Other calc
(Al1)(Pt1)	-92.784	-100.420	-	0.309571	-	-
(Al16)(Pt16)	-92.809	(B20 structure is stable phase)	-	0.310333	-	-
(Al1)(Ru1)	-68.515	-62.05 [*]	-74.113[*]	0.300889	0.303 [*]	0.303 [*]
(Al16)(Ru16)	-68.423		-47.32 [*]	0.301679	0.295 [*]	0.298 [*]
			-95.510 [*]		0.29916 [*]	0.296 [*]
			-47.4 [*]		0.3036 [*]	0.3005 [*]
					0.29914 [*]	
					0.29884 [*]	
(Pt1)(Ru1)	33.905	-	-	0.3123371	--	-
(Pt16)(Ru16)	45.488			0.3231449		
(Al1)(Al1)	-347.080	-	-346.98 [30]	0.3240399	-	0.3244 [30]
(Pt1)(Pt1)	-573.790	-	-574.61 [30]	0.3175766	-	0.3175 [30]
(Ru1)(Ru1)	-823.155	-	-823.164 [30]	0.3031952	-	3.068 [30]
(Al16)(Ru15,Al1)	-65.33	-	-	0.30221	-	-
(Al16)(Ru15,Va1)	-65.43	-	-	0.30124	-	-
(Ru16)(Al15,Ru1)	-61.24	-	-	0.302206	-	-
(Ru16)(Al15,Va1)	-56.43	-	-	0.313417	-	-
(Al16)(Ru15,Pt1)	-72.09	-	-	0.3019	-	-
(Ru16)(Al15,Pt1)	-62.48	-	-	0.30256	-	-

[*] For original references, refer to [31]

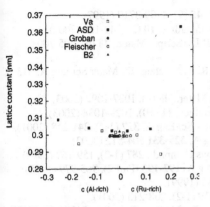

Figure 2. Comparison of the lattice constants for B2-Al-Ru.

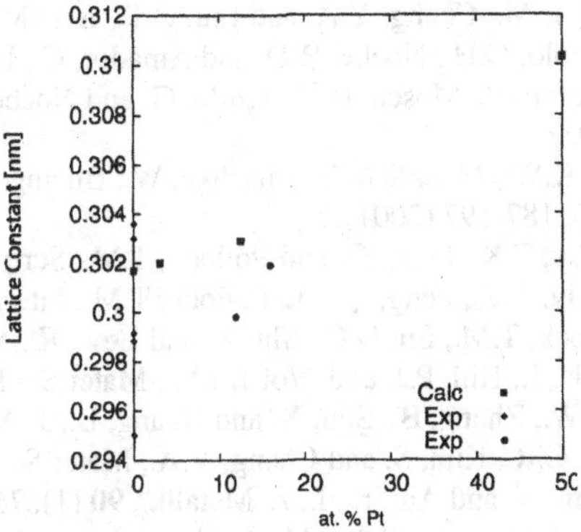

Figure 3. Comparison of experimental and calculated lattice constants for RuAl with Pt additions.

The effect of Pt as a ternary dopant was also studied. In Figure 3, the calculated lattice parameters for RuAl with Pt additions are compared to experimental results. The lattice parameter of RuAl increases with increase in Pt additions, which is in agreement with experimental lattice parameter measurements of alloys in the Al-Pt-Ru ternary system [21].

CONCLUSIONS

Ab initio calculations can be used to calculate order and disorder formation energies for bcc B2 and A2 phases, respectively. These values can be used as meaningful starting values in CALPHAD assessments where little or no thermodynamic data are available due to unstable end-member phases. This allows then for the assessment of only the interaction parameters where there are mixing on the sublattices. It has also been shown that defect structures on B2 phases can be calculated to predict the stable defects structure which can be expected in the phase. The effect of the defects and ternary doping on the lattice parameters have also be determined. The results are in good agreement with initial experimental results from the Al-Pt-Ru ternary system.

ACKNOWLEDGEMENTS
The work is funded by the NSF (USA) through grants DMR-9983532 and DMR. SNP also acknowledges the financial support of the CSIR-NML, the Platinum Development Initiative and NSF-IF-1012FP (South Africa).

REFERENCES
1. Felten, E.J. and Pettit, F.S., Oxi. Metals, **10** (3), 189-223 (1976).
2. Fleischer, R.L., J. Mater. Sci., **22** (7), 2281-8 (1987).
3. Tryon, B., Pollock, T.M., Gigliotti, M.F.X. and Hemker, K., Scripta Mater., **50** (6), 845-848 (2004).
4. Fleischer, R.L., Acta Mat., **41** (3), 863-869 (1993).
5. Kao, C.R., Pike, L.M., Chen, S.-L. and Chang, Y.A., Intermetallics, **2** (4), 235-247 (1994).
6. Pike, L.M., Chang, Y.A. and Liu, C.T., Intermetallics, **5** (8), 601-608 (1997).

7. Pike, L.M., Chang, Y.A. and Liu, C.T., Acta Mat., **45** (9), 3709-3719 (1997).
8. Bozzolo, G.H., Noebe, R.D. and Amador, C., Intermetallics, **10** (2), 149-159 (2002).
9. Gargano, P., Mosca, H., Bozzolo, G. and Noebe, R.D., Scripta Mater., **48** (6), 695-700 (2003).
10. Liu, K.W., Mucklich, F., Pitschke, W., Birringer, R. and Wetzig, K., Mater Sci Eng A, **313** (1-2), 187-197 (2001).
11. Nandy, T.K., Feng, Q. and Pollock, T.M., Scripta Mater., **48** (8), 1087-1092 (2003).
12. Nandy, T.K., Feng, Q. and Pollock, T.M., Intermetallics, **11** (10), 1029-1038 (2003).
13. Pollock, T.M., Lu, D.C., Shi, X. and Eow, K., Mater Sci Eng A, **317** (1-2), 241-248 (2001).
14. Hohls, J., Hill, P.J. and Wolff, I.M., Mater Sci Eng A, **329-331** 504-512 (2002).
15. Hu, W., Zhang, B., Shu, X. and Huang, B., J. Alloys Compnd., **287** (1-2), 159-162 (1999).
16. Kao, C.R., Kim, S. and Chang, Y.A., Mater Sci Eng A, **192-193** (Part 2), 965-979 (1995).
17. Dupin, N. and Ansara, I., Z. Metallk., **90** (1), 76-85 (1999).
18. Hillert, M. and Selleby, M., J. Alloys Compnd., **329** (1-2), 208-213 (2001).
19. Ansara, I., Dupin, N., Leo Lukas, H. and Sundman, B., J. Alloys Compnd., **247** (1-2), 20-30 (1997).
20. Prins, S.N., Cornish, L.A., Stumpf, W.E. and Sundman, B., Calphad, **27** (1), 79-90 (2003).
21. Prins, S.N., M.Sc., University of Pretoria, 2003.
22. Kresse, G. and Furthmuller, J., Comp Mater Sci, **6** (1), 15-50 (1996).
23. Kresse, G. and Hafner, J., Phys. Rev. B, **6** (40), 8245-57 (1994).
24. Kresse, G. and Hafner, J., Phys. Rev. B, **48** (17), 13115-18 (1993).
25. Kresse, G. and Hafner, J., Phys. Rev. B, **47** (1), 558-61 (1993).
26. Zunger, A., Wei, S.H., Ferreira, L.G. and Bernard, J.E., Phys. Rev. Let., **65** (3), 353-6 (1990).
27. Jiang, C., Wolverton, C., J, S., Chen, L.-Q. and Liu, Z.-K., Phys. Rev. B, **68B** 214-220 (2004).
28. Gobran, H.A., Liu, K.W., Heger, D. and Mucklich, F., Scripta Mater., **49** (11), 1097-1102 (2003).
29. Obrowski, W., Naturwissenschaften, **47** 14 (1960).
30. Wang, Y., Curtarolo, S., Jiang, C., Arroyave, R., Wang, T., Ceder, G., Chen, L.Q. and Liu, Z.K., Calphad, **28** (1), 79-90 (2004).
31. Mucklich, F. and Ilic, N., Intermetallics, **In Press, Corrected Proof**

Mater. Res. Soc. Symp. Proc. Vol. 842 © 2005 Materials Research Society   S5.13

B2 Phases and their Defect Structures: Part II. Ab initio Vibrational and Electronic Free Energy in the Al-Ni-Pt-Ru System

Raymundo Arroyave[1], Sara Prins[1,2] and Zi-Kui Liu[1]
[1]Dept. Mat. Sci. & Eng., Pennsylvania State University, University Park, PA 16802, USA
[2]CSIR-NML, P.O. Box 395, Pretoria, 0001, South Africa

ABSTRACT

In this work, we calculate the finite temperature thermodynamic properties of the binary B2 phases in the Al-Ni-Pt-Ru system, particularly the B2 RuAl phase in the Pt-Al-Ru ternary, through the incorporation of the vibrational and electronic contributions to the total free energy. The harmonic approximation is used to consider the atomic vibrations, with the quasi-harmonic correction to account for volume expansion effects on the vibrational entropy as the temperature increases. The vibrational entropy calculations are incorporated through the supercell approach. The calculated phonon dispersion curves show that the B2 PtRu structure is mechanically unstable at low temperatures, while B2 PtAl is marginally stable. The thermal electronic contribution is added to the total free energy. Finally, the formation enthalpies and entropies of B2 RuAl are calculated as a function of temperature.

INTRODUCTION

The B2 phases in the Al-Ni-Pt-Ru system are important in Thermal Barrier Coating (TBC) applications for jet engine turbine blades. TBCs consist of three layers: a ceramic top coat, a thermally grown oxide (TGO), and a bond coat (BC) [1]. The BC layer is used to match the thermal expansion coefficients of the ceramic thermal barrier and the super alloy substrate and also serves as an Al source so the TGO formed during operation is Al_2O_3 instead of other oxides. The most commonly used bond coat material is the B2 NiAl phase. Despite the success of the TBC technology, it is necessary to improve the durability of the TBC, which is affected by the formation and growth of the TGO layer, as well as by the stability of the BC [1]. Besides the binary NiAl B2 phase, additions of Pt[2] and Ru [3] have been considered as there seems to be evidence that they improve the thermodynamic stability and mechanical properties of the BC.

Improvement of current TBC technologies depends on the understanding of the mechanisms behind the decomposition and transformation of the B2 BC layers as well as their interaction with the TGO and the alloy substrate. Using thermodynamic models based on the CALPHAD approach [4], for example, it may be possible to predict likely reaction sequences and phase equilibria resulting from the interactions of the B2 phases with their surroundings.

In part I, *ab initio* techniques were used to calculate the formation energies of the stoichiometric B2 phases in the Pt-Al-Ru system as well as the energetics of the possible point defects that could exist in these structures. In this work, we complement the results at 0 K with calculations of the thermochemical properties of the perfect stoichiometric B2 phases in the Pt-Ru-Al system at finite temperatures. Entropic contributions to the total free energy have been assumed to result from the vibrational and electronic degrees of freedom (DOF) of the structure. Configurational contributions have been neglected. The calculated phonon dispersion curves, as well as the resulting thermochemical properties will be presented and the mechanical instability of PtRu and its implication to the stability of BC coats will be discussed. The calculated entropies of formation and specific heats will eventually be used to refine existing CALPHAD models describing the B2 phases in the Al-Ni-Pt-Ru quaternary system.

THEORY

Ignoring magnetic and configurational DOF, the total free energy of a phase can be expressed as:

$$F(V,T) = E_0(V) + F_{vib}(V,T) + F_{el}(V,T)$$

$$F_{vib}(V,T) = k_B T \int_0^\infty \ln\left[2\sinh\left(\frac{h\nu}{2k_B T}\right)\right] g(\nu,V)$$

$$F_{el}(V,T) = E_{el}(V,T) - TS_{el}(V,T) \tag{0.1}$$

$$E_{el}(V,T) = \int n(\varepsilon,V) f\varepsilon d\varepsilon - \int^{\varepsilon_F} n(\varepsilon,V)$$

$$S_{el}(V,T) = -k_B \int n(\varepsilon,V)\left[f\ln f + (1-f)\ln(1-f)\right]d\varepsilon$$

where $E_0(V)$ is the energy of the structure at 0 K, $F_{vib}(V,T)$ is the vibrational free energy and $F_{el}(V,T)$ is the electronic free energy, $g(\nu,V)$ is the phonon density of states (DOS) and is explicitly made a function of the volume of the structure, $n(\varepsilon,V)$ is the electronic DOS, ν is the vibrational frequency, ε is the energy of the electrons close to the Fermi level, f is the Fermi function and k_B is Boltzman's constant.

Currently there are two main approaches to obtain the phonon DOS. The first one is linear response theory (LRT) and consists on the direct evaluation of the second order derivatives of the crystal potential with respect to atomic displacements [5]. The other method is called the supercell (SC) approach and consists on the evaluation of the force constant through the calculation of the forces corresponding to atomic displacements and solving the resulting system of equations [6]. Given its simplicity and ease of implementation, in this work the SC method is used. A further correction to the harmonic approximation is to make the phonon DOS volume-dependent. This is usually done by calculating the harmonic vibrational free energy at different volumes. By minimizing the resulting total free energy $F(V,T)$ at each temperature with respect to volume, we obtain the free energy with respect to T alone, $F(T)$ [7], evaluated at the minimizing volume, $V(T)$.

METHODOLOGY

First principles calculations using Density Functional Theory (DFT) [8], within the Generalized Gradient Approximation with Projected Augmented Waves as implemented in the VASP code [9] were performed. The energy cutoff used for all the structures was 350 eV and the calculations were performed on a uniform mesh of at least 5000 k-points per unit cell. For hcp Ru, we used a Gamma centered mesh. The vibrational properties of the structures were calculated using the *fitfc* program in the ATAT package[10]. Five different volumes were considered for each structure. Two different supercell sizes were used for each structure and the interatomic forces resulting from the perturbations in the atomic positions were calculated with VASP. For each of the volumes considered, the electronic free energy was calculated using the *felec* utility in the ATAT package. The 0K energy for each volume was added to the vibrational and electronic free energy. Once the free energy was obtained as a function of volume and temperature, the free energy was minimized with respect to volume and an expression for $F(T)$ was obtained.

RESULTS AND DISCUSSION

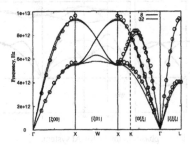

Figure 1. Calculated phonon dispersion curve for fcc Al compared with experiments [12].

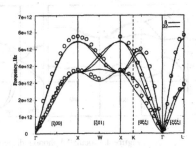

Figure 2. Phonon dispersion for fcc Pt compared with experiments [11].

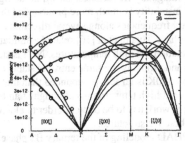

Figure 3. Phonon dispersion curve for hcp Ru compared with experiments [13].

Figures 1-3 show the calculated and experimental phonon dispersion curves for fcc Al, Pt and hcp Ru for two different supercell sizes for each structure. For the case of fcc Al (Figure 1), it can be seen that even with the smallest supercell size (8 atoms) the agreement between calculations and experiment is very good. For fcc Pt (Figure 2), the agreement is not as good as for fcc Al, even with the largest supercell. The 32 atoms calculation, however, is able to reproduce the softening of the lowest TA branch at the W point of the Brillouin zone. Another feature that is well reproduced is the change in slope of the lowest TA branch along the [0ξξ] direction, which is due to a Kohn anomaly [14]. For the case of hcp Ru (Figure 3), the agreement between the available experiments and calculations is better when the largest supercell (36 atoms) is used. The 8-atom supercell underestimates the frequency for the L branches at the A point by about 0.8 THz. With 36 atoms, this frequency is reasonably reproduced.

Figure 4 shows the calculated phonon dispersion curves for the B2 phases in the Pt-Ru-Al system. Phonon dispersion relations can provide an insight into the stability of a structure and in this case it shows that, while the RuAl phase is clearly stable--the B2 phase is a ground state in the Ru-Al binary [15]—the other two B2 phases are not stable at 0K. B2 PtAl for example, is marginally stable –its ground state is B20 [15]-, with the lowest TA mode having 0 frequency at the M point of the Brillouin zone.

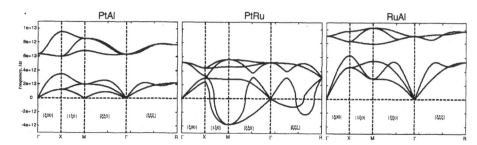

Figure 4. Calculated phonon dispersion curves for B2 phases in the Pt-Ru-Al system.

It was observed that by expanding the lattice by small amounts this branch in the B2 PtAl phase became unstable, yielding imaginary frequencies, shown in the figure as negative. For the B2 PtRu phase, the extent of the instability is much more pronounced that for B2 PtAl, with several branches having imaginary frequencies along several directions. In their study of the lattice instability of the B2 NiTi phase, Huang et al [16] related a similar instability to the $B2 \rightarrow B19'$ martensitic transformation. At high temperatures, NiTi is stabilized by phonon-phonon anharmonic interactions and probably the same entropic stabilization can take place in the case of PtAl, whose instability at 0K is far less pronounced than that of B2 NiTi. In fact, the B2 PtAl phase appears to be stable at high temperatures although at an off-stoichiometric composition, rich in Pt [17]. More work has to be done in order to elucidate the actual reasons for the instabilities observed in these B2 structures.

For the purpose of the CALPHAD modeling of the ternary B2 phase in the Pt-Ru-Al system it is desirable to obtain an expression for the entropies and specific heats of the three end members of the sublattice model [4], namely, PtAl, PtRu and RuAl. Unfortunately, the entropy of the unstable phases is not defined and although there have been some efforts regarding the incorporation of unstable modes in the total free energy of crystal structures [18], the validity of such an approach has not been rigorously corroborated. In light of this, we limit the thermodynamic analysis to the stable B2 RuAl phase.

Figure 5 and Figure 6 show the calculated and tabulated [19] enthalpies of the fcc Al and hcp Ru structures, respectively. The calculations considered four approximations. First the harmonic approximation with and without thermal electronic contributions was considered. Afterwards, the quasi-harmonic approximation was used. As can be seen, the high temperature properties of these phases cannot be reproduced unless volume effects on the phonon DOS as well as electronic DOF are considered. Corrections including the electronic DOS are most important for structures that have a high density of states at the Fermi level. hcp Ru has a higher DOS at the Fermi level than fcc Al and, as one would expect that electronic contributions to the total Free Energy would be larger for Ru. Figure 6 shows that this is the case. In fact, only by including electronic contributions to the quasi-harmonic approximation it is possible to accurately predict the correct temperature dependence of the enthalpy of hcp Ru.

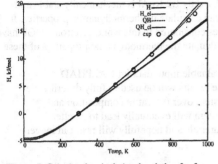

Figure 5. Calculated and tabulated enthalpy for fcc Al.

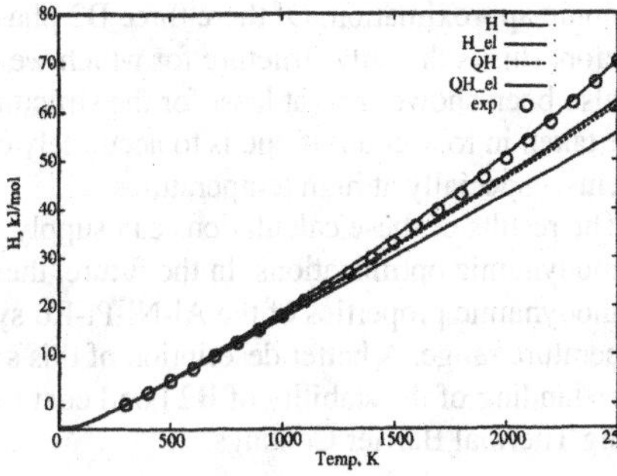

Figure 6. Calculated and tabulated enthalpy for hcp Ru.

Using the calculated thermodynamic properties of fcc Al and hcp Ru, as well as those of B2 RuAl, we proceeded to calculate the formation enthalpy and entropy of this intermetallic compound using the harmonic and quasi-harmonic+electronic approximations. As expected (Figure 7), the formation entropy with only harmonic contributions is essentially constant once the Debye temperature is reached. Adding volume effects plus electronic DOF causes the formation entropy to become more negative as temperature increases. Similar calculations for B2 NiAl show basically the same trend. The temperature dependence of the formation entropy also plays a role in determining the behavior of the formation enthalpy and, as can be seen in Figure 8, the formation enthalpy of B2 RuAl has a slight negative deviation as temperature increases. Although no experiments are available, this behavior has been observed in B2 NiAl [20] and is likely then that the present calculations actually represent the real thermodynamic behavior of B2 RuAl.

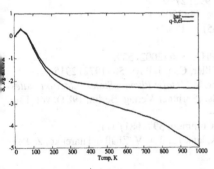

Figure 7. Calculated formation entropy of B2 RuAl

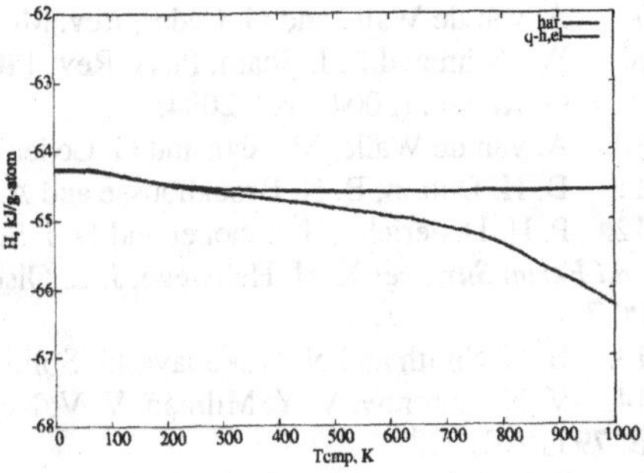

Figure 8. Calculated enthalpy of formation of B2 RuAl.

CONCLUSIONS

In this work, it has been shown how we can use ab initio calculations to determine, with a great degree of confidence, the thermodynamic properties of intermetallic phases, particularly, the B2 phase in the RuAl binary system. The harmonic approximation to the crystal potential has been used to obtain the vibrational behavior of fcc Al, fcc Pt and hcp Ru, as well as the B2 structures of the PtAl, PtRu and RuAl binaries. To account for the effects of volume expansion on the vibrational free energy, we have further refined the calculations by using the quasi-harmonic approximation. Of there three B2 phases, only B2 RuAl is mechanically stable and therefore this is the only structure for which we can calculate its thermodynamic properties. It has also been shown that, at least for the structures considered in this work, electronic DOF have to be taken in to account if one is to accurately calculate the thermodynamic properties of these systems, especially at high temperatures.

The results of these calculations can supply valuable input data for CALPHAD thermodynamic optimizations. In the future, these results will be used to fully describe the thermodynamic properties of the Al-Ni-Pt-Ru system over the entire composition and temperature range. A better description of this system will eventually lead to a better understanding of the stability of B2 bond coat materials and hopefully will result in longer-lasting Thermal Barrier Coatings.

ACKNOWLEDGEMENTS

NSF financial support under grants DMR-9983532 and DMR-0209624 is acknowledged. SNP also acknowledges the financial support of the CSIR-NML, the Platinum Development Initiative and NSF-IF-1012FP (South Africa).

REFERENCES

1. M. W. Chen, R. T. Ott, T. C. Hufnagel, P. K. Wrigth and K. J. Hemker, Surf. Coat. Technol. **163-164** (2003) 25.
2. V. K. Tolypgo and D. R. Clarke, Acta Mater. **48** (2000) 3283.
3. B. Tryon, T. M. Pollock, M. F. X. Gigliotti and K. J. Hemker, Scripta Materialia **50** (2004) 845.
4. N. Saunders and A. P. Miodownik, *CALPHAD (Calculation of Phase Diagrams): A Comprehensive Guide* (Pergamon, Oxford ; New York, 1998).
5. X. Gonze, Phys. Rev. B **55** (1997) 10337.
6. S. Wei and M. Y. Chou, Phys. Rev. Lett. **69** (1992) 2799.
7. A. van de Walle and G. Ceder, Rev. Mod. Phys. **74** (2002) 11.
8. W. Kohn and L. J. Sham, Phys. Rev. **140** (1965) A1133.
9. G. Kresse. (2004), vol. 2004.
10. A. van de Walle, M. Asta and G. Ceder, CALPHAD **26** (2002) 539.
11. D. H. Dutton, B. N. Brockhousse and A. P. Miiller, Can. J. Phys. **50** (1972) 2915.
12. P. H. Dederichs, H. Schober and D. J. Sellmyer, in *Metals: Phonon States, Electron States and Fermi Surfaces* K. H. Hellwege, J. L. Olsen, Eds. (Springer-Verlag, Berlin, 1981), vol. 13a, Pt. 7.
13. H. G. Smith and N. Wakabayashi, Solid State Commun. **39** (1981) 371.
14. V. N. Antonov, V. Y. Milman, V. V. Nemoshkalenko and A. V. Zhalko-Titarenko, Z. Phys. B. **79** (1990) 223.
15. P. Villars. (2004), vol. 2004.
16. X. Huang, C. Bungaro, V. Godlevsky and K. M. Rabe, Phys. Rev. B **65** (2001) 141081.
17. A. J. McAlister and D. J. Kahan, Bull. Alloy Phase Diagrams **7** (1986) 47.
18. D. C. Swift, G. J. Ackland, A. Hauer and G. A. Kyrala, Phys. Rev. B **64** (2001) 2141071.
19. O. Knacke, O. Kubaschewski and K. Hesselmann, *Thermochemical Properties of Inorganic Substances* (Springer-Verlag, New York, ed. Second Edition, 1991), vol. I.
20. Y. Wang, Z.-K. Liu and L.-Q. Chen, Acta Mater. **52** (2004) 2665.

Mater. Res. Soc. Symp. Proc. Vol. 842 © 2005 Materials Research Society   S5.15

Thermal Analysis of Relaxation Processes of Supersaturated Vacancies in B2-Type Aluminides

Ryusuke Nakamura, Kyosuke Yoshimi, Akira Yamauchi and Shuji Hanada
Institute for Materials Research, Tohoku University,
Katahira 2-1-1, Sendai 980-8577, Japan.

ABSTRACT

Relaxation behavior of supersaturated vacancies in water-quenched and rapidly solidified B2 type CoAl and NiAl with the stoichiometric composition was studied by differential scanning calorimetry (DSC). In both CoAl and NiAl quenched from 1773 K, a single exothermic peak due mainly to the recovery of supersaturated vacancies appeared near 950 K. On the other hand, the exothermic peak was also observed in as-spun rapidly solidified CoAl and NiAl ribbons, and the temperature range of the peak was much broader than that of quenched-in bulks, suggesting that the peak in the rapidly solidified ribbons results from not only a single recovery process of vacancies but also those of some other lattice defects. TEM observation showed that a large amount of dislocations were introduced in the NiAl ribbon and that comparatively many voids as well as an amount of dislocations were introduced in the CoAl ribbon.

INTRODUCTION

Vacancies in B2 type intermetallic compounds have anomalous properties compared to those in metals and alloys. For example, thermally activated vacancies in the compounds are generated with a 1% order concentration at high temperature near their melting point and they are readily frozen by a operation of water-quenching from high temperature. These phenomena can be explained mainly by the properties of a low vacancy formation enthalpy and a high vacancy migration enthalpy in the B2 intermetallic compounds. Recently, Yoshimi et al. [1-3] have actively investigated the clustering behavior of supersaturated vacancies in both rapidly solidified and quenched-in polycrystalline FeAl. Because the vacancy clustering behavior is strongly related to vacancy migration, it is essential to understand the kinetics of the relaxation process of supersaturated vacancies. So far, the recovery processes of quenched-in vacancies in B2 type intermetallic compounds such as NiAl [4], CoGa [5] and FeAl [6] have been studied by thermal analysis using a differential scanning calorimeter. However, the information on the kinetics of supersaturated vacancies is still scarce, whereas there is an amount of knowledge accumulated for many years to grasp static point defects characters.

In the present work, a thermal analysis by differential scanning calorimetry (DSC) is applied to investigate the relaxation process of supersaturated thermal vacancies frozen from high temperature in B2-type CoAl and NiAl. Furthermore, the DSC measurements are extended to the rapidly solidified ribbons in which the concentrations of supersaturated vacancies are considered to be much higher than those in quenched bulk specimens.

EXPERIMENTAL PROCEDURE

Ingots of nominally stoichiometric CoAl and NiAl alloys were produced from commercially available high purity nickel, cobalt and aluminum by arc-melting and induction melting techniques. To obtain polycrystalline bulk specimens with large grains and low dislocation density, the ingots were annealed at 1773 K for 86.4-172.8 ks and then furnace-cooled with the rate of 0.5-1 K/min. The obtained grain size was 2 to 4 mm. After the heat treatment, the pieces of the alloys were capsulated in a silica tube with high purity argon gas and annealed at 1773 K for 3.6 ks, followed by quench into ice water. The capsules were crushed immediately after the water-quench in the ice water. On the other hand, the procedure to produce rapidly solidified ribbons is as follows. The ingots made by arc-melting above were re-melted in the arc-melting furnace and drop-cast into a rod-shape Cu mold of 6 and 11 mm in diameter. Rapidly solidified ribbons were produced by melting the drop-cast rods of the alloys using a twin-roll melt-spinning apparatus with Cu rolls at the rotating speed of 3600 rpm.

Different scanning calorimetry (DSC) measurement was performed for both the quenched-in bulk and as-spun rapidly solidified ribbons in a Netzsch DSC 404. The samples of 200 ~ 250 mg in weight were loaded into an alumina pan and the measurement was carried out in an atmosphere of high purity argon gas flow in the temperature range from room temperature to 1400 K at the heating and cooling rates of 30 K/min.

Thin foils for transmission electron microscopic (TEM) observation were prepared by twin-jet electropolishing using a Struers Tenupol-5 in a 33% nitric acid-methanol solution at 7-8 V and 246 K. The thin foils were examined using a JEM-2000EX II operating at 200 kV.

RESULTS AND DISCUSSION

Figure 1 shows DSC heating curves of stoichiometric CoAl bulk furnace-cooled and quenched from 1773 K. An exothermic peak in the temperature range between 800 and 1100 K appears in the first heating step for the quenched-in CoAl, but no peak can be seen in the second heating step. On the other hand, the curves of the first and second heating steps of the furnace-cooled CoAl almost overlap each other. The irreversible exothermic peak appearing in the quenched CoAl would result from the recovery of supersaturated vacancies frozen by quenching from the high temperature.

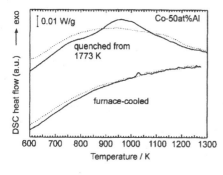

Figure 1. DSC heating curves of Co-50 at%Al bulks furnace-cooled and quenched from 1773 K measured at the heating rate of 30 K/min. Solid and dotted lines correspond to the first and second heating steps, respectively.

Hereinafter, the difference in calorie between first and second step curves is plotted against temperature as the DSC curve because the second cycle curve can be regarded as the base line. The DSC curves of quenched bulk and rapidly solidified ribbons for both the stoichiometric CoAl and NiAl are shown in figures 2 and 3, respectively. In not only CoAl but also NiAl quenched from 1773 K, a single exothermic peak is observed in the temperature range between 800 and 1100 K. On the other hand, a single, larger exothermic peak is observed in rapidly solidified CoAl and NiAl ribbons. Although the onset temperature of the exothermic peak is about 800 K as is the case with the quenched bulk sample, the end temperature is about 200 K higher than that of the quenched bulk. These results suggest that the peak observed in the as-spun ribbons includes several, more complicated recovery processes of supersaturated vacancies, motivating us to investigate microstructure of rapidly solidified ribbons.

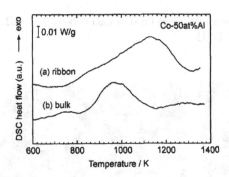

Figure 2. DSC curves of Co-50at%Al (a) rapidly solidified ribbons and (b) bulks quenched from 1773 K measured at the heating rate of 30 K/min.

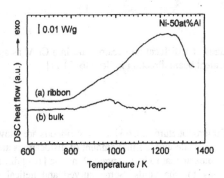

Figure 3. DSC curves for Ni-50at%Al (a) rapidly solidified ribbons and (b) bulks quenched from 1773 K measured at the heating rate of 30 K/min.

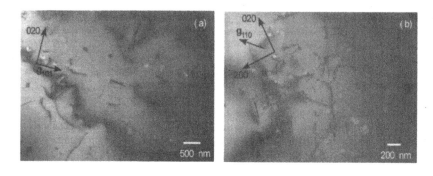

Figure 4. Bright field images obtained in CoAl as-spun ribbons. The incident beam direction is (a) $B \approx [\bar{1}01]$ and (b) $B \approx [001]$.

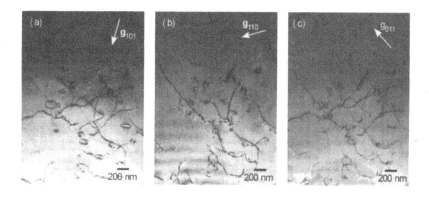

Figure 5. Bright field images of a dislocation substructure in a CoAl as-spun ribbon taken with different g vectors. The incident beam direction is close to $[\bar{1}11]$.

Bright field images of microstructure in CoAl as-spun ribbons are shown in figures 4 and 5. Many voids on a nano to meso-scale with a small amount of dislocations are observed in figure 4. It should be noted that the voids appear to be arranged along the [101] direction. There may be a preferential arrangement way for the voids. Some curved and helical dislocations are also observed as well as prismatic loops in figure 5. From these images taken with three different g vectors, it is clarified that the observed dislocations and prismatic loops in the CoAl ribbon have the Burgers vectors parallel to <100> directions. Figures 6 and 7 show the bright field images of dislocation substructure in the NiAl as-spun ribbons. A large amount of dislocations, which have the <100> type Burgers vector, can be seen in NiAl as-spun ribbons, as shown in figure 6, whereas the density of voids is much smaller than that in CoAl as-spun ribbons, as shown in figure 7. Therefore, it can be considered that the broadening of the exothermic peak in the DSC curves of rapidly solidified ribbons would result from complex recovery processes of dislocations, prismatic loops, voids and so on in addition to supersaturated vacancies.

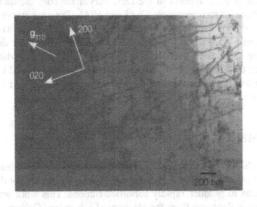

Figure 6. A large amount of dislocations introduced in as-spun NiAl ribbons. $B \approx [001]$.

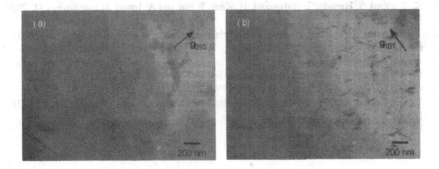

Figure 7. Bright field images of a NiAl as-spun ribbon. The incident beam direction is close to $[\bar{1}01]$.

It is important to estimate the activation energy of vacancy migration in quenched bulks by Kissinger's method although it is difficult to obtain in as-spun ribbons because of the complex relaxation processes for several kinds of lattice defects as mentioned above. The activation energy of vacancy migration is now under investigation.

CONCLUSIONS

A single exothermic peak appears in the DSC curves for both the quenched-in bulks and rapidly solidified ribbons of stoichiometric CoAl and NiAl. The peak width is much broader in the rapidly solidified ribbons than that in the quenched-in bulks. In CoAl as-spun ribbons, there are many nano- to meso-scale voids with a small amount of <100>-type dislocations. In NiAl as-spun ribbons, there are a large number of <100>-type dislocations, whereas the density of voids is much smaller than that in CoAl as-spun ribbons. The existence of the large amounts of the lattice defects as well as supersaturated thermal vacancies would lead to the broadening in the DSC curves of the as-spun ribbons.

ACKNOWLEDGEMENT

The authors would like to thank Prof. S. Tsurekwawa and Mr. T. Matsuzaki of Department of Nanomechanics, Graduate School of Engineering, Tohoku University for the use of a twin-roll melt-spinning apparatus to produce rapidly solidified ribbons. This work was supported by the Grant-in-Aid for Science Research from the Ministry of Education, Culture, Sports, Science and Technology of Japan under Contract No. 16360339.

REFERENCES

1. K. Yoshimi, S. Hanada, T. Haraguchi, H. Kato, T. Itoi and A. Inoue, *Mater. Trans.*, **11**, 2897, (2002).
2. T. Haraguchi, K. Yoshimi, H. Kato, S. Hanada and A. Inoue, *Intermetallics*, **11**, 707, (2003).
3. K. Yoshimi, T. Kobayashi, A.Yamauchi, T. Haraguchi and S. Hanada, *Philosophical Magazine*, in press.
4. M. Shimotomai, T. Iwata and M. Doyama, *Phil. Mag. Let.*, **51**, L49, (1985).
5. O. Sassi, J. Aride, A. Berrada and G. Moya, *Phil. Mag. Let.*, **68**, 53, (1993).
6. S. Zaroual, O. Sassi, J. Aride, J. Bernardini and G. Moya, *Mater. Sci. Eng. A*, **279**, 282, (2000).

Mater. Res. Soc. Symp. Proc. Vol. 842 © 2005 Materials Research Society   S5.28

Ab initio calculation of point defect energies and atom migration profiles in varying surroundings in L1$_2$-ordered intermetallic compounds

Doris Vogtenhuber[1], Jana Houserova[2], Walter Wolf[3], Raimund Podloucky[4], Wolfgang Pfeiler[1] and Wolfgang Püschl[1]

[1]Institut für Materialphysik, University of Vienna, Strudlhofgasse 4, A-1090 Vienna, Austria
[2]Institute of Physics of Materials, Academy of Sciences of the Czech Republic, Zizkova 22, CZ-616 62 Brno, Czech Republic
[3]Materials Design s.a.r.l., 44 av. F.-A. Bartholdy, F-72000 Le Mans, France
[4]Institut für Physikalische Chemie, University of Vienna, Liechtensteinstrasse 22a, A-1090 Vienna, Austria

ABSTRACT

Formation energies of antisite defects and vacancies were derived for the L1$_2$-ordered intermetallics Ni$_3$Al, Ni$_3$Ga, Pt$_3$Ga, and Pt$_3$In by a supercell ab initio approach. A thermodynamic treatment of point-like defects was then used for the calculation of temperature-dependent defect properties. Energy profiles for atom jumps in Ni$_3$Al in systematically varied atomic neighborhoods were calculated by statically displacing the jumping atom or by using a nudged elastic band method. It is discussed how a kinetic Monte-Carlo model can be modified so that the jump barrier height reflects the strongest neighborhood influences.

INTRODUCTION

L1$_2$-ordered intermetallic compounds due to high-temperature strength and favorable corrosion behavior have been regarded as promising structural materials for some time now [1,2]. The desired properties being linked to the long-range ordered state, its stability and rate of change are of great practical significance. Ordering kinetics in various long-range ordered alloys has been investigated experimentally by some of the authors using residual electrical resistivity as a very sensitive indicator [3]. In recent work on this class of materials, special attention was devoted to Ni$_3$Al. Residual resistivity measurements of ordering kinetics were supplemented by Monte-Carlo (MC) simulation [4,5]. In order to properly understand the meaning of the activation parameters found experimentally, and to correctly describe the elementary process in the MC simulations, ab initio calculations of defect energies have been undertaken [6]. In the present work, the scope is extended to analogous L1$_2$ compounds and to atom jump behavior in different atomic neighborhoods in Ni$_3$Al.

AB INITIO CALCULATION

For all ab initio density functional calculations the Vienna Ab initio Simulation Package (VASP) [7] was applied, and for exchange and correlation the generalized gradient approximation (GGA) was adopted. Point defects were modeled by embedding them in supercells of 32 atoms which corresponds to 2x2x2 L1$_2$ unit cells. The selected cell size has been shown to be sufficient by comparison to results based on larger unit cells. Relaxations of two coordination shells of neighboring atoms around a point defect were calculated by atomic force minimization, but the influence of volume changes is neglected. The formation energies of defect supercells are entered into a thermodynamic treatment, defects are assumed to be sufficiently dilute and therefore non-interacting. Only configuration entropy contributions are considered. Invoking a grand canonical ensemble where the number of particles is regulated by chemical potentials which must adjust to the proper values necessary to preserve stoichiometry [8], . temperature dependent effective formation energies of defects are derived.

Energy profiles of jumping atoms have been calculated by statically displacing the atom along the jump path and relaxing the surrounding atom positions at each step.

RESULTS AND DISCUSSION

In Table I the results for the effective point defect formation energies in the compounds considered are given in units of eV for a temperature of 1000 K. The temperature dependence is weak, and energies are affected by typically less than 10% in the remaining temperature range up to the melting point. It is noted that the antisite formation energies are exactly the same for majority and minority atoms. This follows from the requirement to maintain the composition of the alloy and is true for ideal stoichiometry only. For all compounds studied, antisite formation takes distinctly less energy than vacancy formation. The tendencies in defect formation energies can largely be made plausible by atom size arguments as discussed in detail in [9].

Table I. Defect formation energies (relaxed) [eV].

Compound	vacancy Ni/Pt	vacancy Al/Ga/In	antisite Ni/Pt	antisite Al/Ga/In
Ni_3Al	2.00	2.48	0.51	0.51
Ni_3Ga	1.62	2.35	0.46	0.46
Pt_3Ga	1.74	1.48	0.66	0.66
Pt_3In	1.37	1.88	0.71	0.71

An investigation was made of the influence of the specific atomic environment of the jumping atom on the jump rate, in view of a more accurate modeling of atom kinetics by a kinetic Monte Carlo algorithm. Only ground-state energies are considered in this analysis, and Ni_3Al has been taken as a representative alloy. The barrier heights (E^{barr}) substantially depend on the atomic species in the shells around the jumping atom, via their influence on local lattice relaxations, chemical interactions along the diffusion path and relative stabilization of initial or final states of an atomic jump. In general, (i) a decrease of E^{barr} facilitates diffusion by speeding up the jump rates and (ii) a decrease of the total energy along the jump path stabilizes the final state with respect to the initial state.

To study the basic effects of changes in the local environment of a jump which changes the degree of order, we modeled selected antisite defects in the nearest neighbor (NN) and next-nearest neighbor (NNN) shell of atoms jumping in Ni_3Al. In a first step jump processes were assumed as jumps of Ni atoms into NN Al vacancies, giving as result a Ni antisite defect and a Ni vacancy. As the atomic volume of Al is about 14% larger than the atomic volume of Ni, size effects are expected to play a significant role. Hence, a substitution of Ni by Al in the cell invariably increases the local pressure in the unit cell, whereas a substitution of Al by Ni corresponds to some local relaxation.

The configurations considered in our calculations are listed in Table II. The respective jump profiles are depicted in Figs. 1a and 1b. As might be expected, the strongest influence on jump barriers is exerted by the number of antisites in a four-atom window of nearest neighbors through which an atom jumping to a NN vacancy site has to pass (configurations I0 to K1 in Table II). In terms of the 'pure' barrier height E_0^{barr} (arithmetic mean of barriers for jump and back-jump, see below), the following general picture emerges: For neighboring antisite atoms outside the four-atom window E_0^{barr} only comparatively small changes occur (below ±10%). If the antisite belongs to the window (Al instead of Ni), the changes in E_0^{barr} with respect to the perfectly ordered $L1_2$ lattice (A in Table II) are about 18% for one antisite (I0), about 30-40% for antisites in adjacent position (I2, J0, K0), and about 50-60% for diagonally opposite antisites (I1, J1, K1). In Fig. 1a we have plotted the jump profiles for neighborhoods where the replaced atoms do not belong to the window (A-H), Fig. 1b shows the jump profiles that result when window atoms are among the replaced atoms (I0-K1).

First results of calculations with the nudged elastic band method [10] seem to indicate that the deviations in energy from the results with straight jump paths are not significant. Investigation of jumps of Al into Ni vacancies and of Ni atoms within their own sublattice for different surroundings are in progress.

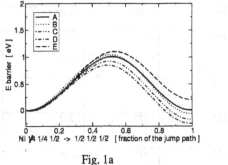

Fig. 1a

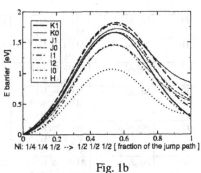

Fig. 1b

Figure 1. Energy profiles for Ni jumping into an Al vacancy; jump coordinates are in fractional units of the supercell. (a) Neighborhood Al atoms outside the four atom-window are successively replaced by Ni (sequence A-D; E refers to an antisite pair placed in the neighborhood). (b) Window atoms are now among the replaced ones.

Table II. Barrier energies for Ni jumps. Left to right: Designation of configuration, stoichiometry of the supercell, positions of exchanged neighboring atoms in the supercell in fractional coordinates of the supercell. E_{in}^{barr}: Energy barrier for the jump in the specified environment, E_{out}^{barr}: Barrier for the back-jump, $E_0^{barr} = \left(E_{in}^{barr} + E_{out}^{barr}\right)/2$ 'pure barrier height'. All energies are given in eV.

Ni jumps ($\frac{1}{4}$, $\frac{1}{4}$, $\frac{1}{2}$) $\rightarrow$ ($\frac{1}{2}$, $\frac{1}{2}$, $\frac{1}{2}$)					
Configuration	Stoichiometry	Position of replaced atoms	E_{in}^{barr}	E_{out}^{barr}	$E_0^{barr} = \frac{1}{2}\left(E_{in}^{barr} + E_{out}^{barr}\right)$
Exchanges of atoms outside the window					
A	Ni$_{24}$Al$_7$	-	1.004	0.996	1.000
B	Ni$_{25}$Al$_6$	Ni ($\frac{1}{2}$, $\frac{1}{2}$, 0)	1.040	1.100	1.070
C	Ni$_{26}$Al$_5$	Ni ($\frac{1}{2}$, $\frac{1}{2}$, 0) Ni (0, $\frac{1}{2}$, 0)	0.922	1.096	1.009
D	Ni$_{27}$Al$_4$	Ni ($\frac{1}{2}$, $\frac{1}{2}$, 0) Ni (0, $\frac{1}{2}$, $\frac{1}{2}$) Ni ($\frac{1}{2}$, 0, $\frac{1}{2}$)	0.849	1.095	0.972
E	Ni$_{24}$Al$_7$	Ni (0, $\frac{1}{2}$, $\frac{1}{2}$) Al ($\frac{1}{4}$, $\frac{3}{4}$, $\frac{1}{2}$)	1.099	0.892	0.996
F	Ni$_{23}$Al$_8$	Al ($\frac{3}{4}$, $\frac{3}{4}$, $\frac{1}{2}$)	0.947	0.872	0.910
G	Ni$_{23}$Al$_8$	Al ($\frac{3}{4}$, $\frac{1}{4}$, $\frac{1}{2}$)	1.105	0.826	0.965
H	Ni$_{23}$Al$_8$	Al ($\frac{3}{4}$, $\frac{1}{2}$, $\frac{1}{4}$)	1.066	0.738	0.902
Window atoms (*) among the exchanged atoms					
I0	Ni$_{22}$Al$_9$	Al ($\frac{1}{4}$, $\frac{1}{2}$, $\frac{3}{4}$) * Al ($\frac{3}{4}$, $\frac{1}{2}$, $\frac{1}{4}$)	1.451	0.905	1.178
I1	Ni$_{22}$Al$_9$	Al ($\frac{1}{4}$, $\frac{1}{2}$, $\frac{3}{4}$) * Al ($\frac{1}{2}$, $\frac{1}{4}$, $\frac{1}{4}$) *	1.794	1.362	1.578
I2	Ni$_{22}$Al$_9$	Al ($\frac{1}{4}$, $\frac{1}{2}$, $\frac{3}{4}$) * Al ($\frac{1}{2}$, $\frac{1}{4}$, $\frac{3}{4}$) *	1.462	1.165	1.314
J0	Ni$_{21}$Al$_{10}$	Al ($\frac{3}{4}$, $\frac{1}{2}$, $\frac{1}{4}$) Al ($\frac{1}{4}$, $\frac{1}{2}$, $\frac{1}{4}$) * Al ($\frac{1}{4}$, $\frac{1}{2}$, $\frac{3}{4}$) *	1.665	1.096	1.381
J1	Ni$_{21}$Al$_{10}$	Al ($\frac{3}{4}$, $\frac{1}{2}$, $\frac{1}{4}$) Al ($\frac{1}{4}$, $\frac{1}{2}$, $\frac{1}{4}$) * Al ($\frac{1}{2}$, $\frac{1}{4}$, $\frac{3}{4}$) *	1.822	1.222	1.522
K0	Ni$_{20}$Al$_{11}$	Al ($\frac{3}{4}$, $\frac{1}{2}$, $\frac{1}{4}$) Al ($\frac{1}{4}$, $\frac{1}{2}$, $\frac{1}{4}$) * Al ($\frac{1}{4}$, $\frac{1}{2}$, $\frac{3}{4}$) * Al ($\frac{3}{4}$, $\frac{1}{2}$, $\frac{3}{4}$)	1.723	0.875	1.299
K1	Ni$_{20}$Al$_{11}$	Al ($\frac{3}{4}$, $\frac{1}{2}$, $\frac{1}{4}$) Al ($\frac{1}{4}$, $\frac{1}{2}$, $\frac{1}{4}$) * Al ($\frac{1}{2}$, $\frac{1}{4}$, $\frac{1}{4}$) * Al ($\frac{1}{2}$, $\frac{1}{4}$, $\frac{3}{4}$) *	1.660	1.377	1.518

KINETIC MONTE CARLO ALGORITHM WITH VARIABLE JUMP BARRIERS

From our simulations it is clear that the exact local atom configuration in the vicinity of the jumping atom exerts a strong influence on barrier height and thus on jump probabilities. This is especially the case for the four-atom "window" which has to be passed in an fcc (or L1$_2$) structure by a jumping atom. In existing kinetic Monte Carlo algorithms barrier heights are either not considered at all (as they simply multiply the overall time scale by some factor) or constructed by superimposing a constant contribution E_0^{barr} to an energy level which is the arithmetic mean between the initial and final position of the jumping atom. We want to make a more realistic ansatz with the barrier height more sensitively dependent on the atomic surroundings, as reflected in the results of the ab initio calculations. What are the conditions the jump probabilities have to fulfill? Let us assume that our system contains N atoms and is part of a canonical ensemble. Following the usual approximations in atom kinetics we may take transition state theory [11] to be valid. This gives a transition probability per unit time of a jumping atom of the form

$$\Gamma_{ij} = \Gamma_{ij}^* \exp\left(-\frac{\Delta E_{is}}{kT}\right)$$

Γ_{ij}^* is an *effective* attempt frequency derived from normal frequencies of the atom potential E in the vicinity of the initial equilibrium state and the saddle point state. ΔE_{is} means the potential difference between the initial and the saddle point state. Any kinetic ansatz must lead to the correct thermodynamic equilibrium. A sufficient condition is the principle of detailed balance:

$$f_i \Gamma_{ij} = f_j \Gamma_{ji},$$

where f_i is the number of systems of our ensemble being in equilibrium in state i. Since $f_i \propto \exp\left(-\frac{E_i}{kT}\right)$, this condition is fulfilled provided $\Gamma_{ij}^* = \Gamma_{ji}^*$. Formally, this does not seem to be guaranteed in transition state theory. In a real system in stable equilibrium, detailed balance must however be fulfilled. Considering the approximations of transition state theory and the fact that the effective potential cannot be calculated practically anyhow, we feel justified in adopting a symmetrical attempt frequency $\Gamma^* = \Gamma_{ij}^* = \Gamma_{ji}^*$. We therefore set

$$\Delta E_{is} = \frac{1}{2}\Delta E_{ij} + E_0^{barr}\{s(ij)\}.$$

The energy difference between final and initial state ΔE_{ij} cannot be calculated easily ab initio for all possible configurations and therefore shall be determined by a pair interaction model as usual. The 'pure' barrier height E_0^{barr} is now assumed to depend on a set of surrounding atom occupations. According to the results shown in Table II we may compute it by a simple approximate rule: Take 0.18eV for the first Al atom in the window, add another 0.18eV for an adjacent Al atom in the window, add 0.36eV if the second atom in the window is diagonally opposite. First results with the modified MC algorithm will be given in a forthcoming paper.

CONCLUSIONS

(i)   We have been able to calculate effective point defect formation energies by subjecting the results of a supercell ab initio method to a grand-canonical formalism, fulfilling thus the requirement of stoichiometry.

(ii)   The results so obtained depend weakly on temperature (listed here for a typical annealing temperature of 1000K) and contain energy contributions of more than one defect type. Generally, antisite defects are easier to form (~0.5-0.7 eV) than vacancies (~1.4-2.5 eV).

(iii)   The influence of atomic environment on atom jumps in Ni_3Al can be reassumed as follows: The barrier of antisite-generating jumps is markedly raised as (smaller) Ni atoms in the neighborhood are replaced by (larger) Al atoms. For the inverse case a lowering of the barrier is observed, although to a lesser extent.

(iv)   If Ni atoms are replaced by Al atoms in a four-atom window through which the jumping atom has to pass the barrier height can be increased significantly.

(v)   We propose how to incorporate variable barrier height into a kinetic Monte Carlo algorithm.

ACKNOWLEDGMENT

The support of the Austrian Fonds zur Förderung der Wissenschaftlichen Forschung (FWF) under grant P-12538-N03 is gratefully acknowledged.

REFERENCES

1. J.H. Westbrook and R.L. Fleischer, in *Intermetallic compounds-principles and practice*, Vol.1 and 2 (Wiley, Chichester, 1994).
2. G. Sauthoff, *Intermetallics*. (VCH Verlagsgesellschaft, Weinheim, 1995).
3. W. Pfeiler, *JOM* **52**, 14 (2000)
4. P. Oramus, R. Kozubski, V. Pierron-Bohnes, M. C. Cadeville, C. Massobrio and W. Pfeiler, Mat. Sci. Eng. A **324**, 11 (2002).
5. P. Oramus, R. Kozubski, V. Pierron-Bohnes, M. C. Cadeville and W. Pfeiler, Phys. Rev. B **63**, 174109 (2001).
6. H. Schweiger, O. Semenova, W. Wolf, W. Püschl, W. Pfeiler, R. Podloucky and H. Ipser, Scripta Materialia **46**, 37 (2001).
7. G. Kresse and J. Furthmüller, Phys. Rev. B **54**, 11169 (1996); Comput. Mater. Sci. **6**, 15 (1996).
8. M. Rasamny, M. Weinert, G.W. Fernando and R.E. Watson, Phys. Rev. B **64**, 144107 (2001).
9. J. Houserova, D. Vogtenhuber, W. Wolf, R. Podloucky, W. Pfeiler, and W. Püschl, Defect and Diffusion Forum, in the press (2004).
10. G. Mills, H. Jónsson, and G. Schenter, Surface Sci. **324**, 30 (1995).
11. G. H. Vineyard, J. Phys. Chem. Solids **3**, 121 (1957)

AUTHOR INDEX

Acosta, Manuel, 383
Akiba, Etsuo, 339
Appel, Fritz, 103, 157, 223
Arroyave, Raymundo, 523, 529
Arzt, E., 73, 127
Au, Peter, 199

Baier, F., 439
Baker, I., 35
Bartels, Arno, 121, 127, 157, 187, 193
Baudin, C., 27
Bei, H., 21, 67
Bewlay, Bernard P., 273, 297
Boehlert, Carl J., 235
Bonnentien, Jean-Louis, 145
Bourke, M.A.M., 21
Brokmeier, H.-G., 223
Brown, D.W., 21
Bystrzanowski, S., 121, 127, 193

Cawkwell, Marc J., 309
Chang, Y.A., 297
Chen, Yali, 211, 217
Chevalier, Jean-Pierre, 145
Chiu, Yu-Lung, 253
Chladil, Harald F., 187, 193
Clemens, Helmut, 73, 121, 127, 157, 187, 193
Collins, Peter C., 285
Couret, Alain, 151, 205
Cowen, Christopher J., 235
Cretegny, Laurent, 273

Dehm, G., 73, 121, 127
Dehoff, Ryan R., 285
Demura, Masahiko, 85, 431
Deshpande, Anirudha R., 377, 389

Evans, N.D., 163

Fischer, F.D., 157
Fraser, Hamish L., 285
Freeman, Arthur J., 511
Frommeyer, Georg, 41

Fu, C.L., 505
Fujimoto, Towako, 371
Fujio, S., 491
Fujishima, Kousuke, 365

George, E.P., 21, 67
Gerling, Rainer, 121, 127, 187, 193
Gomi, Nobuyuki, 461
Gornostyrev, Yuri N., 511
Greenberg, B.A., 241

Hagihara, Koji, 91
Hagiwara, T., 53
Halford, Timothy P., 175, 229
Hanada, Shuji, 315, 449, 535
Haneczok, G., 127
Hanna, J.A., 35
Hara, Toru, 365
Hata, S., 259
Hayashi, Koutaro, 253
Hayashi, Yoshinori, 407
Higo, Yakichi, 175, 229
Hino, Takehisa, 413
Hirano, Toshiyuki, 85, 431
Hosoda, Hideki, 347, 353
Hotta, Y., 181
Houserova, Jana, 541
Hu, D., 139

Inamura, Tomonari, 347
Inoue, Akihisa, 133
Inoue, H., 181
Inoue, Kanryu, 353
Inui, Haruyuki, 407, 419, 425, 491
Ishibashi, Hiroki, 401
Issro, Chaisak, 383
Itakura, M., 259
Itoh, Yoshiyasu, 413

Jackson, Kamili M., 497
Jackson, Melvin R., 273, 297
Jiang, Chao, 523
Jiang, H., 139
Johansson, Börje, 517
Johnson, David, 473

Jones, I.P., 139

Kabra, S., 21
Kameoka, Satoshi, 431
Kaneno, Y., 181, 291
Kase, Takashi, 15
Kathrein, Martin, 467
Kazantseva, N.V., 241
Kim, Han-Sol, 59
Kim, Won-Yong, 59
Kimura, Yoshisato, 395
Kishida, Kyosuke, 85, 431
Kobayashi, Satoru, 47, 247
Kobayashi, Toshiro, 247
Kogachi, Mineo, 401
Koizumi, Y., 53
Koneva, N.A., 485
Konrad, Joachim, 41
Kontsevoi, Oleg Y., 511
Korzhavyi, Pavel A., 517
Kozeschnik, Ernst, 187
Kozlov, E.V., 485
Kozubski, R., 97
Kraft, O., 73
Krcmar, M., 505
Kruzic, J.J., 303
Kuwano, N., 259

Lamirand, Mélanie, 145
Lang, Candace I., 497
Lapin, Juraj, 115
Leitner, Harald, 187, 193
Lenhart, Helmut, 467
Li, Dingqiang, 235
Liu, C.T., 163
Liu, Zi-Kui, 523, 529
Lorenz, U., 223
Louzguina-Luzgina, Larissa V., 133
Louzguine-Luzgin, Dmitri V., 133

Malaplate, Joel, 205
Marketz, Wilfried T., 187
Maruyama, Kouichi, 109, 199
Matsuda, Jun, 109
Matsuda, Mitsuhiro, 365

Matsuda, Mituhiro, 371
Matsuo, Takashi, 461
Mayrhofer, Paul H., 467
Maziasz, P.J., 163
Mills, Michael J., 285
Minamino, Y., 53
Mishima, Yoshinao, 395
Mitterer, Christian, 467
Miura, Seiji, 279, 455
Miyazaki, Shuichi, 347, 353
Mohri, Tetsuo, 279, 455
Molénat, G., 151
Morita, Shigeki, 247
Morita, Sumio, 461
Morizono, Yasuhiro, 365, 371
Morris, D.G., 27
Morris, J.R., 505
Mrovec, Matous, 309
Muñoz-Morris, M.A., 27
Munroe, P.R., 35

Nagase, Takeshi, 347
Nakamura, Ryusuke, 449, 535
Nakano, Takayoshi, 91, 253, 259, 347
Nakanura, Yumiko, 339
Nakashima, Hideharu, 371
Nakatani, Yujiro, 413
Nazmy, Mohamed, 115
Nguyen-Manh, Duc, 309
Nishida, Minoru, 365, 371
Nxumalo, Silethelwe, 497
Nzula, Miyelani P., 497

Oehring, M., 223
Oh, Min-Wook, 425
Ohira, K., 291
Ohkubo, Kenji, 279, 455
Ohtsuka, Hideyuki, 395
Okamoto, Norihiko L., 419
Okuno, Hiroyuki, 455

Palm, Martin, 3
Partyka, E., 97
Perepezko, John H., 321
Pettifor, David G., 309

Pfeiler, Wolfgang, 383, 541
Pierron-Bohnes, Véronique, 383
Pilugin, V.P., 241
Podloucky, Raimund, 541
Pope, David P., 211, 217
Prins, Sara, 523, 529
Püschl, Wolfgang, 383, 541

Ra, Tae-Yeub, 59
Raabe, Dierk, 41
Reynolds, Todd, 473
Riethmüller, J., 73
Ritchie, R.O., 303
Ritter, Ann M., 273
Rogl, Peter F., 383
Rudinal, D., 175

Saada, G., 151
Sakaino, Yuki, 169
Sakaki, Kouji, 339
Sakamoto, H., 491
Sakidja, Ridwan, 321
Sarosi, Peter M., 285
Sato, K., 439
Sauthoff, Gerhard, 3
Schaden, T., 157
Schaefer, H.-E., 97, 439
Schimansky, F.-P., 121, 193
Schneibel, J.H., 303
Schneider, André, 3, 41, 47
Schryvers, Dominique, 329
Sekido, Nobuaki, 321
Selleby, Malin, 517
Seo, Dongyi, 199
Sepiol, Bogdan, 383
Shibuya, S., 181
Shindo, Kentaro, 247
Shindo, Takahiko, 413
Shiraishi, K., 259
Simkin, Benjamin A., 407
Singh, J.B., 151
Soffa, William A., 383
Solov'eva, Yu.V., 485
Song, Shunzi, 169
Sprengel, W., 97, 439
Starenchenko, V.A., 485

Staubli, Marc, 115
Stein, Frank, 3
Sundararaman, M., 151
Sundman, Bo, 517
Suzuki, Akane, 479

Tadachi, Nobutoshi, 401
Takahashi, Tohru, 169
Takahashi, Yohei, 347
Takaku, Reki, 413
Takashima, Kazuki, 175, 229
Takasugi, T., 181, 291
Takeyama, Masao, 175, 247, 461,
479
Tanaka, K., 359, 491
Tanaka, Tetsunori, 91
Taniguchi, Shigeji, 265
Terada, Eiji, 425
Tetsui, Toshimitsu, 181, 247
Tirry, Wim, 329
Tomokiyo, Y., 259
Tsai, An-pang, 431
Tsuji, N., 53
Tsurekawa, Sadahiro, 449

Umakoshi, Yukichi, 15, 79, 91,
253, 259, 347

Veyssière, Patrick, 151, 253
Vitek, Vaclav, 309
Vogtenhuber, Doris, 541

Wakashima, Kenji, 347, 353
Wee, Dang-Moon, 425
Wei, Fu-Gao, 395
Weller, M., 127
Wellner, P., 73
Wiezorek, Jörg M.K., 377, 389
Wittmann, M.W., 35
Wolf, Walter, 541

Xu, Ya, 431

Yamamoto, Y., 163
Yamauchi, Akira, 315, 449, 535
Yanai, Toshifumi, 79

Yang, Y., 297
Yang, Zhiqing, 329
Yasuda, Hiroyuki Y., 15, 79
Yeo, In-Dong, 59
Yoshihara, Michiko, 265
Yoshimi, Kyosuke, 449, 535

Yoshimi, Kyousuke, 315

Zaefferer, Stefan, 41, 47
Zhang, X.Y., 439
Zhu, Hanliang, 109, 199

SUBJECT INDEX

ab initio, 529
    calculations, 505, 517
    enthalpy, 523
actuator, 359
AFM, 211, 217
alloy design, 47
Al-Ni-Pt-Ru, 523
Al_5Ti_3, 253
APB, 253
atom migration, 541
atomic processes, 439
atomistic modeling, 309
austenite, 461

B2, 523
    phases, 529
bismuth tellurium thermoelectric
    module, 413
bond order potentials, 309

catalyst, 431
chirality, 491
chromium, 497
clathrate, 419
coating, 467
cobalt(-)
    aluminide, 395
    and nickel aluminides, 535
    iron-aluminium alloys, 401
coercivity, 377
coincide site lattice, 371
cold rolling, 85
combined reaction processing, 389
composite, 67, 273, 315
composition dependence, 169
creep, 115, 169, 199, 205
    behavior, 121
    resistance, 109
    strength, 27
CuAu, 455

$D0_{22}$, 479
defect, 339
    formation energies, 541

deformation, 211
    behavior, 157
    twin, 371
differential-scanning-calorimetry,
    187
diffusion, 127
directional solidification, 67
dislocation(s), 205, 253
    structure, 485, 511
DS solidification, 297
DSC and TEM, 535

EBSD, 41, 279
electron diffraction, 491
enantiomorph, 491
environmental embrittlement, 181
extrusion, 223

fatigue, 139, 229, 303
Fe_3Al, 21
Fe_2Nb, 461
FePd, 383
ferromagnetism, 395, 401
finite element method, 413
fracture toughness, 175, 303, 473

gamma
    alloy, 247
    Ti-Al, 145
    titanium aluminide, 223
gold aluminide, 467
grain growth, 377

HAADF-STEM, 425
heat treatment atmosphere, 365
heterogeneous nucleation, 321
Heusler, 35
    alloy, 401
Hf-Ti-Si, 297
high-temperature oxidation, 265,
    315
hot-worked, 247
hydrogen, 339
    absorbing alloy, 339
    embrittlement, 181

impact resistance, 247
intermetallic(s), 235, 241, 285, 449, 485
    compound, 133
internal
    friction, 127
    stress, 455
interstitial elements, 145
ion implantation, 265
iron(-)
    aluminide(s), 3, 15, 41, 53, 59
        based on Fe_3Al, 27
    aluminum, 3, 35
    cobalt, 505

$L1_0$ ordered
    FePd, 377
    intermetallics, 389
$L1_2$ ordered intermetallics, 541
lamellar
    γ-TiAl, 157
    structure, 279
    TiAl, 151
lattice defects, 535
long-
    period superstructure, 259
    range order, 383

magnetism, 79
martensitic transformation, 347, 353
massive transformation, 193
measuring techniques, 439
mechanical property(ies), 133, 235, 291, 497
melt spinning, 347
metal matrix composites, 285
methanol decomposition, 431
microsized, 229
    testing, 175
microstructural stability, 121
microstructure, 59, 73, 109, 115, 139, 145, 199, 291, 449
misfit strain, 479
molybdenum silicides, 303, 309, 315

multistage martensitic transformation, 365

Nb silicide, 273, 279
neutron diffraction, 21
NiAl, 73, 67
Ni_3Al, 85, 431
Ni-Al-Pt-Ru, 529
Ni-based compound, 91
nickel aluminide, 79
niobium silicides, 285
Ni-Ti, 329
Ni_3V, 479
Ni_3X, 291
numeric modeling, 157

order disorder phase transformation, 359
ordering, 53
    of carbon, 395
orientation, 407

phase
    diagram calculation and solidification simulation, 297
    transformation, 241, 439
planar defects, 389
plastic
    behavior, 91
    flow, 217
platinum, 497
    group alloys, 511
positron lifetime spectroscopy, 97
powder x-ray diffraction, 461
precipitation, 47, 103, 321
pseudoelasticity, 15, 21

QHRTEM and EELS, 329
quenching dilatometer, 193

recrystallization, 47
rolling, 41
RuAl, 473
Ru_2Si_3, 407
ruthenium, 473

Schmid law violation, 151
severe deformation, 241
shape memory
    alloy, 365
    effect, 359
sigma
    combination rule, 371
    phases, 517
slip transmission, 151
strain and concentration gradient
    fields, 329
strength, 273
strengthening, 3
    mechanisms, 27
structural stability, 505

TEM, 259
temperature anomaly, 485
ternary element, 353
texture, 223
thermal
    stability, 163, 467
    strain, 413
    stress, 73
    vacancies, 97
thermodynamic
    modeling, 517
    simulations, 187
thermoelectric, 407
    material, 419

thin film, 383
TiAl, 103, 115, 139, 175, 181, 187,
    199, 211, 217, 229, 259
    alloys, 109
    based alloys, 121
    intermetallics, 205
Ti_3Al, 169
Ti-based alloys, 133
TiNi, 353
TiNiPt, 347
titanium, 235
    aluminide(s), 127, 163, 265
TNB, 193
transition metal disilicide, 425
transmission electron
    microscope, 53, 321
    microscopy, 15, 35, 79, 85,
    163
twin, 455
twinning, 103

vacancy, 425, 449

wear resistance, 59

x-ray diffraction, 419

yield stress
    anomaly, 91
    behavior, 511

Printed in the United States
By Bookmasters

Printed in the United States
By Bookmasters